Springer Series in Statistics

Springer Series in Statistics

Measures of Association for Cross Classifications
Leo A. Goodman and **William H. Kruskal**
1979 / 146 pp. / cloth
ISBN 0-387-**90443**-3

Statistical Decision Theory: Foundations, Concepts, and Methods
James Berger
1980 / 425 pp. / 20 illus. / cloth
ISBN 0-387-**90471**-9

Simultaneous Statistical Inference, Second Edition
Rupert G. Miller, Jr.
1981 / 299 pp. / 25 illus. / cloth
ISBN 0-387-**90548**-0

Point Processes and Queues: Martingale Dynamics
Pierre Brémaud
1981 / approx. 384 pp. / 31 illus. / cloth
ISBN 0-387-**90536**-7

Non-Negative Matrices and Markov Chains
E. Seneta
1981 / 304 pp. / cloth
ISBN 0-387-**90598**-7

Statistical Computing with APL
Francis John Anscombe
1981/426 pp./70 illus./cloth
ISBN 0-387-**90549-**9

Concepts of Nonparametric Theory
John Pratt and **Jean D. Gibbons**
1981/469 pp./23 illus./cloth
ISBN 0-387-**90582-**0

Francis John Anscombe

Computing in Statistical Science through APL

Springer-Verlag
New York Heidelberg Berlin

Francis John Anscombe
Department of Statistics
Yale University
Box 2179, Yale Station
New Haven, Connecticut 06520
USA

With 70 illustrations

Library of Congress Cataloging in Publication Data
Anscombe, Francis John, 1918–
Computing in statistical science through APL
(Springer series in statistics)
Bibliography: p.
Includes index.
1. Statistics—Data processing. 2. APL (Computer program language) I. Title. II. Series.
QA276.4.A45 519.5′028′5424 81-978 AACR2

Typeset by Computype, Inc., St. Paul, Minn. Printed and bound by Halliday Lithograph, West Hanover, Mass.

Printed in the United States of America.

9 8 7 6 5 4 3 2 1

ISBN 0-387-90549-9 Springer-Verlag New York Heidelberg Berlin
ISBN 3-540-90549-9 Springer-Verlag Berlin Heidelberg New York

Preface

At the terminal
seated, the answering tone:
pond and temple bell.

TODAY as in the past, statistical method is profoundly affected by resources for numerical calculation and visual display. The main line of development of statistical methodology during the first half of this century was conditioned by, and attuned to, the mechanical desk calculator. Now statisticians may use electronic computers of various kinds in various modes, and the character of statistical science has changed accordingly. Some, but not all, modes of modern computation have a flexibility and immediacy reminiscent of the desk calculator. They preserve the virtues of the desk calculator, while immensely exceeding its scope. Prominent among these is the computer language and conversational computing system known by the initials APL.

This book is addressed to statisticians. Its first aim is to interest them in using APL in their work—for statistical analysis of data, for numerical support of theoretical studies, for simulation of random processes. In Part A the language is described and illustrated with short examples of statistical calculations. Part B, presenting some more extended examples of statistical analysis of data, has also the further aim of suggesting the interplay of computing and theory that must surely henceforth be typical of the development of statistical science.

No previous knowledge of computing is expected of the reader. Familiarity with some statistical concepts and theory is assumed—different amounts in different parts of the book, according to the nature of the subject matter. This is not a primer of statistics; it is not a textbook for a course. In fact, four distinct courses could be built around or developed out of material in the book, courses on statistical computing (Chapters 1–8), on time series (Chapter 10), on regression analysis (Chapters 9 and 11), on categorical variables (Chapter 12). There are some sets of exercises. A few exercises suggest methods of studying the material presented in the text; most are concerned with complementing the text; there is no drill.

There are many aspects to computing in statistics. Only some are considered here. The excellent book on statistical algorithms by Chambers (1977) is different in subject-matter and intent, as are most works presenting program packages.

The word "statistics" has many uses and misuses. The oldest surviving meaning is factual numerical information about the condition of society, and hence factual numerical information about anything. By "statistical science" is meant the methodology and theory of gathering and interpreting statistical information—of the design of statistical investigations and of the analysis of material obtained.

The book has been long in writing. I first had access to an APL system, at IBM's Thomas J. Watson Research Center, in the summer of 1967. The following year a duplicated technical report on the use of APL for statistical computing was issued (Anscombe, 1968b), and in 1969 I embarked on the project of preparing a book. A draft of the first nine chapters was duplicated, but not widely circulated, late in 1970. Some of these chapters, particularly Chapter 7, appear now with little change; others, particularly Chapter 8, have been largely rewritten. The later parts of the book, begun after 1970, have turned out more ambitious than originally planned. Three further long chapters have been written, Chapter 10 on time series (essentially completed in 1977), Chapter 11 on regression analysis (essentially completed in 1973), Chapter 12 on contingency tables (essentially completed in 1974, except for the material on alternatives to χ^2, added in 1979). Finally, the four sets of exercises have been added; and three appendices present supplementary material, including a collection of programs.

During these years that I have explored statistical computing with APL, I have found my interests changing. It is perhaps not untypical of others' experience that my first desire, upon gaining access to an enjoyable and powerful computing facility, was to perform substantial tasks. I had just been involved in a study of regression methods (Anscombe, 1967), both what is now generally called the modified Gram-Schmidt method for fitting linear relations by least squares, and also an "adaptive" method for "robust" fitting of linear relations. I produced forerunners of the regression programs described in Appendix 3 in *HOWREGRESSION*; and then immediately delved into the quite heavy adaptive fitting described in *HOWT7*. I have carried out the adaptive fitting on sets of data only a few times; the programs are reproduced for the record, but perhaps no one else will ever run them. Later I came to value the computer for lighter tasks, particularly for making scatterplots and other graphical displays, which so often afford rapid and vivid insight into the questions being studied. More recently I have occasionally resorted to heavier computing, as in the tests of residuals described in *HOWTESTS*; but this with some reluctance, since quick graphical procedures often tell all we need to know, in the clearest way.

A stylistic feature of the book calls for apology. I hope women readers will judge outmoded rather than offensive my references to the reader, the computer user, the scientist, and other such personages, as "he" rather than "she" or "she or he" or "it". It's *homo*, not *vir*, every time. An obscure lawyers' phrase in the Bye-laws of the Royal Statistical Society runs: "Words importing only the masculine gender include the feminine gender." I mean something like that.

The studies reported here and the writing of the book have been liberally supported both by Yale University, particularly in three semesters of leave, and by the U.S. Office of Naval Research in successive contracts (most recently contract N00014-75-C-0563, task NR 042-242). Without such support nothing could have been done.

I am greatly indebted to Beat Kleiner (Bell Telephone Laboratories) and Christine Waternaux (Harvard University, Department of Biostatistics) for their painstaking renderings of the German and French abstracts, which we hope will be helpful to some readers.

Portions of the book in draft have been read and commented on by many colleagues and friends during the years of preparation—too many to list fairly. I do, however, particularly wish to thank three persons who, while students at Yale, collaborated substantially in these studies: Diccon R. E. Bancroft, William J. Glynn, Gary W. Oehlert. I thank Barbara Amato for skilful typing of almost the whole of the text.

I am aware of a very special debt to two contemporaries, Kenneth Eugene Iverson and John Wilder Tukey. That Tukey's name should be mentioned can surprise no one. So many of the things statisticians think about today, the questions they ask, the goals they try to reach, have been deeply and permanently changed by what John Tukey has done. Iverson's name is less familiar in the statistical world than it surely will be eventually —his efforts have been directed broadly towards computing and mathematics, not specifically towards statistics. Anyone venturing to write about computing and statistics can hardly avoid having been greatly influenced, directly or indirectly, by both men. Though in many respects this book does not reflect the actual opinions and attitudes of either person, I like to think that it is somewhat a festivity in their honor.

Zusammenfassung

Statistisches Computing mit Hilfe von APL

Gegenstand von Kapitel 1 ist ein Vergleich zwischen der Rolle des Computers in der Statistik mit der der Mathematik. Diverse Gesichtspunkte des statistischen Computings werden behandelt, und der weitverbreitete Gebrauch statistischer Programmpakete wird mit der direkten Programmierung in einer Allzweck-Programmiersprache verglichen. Die interaktive Programmiersprache APL ("A Programming Language") von K. E. Iverson bietet sich in diesem Zusammenhang als speziell geeignetes Hilfsmittel für den Statistiker an, denn dieser sollte ebensowenig ein Computerprogramm verwenden, welches er nicht lesen und verstehen kann, wie sich auf einen mathematischen Satz stützen, den er nicht selber beweisen kann.

Kapitel 2–8 bilden Teil A des Buches. Dieser befaßt sich mit statistischem Computing mit Hilfe von APL, beschreibt die wesentlichen Eigenschaften dieser Computersprache und demonstriert ihre Verwendbarkeit anhand von typischen statistischen Beispielen. Kapitel 2 führt den Begriff des Computer-Algorithmus ein und behandelt die Notation in APL, die ursprünglich eingeführt worden ist, um die einzelnen Schritte eines Algorithmus in kurzer, Menschen (und nicht Maschinen) eingängiger Form darzustellen. Kapitel 3 enthält die einfachen Skalarfunktionen und erklärt die syntaktische Grundregel, nach der in einem zusammengesetzten Ausdruck alle Operationen streng von rechts nach links der Reihe nach ausgeführt werden. Kapitel 4 führt den Begriff des Arrays ein (der sowohl Vektoren als auch zwei- und mehrdimensionale Matrizen umfaßt) und behandelt das Dimensionieren von Arrays, sowie den Umgang mit Indices. Kapitel 5 beschreibt Anwendungen einfacher Skalarfunktionen auf Arrays wie Operationen, die auf jedes Element einzeln einwirken (z. B. die Addition zweier Matrizen), die sog. "Reduktion" (der Dimension eines Arrays) und das äußere und innere Produkt. Der APL-Zufallszahlengenerator wird ebenfalls in diesem Kapitel behandelt und durch Beispiele illustriert. In Kapitel 6 werden definierte Funktionen, d.h. Programme, diskutiert. Kapitel 7 enthält diverse Beispiele. Diese umfassen den Vergleich zwischen erwarteten und beobachteten Häufigkeiten in einfachen Glücksspielen, Streuungszerlegungen von Zwei- und Mehr-Weg-Tafeln, die Simulation von Zufallsprozessen und die Modellierung eines binären Prozesses durch eine stationäre Markoff-Kette.

Kapitel 8 bringt die Anwendung von APL im statistischen Computing. Die Struktur von APL macht es bedeutend attraktiver, Algorithmen mit Hilfe von Vektor- und

Matrizen-Operationen zu definieren als mit Schleifen, die jedes Element einzeln behandeln. Der Umstand, daß APL Arrays, die aus Zeichen bestehen, sowie numerische Arrays gleich behandelt, führt zu einer wesentlichen Vereinfachung des Programmierens von Streudiagrammen und anderen graphischen Hilfsmitteln. Die Leistungsfähigkeit von APL wird anhand eines Beispieles in Text-Processing demonstriert. APL kann, unabhängig von seiner Funktion als Computersprache, auch als mathematische Kurzschrift verwendet werden. Der erste Teil des Buches schließt mit einer Reihe von Übungen, die teilweise das Verständnis des Stoffes vertiefen, vor allem aber Ergänzungen zum Inhalt der Kapitel sind. Diese umfassen Tukeys "median polish" zum Anpassen eines additiven Modelles für eine Zwei-Weg-Tafel, die Unterteilung eines Textes in einzelne Paragraphen und das Formatieren einer Tabelle. Als Beispiel für die Verwendung von APL als algebraische Sprache wird die Multiplikation zweier Polynome diskutiert.

Kapitel 9–12 bilden Teil B des Buches, welcher der statistischen Datenanalyse gewidmet ist und sich speziell mit dem Zusammenhang zwischen statistischer Theorie und statistischem Computing befaßt. Dieser Teil bezieht sich nur selten auf eigentliche Computer-Methoden, und kann deshalb ohne APL-Kenntnisse gelesen werden. Kapitel 9 behandelt Sinn und Zweck der statistischen Datenanalyse mit Kritik an einer Analyse von R. A. Fisher, der in seinem Buch "Statistical Methods for Research Workers" Daten des Broadbalk-Weizenexperiments benützt hat, um eine einfache Regression zu illustrieren.

Kapitel 10 behandelt Zeitreihen. Als Beispiel dient die 180-jährige Zeitreihe der Studentenzahl an der Yale Universität in den Jahren 1796–1975. Es zeigt sich, daß die Logarithmen der Studentenzahlen einen linearen zeitlichen Trend aufweisen, und daß sich die Studentenzahl ca. alle 30 Jahre verdoppelt. Die Residuen können als Prozess mit "rotem Rauschen" angesehen werden. Mögliche Beziehungen zwischen der Studentenzahl und der Wirtschaftslage in den Vereinigten Staaten werden durch Regressionen im Frequenzbereich auf drei ökonomische Zeitreihen untersucht: dem Wert (in Millionen von Dollars) der jährlichen Warenimporte in die USA seit 1790, dem durchschnittlichen Butterpreis in New York seit 1830 und der Anzahl jährlich neuerstellter Wohnungseinheiten in den USA seit 1889. Ein Großteil des Kapitels ist einem Überblick über die harmonische Analyse von Zeitreihen gewidmet. Die Fourier-Transformation einer Zeitreihe wird beschrieben, sowie das daraus resultierende rohe (d.h. ungeglättete) Spektrum und dessen Phase; auch die Effekte von Tapering und der Anwendung eines linearen Filters werden besprochen. Wir betrachten Tapering als eine Methode, die eine Zeitreihe mit überlappenden Enden auf einen Kreis aufspannt und die Anzahl der Fourierfrequenzen entsprechend reduziert. Eine Regression zwischen zwei Zeitreihen im Frequenzbereich kann als Suche nach demjenigen linearen Filter interpretiert werden, der, auf eine Reihe angewandt, eine möglichst gute (im Sinne kleinster Quadrate) Approximation an die andere Zeitreihe ergibt; zuerst werden beide Serien in den Frequenzbereich transformiert und dann wird für jedes schmale Frequenzband eine iterative nicht-lineare Regression durchgeführt. Dies entspricht dem visuellen Anpassen einer Trendkurve in einem Diagramm des Kreuzspektrums zwischen den Zeitreihen. Die Anwendung dieser Methode auf die obenerwähnten Daten ergibt mehrere Frequenzbänder, in denen die Zeitreihen signifikant voneinander abhängen. Diese Resultate können im Zeitbereich überprüft werden, indem man zuerst die Zeitreihen einen linearen Filter passieren läßt, sodaß praktisch nur noch die in Frage stehenden Frequenzkomponenten vertreten sind, und anschließend ein Streudiagramm der (entsprechend gegeneinander versetzten) gefilterten Zeitreihen konstruiert. Die aus der harmonischen Analyse gefundenen Beziehungen können zum größten Teil durch grosse Schwankungen während Kriegszeiten erklärt werden. Dies führt zu der Schlußfolgerung, dass kein langfristiger Zusammenhang zwischen der Studentenzahl in Yale und der Wirtschaftslage in den Vereinigten Staaten besteht. Zusätzlicher Stoff ist in den Übungen am Ende des Kapitels enthalten. Er umfaßt u.a. die genaue Verteilung des empirischen Spektrums einer Realisierung einer normalverteilten stationären Zufallsfolge, die Schätzung von Saisonschwankungen und das Anpassen eines autoregressiven Modelles oder eines Modelles mit gleitenden Durchschnitten auf Grund des Spektrums einer Zeitreihe.

Kapitel 11 behandelt die gewöhnliche Regression anhand einer Studie, in der die öffentlichen Pro-Kopf-Ausgaben für das Schulwesen in jedem Staat der USA mit drei unabhängigen Variablen in denselben Staaten verglichen werden, nämlich Pro-Kopf-Einkommen, prozentualer Bevölkerungsanteil der Jugendlichen unter 18 Jahren und Prozentsatz der Bevölkerung, der von der Volkszählung als städtisch eingestuft worden ist. Die Berechnungen verwenden das Orthogonalisierungsverfahren von Gram-Schmidt, welches Schritt für Schritt durchgeführt und erläutert wird. Ferner werden verschiedene Arten von Streudiagrammen besprochen, die eine Beurteilung der Resultate von Regressionsanalysen erlauben. Der zweite Teil des Kapitels befaßt sich mit diversen technischen Fragen: Anpassungstests, die auf Residuen basieren; das erwartete Verhalten der Diagonalelemente der Projektionsmatrix, die die Werte der abhängigen Variablen auf den Vektor der Residuen projiziert; das Anpassen einer Regression, wenn die Fehler eine Verteilung vom Typ Pearson VII oder II aufweisen und die robuste Anpassung einer Regression nach der Methode von P. J. Huber. Die Übungen am Schluß des Kapitels enthalten wiederum viel zusätzliches Material, darunter normale Wahrscheinlichkeitsdiagramme, stufenweise Regression, Kovarianzanalyse, Anpassen additiver und multiplikativer Modelle für Zwei-Weg-Tafeln, die Zerlegung einer Matrix auf Grund ihrer Singulärwerte und die APL-Darstellung von Daten, die nicht durch rechteckige Matrizen dargestellt werden können.

Kapitel 12 behandelt kategorische Variablen und Kontingenztafeln. Der χ^2*-Test von Pearson für Unabhängigkeit wird diskutiert. Da dieser nicht benützt werden sollte, wenn die Erwartungswerte einiger Felder sehr klein sind, wird mitunter ein Likelihood-Quotienten-Test an seiner Stelle verwendet. Ein weiterer Test basiert auf der Wahrscheinlichkeit der beobachteten Werte unter der Null-Hypothese unabhängiger Zeilen und Spalten; dieser Test scheint den anderen vorzuziehen zu sein. Er besitzt auch die bemerkenswerte Eigenschaft, daß seine Verteilungsfunktion unter der Null-Hypothese bedeutend besser durch stetige Verteilungsfunktionen wie die tabellierte* χ^2*-Verteilung approximiert werden kann als die anderer Test-Statistiken. Der zweite Teil des Kapitels befaßt sich mit der Messung der Stärke des Zusammenhangs zwischen geordneten kategorischen Variablen durch die Anpassung einer Wahrscheinlichkeitsverteilung mit konstanten Kreuzproduktverhältnissen. Solche Verteilungen sind von R. L. Plackett untersucht und auch von K. Pearson kurz betrachtet worden. Das Verfahren wird auf mehrere publizierte Kontingenztafeln angewendet. Die Übungen am Schluß des Kapitels umfassen u.a. ein alternatives Verfahren zur Messung der Stärke* *des Zusammenhangs zwischen geordneten kategorischen Variablen, das von L. A. Goodman entwickelt worden ist.*

Das Buch enthält drei Anhänge. Der erste besteht aus zwei kurzen Aufsätzen über den Stand der statistischen Wissenschaften. Der zweite Anhang gibt einen Überblick über Anpassungstests, die auf Residuen einer Regression basieren, und über Tests für die Größe der Kurtosis des Voraussagefehlers eines Zeitreihenmodelles. Auch eine Regel von J. W. Tukey wird diskutiert, welche entscheidet, wieviele Parameter in ein Modell aufgenommen werden sollten, um möglichst informative Residuen zu erhalten. Der dritte Anhang enthält die statistischen APL-Programme des Autors.

B. K.

Résumé

Les calculations statistiques à l'aide d'APL

Le premier chapitre est consacré au rôle des calculations en statistique, qui est comparé au rôle des mathématiques. Des aspects divers des calculations statistiques sont indiqués. L'usage répandu des collections de programmes statistiques est mis en contraste avec la programmation directe dans un langage d'utilité générale. Il est aussi imprudent pour un statisticien de se servir d'un programme qu'il ne peut pas lire que de s'appuyer sur un théorème mathématique qu'il ne peut pas démontrer. On suggère que le langage de programmation et système interactif, que l'on appelle APL d'après le livre de K. E. Iverson: "A Programming Language", est particulièrement approprié.

Les chapitres 2 à 8, qui constituent la partie A de ce livre, s'adressent aux calculations statistiques en APL. Certains des traits principaux du langage sont décrits et son usage est démontré pour des calculations statistiques typiques. Le deuxième chapitre s'adresse aux méthodes de calculation, ou algorithmes, et à l'origine d'APL dans la notation d'Iverson pour une communication concise et sans ambiguité d'algorithmes aux lecteurs humains. Le troisième chapitre présente les fonctions scalaires d'APL et la règle de syntaxe qui impose que dans les expressions composées la fonction la plus à droite est exécutée d'abord. Le quatrième chapitre s'adresse aux matrices, leurs indices et leurs vecteurs de dimension. Le cinquième chapitre montre comment les fonctions scalaires sont appliquées aux matrices. La génération des variables pseudo-aléatoires dans ce système est décrite et illustrée. Le sixième chapitre traite des fonctions définies, c'est à dire des programmes. Dans le septième chapitre on présente des exemples de calculations statistiques et en particulier la comparaison des fréquences observées et espérées dans des jeux simples de hasard, l'analyse de variance pour des données sous forme de tableaux à deux dimensions, la simulation de processus aléatoires, l'analyse d'une suite alternée en terme d'une chaîne de Markov stationnaire et l'analyse de variance pour des données sous forme de tableau avec de nombreuses dimensions.

Dans le huitième chapitre se trouvent des remarques diverses à propos de l'usage d'APL pour les calculations statistiques. Le langage APL facilite l'expression d' algorithmes en fonction d'opérations matricielles plutôt qu'en termes d'itérations dans lesquelles interviennent les éléments individuels. Le traitement identique des matrices à éléments numériques et à éléments alphabétiques facilite la programmation des figures et autres représentations graphiques. L'efficacité du langage APL est discutée, en partie par un exemple de traitement et édition d'un texte. APL peut aussi servir comme une notation mathématique indépendamment des calculations. La partie A se termine avec une série d'exercices. Quelques-uns de ces exercices suggèrent des méthodes pour étudier la matière présentée dans les chapitres; la plupart sont des compléments. Dans les sujets abordés on trouve: la méthode appelée "median polish" de J. W. Tukey pour ajuster une structure additive dans un tableau à deux dimensions, l'arrangement d'un texte en paragraphes et la formation de tables, et la multiplication de polynômes comme un exemple de calculations algébriques automatisées.

Les chapitres de 9 à 12, qui forment la partie B de ce livre, sont consacrés à certaines analyses statistiques de données et en particulier à l'interaction entre les calculations et les raisonnements théoriques. Les références aux techniques de calculation sont rares et il n'est pas nécessaire que le lecteur sache programmer en APL. Dans le neuvième chapitre on présente une discussion des buts de l'analyse statistique, avec une critique de l'utilisation par R. A. Fisher dans "Statistical Methods for Research Workers" de certaines données venant de l'expérience Broadbalk sur le blé pour illustrer la régression simple.

Le dixième chapitre, sur l'analyse des séries temporelles, commence par l'examen d'une série sur 180 années de l'enrôlement total annuel des étudiants à l'université de Yale de 1796 à 1975. Le logarithme de l'enrôlement semble avoir une composante à tendance linéaire qui prédirait que l'enrôlement double tous les 30 ans. Les résidus semblent constituer un processus léthargique de "bruit rouge". On se demande alors si l'on peut détecter une corrélation entre l'enrôlement à Yale et l'état de l'économie aux Etats-Unis. On se sert d'une série économique longue: le total annuel des importations de marchandises (en millions de dollars) depuis 1790. Deux séries économiques plus courtes sont aussi considérées: le prix moyen du beurre à New York depuis 1830, et le nombre de logements neufs commencés chaque année depuis 1889. On essaie de rapporter les séries économiques et la série des enrôlements à Yale par la méthode de régression dans le domaine des fréquences. Une grande partie de ce chapitre est consacrée à une esquisse de l'analyse harmonique des séries temporelles. On y décrit la transformation de Fourier d'une série (le spectre et les phases) et on y examine l'effet d'un filtre linéaire et du "tapering". Le tapering consiste à circulariser une série par recouvrement des extrémités, après quoi l'ensemble des fréquences de Fourier est réduit. La régression d'une série sur l'autre signifie que l'on cherche le filtre linéaire qui porte la seconde série près de la première au sens des moindres carrés. Après avoir transformé les séries par la méthode de Fourier, on effectue une régression non linéaire avec itérations dans une bande étroite de fréquences, ce qui correspond à placer à l'oeil une tendance sur la représentation graphique du cospectre. Quand cette méthode est appliquée aux séries données, on trouve quelques bandes de fréquences pour lesquelles il semble y avoir une association réelle entre les deux séries. Ce résultat peut être vérifié dans le domaine du temps par l'application d'un filtre linéaire qui favorise la bande de fréquences en question, suivie d'une représentation graphique d'une série contre l'autre avec un décalage approprié. Les associations révélées par l'analyse harmonique semblent être principalement causées par les grandes perturbations apportées aux séries en temps de guerre. On en conclut que l'enrôlement à Yale n'a pas d'association permanente avec l'économie des Etats-Unis. Des exercices à la fin du chapitre ajoutent du matériel supplémentaire. Parmi les sujets abordés on trouve: la distribution exacte du spectre empirique d'un échantillon d'un processus aléatoire stationnaire à loi normale, l'estimation des variations saisonnières dans une série donnée, et l'ajustement d'une structure à moyennes mobiles ou d'une structure auto-régressive par l'examen du spectre.

Dans le onzième chapitre, sur la régression ordinaire, on effectue une régression des dépenses dans les écoles publiques (par personne) dans chacun des états des Etats-Unis sur trois variables explicatives: le revenu par personne, la proportion de la population en dessous de 18 ans, et la proportion de population classifiée urbaine dans le recensement pour chaque état. Les calculs sont effectués étape par étape en utilisant la méthode d'orthogonalisation de Gram-Schmidt. Plusieurs représentations graphiques sont fournies pour permettre de porter un jugement sur la validité de l'analyse. Dans la dernière partie du chapitre, on traite certaines questions techniques: les tests d' ajustement sur les résidus, les propriétés espérées des éléments diagonaux de la matrice Q qui projette le vecteur des observations sur le vecteur des résidus, l'ajustement d'une régression quand les erreurs ont une distribution de Pearson du type VII ou II, la méthode d'ajustement robuste de P. J. Huber. Les exercices à la fin du chapitre ajoutent beaucoup de matériel supplémentaire. Parmi les sujets abordés on trouve: les diagrammes à échelle fonctionelle normale, la régression en plusieurs étapes et l'analyse de covariance, l'ajustement de modèles additifs et multiplicatifs sur les tableaux à deux dimensions, l'expression d'une matrice en fonction de ses valeurs singulières, la représentation de tableaux non rectangulaires en APL.

Dans le douzième chapitre, on traite des variables catégoriques et des tableaux de contingence. On y considère le test d'association du χ^2 *de Pearson. Le test ne se comporte pas bien quand les valeurs espérées dans certaines cellules d'un tableau de contingence sont très petites. Un test du rapport de vraisemblance est souvent recommandé à sa place; la probabilité des données, sous l'hypothèse qu'il n'y a pas d' association, peut aussi servir de statistique pour le test et semble être préférable. Cette statistique a la propriété remarquable d'avoir une distribution relativement lisse, pour laquelle l'approximation du* χ^2 *est meilleure que pour d'autres statistiques. La deuxième partie du chapitre est consacrée au mesurage de l'association entre des variables catégoriques dont les catégories ont une relation d'ordre. Dans ce but on ajuste une distribution de probabilités pour laquelle un rapport quadruple de probabilités est constant. De telles distributions ont été étudiées par R. L. Plackett et furent brièvement considérées par K. Pearson. La méthode est utilisée pour certaines tables publiées. Des exercices à la fin du chapitre contiennent du matériel supplémentaire, y compris une approche alternative pour mesurer l'association de variables à catégories ordonnées due à L. A. Goodman.*

Il y a trois appendices. Le premier contient deux articles courts sur l'état de la science statistique. Le deuxième présente un résumé de résultats au sujet des tests d'ajustement effectués sur les résidus d'une régression et des tests sur la forme de la distribution des innovations d'une série temporelle. On y discute une règle due à J. W. Tukey pour décider combien d'effets doivent être estimés de façon à obtenir des résidus instructifs. Le troisième appendice présente l'ensemble des programmes statistiques en APL écrits par l'auteur.

C. W.

Contents

Chapter 1

Statistical Computing

Ever since high-speed stored-program digital computers became available round about 1950 they have been applied to statistical work. Their influence has grown gradually. The computer did not immediately have the vital part in statistics that it had in, say, space exploration. Yet we now see that the computer is exercising a profound effect on the science of statistics, transforming not only methods of operation but also basic ideas and understanding and objectives. And the transforming is still far from complete.

The influence of the computer on statistics is as great as the influence of mathematics. There is general recognition today that a professional statistician needs a substantial mathematical training, even if his main kind of activity should seem nonmathematical. Eminent statisticians of the past have succeeded without much mathematics. John Graunt was no mathematician, and even if he had been one there was little in the mathematics of his day that could have helped him; yet some of his work still seems to be of very high quality. It is nevertheless true that mathematics is now advisable. Even in apparently nonmathematical contexts a little mathematical insight often proves valuable; contexts sometimes become mathematical suddenly; and a statistician with inadequate mathematics is denied access to much of the modern literature of his subject—not all of which literature is irrelevant or pointless.

There is less general recognition that a professional statistician needs to be able to compute, even if his main interests are strongly mathematical and theoretical. Training programs nowadays usually include considerable exposure to computing, and many of the younger professionals have high mathematical and computing expertise. The mathematical statistician who does not compute is still with us, but seems destined soon to become rare if not extinct.

The slowness with which statistics has embraced the computer has not been without good reason. As attempts were made to apply the computer to statistical purposes, difficulties were encountered, manifold and serious difficulties that have taken time to be perceived and resolved.

The diversity in the problems of statistical computing comes not only from the diversity of tasks addressed, which is the diversity comprehended by the word "statistics" itself, but also from the diversity of persons addressing these tasks. By "statistics" may be meant organization, presentation or evaluation of statistical data, or study of random processes, or the theory of decision-making under uncertainty, or many other kinds of mathematical theory considered to be statistical. Computers may be used to perform tasks of arrangement or analysis of data, to simulate random processes, and to aid mathematical investigation in various ways.

The persons who may seek to use a computer for some statistical purpose vary greatly in their interest in, and knowledge of, both statistics and computing. Three very different sorts of person, in these respects, are:

(i) the professional statistical theorist, interested in developing new ideas or methods, interested in particular statistical problems mainly as examples of general ones, demanding, if he comes to the computer, to understand fully what is going on—preferably his own programmer, so that he may do what he wishes rather than what someone else has wished;
(ii) the scientist in a field that yields statistical problems, himself not primarily a statistician, who wants to use good statistical tools to better his understanding of his field, but has no interest in statistical methods for their own sake nor any desire to experiment with them if he can help it—who prefers to be concerned as little as possible with the technicalities of statistical calculations;
(iii) the student or other beginner in statistics, desiring to understand the main ideas without being submerged by details, generally preferring simple examples to raw reality.

Granted the kind of statistical task to be addressed and the kind of user to be catered to, there are several quite different aspects of computing that call for attention. They are so different that progress on one has little or no bearing on the others; they might be called orthogonal. Some of these aspects, problem areas or challenges in statistical computing, can be listed as follows.

(*a*) *Immediacy*. How can the statistician control a computer, so that it becomes an extension of himself, as the piano is to the pianist? Often in the past, informing the computer of the user's desires has been anything but quick and easy. How can the immediacy of the desk calculator be recaptured?

(*b*) *Safety*. How can statistical calculations be made safe from error and misunderstanding? Two sorts of safety to be aimed at are: (i) numerical stability, freedom from perceptible (possibly devastating) effects of rounding in the computations; (ii) comprehensibility of the output, so that the bearing of the calculations on the topic of study will be correctly understood. Yet another sort of safety is of stored material from damage or loss caused by malfunctioning of the equipment.

(*c*) *Standardization*. Most or all features of statistical computation—computer hardware, software systems, coding, languages, symbols, terminology, procedures—have much to gain from elimination of pointless variations, redundancies and confusion. Yet pointlessness is not always easy to judge. The only quite satisfying rule of standardization is that you adopt my standards.

(*d*) *Simulation*. Problems that arise when random processes are simulated include: the design of random number generators, methods for generating samples of specified distributions, and the many devices for economization of effort sometimes referred to as the Monte Carlo method.

(*e*) *Great size*. Peculiar management problems arise in large studies. For example, a very large data file, listing the raw material of a census or of a medical or economic survey, should be so organized that interesting portions can be retrieved rapidly, without a scan of the whole file. Very large computations can be encountered in combinatorial or Monte Carlo work, necessitating great ingenuity in coding and arrangement. Presentation of very large amounts of data calls for special statistical procedures.

This book is concerned with aspect (*a*) above, with turning the computer into a flexible and ready instrument for statistical work, and with some of the effects such an instrument has on statistics. Other aspects may be glimpsed from time to time, but will not be addressed systematically.

Two different approaches towards making computers accessible to users have been followed, one directed narrowly to a particular class of users with similar interests, the other much broader. The narrow approach aims to provide users with programs that will permit some desired task or set of tasks to be accomplished. The programmer may be addressing, for example, multiple linear regression, and aiming to permit the user to make (possibly) the sort of calculations and plots discussed below in Chapter 11; or he may be addressing harmonic analysis of time series (as in Chapter 10), or analysis of results of experiments having any kind of balanced factorial design, or singular-value decomposition and factor analysis of data matrices—to mention some prominent areas of statistical analysis of data. Most of the computing to be done can conveniently be arranged in a single program, or perhaps in two or three programs to be used in succession. Any such program is likely to be rather complicated, and much ingenuity may be expended on making the program fast in execution, numerically stable, yielding well arranged output. Sometimes a single program seems (at first) to do all that is needed, but before long the user is tentatively inquiring about modifications. Circumstances suggest minor variations in what the program should do. Variations can be accommodated by options in the program. The more options there are, the more versatile the program, the longer and less penetrable the instructions that go with it. Unfortunately, the programmer finds composing and testing a program far more fun than writing an intelligible description of what the program does and how to run it.

As the user gains familiarity with the program and its output, and is encouraged to use it frequently, he becomes aware of numerous small problems. Even if what the program does is entirely satisfactory, little things need to be done to the input before the program can be run as he wants it, and perhaps other little things need to be done to the output. For example, the raw data need to be checked: maxima and minima of each variable should be seen, as a precaution against grossly wrong entries, or the average may be compared with a known value; differences or ratios of variables that ought to be closely related should be examined. If variables are stacked as columns in a data matrix, some commonly needed operations are: to delete some columns, to transform all the entries in one of the columns to logarithms, to insert a new column consisting of the elements of a previously defined vector with some elements dropped from one end, to delete any rows for which the value in one of the columns lies in a certain range. Each of these operations seems to be simple, but there is no knowing how many of them will be needed and whereabouts in the proceedings. They cannot well be accommodated in the program as options. Instead, the programmer expands his offering from a single substantial program, or two or three programs, to a whole package or system, incorporating many further programs that can be run successively to accomplish the little editing and arranging and modifying jobs that are called for. If a high degree of flexibility is achieved in this way, something close to a general-purpose language has been devised, and the number of individual programs or commands will be closer to a hundred than to the original one or two or three. However, the general-purpose language has been invented unintentionally—it has entered by the back door, and is almost certainly not very good. Anywhere near as many as a hundred commands are difficult to remember and use accurately, unless they are very cleverly designed.

A variant of the special-purpose system is possible with conversational computing. The computer "takes charge" of the whole procedure, interrogating the user about what material he wants analyzed (or what kind of process he wants simulated, or whatever sort of thing the system is intended for), explaining to him what procedures are available, and asking him at each step what he would like done next. Such an interrogatory system can give assistance and guidance to the user, while permitting him considerable freedom of choice. The problem of design is not only statistical but psychological. How can the interrogation be conducted, so that misunderstanding is unlikely, guidance is offered to users who need it, and the system does not seem excessively clumsy to experienced users who know how they wish to proceed?

There is a different sort of difficulty with a comprehensive statistical package or system that is hard to avoid, namely that the instructions offered to the user may not explain and fully specify everything that the programs do, and the programs themselves are usually not open to direct inspection. A user whose interest in statistics is only secondary will perhaps not be too concerned about a little mystery in the output—he may consider statistical work altogether mysterious in any case. A more critical user may well find the uncertainties intolerable. It is just as unwise for a professional statistician to make use of a computer program that he cannot read as it is for him to rely on a mathematical theorem that he cannot prove.

The other way of making computers usable to the statistical public is to have a general-purpose language implemented, suitable for a broad class of computations, perhaps for all possible kinds of computations; and then persuade likely users that learning this language will *in the long run* prove the easiest and most satisfying route to control of the computer. Before, say, 1970 the run was indeed likely to be long, because the available languages were rather hard to learn and remember, containing numerous arbitrary quirks and restrictions, and rather clumsy to use. Almost all computing is done on arrays of entities, but most computer languages are designed to operate primarily on single entities, arrays being negotiated by loops—loops, and loops within loops, initialized, ranged and fussed over. So although today many persons concerned with statistics can wield Fortran or other similar language effectively, many cannot. The latter feel perhaps that they should not invest the time needed to acquire skill. They prefer, if computing seems inevitable, to employ a programmer—a course often both slow and infuriating.

One purpose of this book is to suggest that the computer language and conversational system known as APL has changed the situation. Though the language was not developed specifically for statistical work, it is in fact very suitable for the purpose. Four reasons are:

(i) APL was designed at the outset to handle scalars, vectors, matrices, and rectangular arrays in any number of dimensions, either numerical arrays or character arrays. Many basic operations can be specified for arrays just as well as for scalars, without any loop written in the program. Programs in APL therefore tend to contain few loops. The user is encouraged to think of array operations without a logically irrelevant internal sequence; this is aesthetically pleasing and often illuminating.

(ii) There is a high degree of consistency in APL, resulting from a high

degree of generality in the definitions. Syntax is governed ruthlessly by a few simple rules. Once the vocabulary is learned, the language is easy to remember. There is a remarkable absence of arbitrary features that require frequent reference to the manual. The language therefore has a peculiar dignity and reasonableness. It is worth learning.
(iii) Because character arrays are first-class citizens, addressed as easily as numerical arrays, scatterplots and other graphical displays can be programmed as readily as numerical calculations.
(iv) The implementation of the language as a conversational computing system affords the flexibility that is badly needed but hard to find in special-purpose packages.

To do one's own programming in APL compares well in effort with using the more versatile of the special statistical systems. That is, a person can get a general idea and begin to make use of APL for statistical work about as quickly as he can make similar progress with a special system; a two-hour introduction will set him on the road. To gain a mature understanding of APL, to take proper advantage of its power, to unlearn inappropriate habits of thought acquired from other computer languages—all that takes far longer, naturally, but the going is pleasant. Mature understanding of anything interesting takes a long time.

What will the relative usefulness of APL for statistical computation prove to be, compared with special statistical systems? The scientist, not primarily a statistician, who encounters statistical problems in his work, is likely to prefer a program package or interrogatory system requiring him to do nothing that would ordinarily be called programming. The better the user's understanding of statistics and the more he is interested in experimenting or exploring new directions in statistics, the more directly he will wish to control his computing and the greater the variety of computing he will wish to do. He will, as it seems, be best served by the best general-purpose computing system, which is now (by an order of magnitude) APL. A ready-made statistical system may be useful to him if composed in APL and open to inspection, with free-lance operation not excluded.

If this conclusion about working with APL ultimately receives general assent, the situation in regard to statistical software will resemble that for hardware. In the later 1940's, before computers were generally available, discussions were held between statisticians and computer men about statistical computing. Would it be possible to develop a machine for analysis of variance, say, having special facilities for summing squares and products? The experts advised against this; the effort would be better expended on a general-purpose machine. We see now that their judgment was right. If a good-enough general-purpose machine is available, no-one wants to be handicapped with a special-purpose machine. The situation seems to be similar with software.

The rest of this book is in two parts. Part A presents APL, showing how it can be applied to statistical calculations. Part B is devoted to examples of statistical analysis in the presence of flexible computing. A reader who does not want to study the details of APL may go straight to Part B, skipping there the few passages referring specifically to APL. Appendix 1 contains some further evaluative remarks, in the vein of this chapter. Appendix 2 is a supplement to Chapter 11. Appendix 3 lists a set of statistical programs that some readers may take as the starting point for their own APL library.

Notes for Chapter 1

The development of interest in statistical computing during the 1960's can be traced in three symposia listed in the bibliography under IBM (1965), Nelder and Cooper (1967), Milton and Nelder (1969). From the first to the third of these is a big jump.

The earliest statement that I have seen concerning the appropriateness of APL for statistical calculations is one by Hunka (1967), unfortunately laced with misprints. Smillie issued an extensive set of APL programs for statistical analysis and linear programming, *STATPACK*1 (1968) and a second edition called *STATPACK*2 (1969). Other writings of the same author include an expository article (1970), later developed into a book (1974). Smillie's programs contain many elegant and ingenious features, and one's programming skill is improved by reading them. However, I have purposely refrained from ever executing any of his (or anyone else's) statistical programs, preferring to consider the statistical problems afresh. How can I blame anyone who declines to use my programs?

For a thoughtful assessment of some problems in the design of interactive systems for statistical analysis see Ling (1980).

PART A
DESCRIPTION OF APL

Chapter 2

Origins

Computers perform logical tasks—numerical calculation and symbol arranging. A procedure for carrying out such a task is sometimes called an *algorithm*. (The dictionaries say the word ought to be "algorism" and it means the Arabic decimal system of arithmetic, named after an Arabic mathematician of the ninth century A.D.; but with spelling and meaning just stated the word is widely used in computing circles.) A precise specification of an algorithm, especially one so coded that it may be followed by a computer, is also called a *program*. A program consists of a sequence of *statements* or *commands*; some statements specify one or more entities in terms of operations on other entities; other statements, known as *branches*, specify which statement shall next be executed. A set of statements that are executed repeatedly, because a branch at the end returns execution to the beginning of the set, is called a *loop*.

For example, a well-known procedure for finding the square root y of a given positive number x may be expressed informally thus: guess a value for y; find the average of y and the quotient $(x \div y)$; this is a better guess; repeat until the better guess is equal to the previous guess to however many decimal digits you want. The procedure might be set out in numbered statements more formally thus:

Procedure to find an approximate square root y of a given number x.

1. Set $y = 1$.
2. Set $y' = y$.
3. Replace the previous value for y by

$$\tfrac{1}{2}\left(y' + \frac{x}{y'}\right).$$

4. If $|y - y'| \geqslant 10^{-6} \cdot y$, go back to statement 2; otherwise stop.

Here statements 1–3 are specification statements, statement 4 is a branch, statements 2–4 constitute a loop. Provided the given number x is positive and the calculations called for in statements 3 and 4 can actually be carried out with adequate precision, this procedure is feasible and terminates after a finite number of steps. In fact, the loop will be executed not more than $5 + \frac{1}{2}|\log_2 x|$ times.

This is not the only possible procedure, or algorithm, for finding a square root. A very different method is to take the logarithm of the given number (to any convenient base), halve it, and then take the antilogarithm. This method requires access to the logarithmic function and its inverse; the precision with which the functions are evaluated controls the precision of the result. There is no loop in the procedure as just stated; conceivably there could be loops in the invoked procedures for finding logarithms and antilogarithms. A third method for calculating square roots is a modification of ordinary long division, in which as each successive digit in the answer is determined the divisor is changed. The procedure is none too easy to describe precisely, although school-children have been known to carry it out readily. Which of these three procedures is best depends on the relative ease with which the required operations (division, taking logarithms, etc.) can be executed as well as on the number of steps to be performed. The best procedure for human calculation with paper and pencil is not necessarily the best for a computer.

Even the most rudimentary of arithmetical operations can be performed in more than one way, and a precise description of the procedure is harder to give than anyone might expect who had not done it. Consider, for example, how to decide which of two given positive numbers is the greater. Let us suppose that each number is expressed in decimal notation with a finite number of digits. If we are considering how a person does this task, we are tempted to say: look and see. Anyone can tell at a glance that 37 is greater than 6.4. But which of these numbers is the greater: 7933567002 and 894317114.75? The glance must be more careful. A possible procedure can be set out (informally, imprecisely) as follows:

1. Determine for each number how far to the left (or right) of the decimal point (or right end of an integer if no decimal point is shown) is the leftmost nonzero digit, called the leading digit. If the two leading digits are not equally far to the left (or right) of the decimal point, that settles the matter—as for the numbers above.
2. If the leading digits are equally far to the left (or right) of the decimal point, compare them. If they are unequal, that settles the matter.
3. If they are equal, move one place to the right in both numbers and again compare the digits. If they are unequal, that settles the matter. Otherwise, go back to the beginning of this statement.

The procedure either terminates with a decision or else stalls because no digits are left to be compared, in which case the two numbers are equal.

Something close to this is how any human would perform the task, whenever the numbers were too long to be taken in at a glance. Expressed with due precision, the procedure can be implemented on a computer, but will not necessarily seem the most attractive method. Subtraction will have been carefully implemented, and a good computer procedure may therefore be to subtract the second number from the first and note whether the difference is positive, zero or negative—that is, roughly, subtract and see the sign. No human would dream of doing the task that way.

Before the advent of computers, the word "algorithm" was best known in "Euclid's algorithm" for finding the highest common factor of two given positive integers. Replace the larger number by its remainder after division by the smaller number. Repeat this operation on the two numbers now in hand until a remainder of 0 is encountered, whereupon the other number (the last divisor) is the desired highest common factor.

All these algorithms seem simple to anyone familiar with them. Yet it is not so simple to describe them briefly but unambiguously, without leaving some details to the common sense of the reader.

Algorithms (as the word is currently understood) are not necessarily concerned with numerical problems. Consider the nonnumerical problem of arranging a list of names in alphabetical order. Let us suppose that the names are single words made out of the 26-letter alphabet, with no distinction between upper and lower case, no hyphens, apostrophes, accents or other complications, and that every name has at most ten letters. Names having fewer than ten letters may be thought of as augmented by one or more blank characters at the end, so that, with these blanks counted, all have ten letters. The task of arranging ten-letter names in alphabetical order is close to that performed by the author of a book who arranges the index in alphabetical order. When the task is done by hand, the names are written singly on cards, so that they can be rearranged easily.

A basic operation that may be carried out is to sort the names according to the letter that appears in the jth place, for any chosen j ($1 \leqslant j \leqslant 10$). That is, the names are arranged in 27 sets (or piles of cards, some possibly empty): the jth character of every name in the first set is blank, of every name in the second set is A, . . . , of every name in the last set is Z. Within any set, let the order of the names be the same as it was before the sorting. If the sorting is done with cards by hand, preservation of order within sets can be obtained by placing each card face down on the appropriate pile. Before computers there were mechanical devices for sorting punched cards on any column, with preservation of order within sets.

Suppose that, in some way or another, such sorting is available. How can it be used to order the names? The author ordering his index will almost certainly (if he is not a computer scientist) sort his cards according to the first letter, and get possibly as many as 26 piles of names beginning with A, names beginning with B, . . . , names beginning with Z (presumably there are no names beginning with a blank). If any of these piles contains only one card, no further sorting of it is needed; and if the pile contains just a few cards, but more than one, he will probably rearrange it somehow into alphabetical order. But any pile that contains more than just a few cards will be sorted according to the second letter, becoming possibly as many as 27 new piles. And if any of these contains more than just a few cards it will be sorted on the third letter, and so on. If the names to be ordered do not number more than a few hundred altogether, this method works well. It would, however, be bad for arranging the entries in a big-city telephone directory. Until all cards in a pile have been arranged in correct order, the piles must be kept separate, and the number of letters on which each pile has already been sorted should be known. The bookkeeping is too complicated.

The ordering of ten-letter names becomes much simpler in principle when the procedure is reversed, because then the piles need not be kept separate. Let the first sort be on the tenth letter. Let the resulting sets (piles) be put together into one set of names, which are now ordered in their last letter. Then the second sort is on the ninth letter, and again the sets are put together so that the names are ordered in their last two letters; and so on, the last sort being on the first letter. In this procedure, possibly more sorting is done than was really needed—perhaps the names would have appeared in correct alphabetical order had they been sorted on only their first five letters instead of on all ten. But the procedure is administratively simple.

Yet another way of arranging the names in alphabetical order is as follows. Associate with each name a vector of ten integers, showing the ranking that each of the ten letters has in the alphabet. A has rank 1, M rank 13, Z rank 26; and let us say that a blank has rank 0. For the four-letter name ADAM (followed by six blanks) we get the vector

$$1\ 4\ 1\ 13\ 0\ 0\ 0\ 0\ 0\ 0.$$

Regarding the vector as digits of an integer expressed in the scale of 27, calculate the number in whatever scale you prefer. From ADAM we obtain the decimal number

$$(1 \times 27^9) + (4 \times 27^8) + (1 \times 27^7) + (13 \times 27^6) = 8770812450471.$$

Now use a suitable algorithm (many exist) to arrange these numbers in nondecreasing order.

No human in his right mind would attempt to use that procedure by himself unaided. For some computers, however, the method is not only feasible but good.

The coming of stored-program digital computers gave an immense stimulus to the development, comparison and study of algorithms. Unambiguous expression has often been achieved by writing the algorithms in the "machine language" of a particular computer; but various "high-level" languages for expressing algorithms have also been devised that are more or less machine-independent and conform more or less to natural ways of thinking about the problems. In particular, the language known as Algol, developed by an international committee during the years 1958–1960, has been widely adopted as a means of communicating procedures for numerical computation—both communication with human readers and instruction of computers. Algol's immediate ancestor, Fortran, is still the most widely implemented high-level language for numerical computation.

For the purpose of communicating algorithms to human readers, it would be natural to turn to ordinary mathematical notation, which is certainly precise and concise. Unfortunately mathematical notation depends on context. That is, the same notation can have different meanings in different contexts. Suppose a mathematician refers to an entity called AB. If the context is elementary geometry, A and B are probably names of points and AB denotes the line segment between them. But if the context is algebra, AB is probably some kind of product—thus A and B may denote matrices and AB their inner product. Far from being ashamed of this ambiguity of notation, mathematicians rejoice in it. If, in a new field, a mathematician wishes to name an entity derived in some way from two other entities, A and B, and if he sees an interesting parallel with a geometric notion of line segment or with an algebraic notion of product, he will call attention to the analogy by choosing the name AB. "Suggestive notation" is esteemed. However, computing procedures are liable to overlap what are traditionally regarded as different branches of mathematics, and the confusion in customary notation can be undesirable.

In his book, *A Programming Language*, published in 1962, K. E. Iverson described a new language for precise and concise expression of algorithms, and applied it to various topics in computing method and design. He wrote:

> "Most of the concepts and operations needed in a programming language have already been defined and developed in one or another branch of mathematics. Therefore, much use can and will be made of existing notations. However, since most notations are specialized to a narrow field of discourse, a consistent unification must be provided. For example, separate and conflicting notations have been developed for the treatment of sets, logical variables, vectors, matrices, and trees, all of which may, in the broad universe of discourse of data processing, occur in a single algorithm." (Iverson, 1962, pp. 1–2.)

Iverson's language differed from Fortran and Algol in many respects. It was intended for all kinds of algorithms, not only numerical ones. It was fundamentally simpler and more mathematicianly; it was richer and more powerful. Because its purpose was to inform humans, and not directly to instruct a computer, no attention was paid, in formulating the language, to the existing peculiarities of computer hardware and organization, nor to the existing meagerness of key-punch character sets.

Later, Iverson with A. D. Falkoff and other colleagues at IBM's Thomas J. Watson Research Center implemented a modified version of the language as a high-level coding language for computation in conversational mode through typewriter terminals. The modified language is known by the initials of the title of the book, APL. The whole interactive system, as implemented on computers of IBM's 360 series, was known as *APL*\360. (The backward slash is an APL symbol meaning extension of what follows by what precedes.) There have been various further implementations. The programs listed in Appendix 3 of this book are in a more recent dialect known as *VS APL*.

There is a good deal of difference between APL and Iverson's language of 1962. Part of this is merely superficial. APL is transmitted through a typewriter having fewer printing devices than were called for in the book. Expressions in APL are typed on one line, without subscripts, superfixes or other vertical displacements; and only one style for letters of the alphabet is used (Italic capital). But in addition to superficial differences there are substantial differences of principle, resulting from further experience in the expression of algorithms, from adaptation to a reduced character set, from further logical analysis, and from the challenge of implementation in a variety of computing environments. APL is described in the manual by A. D. Falkoff and K. E. Iverson (*APL*\360 *User's Manual*, 1968), and in the current anonymous successor, *APL Language* (IBM, 1975). One of these, and not the 1962 book, should be consulted for information about APL.

We proceed now to a partial description of APL, as it can be used in any of the current implementations, for statistical computation. The complete facilities available are stated in the relevant manual. This partial description is intended to convey the feel of the language—to interest the reader rather than give full instruction. Emphasis is on the language as a means of expressing what is to be done, and not on management of the computing system.

Notes for Chapter 2

A wealth of information about algorithms is given by Knuth (1968, 1969, 1973). Perlis (1975) compares various computer languages, notably Algol60 and APL. Dijkstra (1962) describes Algol60.

Although Iverson's original book (1962) is not recommended as a source of information about APL, it is interesting, together with the companion book by Brooks and Iverson (1963), as showing the kinds of ideas out of which APL grew. The development of APL has been described by Falkoff and Iverson (1973); see also McDonnell (1979).

Numerous books about APL have appeared, notably Gilman and Rose (1970) and Polivka and Pakin (1975). A standard for APL has been published (Falkoff and Orth, 1979).

Chapter 3

Primitive Scalar Functions

One communicates with an APL system through a computer terminal. There is a keyboard much like that of an ordinary typewriter, through which the user gives instructions, and there is a display device permitting the user to see both what he has typed and what response the computer makes. Some terminals are indeed typewriters, and the input and the output appear typed on a roll of paper, which can be kept afterwards as a record of the proceedings. The printing is done by a typeball or other printing device, on a carrier that moves across the page to type a line. Other terminals show input and output on a cathode-ray tube (television screen). These have some advantages over printing terminals: output comes fast, and they are relatively inexpensive; but there is no record to take away; an auxiliary device must be used to obtain a permanent copy. All the illustrations for this book have been produced on a (quite slow) printing terminal with typeball, which yields the most legible record; and that is the kind of terminal that will be referred to.

A terminal connected to an APL system can be used as a desk calculator. If the user types "2 + 3" and then touches a key to transmit this message, the computer responds by causing "5" to be typed on the next line. When the keyboard of the terminal is unlocked and ready to receive a new command, the typeball carrier is in such a position that the user's command is indented six spaces, whereas the machine's reply is usually not indented. Thus a reader can tell which party typed what.

Figure A:1 reproduces a desk-calculator-like dialog between the user and the computer. The APL dialog is on the left, and some notes and numbering of lines for ease of reference have been added afterwards on the right. The display has been reduced in size a little in reproduction.

Subtraction is denoted by the ordinary minus sign, as in line 3. Negative numbers are represented with a "high minus" negative sign in front, which is an inseparable part of the number just like the numerals and decimal point (if any)—see line 4.

```
        2+3                                   Sum                      1.
5                                                                      2.
        2-3                                   Difference               3.
¯1                                                                     4.
        2×3                                   Product                  5.
6                                                                      6.
        2÷3                                   Quotient                 7.
0.6666666667                                                           8.
        2*3                                   Power                    9.
8                                                                     10.
        10⍟2                                  Logarithm               11.
0.3010299957                                                          12.
        2⌈3                                   Greater of (max)        13.
3                                                                     14.
        2⌊3                                   Lesser of (min)         15.
2                                                                     16.
        3!7                                   Binomial coefficient    17.
35                                                                    18.
        3|7.5                                 Residue                 19.
1.5                                                                   20.
        2○3                                   Circular function       21.
¯0.9899924966                                                         22.
        2=3                                   Equal                   23.
0                                                                     24.
        2≤3                                   Not greater than        25.
1                                                                     26.
        'A'≠' '                               Unequal                 27.
1                                                                     28.
```

Figure A:1

Multiplication and division are denoted by their old-fashioned standard signs. (Multiplication is *not* represented by juxtaposition, nor by a period; division is not represented by a solidus.)

The computer follows some standing rules of "format" when output is displayed through a terminal. Various methods are available for altering or by-passing the rules, so that the format of output (number of digits, spacings, indentations, etc.) can be minutely controlled in accordance with the user's wishes. Otherwise, the standard rules apply, as in Figure A:1. Integers in output are usually represented in ordinary decimal notation without a decimal point, as in lines 2, 4, and 6. Fractional numbers are usually shown as decimal fractions with ten significant digits, except that trailing zeroes after the decimal point are omitted. Thus in line 8 the number two-thirds is shown to ten decimal places, and lower down the page at line 20 the number one-and-a-half is shown to one decimal place. A comma or blank space is never used inside the decimal representation of a number to mark the thousands or millions.

The four basic arithmetic operations of addition, subtraction, multiplication and division, are indicated in APL by familiar signs. Each function has two arguments, and the sign for the function is placed between the arguments. Functions that are represented in APL by a single special symbol are called *primitive*; they are part of the given vocabulary of the language. A function that has two arguments is called *dyadic* (or one argument, *monadic*). The four basic arithmetic functions are defined for scalar arguments and yield scalar values. Thus these four functions are described as *primitive dyadic scalar functions*. There are seventeen other primitive dyadic scalar functions in APL, and some of them are shown lower down in Figure A:1.

In line 9 we see the power function denoted by a star. The ordinary mathematical notation for 2 to the power 3 is 2^3. That notation requires a superscript, which is not available on the terminal; but in any case it would not be satisfactory in APL because the function name is omitted and replaced by the special convention of superscripting. APL abhores special conventions!

Next in line 11 we see the inverse of the power function, the logarithm function represented by a star inside a circle. The first argument is the base, the second the number whose logarithm is to be found.

The function in line 13 yields the greater of the two arguments, that in line 15 the lesser. In line 17 we see the binomial coefficient function. The usual mathematical notation for line 17 would be $\binom{7}{3}$, about which the same comment can be made as for powers.

In line 19 there is the residue function, yielding the smallest nonnegative number that can be obtained by subtracting an integer multiple of the first argument from the second.

In line 21 is a circular function, denoted by a circle. The first argument must be an integer and indicates which circular function is to be applied to the second argument. If the first argument is 1 we get the sine function, if 2 the cosine function, if 3 the tangent. Other choices for the first argument yield other functions, natural or hyperbolic, direct or inverse. So line 21 asks for cosine 3. The second argument is in radians.

The last three functions shown in Figure A:1 are concerned with relations. They yield the value 1 or 0 according as the relation indicated is true or false. Thus their value is a truth value. The functions $=$ and $\neq$ can be applied not only to numerical arguments but also to characters. Line 27 asks for the truth value of the proposition that the character *A* differs from the blank character (a space).

In all these instances, the symbol representing a dyadic function is placed between its arguments. As for $+$, $-$, $\times$, $\div$, and the relations $=$, $\leqslant$, etc., so for the others. This is a syntactic rule in APL, and applies not only to all primitive dyadic functions, whether scalar or not, but also (as we shall see) to dyadic functions defined by the user.

The examples in Figure A:1 all involve a single dyadic function. Often we need to have several such functions in the same expression. Let us consider for a moment the state of affairs with ordinary mathematical notation for algebraic expressions involving scalar numerical-valued quantities a, b, c. Because addition is associative,

$$a + b + c$$

has a unique meaning, the same whether we consider the expression to mean

$$(a + b) + c$$

or

$$a + (b + c).$$

Subtraction is not associative, and execution of compound expressions is from left to right. Thus

$$a - b + c$$

means

$$(a - b) + c$$

and not

$$a - (b + c).$$

Multiplication is understood to have priority over addition and subtraction, so that

$$ab + c$$

means

$$(a \times b) + c$$

and also, because addition is commutative and multiplication has priority,

$$c + ab,$$

but not

$$a(b + c).$$

We reach confusion when we come to division. Authorities assert that

$$a + b/c$$

means always

$$a + \frac{b}{c};$$

but in fact mathematicians sometimes write the former expression when they mean

$$\frac{a + b}{c},$$

relying on context to keep the reader with them.* Everyone agrees that

*When as editor of a mathematical journal I pointed out to an author that his notation $n + 1/2$ properly meant $n + 0.5$, he countered by suggesting that no mathematician would misunderstand his intention in that way.

raising to a power has priority over multiplication, so

$$ab^2$$

does not mean

$$(ab)^2.$$

APL has many primitive functions, many more than elementary algebra. A table of priorities for their execution would be long, difficult to remember, and a source of annoyance. The system of inherent priorities has therefore been abandoned, and replaced by the simple rule that when an expression involving several functions is evaluated the rightmost function is executed first, then the next rightmost, and so on, except insofar as parentheses indicate a different order of execution. The rule is pleasantly unambiguous. Sometimes it does violence to custom.

In Figure A:2, line 1 means what the uninitiated would expect. On the other hand, lines 3 and 5 do not. It is always permissible to insert unnecessary parentheses to make the order of execution clear, as in the alternative expressions shown on the right. In lines 7 and 9 the parentheses in the expressions on the left are indispensable if they are intended; they can be omitted only after rearrangement of terms, as shown on the right.

```
      2+3÷4                            Or:     2+(3÷4)              1.
2.75                                                                2.
      2×3+4                            Or:     2×(3+4)              3.
14                                                                  4.
      2-3-4                            Or:     2-(3-4)              5.
3                                                                   6.
      (2×3)+4                          Or:     4+2×3                7.
10                                                                  8.
      (2-3)-4                          Or:     ¯4+2-3               9.
¯5                                                                 10.
      -2+3                             Or:     -(2+3)              11.
¯5                                                                 12.
      ¯2+3                             Or:     (¯2)+3              13.
1                                                                  14.
      -÷3                              Monadic   -   and   ÷       15.
¯0.3333333333                                                      16.
      |3-!4                            Monadic   |   and   !       17.
21                                                                 18.
      0.01×⌊0.5+100×*1                 Monadic   ⌊   and   *       19.
2.72                                                               20.
      1○○÷3                            Monadic   ○   and   ÷       21.
0.8660254038                                                       22.
      1÷2÷3÷4                          Or:     (1×3)÷(2×4)         23.
0.375                                                              24.
      0=0=0                                                        25.
0                                                                  26.
      17×0=0=0=0                                                   27.
17                                                                 28.
```

Figure A:2

We are familiar in ordinary mathematical notation with a monadic use of "+" and "−". "+3" means just 3, "−3" means negative-3. In APL, most of the symbols for primitive dyadic functions do double service and represent monadic functions whenever they have no preceding argument. Monadic "−" appears in line 11, which should be compared with line 13, where the "high minus" is a part of the notation for the number negative-2 and is not a function. Monadic "−" appears again in line 15 together with monadic "÷", which denotes the reciprocal function. In line 17 appear monadic "!", representing the factorial function, and monadic "|", denoting the absolute-value function. In line 19 the monadic star denotes the exponential function; and the symbol whose dyadic meaning is "lesser of" has monadic meaning "greatest integer not exceeding". Thus line 19 asks for the number usually denoted by e, rounded to two decimal places. In line 21 we see monadic circle, which just multiplies its argument by π, so this line asks for sin 60°. All monadic functions in APL, whether primitive or defined, obey the same syntactic rule that the name of the function precedes the one argument.

Each of the commands illustrated in Figures A:1 and A:2 has called for a calculation to be made and the answer immediately returned through the terminal. Nothing has been stored. Ordinarily the user prefers the result of a calculation to be stored so that it can be used later without having to be reentered from the keyboard. This is done by assigning the answer to a named variable.

Various objects in APL can be given names by the user—variables, defined functions, groups of these, workspaces. All such names consist of a string of one or more letters or digits, starting with a letter. Any of the letters may optionally be underlined. A name may not contain any blank space nor symbol representing a primitive function.

Assignment of a value (or other content) to a named variable is denoted by a left arrow. In line 1 of Figure A:3, the value 1.7 is assigned to a variable named X; that is, 1.7 is stored and can later be recalled by referring to it as X. When the computer has made this assignment, the keyboard unlocks to receive a new command, no other reply being needed.

The next few lines of Figure A:3 consist of an attempt to approximate the square root of 3. The Newton-Raphson iterative procedure mentioned in Chapter 2 is followed (more or less), with 1.7 as the initial value. In line 2 the difference between 3 and the square of X is called for (the answer being returned in line 3). Calling for the value of X in this way does not disturb X—the number 1.7 is still stored under the name X. In line 4, X is added to the quotient of 3 by X, the sum is halved and put back under the name X. So at this point X changes its value; the old value, 1.7, is blotted out by the new. The two leftmost symbols in line 4 call for the new X to be displayed. The square symbol, known as "quad", means the terminal.

```
      X←1.7                                             1.
      3-X×X                                             2.
0.11                                                    3.
      ⎕←X←0.5×X+3÷X                                     4.
1.732352941                                             5.
      3-X*2                                             6.
¯0.001046712803                                         7.

      3-X×X←⎕←0.5×X+3÷X                                 8.
1.732050834                                             9.
¯9.126879674E¯8                                        10.
      RT3←⎕←3*0.5                                      11.
1.732050808                                            12.
      |1-X÷RT3                                         13.
1.521146586E¯8                                         14.

      ⍝ SOME POISSON PROBABILITIES                     15.

      ⎕←MU←17×0.24                                     16.
4.08                                                   17.
      ⎕←P3←(MU*3)×(⎕←P0←*-MU)÷!3                       18.
0.01690746565                                          19.
0.1913849366                                           20.
      P3÷1-P0                                          21.
0.1946764216                                           22.
      P3÷ ⎕←  P4← P3 × MU÷4                            23.
0.1952126354                                           24.
0.9803921569                                           25.
```

Figure A:3

"Assign to quad" means "display". So line 4 is a double specification, a new value being assigned to X and then this value assigned to quad. Whenever an APL command calls for something to be calculated but does not assign it anywhere, assignment to quad is understood by default. The same two leftmost symbols in line 4 could optionally have been inserted on the left of all the commands in Figures A:1 and A:2 and also on the left of line 2 in Figure A:3. With this understanding concerning assignment to quad by default, we may say that all the commands in Figures A:1–3 are specification statements.

Instead of making the double specification in line 4, we could have omitted the first two symbols of line 4 and inserted an immediately following line consisting just of the command "X", which would have caused the display shown in line 5 of Figure A:3.

In line 6, the square of the new X is compared with the number 3. Since X multiplied by itself is the same as X squared, lines 2 and 6 are equivalent; the answers differ only because X has different values.

Another cycle of iteration is asked for in line 8. The new approximation is assigned to quad (displayed in line 9) and also put back under the name X, which is then multiplied by itself and subtracted from 3. The answer, not being assigned explicitly, is displayed (line 10).

Line 10 introduces us to "floating-point" notation, used to avoid excessively many zeroes in the output. The number before the letter *E* is understood to be multiplied by 10 to the power of the integer following the *E*. "1*E*6" means a million. The entry in line 10 means ¯0.0000000913 approximately.

In line 11 the square root of 3 is obtained from the power (star) function, displayed (line 12) and also given a name (*RT*3). The absolute value of the proportional difference between the last value of *X* and the true square root of 3 is asked for in line 13.

The remainder of Figure A:3 is concerned with some probabilities in a Poisson distribution whose mean is specified in line 16. Line 15 contains a note intended for a reader, not for the computer; it begins with a lamp sign. Any input line beginning this way is ignored by the computer (which would otherwise search for an object named *PROBABILITIES* and report failure in an error message). In line 18 two variables are defined (the probabilities of no and of three events) that in ordinary mathematical notation might have been expressed like this:

$$p_0 = e^{-\mu}.$$

$$p_3 = \frac{\mu^3 p_0}{3!}.$$

Line 21 asks for the conditional probability of three events, given that at least one has occurred. Line 23 asks for the probability of four events and also the ratio of probabilities of three and four events.

Line 23 illustrates a detail concerning the format of input, namely that any number of unnecessary spaces may optionally be inserted anywhere, except inside a name. This is the only input line in Figures A:1–3 that is not typed as compactly as possible.

The peculiar *convenience* of APL for statistical computing (as indeed for other kinds of computing) does not begin to appear until we come to arrays. But already we see several characteristic features of the language—the uniform treatment of functions, the priority in execution based only on position (execute the rightmost executable function first), and a quality that can be called permissiveness. As soon as the reader understands the multiple specifications in lines 4 and 8 of Figure A:3, he will see that the more tortuous-looking line 18, with its first two executed specifications enclosed in parentheses, is similar; the meaning is clear, and the computer accepts it.

For readers already familiar with another computer language, like Fortran or Algol, a few further comments may be helpful. APL does not require any preliminary declaration about the type of a variable (nor, when we come to arrays, a preliminary declaration of dimension). The same named variable may change its type (and dimension) as the computation proceeds. For most purposes we do not need to be concerned whether a numerical-valued variable has only the "logical" values 0 and 1, or whether it has other integer values, or whether it has fractional values. Changes of type are made automatically by the system when needed. Note in particular that logical variables are numerical and not relegated to a separate "Boolean" category. (Line 27 of Figure A:2 may seem to be only a joke, but it is a permissible joke—the integer 17 *can* be multiplied by a truth-value of 1.)

Numbers are represented in the computer in the binary scale. A binary digit is called a *bit*, and 8 bits are called a *byte*. In regard to accommodation, three sorts of numbers are recognized in *APL*\360 and some other APL systems, namely (i) logical numbers equal to either 0 or 1, which can be stored as single bits, (ii) signed integers less than 2^{31} (roughly 2.1×10^9, or, if you prefer, 2.1*E*9) in magnitude, which can be stored in 32 bits or 4 bytes, (iii) signed floating-point numbers whose magnitude lies between (roughly) 7.2*E*75 and the reciprocal of that and whose precision is equivalent to about 16 decimal digits—such numbers can be stored in 64 bits or 8 bytes. Characters are stored in single bytes. Knowledge of these facts can sometimes be turned to account in devising ways to store bulky material economically. The size of the "active workspace" in which computations take place depends on local conditions. Most of the illustrations in this book have been executed in a workspace whose available working area has size approximately 32000 bytes.

Chapter 4

Arrays

This chapter is concerned with description and formation of arrays. Three new symbols representing primitive nonscalar ("mixed") functions are introduced. The following chapter will deal with extension to arrays of the primitive scalar functions that we have already met, and will introduce some further primitive mixed functions.

A named variable in APL stands in general, not necessarily for a single item which we refer to as a scalar, but for an array of items. The members of such an array must be either all numbers or all characters (letters, digits, or other symbols of the language). The permitted kinds of array are: a scalar; a vector; a rectangular matrix; a rectangular array in three or more dimensions. There is no restriction in principle on the number of dimensions—whatever restriction there is comes merely from the size of the workspace in which computations take place. All arrays, if not scalars or vectors, must be perfectly rectangular. Other kinds of array, such as, for example, a two-dimensional array consisting of rows of unequal length, or a list having a tree structure, or an array of any shape having partly numerical and partly character members, can be readily handled in APL, but a little less directly, after expression in terms of permissible variables.

Sharp distinction is made between the possible structures of variables in APL. Whereas a scalar is a single element, number or character, not indexed, a vector is a list or concatenation of elements indexed by one indexing variable taking consecutive positive integer values; a matrix is a set indexed by two such indexing variables; and so on. A single number or character entered through the terminal is ordinarily taken to be a scalar, though by special action it can be made into a vector of unit length, or a matrix with one row and one column, etc. Similarly, a simple list of numbers or characters is ordinarily taken to be a vector, though if we wish we can make from it a matrix having either one row or one column, these being the "row-vector" and "column-vector" familiar in matrix algebra. Vectors in APL are neither row-vectors nor column-vectors, but just vectors!

A numerical vector is represented, in input and output, as a string of numbers separated by spaces. At least one space between consecutive numbers is essential; for the numbers in their decimal representation are not of fixed length, and without spaces or some other marker there would be no way to tell where one number ended and the next began. By contrast, in character vectors a space is one of the characters, and all characters have the same (unit) width, so extra spaces or other markers to distinguish the elements of the vector are not needed. In input, the elements of a character vector must be enclosed in quotes, but such quotes do not appear in output.

In line 1 of Figure A:4 we see the variable *VA* specified as a numerical vector, and in line 2 the variable *VB* is specified as a character vector. In line 3 a display of *VA* is asked for, appearing in line 4, which is like what was entered in line 1 except that the elements of the vector are separated by two spaces each and trailing zeroes are omitted after the decimal point.* In line 5 a display of *VB* is called for, appearing in line 6 exactly as it was entered except for omission of the quotes.

A further example of a character vector is entered in line 7. This line of literary text has quotes in it. To distinguish a quote (or apostrophe) that is "really there" from the quotes needed to enclose a character array at input, the quotes must be doubled—compare the input in line 7 with the output in line 8. Note that the computer makes no attempt to execute a character array as though it were an APL command—it does not try to execute the !-function, nor in *VB* did it try to execute "2 + 3".

We may call for individual elements of a vector by reference to the indexing variable, which cannot be shown as a subscript and so is enclosed in square brackets. (In APL square brackets always enclose something that qualifies like a subscript what is immediately to the left of the left bracket.) In line 9 the first element of *VB* is asked for; the result, displayed in line 10, is a scalar. In line 11 several elements of *VB* are asked for, corresponding to a vector of index values; the result, displayed in line 12, is a vector.

In this book, indexing variables always takes values from 1 upwards. That is called 1-origin indexing. An option is open to the user to designate 0 instead of 1 as the place where counting begins—called 0-origin indexing. Several of the functions mentioned in this and the next chapter, being concerned with indexing, behave differently when 0-origin indexing is specified. No further reference to 0-origin indexing will be made here.

The Greek rho (ρ) is the symbol for two primitive functions concerned with the structure of arrays, one function monadic, the other dyadic.

*When the format of numerical output is not directly specified by the user, standard rules apply. The way the members of a numerical array are separated by spaces has varied a little in the different implementations of APL. Some implementations would show the numbers in line 4 separated by a single space, not two.

```
       VA←39.4 1.7 1.3 5.0 5.2 9.9 5.7 6.7 2.2 2.0 0.6 0.1 2.3 15.9    1.
       VB←'LINE 1 OF FIG. A:1 READS:  2+3'                              2.
       VA                                                                3.
39.4  1.7  1.3  5  5.2  9.9  5.7  6.7  2.2  2  0.6  0.1  2.3  15.9      4.
       VB                                                                5.
LINE 1 OF FIG. A:1 READS:  2+3                                           6.

       ⎕←CTA3740←'''TEHEE!'' QUOD SHE, AND CLAPTE THE WYNDOW TO,'       7.
'TEHEE!' QUOD SHE, AND CLAPTE THE WYNDOW TO,                             8.

       VB[1]                                                             9.
L                                                                       10.
       VB[6 28 27 3 8 8 3 14]                                           11.
12 NOON.                                                                12.
       ρVA                                                              13.
14                                                                      14.
       ρVB                                                              15.
30                                                                      16.
       ρρVA                                                             17.
1                                                                       18.
       ρρVB                                                             19.
1                                                                       20.
       ρ17                                                              21.
                              (blank line returned by the computer)     22.
       ρρ17                                                             23.
0                                                                       24.
       ρρρ17                                                            25.
1                                                                       26.
```

Figure A:4

Monadic ρ yields a vector, called the *dimension vector* or *shape*, showing the size of its argument. For any array X, ρX is a vector of length equal to the number of indexing variables of X; it lists the number of values that each of these indexing variables takes. In front of a vector, ρ gives a one-element vector equal to the length of the argument—see lines 13–16. If we place another ρ in front, we get a one-element vector equal to 1, which is the dimensionality or *rank* of the original argument, the number of indexing variables that it had. Monadic ρ in front of a matrix yields a vector with two elements, the number of rows and the number of columns; and if we place another ρ in front we get a vector consisting of one element equal to 2, the number of indexing variables of the original matrix. Similarly, monadic ρ in front of a three-dimensional array yields a three-element vector, the number of planes, the number of rows and the number of columns; and ρ in front of that gives a one-element vector equal to 3. A scalar is an array with no indexing variable. Monadic ρ applied to a scalar therefore yields an empty vector (having no elements, displayed as a blank in line 22), and ρ applied to that counts the number of elements in its argument and yields a one-element vector equal to 0 (line 24). Three or more ρ's in front of any array always yields a one-element vector equal to 1 (lines 25–26).

Leaving dyadic ρ aside for the moment, let us consider another symbol concerned with the structure of arrays, namely the comma (,), which again stands both for a monadic and for a dyadic primitive function. Monadic comma transforms its argument into a vector composed of the same elements. If the argument is already a vector, monadic comma leaves it unchanged. If the argument is scalar, monadic comma makes it into a one-element vector. (Compare lines 21–24 of Figure A:4 with lines 1–4 of Figure A:5.) If the argument is a matrix, monadic comma strings it out as a vector, the elements of the first row followed by those of the second row, and so on; the length of the resulting vector being equal to the number of elements in the original matrix (product of number of rows and number of columns). Monadic comma is referred to as the *ravel* function. The notation is unintuitive.

Dyadic comma is used to join arrays together in various ways. We consider here what happens when each argument of dyadic comma is either a vector or a scalar. The result is the *catenation* or stringing together of the two arguments to form a vector. For example, "3,4" means the catenation of the scalars "3" and "4" to form the two-element vector "3 4"; and "3 4,5" means the catenation of the two-element vector "3 4" with the scalar "5" to make the three-element vector "3 4 5". It is assumed here, as usual when the meaning of APL functions is being illustrated, that the expressions mentioned are isolated. It is not true that in all contexts "3,4" means the same as "3 4", because in compound expressions functions must be executed from the right. Thus "3, 4 × Y" means that 3 is to be catenated onto (4 × Y), but "3 4 × Y" means that the vector (3 4) is to multiply Y. These expressions may or may not be meaningful, as we shall see, but in any case they are different. The standard representation for a numerical vector is a string of numbers separated by spaces, *not* commas. Such a string is regarded as an unbreakable whole, as though enclosed in parentheses. "3 4 × Y" means "(3 4) × Y", not "3 (4 × Y)". The latter is meaningless because the space is not a function, but either a character or a distinguishing mark to be used in some recognized contexts.

We may now return to ρ and consider the dyadic function, as in "$X\rho Y$". The first argument (X) must be a vector of integers—or possibly, by typical permissiveness, a scalar integer, which is interpreted as a one-element vector. The second argument (Y) can be any array, which will be interpreted as if it had monadic comma in front of it and was therefore strung out as a vector. What is yielded is an array whose dimension vector is X and whose content is made out of the elements of Y. As many elements of Y as needed are taken; if Y does not have enough elements to go round, its elements are used over again cyclically. Thus $X\rho Y$ is always equivalent to

$$X\rho(,Y),(,Y),(,Y),\ldots$$

```
      ρ,17                                                          1.
1                                                                   2.
      ρρ,17                                                         3.
1                                                                   4.

      VB[6 29]←'3-'                                                 5.
      2 19 ρVB,8ρ' '                                                6.

LINE 3 OF FIG. A:1                                                  7.
READS:  2-3                                                         8.

      ⎕←I←3 3 ρ1 0 0 0                                              9.

 1 0 0                                                              10.
 0 1 0                                                              11.
 0 0 1                                                              12.

      ρI                                                            13.
3  3                                                                14.
      ρρI                                                           15.
2                                                                   16.
      ,I                                                            17.
1  0  0  0  1  0  0  0  1                                           18.
      (ρ,I),ρρ,I                                                    19.
9  1                                                                20.

      2 2 6 ρι20                                                    21.

   1   2   3   4   5   6                                            22.
   7   8   9  10  11  12                                            23.

  13  14  15  16  17  18                                            24.
  19  20   1   2   3   4                                            25.

      2 4 ρVA                                                       26.

    39.4            1.7            1.3            5                 27.
     5.2            9.9            5.7            6.7               28.

      I[ι2;ι2]                                                      29.

 1 0                                                                30.
 0 1                                                                31.

      I[3;]                                                         32.
0  0  1                                                             33.
      ρI[3;]                                                        34.
3                                                                   35.
      I[,3;]                                                        36.

 0 0 1                                                              37.

      ρI[,3;]                                                       38.
1  3                                                                39.
```

Figure A:5

A further primitive function symbol useful in forming arrays is the Greek iota (ι), which as usual stands both for a monadic and for a dyadic function. The dyadic function will not be described here. Monadic ι takes as its argument a nonnegative integer (scalar or one-element vector), say N. Then ιN is a vector of the first N natural numbers (positive integers), $1, 2, \ldots, N$. In particular, $\iota 0$ stands for an empty vector.

In output, a matrix is printed as a rectangular array, as one would expect. If it is a character matrix, there is no spacing between columns (because spaces are characters). For a numerical matrix, the amount of spacing between columns, in standard format, is usually less if all elements are small integers than if some elements have fractional parts. For a three-dimensional array, the indices refer to planes, rows and columns, in that order. In output on two-dimensional paper the planes cannot conveniently be stacked vertically, so they are printed as matrices, one after another, with a blank line between. Examples of such displays are shown in Figure A:5, which is a continuation of Figure A:4 and refers to the vectors VA and VB defined therein. In line 5 two of the characters in VB are changed, and then in line 6 eight blanks are catenated onto VB and the result displayed as a character matrix with two rows and nineteen columns (lines 7–8). An identity matrix is entered in line 9, displayed in lines 10–12, strung out as a vector in lines 17–18, and various dimension vectors are examined. A three-dimensional array is called for in line 21 and displayed in lines 22–25. A matrix of noninteger numbers is formed and displayed in lines 26–28.

To call for individual elements of a matrix, the row and column numbers are appended to the matrix name, separated by a semicolon, inside square brackets. Thus for a matrix M we write "$M[I;J]$". If I and J are both scalar this means the (I,J)th element of M and is scalar. If I is a vector but J is scalar, $M[I;J]$ is the vector of elements in the Jth column and in the rows indexed by I; and similarly the other way round. If I and J are both vectors, $M[I;J]$ is the submatrix of elements in rows I and columns J of M. To obtain all the elements in the row or rows indexed by I we write "$M[I;]$", and similarly to obtain all the elements in column(s) J we write "$M[;J]$". To index a three-dimensional array A, we use two semicolons, thus: "$A[I;J;K]$".

The semicolon acts as a separator, rather like a parenthesis; it is not a function. "$M[1+2;3]$" means "$M[(1+2);3]$", or $M[3;3]$, the third element in the main diagonal of M. It does not mean "$M[1+(2;3)]$", which is meaningless. We shall encounter another use of semicolons later in mixed numerical and character output, where again the semicolon separates like parentheses.

Some things in the foregoing account of arrays have perhaps mystified and exasperated the reader: empty vectors, distinctions between scalars and one-element vectors, empty spaces that are characters and blank lines that are nothing. So let us go back and reconsider the question: what are arrays in APL?

Arrays have both *content* and *structure*. The content is a set of elements —possibly a set of numbers, possibly a set of characters, possibly an empty set containing nothing at all. The structure is the pattern in which these elements are regarded as being arranged. Structure would be represented in ordinary mathematical notation by associating subscripts (indices) with each element. The number of indices is necessarily (you count them, you must get) 0 or 1 or 2 or 3 or whatever, a nonnegative integer. Each index ranges over a set of values—the number of values in each set must necessarily also be a count, 0 or 1 or 2 or . . . , a nonnegative integer. The dimension vector lists the number of values in each index set. When an array X is stored in the computer, both the structure (ρX) and the strung-out content ($,X$) are stored. If we are to understand an array fully, we must know the structure as well as the content. If we ask for a display of X, we shall see (more or less) the content of X, arranged in a way that depends on the structure. But from the display of X alone we cannot infer the structure, because different structures could have given rise to the same arrangement; the display tells us something about the structure, but not all. To see the structure we ask for the dimension vector (ρX).

Before going further, let us pause a moment over the statement just made that asking for a display of X causes us to see (more or less) the content of X. How exactly is the content reproduced? If X is a numerical array whose content consists of not-too-large integers, the display represents those integers exactly. If the numbers are very large integers or have fractional parts, they are displayed with a precision of a prearranged number of decimal digits (10 is the standard choice), though they are held in the computer with a precision of approximately 16 decimal digits.* Then the display almost but possibly not exactly reflects the stored content. If X is a character array, usually in practice a display of X reproduces the characters exactly and unambiguously, but not always. The trouble is that not every possible element in a character array is revealed in output by the printing of a symbol. There is the blank character, which is a perfectly good existent non-empty element of a character array but does not elicit any positive action from the typeball. There are other invisible elements; a carrier return, a linefeed and a backspace can each be accepted in an array, counting as a single character, and it is even possible to have one-byte codes accepted that do not correspond to anything the terminal can do and therefore cannot be put out through the terminal, though they may be worked with by the computer. However, the blank character or forward space is the one non-symbol regularly and inevitably used in character

*This is the precision in the implementations with which the writer is acquainted. In principle, the precision may vary with the implementation.

arrays. In a line of character output, blank characters are only perceived unambiguously if they are followed by a visible character. Trailing blanks are invisible (as in lines 7–8 of Figure A:5). Their presence is not even revealed by the typeball's traveling over them before returning for the next line—when nothing more has to be printed, the carrier returns at once.

A third possible kind of content for an array is an empty set. For most purposes, no distinction is made between empty numerical arrays and empty character arrays—there is only one vacuity. The empty numerical vector ⍳0 is equivalent to the empty character vector `''`; for example, either of these may be catenated (by dyadic comma) onto a non-empty numerical vector or onto a non-empty character vector, the result being the same non-empty vector.* An empty array of any rank is displayed as a blank line. Thus when we see a character array displayed, without also seeing (or knowing) its dimension vector, we may be uncertain whether a lack of printing results from blank characters in the array or from nothingness. But in the display of a numerical array no possible confusion can come from spaces, for all spaces are mere distinguishers in the display and never represent elements of the array.

It is unusual not to know whether a named variable is a numerical array or a character array, but if that were so and a display of the array showed an arrangement of decimal numbers as if a numerical array were being represented, the viewer could not be sure whether the variable was indeed a numerical array or the character representation of such an array. Numbers upon which arithmetical calculations are to be made are held in the computer as numerical arrays in a binary coding (of which the details are invisible to the user). Thus if the computer is to make a calculation with the number eight, it holds this number as a string of bits that might be represented in symbols as 1000. But at input and output, for the convenience of the user, translations are made between numbers in the computer's binary coding and numbers in equivalent decimal notation. Usually numbers are held in named numerical arrays, and the representing character arrays are not named, but for special purposes, such as laying out a mathematical table, a representing character array may be named. (See Exercise 15 at the end of Part A below.) Translations (in either direction) between named numerical arrays and named representing character arrays have always been possible in APL; they are particularly easy in later versions of APL having primitive "format" and "execute" functions.

*It might well have been that APL made no distinction for any purpose between an empty numerical vector and an empty character vector. In fact, the numerical or character origin is preserved when an empty array is defined, and for some purposes it is remembered. When one or more elements are taken from an empty array by the "take" or "expand" functions mentioned in the next chapter, the elements are zeroes if it was an empty numerical array and blanks if it was an empty character array.

To return to structure, one possible structure for an array is scalar. The content is one element (the content cannot therefore be empty), the rank is 0 because there are no indices, and so the dimension vector is empty. All other kinds of array have at least one index, the numbers of values taken by them being listed in the dimension vector (which is therefore not empty). If any element in the dimension vector is zero, the array is empty (has no content). Empty vectors are frequently encountered in APL programs, because (i) the much-used monadic functions ρ and ι both yield empty vectors whenever applied to (respectively) a scalar or zero argument, and (ii) vectors are sometimes built up by successive catenation, and an empty vector may make a convenient starting point. Empty arrays of rank greater than 1 are met with less often.

If all elements in the dimension vector are positive, the array has content. In output, a vector is displayed on one line, unless it is too long, in which case it overflows onto following lines. A matrix with one row is displayed the same way, as also a three-dimensional array whose dimension vector begins with two ones. Thus if an array is displayed on one line and has at least two elements, we infer that it is either a vector, or a matrix with one row, or a three-dimensional array with one plane and one row, etc. If a displayed array has only one element, it could be scalar or a one-element vector or matrix, etc. The computer knows the structure, even if the user doesn't!

It might be suggested that the standard form of display of an array ought to include automatically a listing of the dimension vector, so that full understanding would be available at once. There are several reasons why that would be a nuisance. One is that often the dimension vector of an array is known before the array is generated. The content, not the dimension vector, is what the user doesn't already know and wishes to see. With character arrays it is often precisely the display, not the array, that we wish to examine. When we look at a scatterplot or a news announcement, we usually have no interest in the structure of the array in the computer.

Even when structure is of interest, sometimes the partial information about structure implied by a display of the array is sufficient for our needs. We only need to know the content, and not the structure at all, before using an array as the second argument of dyadic ρ. One-element arrays are regularly interpreted by the computer as though they were either scalars or vectors, whenever the context unambiguously requires one of these. This convenience encourages carelessness. Occasionally the context is not so compelling and the outcome surprises us! It is more satisfying, and more interesting, to be conscious of structure most of the time. Hence the emphasis here.

We have seen that we can always make a vector out of a scalar by preceding it by monadic comma, the ravel function. Conversely, if X is a one-element array, we can make it scalar by writing "$(\iota 0)\rho X$", or, if we know X to be a vector, "$X[1]$", or by a reduction as explained in the next chapter.

The handling of structure in APL, with toleration and consistent treatment of the extreme structures (empty arrays, one-rowed matrices, etc.), though it may seem mystifying at first, is a great strength and convenience of the language, often permitting special cases to be accommodated without special provision.

To sum up this description of arrays: when an array X is stored, it is as though it were coded as a sequence of symbols having the following form and meaning (the actual coding is the business of the implementer and does not concern us):

first, the value of $\rho\rho X$;
next, the members of ρX (if any);
next, an indicator of the kind of content (say **C** for character, **L** for logical number, **I** for 4-byte integer, **F** for floating-point number);
next, the strung-out content (if any).

Thus we can think of the following codings:

	Specification	*Coding of X*
$X \leftarrow 5$	(scalar number)	0**I**5
$X \leftarrow 1.5$	(scalar number)	0**F**1.5
$X \leftarrow ,5$	(one-element vector)	1 1**I**5
$X \leftarrow$'*I AM*'	(character vector)*	1 4**C***IbAM*
$X \leftarrow 3\ 3\ \rho\ 1\ 0\ 0\ 0$	(identity matrix)	2 3 3**L**1 0 0 0 1 0 0 0 1
$X \leftarrow \iota 0$	(empty vector)	1 0**I**
$X \leftarrow 3\ 0\ \rho$'*ABC*'	(empty matrix)	2 3 0**C**

There are no symbols in this coding representing emptiness or any other tenuous notion. Provided the place where the coding begins (the first symbol, always a nonnegative integer) is identified, the extent of the coding can be determined without a marker to show the end of it.

A statistical data set may constitute an array having one of the permissible structures for an APL variable—a vector or a perfect rectangular array in two or more dimensions—and then is naturally represented in APL directly as a variable. Any statistical data set whatever has content and structure. If we know the content, and if we understand and can describe the structure, we can certainly specify both in terms of APL variables. From the fact that perfect rectangular structures are handled particularly elegantly in APL, one should not infer that arbitrary structures are not handled with as much elegance as their arbitrariness permits. We are all

*In this line *b* stands for the blank character or space, represented in the computer by a one-byte code like any other character.

familiar with very convenient notation for some common mathematical functions, such as the usual representation of the product function of arguments a and b by the simple juxtaposition, ab, or of the power function by another juxtaposition, a^b. But ordinary mathematical notation can represent arbitrary functions only a little less concisely, as by the notation $f(a, b)$. Great ease for some purposes does not imply unreasonable clumsiness for other purposes. So it is for the representation of data sets. In fact, data sets having a complicated structure can be represented in various ways in APL, and transformations from one form to another are easily accomplished. For an example of some possibilities, see Exercise 15 at the end of Chapter 11 below.

Chapter 5

Primitive Functions on Arrays

The primitive *monadic* scalar functions that we met in Chapter 3 are extended to non-scalar arrays element by element. If f stands for such a function and X is any array, fX means the array formed by applying f to each element of X; fX has the same size as X.

The primitive *dyadic* scalar functions can be applied to arrays in four ways, namely: element by element, reduction, outer product, inner product. If f stands for such a function and X and Y are any two arrays of the same size, the *element-by-element* application of f is denoted by XfY. It is an array of the same size as X and Y, formed by applying f to corresponding elements. Thus if X and Y are numerical vectors of equal length (or matrices of equal size), $X + Y$ denotes the vector sum (or matrix sum, as the case may be). In general the two arguments must be of the same size, with one exception: either argument may be a scalar or other one-element array, in which case it is associated with each element of the other argument. Thus $X * 2$ denotes an array of the same size as X, consisting of the squares of the elements of X.

Consider the specification statement "$Y \leftarrow 2 = \rho X$". Since the "2" is scalar, the size of Y is equal to the size of ρX. If X is a vector, ρX is a one-element vector equal to the length of X, and so Y is a one-element vector, equal to 1 if X is of length 2, otherwise equal to 0. If X is a matrix, ρX is a vector of two elements and so therefore is Y (for ρX equal to 2 3, Y is equal to 1 0). If X is scalar, ρX is an empty vector and so is Y.

For any primitive dyadic scalar function f and any vector V, the *reduction* of V by f is denoted by f/V. It is a scalar equal to

$$V[1]fV[2]f\ldots fV[N],$$

where N is the length of V (N is the scalar equal to the value of the one-element vector ρV). Thus $+/V$ denotes the sum of the elements of V and $\times/V$ denotes their product. Execution of the compound expression is, as usual, from right to left. Thus $-/V$ denotes the alternating sum,

$V[1] + (-V[2]) + V[3] + (-V[4]) + \ldots$, and similarly $\div / V$ means $V[1] \times (\div V[2]) \times V[3] \times (\div V[4]) \times \ldots$. If V is of length 1, f/V is the scalar equal to the value of V. Thus we may say that N defined above is equal to $+/\rho V$, or equivalently $(\rho V)[1]$. If V is empty, f/V is equal to the identity element of the function f if there is one; these identity elements are explained and listed in the manual. Reduction of an empty vector by $+$ or $-$ yields 0, and by $\times$ or $\div$ yields 1.

If S is scalar, f/S means just S. If A is an array of rank 2 or more, the reduction may be made along any coordinate, and in general the solidus must be "subscripted" to indicate which coordinate is to be reduced. The result is an array of rank one less than A. Thus for a matrix M, the vector of row sums, obtained by reducing over columns, may be written $+/[2]M$; and the column sums obtained by row-reduction are $+/[1]M$. If the subscript is omitted, the last coordinate will be reduced. If, without subscript, the solidus is crossed with a horizontal line (minus sign), the first coordinate will be reduced. Since nearly all reductions in practice are along either the first coordinate or the last coordinate, even for arrays of high rank, we rarely need to use explicit subscripts. The sum of all the elements of a matrix M may be denoted by $+/+/M$, since this means the sum of the elements of the vector of row sums of M. Alternatively we may write $+/,M$. Similarly the sum of all elements of a three-dimensional array T can be denoted by $+/+/+/T$ or by $+/,T$.

Figure A:6 illustrates some of these procedures. Line 1 calls for the first six terms in the Taylor series for $e^{0.03}$. The vector 0 1 2 3 4 5 is labeled N. The factorial sign in front of N gives the vector of factorials of the members of N, that is

$$(1, 1, 2, 6, 24, 120).$$

The expression in parentheses stands for the scalar 0.03 raised to the power of the members of N, that is the vector

$$\left(1, 0.03, (0.03)^2, (0.03)^3, (0.03)^4, (0.03)^5\right).$$

These two vectors are divided, term by term, to give the vector

$$\left(\frac{1}{1}, \frac{0.03}{1}, \frac{(0.03)^2}{2}, \frac{(0.03)^3}{6}, \frac{(0.03)^4}{24}, \frac{(0.03)^5}{120}\right),$$

which is labeled X and displayed in line 2. In line 3 the sum-reduction of X is asked for, approximating the value of $e^{0.03}$, and also the difference-reduction, approximating $e^{-0.03}$. These quantities are asked for directly in line 5 by means of the monadic star (exponential) function; both members of the vector $(3, -3)$ are divided by 100 and then exponentiated. The binomial coefficients of order 5 are asked for in line 7.

```
        ⎕←X←(0.03*N)÷!N←0,⍳5                                                     1.
1  0.03  0.00045  4.5E¯6  3.375E¯8  2.025E¯10                                    2.
        (+/X),-/X                                                                3.
1.030454534  0.9704455335                                                        4.
        * 3 ¯3 ÷100                                   Or:    * 0.03 ¯0.03        5.
1.030454534  0.9704455335                                                        6.
        N!5                                                                      7.
1  5  10  10  5  1                                                               8.

        ⎕←M←(⍳3)∘.×⍳5                                                            9.

    1   2   3   4   5                                                           10.
    2   4   6   8  10                                                           11.
    3   6   9  12  15                                                           12.

        +/M                                           Or:    +/[2]M             13.
15  30  45                                                                      14.
        +⌿M                                           Or:    +/[1]M             15.
6  12  18  24  30                                                               16.

        ÷(⍳3)∘.+¯1+⍳3                                                           17.

      1               0.5             0.3333333333                              18.
      0.5             0.3333333333    0.25                                      19.
      0.3333333333    0.25            0.2                                       20.

        0.05×+/(*-X)÷1+X×X←0.1×,(⍳110)∘.-0.5×1+ 1 ¯1 ×3*¯0.5                    21.
0.6214496235                                                                    22.
```

Figure A:6

Line 9 introduces an *outer product* of two vectors. The name "outer product" has been retained in APL, but the idea has developed greatly from the outer product of matrix algebra and does not necessarily refer to multiplication. If X and Y are any two arrays and f is any primitive dyadic scalar function, the outer product, written $X \circ .fY$, is an array of size $(\rho X), \rho Y$ in which each element of X is combined by f with each element of Y. For example, if X is a matrix and Y a vector, and if the resulting array is called Z, then Z is three-dimensional and $Z[I; J; K]$ is equal to $X[I; J]$ $fY[K]$. In line 9 the outer product of two vectors, using the function $\times$, yields the matrix M. Row sums (reduction along columns) are asked for in line 13, and column sums in line 15. In line 17 a 3-by-3 Hilbert matrix is called for, of which the (i, j)th element is $(i + j - 1)^{-1}$. The outer product of the vectors 1 2 3 with 0 1 2, using the function $+$, is formed and then reciprocals of all elements are taken.

Line 21 shows an example of numerical integration, an attempt at evaluation by quadrature of

$$\int_0^{\infty} \frac{e^{-x}\,dx}{1 + x^2}\,.$$

The method is based on the Gaussian formula*

$$\int_0^h f(t)\,dt \approx \frac{h}{2}\left\{ f\left(\frac{h}{2}\left(1-\frac{1}{\sqrt{3}}\right)\right) + f\left(\frac{h}{2}\left(1+\frac{1}{\sqrt{3}}\right)\right)\right\}.$$

For some suitable h and n, the given integrand is integrated, using this formula, over each of the intervals $(0,h)$, $(h,2h)$, . . . , $((n-1)h, nh)$, and the integral beyond nh is ignored. In line 21, n has been chosen equal to 110 and h equal to 0.1. The result shown in line 22 is in fact correct to six decimal places (that is, 0.621450). One may verify the accuracy by repeating the calculation with other values for n and h—as will be shown in the next chapter. (A good theoretical bound to the truncation error is easily calculated. But as for the accuracy of the Gaussian quadrature, though the fourth derivative of the integrand is obviously bounded, a good bound is troublesome to obtain.)

We now have enough machinery to begin to do interesting things. The reader may at this point go to the first two examples of statistical analysis of data in Chapter 7. (He should then return here!)

The fourth way of applying primitive dyadic scalar functions to arrays is by a generalization of the ordinary (inner) product of two matrices. If X and Y are matrices, with the number of columns of X equal to the number of rows of Y, the (i, j)th element in the product is formed from the ith row of X and the jth column of Y; these are multiplied element by element and then a sum reduction is made. The notation for the product in APL is $X+.\times Y$. In general, if X and Y are any two arrays, conformable in the sense that the last element of ρX is equal to the first element of ρY, and if f and g are any two primitive dyadic scalar functions, there is an *inner product* written $Xf.gY$ whose dimension vector is all but the last element of ρX catenated with all but the first element of ρY. Suppose for example that X is three-dimensional and Y is a matrix, and the resulting array is called Z. Then Z is three-dimensional and $Z[I;J;K]$ is equal to $f/X[I;J;]gY[;K]$.

Inner products using $+$ and $\times$, with arguments of various ranks, are frequently useful in statistical work. Inner products using other functions are occasionally met with. Suppose that the rows of a matrix X specify "locks" and the columns of a matrix Y specify "keys"; and if the ith row of X is identical with the jth column of Y, we say the jth key fits the ith lock. To see which keys fit which locks, we may form the inner product $X\times.=Y$. All elements in this matrix are 0's and 1's, and the 1's indicate the lock-key matches.

The generality of these various methods of applying the primitive scalar

*In mathematical expressions the symbol "$\approx$" is used for approximate equality and "$\sim$" for asymptotic equality.

functions to arrays is impressive. Symbols and devices in APL are introduced to meet common needs, but they are defined broadly, so that many possibilities are brought in "for free", some of which turn out to be useful. The same desire for the general and the simple that mathematicians bring to the content, but not to the notation, of mathematics is here applied to a domain of expression. APL may therefore be described as a mathematicianly language (and not for the trivial reason that it contains special symbols). It may equally fairly be described as an engineer's language, since it has been molded by use.

Any language designed to handle matrices must obviously encompass ordinary matrix addition. In APL we express the sum of two matrices X and Y of the same size by writing $X + Y$. But this notation does not relate specially either to *matrices* or to *addition*. For as we have seen, X and Y may be rectangular arrays of any size and dimensionality (the same for each), or one may be an array and the other a scalar; moreover, "+" may be replaced by any other primitive dyadic scalar function. Many of these possibilities are just as useful as the ordinary matrix sum. And similarly for reduction and the outer and inner products. In particular, the outer product, which may seem relatively unfamiliar because so little used in matrix algebra, is a versatile and common tool in APL.

In addition to the primitive scalar functions, there are a number of primitive "mixed" functions that can be applied to arrays. They are defined and explained in the manual. Some of their names are:

dyadic "take" (↑) and "drop" (↓),
dyadic "compress" (/) and "expand" (\),
monadic "reverse" and dyadic "rotate" (⌽),
monadic and dyadic "transpose" (⍉),
monadic "grade up" (⍋) and "grade down" (⍒).

Their meanings will be explained here briefly when instances are encountered.

A further primitive function symbol of special interest for statistics is the query (*?*), standing for a monadic function known as "roll" and a dyadic function known as "deal". For any positive integer N, "*?N*" means one member drawn at random from the population ⍳N, that is, a random one of the first N natural numbers. This is a scalar function and could have been mentioned in Chapter 3. It is extended to arrays like other primitive monadic scalar functions. Applications of "*?*" are independent. Thus "*?R⍴N*" means a random sample of size R, taken with replacement, from the population ⍳N. Dyadic "*?*" is a mixed function: "*R?N*" means a sample of size R (an R-element vector) taken without replacement from the population ⍳N, where R must be a nonnegative integer not greater than N.

"$N?N$" is a random permutation of ιN. With these two functions we can perform experiments simulating random processes.

In current implementations, a generator of pseudorandom numbers is used. Each time a random number is called for, the value of a stored positive integer known as the "link" is multiplied by 7^5 and then its residue is taken modulo $2^{31} - 1$, a prime number. This process can be expressed in APL thus:

$$LINK \leftarrow 2147483647 \mid 16807 \times LINK$$

Starting with a clear workspace, the initial value of the link is the same as the multiplier, 16807. This number is a primitive element or multiplicative generator of the finite (Galois) field of nonnegative integers under residue arithmetic modulo $2^{31} - 1$. Successive powers generate all $2^{31} - 2$ nonzero elements of the field in a "pseudorandom" order. At any stage the stored value of the link may be displayed and can be changed arbitrarily (though this is not illustrated here).

If what has been called for is "$?N$", where N is an integer less than 2^{31} (and is actually represented internally as an integer rather than a floating-point number), the value returned is the least integer not less than the following amount: N multiplied by the new value of the link, divided by 2^{31}. It will be seen upon reflection that if N is the greatest such integer, namely 2147483647, the value returned is exactly equal to the link. For any integer N not exceeding this maximum, the relative frequencies (or pseudoprobabilities) with which the members of ιN are returned when "$?N$" is called repeatedly are almost equal—they cannot differ by more than about 2^{-30} or 10^{-9}, an inequality far too small for empirical detection. For an N that exceeds $2^{31} - 1$ or is entered in floating-point form, such as "$1E6$" instead of "1000000", the procedure for executing "$?N$" is slightly different and slower, since two new values of the link are generated instead of only one.

All generators of pseudorandom numbers have curious dependences between sets of numbers generated, that can be troublesome for some purposes. The APL generator is considered to be relatively good in this regard. Of course, the user is at liberty to work with any other pseudorandom-number generator that he prefers, when conducting simulation experiments, but the one supplied in the system will run faster. (Primitive functions are generally executed more rapidly than comparable defined functions, as described in the next chapter.)

The first line of Figure A:7 calls, in the manner just suggested, for some successive powers of 7^5 reduced modulo $2^{31} - 1$, starting at the second power. Line 3 calls for a direct verification of the first two entries in line 2. Line 5 calls for twenty simulated observations from the uniform distribution over the unit interval. They are labeled X and displayed in lines 6 and 7 after rounding to two decimal places. The least and the greatest elements

```
      ?6⍴2147483647                                                    1.
282475249  1622650073  984943658  1144108930  470211272  101027544    2.
      2147483647|16807*⍳3                                              3.
16807  282475249  1622650073                                           4.

      0.01×⌊0.5+100×X←1E¯6×?20⍴1000000                                 5.
0.68  0.68  0.93  0.38  0.52  0.83  0.03  0.05  0.53  0.67  0.01      6.
      0.38  0.07  0.42  0.69  0.59  0.93  0.85  0.53  0.09            7.
      (⌊/X),⌈/X                                                        8.
0.007699  0.934693                                                     9.

      ⌊0.5+Y←12×-⍟X                                                   10.
5  5  1  12  8  2  40  35  8  5  58  12  32  10  5  6  1  2  8  29   11.
      (+/Y)÷20                                                        12.
14.12602159                                                           13.
      2 61 ⍴' ∘+*'[1+3⌊+/(0,⍳60)∘.=⌊0.5+Y],61⍴'+---------'            14.

 ++   *∘  *  ∘  +                    ∘   ∘   ∘     ∘             ∘    15.
+---------+---------+---------+---------+---------+---------+        16.

      CZ←((⍳20)∘.≥⍳20)+.×Z÷+/Z←(⌽⍳20)×Y-¯1↓0,Y←Y[⍋Y]                  17.
      (42⍴ 0 1)\' ∘'[1+(⌽0,⍳10)∘.=⌊0.5+10×0,CZ]                       18.

                                            ∘                         19.
                                       ∘  ∘  ∘                        20.
                                     ∘                                21.
                                                                      22.
                                                                      23.
                               ∘  ∘  ∘                                24.
                       ∘  ∘  ∘  ∘                                     25.
               ∘  ∘  ∘  ∘                                             26.
                                                                      27.
   ∘  ∘  ∘  ∘                                                         28.
 ∘                                                                    29.

      0.1×⌊0.5+10×T←300X                                              30.
¯1.6  ¯1.6  ¯0.2  2.6  ¯16.4  ¯0.6  0.1  0.2  ¯10.7  ¯1.7  0  2.6    31.
      0.2  3.8  ¯1.5  ¯3.5  ¯0.2  ¯0.5  ¯11.8  0.3                    32.
      +/(¯18+⍳21)∘.=⌊T                                                33.
1  0  0  0  0  1  1  0  0  0  0  0  0  1  0  4  4  5  0  2  1        34.

      ⎕←M←(⍳4)⌽ 4 4 ⍴'ABCD'[4?4]                                      35.

DACB                                                                  36.
ACBD                                                                  37.
CBDA                                                                  38.
BDAC                                                                  39.

      (8⍴ 0 1)\M[4?4;4?4]                                             40.

 C D B A                                                              41.
 A B C D                                                              42.
 D C A B                                                              43.
 B A D C                                                              44.
```

Figure A:7

of X are called for in line 8. In line 10 X is transformed into a vector Y of twenty observations from the exponential distribution with mean equal to 12; for if x is a random variable uniformly distributed over the unit interval, then y, defined by

$$y = -\lambda \ln x$$

for some positive λ, has the exponential distribution with mean λ and probability differential element

$$\lambda^{-1} e^{-y/\lambda} dy \quad (y > 0).$$

Perhaps it should be explicitly said that in the text of this book mathematical expressions are sometimes written in conventional notation, as above (where an equals sign means assignment and a solidus means division!), and sometimes in APL, whichever seems more appropriate or clearer. Whenever an expression is acceptable in both languages, it means the same in both. Most expressions are recognizably in one language and not the other, and so there should be no confusion.

The elements of Y, rounded to the nearest integer, are displayed in line 11. The average of the elements of Y, expected to be in the neighborhood of 12, the mean of the exponential distribution, is asked for in line 12. Line 14 asks for a graphical display of the values of Y (rounded to the nearest integer). Single values are shown in line 15 as "∘", two coincident values as "+", and three or more as "*". The lower line of the display (line 16) marks a scale: the first "+" stands for 0, the next for 10, the next for 20, etc.

The next few lines are concerned with a test for distribution shape that may be carried out on a set of readings purporting to come from an exponential distribution. If the readings are rearranged in ascending order, their successive differences should be independently exponentially distributed. Specifically, if the readings in ascending order are $y_{(1)}, y_{(2)}, \ldots, y_{(n)}$, and if

$$\begin{aligned} z_1 &= n y_{(1)}, \\ z_2 &= (n-1)(y_{(2)} - y_{(1)}), \\ z_3 &= (n-2)(y_{(3)} - y_{(2)}), \\ &\ldots \\ z_n &= y_{(n)} - y_{(n-1)}, \end{aligned}$$

then the z's should be independently drawn from a common exponential distribution (the same as that of the y's). If now we calculate the cumulative sums of the z's, divided by their total, that is

$$\sum_{r=1}^{i} z_r \Big/ \sum_{r=1}^{n} z_r \quad (i = 1, 2, \ldots, n),$$

we get a sequence of numbers that should plot roughly linearly against their index. Some kinds of departure of the original distribution (of the y's) from the supposed exponential form would be revealed by a bend in the plot of the cumulative z's.

Line 17 calls for the vector Y to be rearranged in ascending order, to be differenced and scaled to form the vector Z, which is then transformed into the scaled cumulative sums CZ. Three of the primitive mixed functions are used in this line. Inside the square brackets on the right, the "grade up" function applied to Y yields a permutation of the index numbers of Y that would rearrange Y in nondecreasing order. This is applied to Y and the result is renamed Y by a specification arrow. Then a zero is catenated onto the front of Y, the last element is dropped from $(0, Y)$ by the "drop" function and the resulting vector is subtracted element by element from Y, then multiplied element by element by the vector $\iota 20$ in reverse order; the result is named Z. Z is now divided by its sum reduction, then multiplied (an inner product) by a matrix consisting of ones on and below the main diagonal and zeroes above, to give CZ.

At this point we could obtain a print-out of CZ, but that would not be easy to assimilate. We wish to see whether the elements of CZ are roughly proportional to their index number. A graphical display is more helpful, so that is called for in line 18. On the right, within the square brackets, zero is catenated onto the front of CZ, the values are multiplied by 10 and rounded to the nearest integer, and then an outer product using the "=" function is formed with the vector of all possible values, namely 10, 9, 8, . . . , 0; the result is a matrix of 11 rows and 21 columns, each column containing ten 0's and one 1. All elements in the matrix are now increased by 1, becoming 1's and 2's; and the result is used to index a two-element character vector consisting of a blank and a small circle. Finally, the whole character matrix is spaced out laterally to make it easier to look at by means of the "expand" function. The left argument of "expand" is a logical vector containing as many 1's as there are columns in the right argument (namely 21), and the 0's in this vector indicate extra blank columns to be inserted between the columns of the right argument.

The bottom leftmost point in the display (line 29) is the origin. One could draw horizontal and vertical lines through it by hand to form axes (or with a little more trouble in the programming the computer could have been persuaded to mark the axes). The "expected" path for CZ is a straight line joining the origin to the top rightmost point (line 19), and that too could be drawn in by hand. The plot is rather coarse; changing the two 10's in line 18 to 25's would improve it, without in fact much altering the impression given. There seems to be no striking deviation from straightness. The maximum vertical displacement between the plotted points and the "expected" straight line could be called for and assessed by a Kolmogorov-Smirnov-type test. Of course, as the data examined here were artificially generated, one would hope that nothing very extraordinary would be revealed!

In line 30 another transformation is applied to our original sample X of twenty readings from the uniform distribution. This time the elements of X are multiplied by π (monadic circle function) and their tangents are taken (dyadic circle function), yielding a vector T that should be a sample from the Cauchy distribution with probability differential element

$$\frac{dt}{\pi(1 + t^2)} \,.$$

The elements of T are shown in lines 31–32, rounded to 1 decimal place. A frequency distribution of the elements of T, grouped in intervals of unit width, is asked for in line 33. Line 34 shows that the T-values have this distribution: one is in the interval (¯17, ¯16), one in (¯12, ¯11), one in (¯11, ¯10), one in (¯4, ¯3), four in (¯2, ¯1), four in (¯1, 0), five in (0, 1), two in (2, 3), one in (3, 4). We expect one half of the readings to fall in (¯1, 1), and have obtained nine out of twenty.

The rest of Figure A:7 is concerned with generating a random 4-by-4 Latin square. What is done is to take a particular 4-by-4 Latin square and apply independent random permutations to the letters, to the rows and to the columns, that being as much randomness as most people care to trouble with. In line 35, the character vector *ABCD* is permuted at random, yielding (as appears later) *BDAC*. Then a matrix is formed, each of whose four rows is this same vector, *BDAC*. Finally the "rotate" function applies a different rotation to each row of the matrix; the left argument (ι4) implies that the first row is to be cyclicly rotated by one step, the second row by two steps, and so on. The rotation of the fourth row by four steps leaves it unchanged. We see the result displayed in lines 36–39. In line 40 random permutations are applied to rows and columns, and the columns are spaced to make the result in lines 41–44 look like a square.

To any reader who has persevered through this account of Figure A:7 some apologies are due. Unless he has already had experience of computing with APL, he is likely to find what has been said only partly comprehensible. No attempt is made in this book to reproduce the systematic detailed description of facilities of an APL system given in the official manual. Figure A:7 shows computations of a kind that some persons engaged in statistical work might like to do, and illustrates how a not-so-simple sequence of computing steps may be expressed in a single line of APL. Only when the user is familiar with most of the vocabulary of APL will such things be done with assurance and ease.

Figure A:7, like all the other figures in Part A, is concerned with the technique of computation. The statistical content of this figure may be described as a statistician's doodling. No suggestions are intended concerning statistical practice. The test for conformity of some readings to an exponential distribution (lines 17–29) is possible, but is in no sense advocated—we see how it could be done if it were wanted. The two graphical displays (lines 15–16 and 19–29) are modest; we shall see more interesting displays later. The obtaining of these displays does, however, illustrate a very important feature of APL, that character arrays are manipulated in exactly the same way as are numerical arrays. A graphical display is usually a print-out of a character matrix. Such a matrix can be indexed, transposed, expanded, and juxtaposed with other matrices, by precisely the same methods as are used for a numerical matrix. Characters and numbers have equal rights. Graphs, charts and tables are just as important to statistical work as numerical computation. They are specified in the same language.

All of Figure A:7 was executed in the order shown, starting with a clear workspace. As long as the random-number generator in the system is not changed, these calculations may be reproduced exactly by anyone interested. But with a different initial value of the link, or possibly with a different sequence of presenting the commands, different calculations will result.

Notes for Chapter 5

The figures in this chapter (and elsewhere in Part A) use only devices that have been available since *APL*\360 was defined and released in 1968. Some later versions of APL would permit greater conciseness in a few places in Figure A:7. In line 5 values of X are called for, rounded to two decimal places; use of the primitive format function would abbreviate the command. Similarly for line 30. Line 14 could be slightly shortened by lamination with the comma function. In line 17, multiplication of Z by a matrix consisting of ones on and below the main diagonal and zeroes above could be replaced by a sum "scan" of Z. The scan operator is another way, besides those mentioned in the text, in which the primitive dyadic scalar functions are extended to arrays.

No matter which the version of APL, the spirit is the same. Appendix 3 shows examples of the facilities available in the more recent dialect known as *VS APL*. Older-style implementations are still in use. If the reader only has access to one of those, he should not feel hopelessly cramped because he is denied a few pleasant and elegant features of up-to-date versions. Any version of APL abounds with pleasant and elegant features.

Chapter 6

Defined Functions

We have seen the use of an APL terminal in the style of a desk calculator. Commands have been executed as soon as entered. If commands are to be executed only once, that is usually the best mode of operation. But if the same, or similar, commands are to be executed repeatedly, or if they are not very simple and brief, it may be preferable to have them in the form of a stored program. For example, the numerical integration in line 21 of Figure A:6 is not a very simple command and it incorporates a choice of values for the two parameters called n and h in the text. If we knew, perhaps from an error analysis, that the chosen values were satisfactory, the calculation need only be done once. But in the absence of a good error bound we should investigate the precision of the result empirically by repeating the calculation with different values for n and h. Then it will be easier to have the procedure entered as a stored program, with explicit parameters, instead of typing it out afresh each time. Similarly, the two graphical displays in Figure A:7 result from not-very-simple commands, the kind the user can just as soon get wrong as get right the first time he tries them. They might be better expressed in a more general form as stored programs that can be tested, corrected if wrong, and then used with some assurance.

There is just one method in APL for storing a program—or a procedure, or a subroutine, or a "macro"—namely, to define a function. Such a function is called a *defined function*, to distinguish it from the primitive functions that are inherent in the system. Whereas a primitive function is denoted by a special dedicated symbol, a defined function is denoted by a name, like a variable. Defined functions are capable of entailing more complicated computations than do any of the primitive functions, and they may have more diverse kinds of behavior. Nevertheless there is much similarity in the ways defined functions and primitive functions are used. That both kinds of object are referred to as "functions" is no accident.

A defined function may have not more than two *explicit arguments*, which can be arrays of any permissible kinds. If there are two arguments, the function is dyadic and when it is to be executed the function name is written between the arguments (necessarily separated from them by at least one space, since otherwise there would be no telling where the function name began and ended). If there is only one argument, the function is monadic and the function name precedes the argument. It is permissible for a defined function to have no explicit argument at all—unlike the primitive functions. A defined function may have any number of what may be called *concealed arguments*, in addition to its (zero or one or two) explicit arguments. That is, a defined function may refer to variables that have been defined before the function is executed. Among the primitive functions we have met, the query functions (roll and deal) have this property, for they refer in execution to the stored value of the link, and so yield an explicit result that does not depend only on the explicit arguments.

As for output, a defined function may yield an *explicit result*, which can be any permissible kind of array, or else no explicit result. (All the primitive functions yield an explicit result.) In addition, a defined function may have other kinds of output or consequence; it may cause material to be displayed, it may change the values of previously defined variables or set up new variables. (The primitive query functions change the stored value of the link, in addition to yielding an explicit result.)

In order to define a function, an upside-down delta called "del" is typed, followed by what is called the *function header*, which gives the name of the function and shows its syntax. When this line is transmitted, the computer goes out of execution mode into function-definition mode. Succeeding lines are automatically labeled on the left by numbers in square brackets. Whatever the user types on these lines is stored as statements in a program, without being executed. Typing another del closes the definition and returns us to normal execution mode. The function may then be called and will be executed (if it turns out to be executable).

The mechanics of function definition are carefully explained in the manual. Here examples will be given to suggest some of the possibilities.

Consider first a function to explore the precision of the numerical integration already discussed. We should have n and h as arguments. They could either be two separate scalar explicit arguments of a dyadic function, or two elements of a single vector explicit argument of a monadic function. The first arrangement seems slightly easier, so we choose that. (A third possibility is that the function has no explicit argument, but the values of n and h are stored as variables that are "concealed arguments" of the function. That is less convenient.) Since we shall only use this function a time or two and then scrap it, we might as well give it a short name. In line 1 of Figure A:8 it is called I (for integral), and the arguments are

```
      ∇ N I H;X                                                   1.
[1]     0.5×H×+/(*-X)÷1+X×X←H×,(⍳N)∘.-0.5×1+ 1 ¯1 ×3*¯0.5        2.
      ∇                                                           3.

        (*-K)÷1+K×K← 10 11 12                                    4.
4.495042551E¯7  1.368991868E¯7  4.23738783E¯8                    5.

        11 I 1                                                    6.
0.6238800424                                                      7.
        22 I 0.5                                                  8.
0.6215404985                                                      9.
        44 I 0.25                                                10.
0.6214542238                                                     11.
        88 I 0.125                                               12.
0.6214497925                                                     13.
        110 I 0.1                                                14.
0.6214496235                                                     15.


      ∇ Y←SQRT1 X;Z                                              16.
[1]     Y←1                                                      17.
[2]     Z←Y                                                      18.
[3]     Y←0.5×Z+X÷Z                                              19.
[4]     →2×1E¯6≤|1-Z÷Y                                           20.
      ∇                                                          21.

      ∇ Y←SQRT2 X;Z                                              22.
[1]     →5×⍳∧/,X>0                                               23.
[2]     Y←''                                                     24.
[3]     'NO GO.'                                                 25.
[4]     →0                                                       26.
[5]     Y←1                                                      27.
[6]     Z←Y                                                      28.
[7]     Y←0.5×Z+X÷Z                                              29.
[8]     →6×∨/,1E¯6≤|1-Z÷Y                                        30.
      ∇                                                          31.

      ∇ Y←SQRT3 X;Z                                              32.
[1]     →2+Y←∧/,X>0                                              33.
[2]     →0,⍴⎕←'NO GO.',Y←''                                      34.
[3]     →3×∨/,1E¯6≤|1-Z÷Y←0.5×Z+X÷Z←Y                            35.
[4]     ⍝  THIS FINDS THE SQUARE ROOT OF EACH ELEMENT OF X.      36.
      ∇                                                          37.

        SQRT3 ⍳6                                                 38.
1  1.414213562  1.732050808  2  2.236067977  2.449489743         39.


      ∇ LATINSQUARE V;N                                          40.
[1]     →2+(2≤N←⍴V)∧1=⍴⍴V                                        41.
[2]     →0,⍴⎕←'NO GO.'                                           42.
[3]     ((2×N)⍴ 0 1)\((⍳N)⌽(N,N)⍴V[N?N])[N?N;N?N]                43.
[4]     ⍝  TO GET A RANDOM 5-BY-5 LATIN SQUARE ENTER:  LATINSQUARE 'ABCDE'
      ∇                                                          45.
```

Figure A:8

called N and H. There is no explicit result. The "; X" at the end of the header is not essential and could be omitted—its purpose will be explained in a moment. The body of the function has only one statement, numbered [1]. This is very similar to line 21 of Figure A:6 and causes an approximate value of the integral to be displayed.

Before we go further, let us decide how large the product nh ought to be. It is easy to see that the integral of the given integrand from nh to ∞ is less than (but close to)

$$\frac{e^{-nh}}{1 + (nh)^2}.$$

This is evaluated for the three values, 10, 11, 12, for nh in lines 4–5. We decide that 11 is good enough for 6-decimal accuracy in the final answer, since the integral will be underestimated by only about 1 in the seventh place because of the truncation. Now we execute our function I several times with values for its arguments having product equal to 11. Each time h is halved and n is doubled, the integration error should be divided by something like 16. The last choice in line 14 is clearly adequate, and this was used in Figure A:6.

The arguments N and H of our function I are what are called *local variables*. They are used when the function is executed, but are then erased. They are initially given whatever values are specified as arguments in the call of the function. In lines 6, 8, 10, 12, and 14 of Figure A:8 numerical values are specified directly. If we had defined A to be equal to 11 and B to be equal to 1, we could have entered "A I B" in line 6 instead of "11 I 1"; the variables A and B would still have the same values after I had been executed, but the temporary use within the function of the names N and H would be forgotten. If a function has an explicit result (which I does not), the name used for it in the function definition is again only a local variable. Our function I causes a variable X to be specified. We have the option of allowing this variable to remain stored after execution of I is completed (in which case it is called a *global variable*), or treating it as a local variable and having it deleted when execution is complete. For the latter option, we add "; X" to the header; for the former we do not. With "; X" in the header, execution of the function I does not interfere with any previously defined global variable. If any of the names N, H, X had already been assigned a global value, that value would be ignored during execution of the function and would resume afterwards.

In Chapter 2 algorithms were introduced by reference to an iterative procedure for finding square roots. There is no practical need to implement this procedure in APL, because square roots are already available in the system. However, such iterative procedures are often encountered, and as this one is so simple it makes a good illustration of the considerations involved. Figure A:8 shows three functions, named *SQRT*1, *SQRT*2, *SQRT*3, each of which expresses essentially the same procedure. *SQRT*1 is a direct translation into APL of the steps listed in Chapter 2. The header shows that the function has one explicit argument *X*, an explicit result *Y*, and an additional local variable *Z*—these correspond to x, y, y' in Chapter 2. In statement [4] there is a branch, and some remarks should be made about that.

Whereas most other computer languages have a considerable vocabulary relating to the flow of an algorithm—specifying conditional branches, loops, etc.—APL has precisely one symbol for branching, the right-arrow (→). Nothing appears to the left of this arrow (except for the statement number in square brackets), and to the right appears an expression having (usually) a nonnegative integer value. The arrow means "go to the statement numbered"—for example, "→ 2" means "go to the statement numbered 2". If there is no statement in the program having the indicated number, the command is interpreted as "stop". In particular, since the statements are numbered from 1 upwards, the command "→ 0" always means "stop". What appears to the right of the arrow is not always a single number; it may be a numerical vector. If *V* is a non-empty vector, "→ *V*" means "go to the statement whose number is the first member of *V*" or "→ *V*[1]". If *V* is an empty vector, "→ *V*" means "go to the next statement", as though the next statement number were automatically catenated onto the end of *V*. A branch is made conditional by the device of computing the statement number to be gone to. A statement that is not a branch is always followed in execution by the next statement.

So what happens when statement [4] of *SQRT*1 is executed is that 10^{-6} is compared with the magnitude of the proportional difference between *Z* and *Y*. If the relation ≤ holds, the statement reads "→ 2", but otherwise it reads "→ 0", or "stop". When execution stops, the current value of *Y* is the desired answer, correct to about 12 decimal digits.

Now if this square-root function were really needed, it would no doubt be stored and referred to frequently, possibly by others than the author. It could be used, therefore, by someone who had forgotten or never knew just what the function did. To guard against breakdown, consideration should be given to any requirement for successful execution. There is one obvious requirement. The argument *X* must be positive; otherwise the iterations will not converge. What about the dimensionality of *X*? We have been thinking about a scalar argument, but could we not perhaps allow *X* to be any kind of numerical array, so that the explicit result *Y* would be the corresponding array of square roots of elements of *X*? For this purpose we should modify statement [4] so that it returns execution to [2] provided any *one* of the ≤ relations holds in the array. If we do not make this modification, we ought to impose a condition that *X* is scalar.

The function *SQRT*2 is like *SQRT*1 except that X is permitted to be any array (and the branch at the end is modified accordingly), and also the condition that every element of X shall be positive is introduced at the outset. In statement [1] the logical array "$X > 0$" is called for, destructured by monadic comma, and reduced by the "and" function. The "and" and "or" functions (∧, ∨) are primitive dyadic scalar functions applicable only to logical arguments equal to 0 or 1; they could have been mentioned in Chapter 3. "∧" could be replaced by "×", but is easier to read, because we know its arguments must be logical numbers. If all elements of X are positive, the reduction yields the value 1, and since "⍳1" has the value 1, the statement finally reads "→ 5", and statement [5] is where the main business begins. But if the reduction yields 0, statement [1] is a branch to an empty vector, and so execution passes to [2]. Here Y is set equal to an empty vector, which is usually quite a good explicit result to be returned when a program breaks down, and something ought to be returned. Then [3] causes a message to be displayed on the terminal, to warn the user that the desired job has not been done, and [4] stops execution. Statements [5], [6], [7], [8] are identical to the statements in *SQRT*1, except for two changes in the last statement, the new labeling of the start of the loop and the insertion of the "or" reduction to allow for a non-scalar argument X.

The function *SQRT*3 is essentially identical to *SQRT*2, but shortened by a kind of cleverness to which APL-men are prone. Statement [1] combines statements [1] and [5] of *SQRT*2, in the sense that if execution does not abort at [2] Y will have been set equal to 1 in [1]. Statement [2] combines [2], [3] and [4] of *SQRT*2, for Y is first changed to an empty vector, then harmlessly catenated onto the character array "*NO GO* ", which is displayed, then counted (by monadic ⍴) and catenated onto 0, so the branch finally reads "→ 0 6", meaning the same as "→ 0" or "stop". The whole of the loop, statements [6], [7], [8] in *SQRT*2, is now in the one statement [3]. There is no appreciable difference in execution time between *SQRT*2 and *SQRT*3. The merit of the shorter program (in addition to having afforded pleasure to its author) is that it is somewhat easier to read, for each statement forms a complete intelligible step: statement [1] tests and initializes, [2] specifies what is to be done if the test fails, [3] is the loop.

A fourth statement has been added onto *SQRT*3, containing a note to anyone who may read a print-out of the program. It starts with a lamp sign and would therefore be ignored by the computer in execution if it were ever reached (which as a matter of fact this statement cannot be, because execution must stop at either [2] or [3]). Such a comment, together with a proper formulation of requirements at the beginning of the program (as here in [1]), is very helpful to a new reader, in the absence of documentation elsewhere. Any number of comments may be inserted anywhere in a program.

Lines 38–39 show a test of *SQRT*3. Identical results would have been obtained by entering instead: "(ι6) ∗ 0.5".

The last item in Figure A:8 is a program for generating a random Latin square of any desired size. The argument *V* should be a character vector of letters (or digits or any other symbols) to be used as the "alphabet" in the square. The method, given in statement [3], is exactly the same as was illustrated in Figure A:7. Statement [1] makes a test of suitability of the argument, namely that it should be of rank 1 (i.e., a vector), with length not less than 2, and the length is named *N*. If *V* does not satisfy these requirements, being a vector of length less than 2, or a scalar, or a matrix, or whatever, then for various reasons the effect of statement [1] is to branch to [2], which stops execution.

Perhaps it should be mentioned that in Figure A:8 and in later figures the displayed definition of a function is always in fact a print-out of a previously stored function, rather than the original entry, from which it differs only in details of format having no interest for present purposes. The command to display the function definition has been suppressed from the figures.

It was remarked above that defined functions and primitive functions were used similarly. Defined functions may appear in compound statements and are executed in accordance with the right-to-left rule, just like primitive functions. They may be invoked in the body of other defined functions. One obvious difference between defined and primitive functions in execution is that whereas primitive functions are always present in the system, a defined function cannot be executed unless its program is actually present in the active workspace. If defined function *F*1 invokes another defined function *F*2, both functions must be in the active workspace before *F*1 can be executed. Another difference between defined and primitive functions is that defined dyadic scalar functions cannot be applied to arrays in the same way as primitive functions for reduction and outer and inner products.

Of all the symbols that can be produced on an APL terminal and are accepted as valid by the system, some denote primitive functions, but most do not—the letters of the alphabet, the digits, the space, decimal point, negative (high minus) sign, small circle, lamp, semicolon. Two symbols present a borderline case. The left-arrow for assignment and the right-arrow for branching are not referred to in the manual as primitive functions, and are therefore not meant to be included when we speak of all primitive functions. However, they behave like functions in that they participate in right-to-left execution. In fact the right-arrow could be described as a monadic function having no explicit result; and in some contexts the left-arrow behaves like a dyadic function yielding its right argument as explicit result. The primary purpose of both of these symbols has to do with management, not with delivering an explicit result, and in this respect they differ from all the objects officially called primitive functions.

This is a convenient point to reconsider the one feature of APL syntax that noticeably ruffles some newcomers, namely, the right-to-left rule. Anyone who has used APL for a while is likely to agree that the principle of priority in execution of functions by position only is very pleasing. Granted such a rule, should it be right-to-left or left-to-right? The matter is worth discussing, as there are arguments on either side. At times the right-to-left rule goes against mathematical convention, and even an experienced user occasionally makes mistakes from this cause. We can consider two questions: (i) would a rigid left-to-right rule seem on the whole more natural than a rigid right-to-left rule? (ii) is there any difference in merit between the rules, apart from naturalness? The answers seem to turn on two topics, monadic functions and reduction.

Functions. Most of us are more familiar in ordinary mathematics with sentences like:

$$\text{“Let } y = f(x).\text{”}$$

than with

$$\text{“Let } x \xrightarrow{f} y.\text{”}$$

Yet the second sentence, read from left to right, correctly shows the sequence of operations in computing. One first gets x, applies the function f to it, and names the result y. The more familiar first sentence, though read from left to right, needs to be executed from right to left. In APL with right-to-left execution we write

$$\text{“}Y \leftarrow F\ X\text{”}.$$

If the rule of execution were left-to-right instead, we should have to write

$$\text{“}X\ F \rightarrow Y\text{”}.$$

Most people would prefer the first way on grounds of familiarity. There is nothing to choose between them on grounds of effectiveness or power.

Reduction. All the primitive dyadic scalar functions, such as were shown in Figure A:1, can be applied to reduce a vector of length 2, that being exactly what was illustrated in Figure A:1. But only eight such functions are ordinarily used to reduce long vectors, namely: $+, -, \times, \div, \lceil, \lfloor, \wedge, \vee$. All but two of these functions are associative (and commutative), and it is immaterial whether a reduction using one of them is executed from right to left or from left to right—they are associative, that is, apart from possible effects caused by round-off error. But the functions "$-$" and "$\div$" are neither associative nor commutative. The order of their arguments in APL follows custom. "$A - B$" means B subtracted from A, not A from B; and

"$A \div B$" means B divided into A, not A into B. In a reduction with either of these functions, left-to-right execution gives generally a different result from right-to-left. The former, with subtraction at least, is the customary interpretation of compound expressions. But the latter is more interesting, for it corresponds to what is customarily written as an alternating sum, or as an alternation of divisions and multiplications. Such alternations frequently occur in numerical computation, and it is convenient to have a concise notation for them. Thus with reduction there is a conflict between custom and effectiveness, and APL has opted here for effectiveness. The same effectiveness could be had with left-to-right execution if the notation for subtraction and division were changed so that the customary order of the arguments was reversed. Then we should have not only the unfamiliar order for all monadic functions but an unfamiliar order for two very common dyadic functions also. So a left-to-right rule of execution, if it were as effective as the present right-to-left rule, would presumably cause at least as much comment.

Notes for Chapter 6

In addition to the del form of function definition described above, later versions of APL permit another procedure of definition through what is called canonical representation. Either procedure has facilities for editing a previously defined function, by adding to it, making deletions or changing it in various ways. Lines in a function definition may optionally be given label names, in addition to the automatic numbering; and then the labels rather than line numbers are used in branches. For all these matters the reader is referred to the manual.

In Appendix 3 will be found a function *INTEGRATE* for evaluating a definite integral of any function of one variable by Gaussian quadrature, in a similar style to the function *I* in Figure A:8. Instructions for use are offered in *HOWINTEGRATE*.

Chapter 7

Illustrations

Expression of statistical calculations in APL will now be illustrated by some short examples. The first two are carried out in simple dialog style, without defined functions. The other examples use defined functions and a slightly richer vocabulary.

Example 1–*Flipped coins.* Figure A:9 relates to observations of the number of heads showing when ten coins have been thrown down. Twenty-five such observations were made by students in an introductory statistics course. Let us see whether the readings conform with the usual theoretical idea of independent binomial variables with probability of heads equal to one half.

The twenty-five readings are entered and named Y in line 1, and counted (lines 2–3). Each reading must necessarily be equal to one of the first eleven nonnegative integers, 0, 1, 2, . . . , 10. The frequencies of each of these values is called for in line 4; an outer product of the value vector and Y, using the function "=", is formed, and the eleven row totals are stored under the name OF (for observed frequencies) and displayed in line 5. At line 6 we calculate the sample mean and at line 7 the sample variance. Of course we know that ρY is equal to 25, and we could just as well have specified 25 as the divisor in line 6 and 24 as divisor in line 7. At line 8 we ask for a display of these quantities and also the sample standard deviation. The mean, variance and standard deviation of the supposed binomial distribution governing the observations are noted in lines 10–11 for comparison—agreement appears satisfactory.

We now do an ordinary χ^2 goodness-of-fit test of the observed frequencies to the supposed binomial expectations. The expected frequencies (EF) are called for, and displayed after rounding to two decimal places. We notice that these expectations are low at both ends of the sequence. Following the old rule of thumb that expectations should not be much less

```
        Y← 6 7 7 3 4 8 5 7 3 2 6 6 6 4 5 7 2 2 5 6 4 7 5 5 3              1.
        ρY                                                                  2.
25                                                                          3.
        ⎕←OF←+/(0,ι10)∘.=Y                                                  4.
0  0  3  3  3  5  5  5  1  0  0                                             5.

        YBAR←(+/Y)÷ρY                                                       6.
        VAR←(+/(Y-YBAR)*2)÷¯1+ρY                                            7.
        YBAR,VAR,VAR*0.5                                                    8.
5  3.166666667  1.779513042                                                 9.
        5,2.5,2.5*0.5                                                      10.
5  2.5  1.58113883                                                         11.

        0.01×⌊0.5+100×EF←(25×2*¯10)×(0,ι10)!10                             12.
0.02  0.24  1.1  2.93  5.13  6.15  5.13  2.93  1.1  0.24  0.02             13.
        EF←A,EF[ 5 6 7 ],A←+/EF[ι4]                                        14.
        OF←(+/OF[ι4]),OF[ 5 6 7 ],+/OF[12-ι4]                              15.
        (+/EF),+/OF                                                        16.
25  25                                                                     17.
        ⍉ 2 5 ρOF,EF                                                       18.

       6                   4.296875                                        19.
       3                   5.126953125                                     20.
       5                   6.15234375                                      21.
       5                   5.126953125                                     22.
       6                   4.296875                                        23.

        +/SR×⎕←SR←(OF-EF)÷EF*0.5                                           24.
0.821618414  ¯0.9393517427  ¯0.4645813314  ¯0.0560678267  0.821618414
2.451474747                                                                26.
```

Figure A:9

than 5 for use in a χ^2 test, we group the first four and last four members of *EF* (line 14) and *OF* (line 15), and reassign the names *EF* and *OF* to these grouped frequencies. Now we wonder if we did that right, so we check that both the expected and the observed grouped frequencies add up to 25 (lines 16–17), and then we display them as a matrix. In fact line 18 calls for the transpose of the 2-by-5 matrix having *OF* as its first row and *EF* as its second row. The reason for transposing the matrix is that without transposition standard output format would require a wider page. (Ordinarily output is displayed to a maximum width of 120 characters, but a considerably smaller width is used here in the figures, so that they can be reproduced on a small page without great reduction in size.) Thus the first column, lines 19–23, shows the grouped observed frequencies, and the second column the corresponding expectations.

Finally we calculate the χ^2 measure of goodness of fit, obtaining the individual standardized residuals *SR* in line 25 and the sum of their squares in line 26. The latter value of 2.45 is entirely unremarkable in relation to the standard probability distribution tabulated as χ^2 with 4 degrees of freedom. Observations seem to agree perfectly with theory.*

Example 2–*Card hands.* Figure A:10 relates to another set of observations made by students in the introductory course. The students were asked to shuffle an ordinary pack of 52 playing cards, deal a hand of 13 cards and note the number of cards in each of the four suits, clubs, diamonds, hearts,

```
      M← 2 5 5 1 3 6 3 1 2 4 6 1 4 4 4 1 3 3 3 4 6 0 3 4 4 5 2 2           1.
      M←M, 4 3 2 4 4 2 3 4 3 5 2 3 3 4 4 2 3 2 4 4 4 3 1 5 3 4 3 3        2.
      M←M, 4 2 6 1 2 4 3 4 4 5 2 2 5 1 4 3 1 4 3 5 3 3 4 3 4 4 1 4        3.
      M←M, 1 3 3 6 4 5 4 0 4 1 5 3 2 2 4 5 3 1 7 2 5 5 1 2 3 5 1 4        4.
      M←M, 3 3 3 4 1 6 4 2 2 4 4 3 7 0 2 4 3 4 1 5 6 2 2 3 1 3 6 3        5.
      M←M, 3 4 4 2 4 2 5 2 3 3 2 5 3 4 3 3 3 4 3 3 3 6 2 2 2 4 2 5        6.
      M←M, 5 2 4 2 4 2 5 2 2 4 4 3 5 6 0 2 8 3 1 1 5 4 2 2 4 3 5 1        7.
      M←M, 3 3 5 2 4 2 2 5 5 4 2 2 2 2 2 7 3 5 4 1 5 2 3 3 1 3 4 5        8.
      M← 60 4 ρM, 3 5 2 3 4 4 4 1 2 5 4 2 3 4 3 3                          9.

      M[ 1 2 59 60 ;]                                                      10.

  2  5  5  1                                                               11.
  3  6  3  1                                                               12.
  2  5  4  2                                                               13.
  3  4  3  3                                                               14.

      +/13≠+/M                           Or:     13+.≠+/M                  15.
0                                                                          16.
      +/⎕←+⌿M                                                              17.
205  207  192  176                                                         18.
780                                                                        19.

      CS←(16÷39)×+/(M-3.25)*2                                              20.
      CSBAR←(+/CS)÷60                                                      21.
      CSBAR,(+/(CS-CSBAR)*2)÷59                                            22.
3.48034188  7.24514128                                                     23.
      (⌈/CS),⌊/CS                                                          24.
13.43589744  0.3076923077                                                  25.

      ⎕←COF←+/ 0.58 1.21 2.37 4.11 6.25 ∘.≥CS                              26.
7  15  20  43  54                                                          27.
      ⎕←OF←(COF,60)-0,COF                                                  28.
7  8  5  23  11  6                                                         29.
      EF← 6 9 15 15 9 6                                                    30.
      +/((OF-EF)*2)÷EF                                                     31.
11.65555556                                                                32.
```

Figure A:10

*Throughout this book the goodness-of-fit statistic suggested by Karl Pearson in 1900 is referred to by his name χ^2. The approximating continuous distribution of the sum of squares of some independent $N(0, 1)$ variables is usually referred to as the tabulated χ^2 distribution.

spades. Each student did this several times, and in all 60 hands were observed. The observations were written down on a sheet of paper in four columns, sixty rows, like this:

C	D	H	S
2	5	5	1
3	6	3	1
2	4	6	1

and so on, finishing

4	4	4	1
2	5	4	2
3	4	3	3

The observations were intended to illustrate the theory of sampling a finite population.

In lines 1–9 of Figure A:10 the observations are entered from the data sheet, row by row, as a vector of length 240, which is immediately restructured as a 60-by-4 matrix M. As a check, the first two and last two rows are asked for in line 10, for comparison with the data sheet. At line 15, a check is made that every row of M sums to 13 (the requested size of the hand)—the number of rows in which the row sum is not equal to 13 is counted and (line 16) found to be 0. The four column totals of M and their sum are asked for at line 17. We note that the sum, 780, is equal to 13×60, as it should be.

At this point in the original treatment of the data, further checks of accuracy in the entering of M were made. The data were entered afresh; each column of the data sheet was entered as a vector and compared with the corresponding column of M. We omit this check now for brevity.

From each row of M a 2-by-4 contingency table can be constructed showing the compositions of the hand and of the balance of the pack. Thus from the first row of M we have

	C	D	H	S
Hand	2	5	5	1
Balance of pack	11	8	8	12

Our purpose is to compare the empirical and the theoretical behaviors of some measure of independence in the sixty such tables. The measure chosen (arbitrarily) is the usual χ^2 criterion. The expected frequencies in each table are

3.25	3.25	3.25	3.25
9.75	9.75	9.75	9.75

and for the first table, quoted above, the χ^2 value is

$$\left(\frac{1}{3.25}+\frac{1}{9.75}\right)\{(-1.25)^2+(1.75)^2+(1.75)^2+(-2.25)^2\}.$$

The multiplier in front of the sum of squares reduces to $16 \div 39$.

At line 20 of Figure A:10 the χ^2 corresponding to each row of M is called for, under the name CS, a vector of 60 elements. The mean and variance of the sixty values is found (lines 21–23), for comparison with the values 3 and 6 respectively for the standard probability distribution tabulated as χ^2 with 3 degrees of freedom. The greatest and least members of CS are called for (lines 24–25).

We now proceed to compare the frequency distribution of the members of CS with this tabulated probability distribution. We must form a list of grouped observed frequencies. Reference to a table of the χ^2 distribution with 3 degrees of freedom gives the following approximate quantiles: median (2.37), lower and upper quartiles (1.21, 4.11), lower and upper 10% points (0.58, 6.25). Cumulative observed frequencies (COF) of members of CS up to these five values are called for at line 26. (It is unnecessary to quote the five values with greater precision, because the actual value set of the χ^2 values is discrete and none is close to any of these theoretical quantiles—as is readily demonstrated by displaying the whole of CS, though this is not done in Figure A:10.) The cumulative observed frequencies are differenced at line 28 to yield the observed frequencies: 7 values of χ^2 are below 0.58, 8 are between 0.58 and 1.21, 5 are between 1.21 and 2.37, . . . , 6 are above 6.25. The expected frequencies for the tabulated χ^2 distribution with 3 degrees of freedom are quoted at line 30 (they are 60 multiplied by 10%, 15%, 25%, . . .). The ordinary χ^2 measure of agreement between lines 29 and 30 is called for at line 31. The value obtained, 11.66, comes at roughly the upper 4% point of the tabulated χ^2 distribution with 5 degrees of freedom. Whether one asserts that the discrepancy between observations and theory is "significant" depends on temperament and philosophy.

In fact, the observed behavior of the χ^2 values CS could differ from the so-called χ^2 distribution for two reasons other than mere chance: (i) the shuffling of the pack and counting were perhaps not always well done; (ii) even if shuffling and counting were perfect, the true probability distribution of the χ^2 criterion, under the specified procedure of observation, is discrete and only represented roughly by the continuous tabulated distribution. In further study of the data this true discrete distribution has been determined and found to agree very satisfactorily with the observed frequencies.

Example 3–*Row-column crossclassification (two-way table).* Let us turn now to analysis of variance. As a simple example, we take a 4-by-6 table quoted by Bliss (1967), relating to an experiment in six blocks comparing the heights of loblolly pines grown from seed from four different sources. The reading obtained from each of the 24 plots is the average height of surviving trees after fifteen years in the plantation.

The first job is to enter the data. This could be done in just the same way as at the top of Figure A:10. However, it is usually advisable to keep sets of data in storage for future reference—we rarely completely dispose of data in one session. A variable like *M* in Figure A:10 can be stored, though for that purpose the name should preferably be more distinctive (*CARDHANDSDATA*, perhaps). Another possibility is to incorporate the data in a defined function, adding a few reference notes.

At the top of Figure A:11 we see a defined function, *LOBLOLLYDATA*, with no explicit argument nor explicit result. It causes a global variable *Y* to be specified, a 4-by-6 matrix of readings, and also causes three lines of notes to be printed. The readings are first entered as integers, then multiplied by 0.1 and restructured as a matrix. Entering the numbers as integers in this way, rather than as numbers having one digit beyond the decimal point, is both slightly easier in typing and more economical of storage space.

When we execute this function (in line 8), *Y* is set up and the three lines of notes are returned by the computer. We can now ask for a display of *Y* (lines 12–16).

We proceed to a standard analysis of variance. Line 17 calls for the general mean (labeled *GM*) to be calculated, displayed (in line 18) and subtracted from each element of *Y* to make a matrix *RY* of residuals. The members of *RY* are squared and summed to give the total sum of squares about the mean (*TSS*). In line 19 are displayed *TSS*, the number of degrees of freedom (23), and the corresponding mean square.

Row effects are called for in line 20. The row means of *RY* are labeled *RE* for "row effects" and displayed (line 21). The elements of *RE* are squared, summed and multiplied by 6 to give the sum of squares for rows (*SSR*). In line 22 are displayed *SSR*, its number of degrees of freedom (3), and the corresponding mean square.

Column effects are similarly called for in line 23. The vector of column effects (*CE*) is displayed in lines 24–25. The sum of squares (*SSC*), degrees of freedom and mean square appear in line 26.

Line 27 is concerned with the residuals after subtraction of row and column effects. The outer product of *RE* and *CE* using the function "+" is subtracted from *RY* to yield the desired matrix of residuals, which is named *RY* again. The elements of the new *RY* are squared and summed to yield the residual sum of squares (*RSS*), displayed in line 28 together with the number of degrees of freedom (15) and the corresponding mean square.

```
    ∇ LOBLOLLYDATA                                                              1.
[1]   Y← 340 293 306 318 340 327 273 276 286 292 302 315 264 250 266           2.
[2]   Y← 4 6 ρ0.1×Y, 252 274 262 248 243 260 265 258 242                       3.
[3]   'Y IS A 4 6 MATRIX OF AVERAGE HEIGHTS OF LOBLOLLY PINES IN FEET'          4.
[4]   'SEED FROM 4 SOURCES (ROWS), 6 BLOCKS (COLUMNS)'                          5.
[5]   'WAKELEY, 1944, QUOTED BY BLISS, VOL. I, TABLE 11.7.'                     6.
    ∇                                                                           7.

      LOBLOLLYDATA                                                              8.
Y IS A 4 6 MATRIX OF AVERAGE HEIGHTS OF LOBLOLLY PINES IN FEET                  9.
SEED FROM 4 SOURCES (ROWS), 6 BLOCKS (COLUMNS)                                 10.
WAKELEY, 1944, QUOTED BY BLISS, VOL. I, TABLE 11.7.                            11.

      Y                                                                        12.

   34        29.3      30.6      31.8      34        32.7                      13.
   27.3      27.6      28.6      29.2      30.2      31.5                      14.
   26.4      25        26.6      25.2      27.4      26.2                      15.
   24.8      24.3      26        26.5      25.8      24.2                      16.

      TSS, 23, (÷23)×TSS←+/,RY×RY←Y-⎕←GM←(+/,Y)÷24                             17.
28.13333333                                                                    18.
210.5133333  23  9.152753623                                                   19.

      SSR, 3, (÷3)×SSR←6×+/RE×⎕←RE←(+/RY)÷6                                    20.
3.933333333  0.9333333333  ¯2  ¯2.866666667                                    21.
171.36  3  57.12                                                               22.

      SSC, 5, (÷5)×SSC←4×+/CE×⎕←CE←(+⌿RY)÷4                                    23.
¯0.008333333333  ¯1.583333333  ¯0.1833333333  0.04166666667                    24.
      1.216666667  0.5166666667                                                25.
17.15833333  5  3.431666667                                                    26.

      RSS, 15, (÷15)×RSS←+/,RY×RY←RY-RE∘.+CE                                   27.
21.995  15  1.466333333                                                        28.

      ⌈/|,CRY←⌊0.5+RY×2×(RSS÷24)*¯0.5                                          29.
4                                                                              30.
      +/(¯5+ι9)∘.=,CRY                                                         31.
1  2  2  5  5  5  2  0  2                                                      32.

      CRY[; 2 3 1 4 6 5 ]                                                      33.

 ¯2 ¯3  4 ¯1  0  1                                                             34.
  0 ¯1 ¯4  0  4  0                                                             35.
  1  1  1 ¯2 ¯1  0                                                             36.
  1  2 ¯1  2 ¯3 ¯1                                                             37.

      RSS1←RSS-⎕←A×⎕←B←(A←+/,RY×RE∘.×CE)×24÷SSR×SSC                            38.
0.1838114556                                                                   39.
4.139219534                                                                    40.
      RSS1, 14, RSS1÷14                                                        41.
17.85578047  14  1.27541289                                                    42.
```

Figure A:11

At this point we can write out the following analysis-of-variance table:

	Sum of squares	D.f.	Mean square
Columns (blocks)	17.16	5	3.43
Rows (seed sources)	171.36	3	57.12
Residual	22.00	15	1.47
Total about mean	210.51	23	9.15

Most of the variation in the given array Y is associated with rows, and so can be ascribed to differences in seed source. The mean tree heights from the four seed sources, equal to $GM + RE$, are roughly: 32, 29, 26, 25 feet; with estimated standard errors about a half a foot for each. The first of the seed sources was close to the site of the experiment in Louisiana, the other three more distant. The experiment was regarded as showing the existence of regional variation in loblolly pines and confirming the general expectation that a local strain would do better than a more distant strain (Wakeley, 1944).

The mean square between blocks (columns) is larger than the residual mean square, as one would expect, but not very much larger; their ratio, 2.34, is only slightly above the upper 10% point, 2.27, of the variance-ratio (F) distribution with 5 and 15 degrees of freedom, so it would be possible to say that a significant block effect had not been established.

It may be noticed that each of the sums of squares, TSS, SSR, SSC, RSS, has been computed directly as a sum, and not indirectly as a difference of sums of squares, as was usually done by desk calculator. Whenever (as always in this book) the total amount of data is sufficiently small that it can be held all at once in the active workspace, the direct method of computing sums of squares is preferable to the difference method, being less sensitive to round-off error. One may easily verify, and should, that $SSR + SSC + RSS$ is equal to TSS to the 10-figure accuracy shown.

This traditional type of analysis of variance of a row-column crossclassification (due to R. A. Fisher) would be perfectly satisfactory if the observations were generated by a random process, such that each observation was the sum of a constant peculiar to the row, another constant peculiar to the column, and an "error", and all the errors were independently drawn from a common normal distribution having variance σ^2, say. If the observations seem reasonably compatible with this ideal theoretical structure, and we see no other cause to be uneasy about it, we shall presumably accept the analysis of variance (with the associated listing of row and column means) as a good summary of the information contained in the data. But if the ideal theoretical structure seems not to fit, we shall presumably look for a more appropriate treatment of the data. The usual method of assessing the suitability of the standard analysis is by examining the residuals. Under the ideal theoretical conditions, the residuals have a multivariate normal distribution, the marginal distribution of each residual being normal with expectation 0 and variance $\nu\sigma^2/n$, where ν is the number of residual degrees of freedom and n is the total number of observations. Here n is only 24 and ν is 15, less than two-thirds of n. Such residuals do not have much information in them. Questions of suitability of the method of analysis will be considered in some detail in Part B. Now a few specimen calculations are illustrated.

In line 29 of Figure A:11 the elements of *RY* are doubled, divided by the square root of an estimate of $\nu\sigma^2/n$, rounded to the nearest integer and named *CRY* for "coded residuals of *Y*". Thus *CRY* expresses the residuals as nearest-integer multiples of a half their standard deviation. The greatest of the absolute magnitudes of the elements of *CRY* is found (line 30) to be 4. Line 31 asks for a frequency distribution of the elements of *CRY*, and line 33 asks for a display of *CRY* after a permutation has been applied to the columns. J. W. Tukey has pointed out that it is interesting to examine residuals in a two-way table after permuting rows and columns so that the row means and the column means are arranged in order of magnitude. Line 21 shows that the row effects are already in decreasing order; but that is not so for the column effects in lines 24–25, so in line 33 the permutation that would rearrange *CE* in increasing order is applied to the columns of *CRY*.

The frequency distribution in line 32 looks close to normal. The display of *CRY* in lines 34–37 also looks reasonably satisfactory; there is no strong suggestion of unequal variances in different rows, for example. There is, however, some slight suggestion that one pair of opposite corners (N.E. and S.W.) contains mainly positive residuals, whereas the other pair (N.W. and S.E.) has mainly negative residuals. That would imply nonadditivity of row and column effects in *Y*. Tukey's "one degree of freedom for non-additivity" provides a test, which is calculated in the remaining lines of the figure.

Inside the parentheses in line 38, the outer product of *RE* with *CE*, using multiplication, is formed, multiplied element by element by *RY*, summed and labeled *A*. The sum of squares of the elements of *RE* ∘.× *CE* is easily seen to be equal to *SSR* times *SSC*, divided by 24. Thus what is displayed at line 39 is the regression coefficient *B* of the elements of *RY* on the corresponding products of *RE* and *CE*. *B* is multiplied by *A* to give Tukey's one-degree-of-freedom term, displayed at line 40. This is subtracted from *RSS* to give a new residual sum of squares, labeled *RSS*1, displayed at line 42 together with the number of degrees of freedom (14) and the corresponding mean square. (We could alternatively have formed a new matrix of residuals after subtracting Tukey's regression, namely *RY* –

$B \times RE \circ . \times CE$; and then $RSS1$ could have been computed directly by summing the squares of the elements of this array.)

Thus the residual line in the above analysis of variance has been decomposed into

Tukey's term	4.14	1	4.14
Remainder	17.86	14	1.28

Evidence of nonadditivity is only weak. Possibly, however, a similar analysis performed on (say) logarithms or reciprocals of Y might be judged slightly more satisfactory.

The foregoing account of an analysis of the loblolly-pine data has been phrased for a reader already well acquainted with the method of analysis of variance. If one were explaining this analysis to someone not so acquainted, it would be helpful to use the terminal to display directly the components into which Y is analyzed, and verify their orthogonality properties. That is, one would show that the given matrix Y can be expressed as the sum of four matrices, all of the same size (4 rows and 6 columns), and these can be displayed in full. The first matrix has all its elements the same, equal to GM. The second matrix has all its columns the same, each being equal to the vector RE. The third matrix has all its rows the same, each equal to CE. The fourth matrix is RY. The elements in the first three matrices are chosen so that the sum of these matrices mimics Y as closely as possible, in the sense of minimizing the sum of squares of discrepancies; and the same for the first matrix alone, and for the first two. Successive residual matrices can be displayed. If now the four component matrices are destructured by monadic comma into 24-element vectors, the inner products $(+.\times)$ of any pair of them can be verified to be 0—or perhaps because of round-off error almost but not quite 0. The analogy with Pythagoras's theorem for right triangles can be shown. In this way a vivid and concrete idea of what is being done can be conveyed. Indeed, such displays may prove illuminating, because unexpected effects are revealed, even for those who are already well acquainted with standard doctrine.

Row-column crossclassifications are frequently encountered. For any particular set of such data, we may possibly wish, not merely to perform a standard analysis of variance, but to repeat the procedure after transforming the readings in various ways. Rather than enter the instructions afresh each time, in the dialog style just illustrated, we may prefer to have the basic procedure available in a defined function. This certainly saves time and trouble. Unfortunately it also saves thought and encourages rigidity. The important thing is to regard no defined function as sacrosanct, however ingeniously and elegantly it may have been written. Such a function, copied from storage, can be altered to suit special needs or in response to a new idea, before it is executed, provided the user understands the program. He should not feel inhibited about making changes.

In devising a function for the analysis of variance, the first question is the degree of generality to be attempted. We should surely make the program refer to any row-column crossclassification, not necessarily of size 4 by 6. It will be convenient to have the output labeled. We may as well throw in Tukey's test, but we shall not incorporate any display of residuals, since that can well be left for subsequent action. If we are more ambitious we may consider a program to handle data in a rectangular array of any number of dimensions, with various possibilities for estimating cross main effects and interactions and examining different sorts of residuals. But first let us try the two-dimensional case.

We shall name the function *ROWCOL* and let it have one explicit argument, the matrix of readings *Y*. Like *LOBLOLLYDATA*, this function will have no explicit result, but will have both displayed and stored output. A possible program for *ROWCOL* is shown in Figure A:12. Some APL-manship has been exercised, but nothing very bizarre—the program is no more tense than *SQRT*3 in the last chapter.

Statement [1] imposes the condition that the explicit argument shall be a matrix having at least two rows and two columns and six elements. Statement [3] and several of the following call for "mixed" output, that is, display of a character vector followed by display of a numerical vector. Here semicolons separate the parts of unlike kind.* In statement [6], to the right of the "÷" sign, the notation means ρY with its first element dropped; an alternative notation would be "$(\rho Y)[2]$". In statement [7], in the expression for the number of degrees of freedom, we see "$1\uparrow\rho Y$", meaning that the first element of ρY is to be taken; "$(\rho Y)[1]$" could have been written instead. Otherwise, the program closely resembles what was done in Figure A:11.

Figure A:12 is a continuation of Figure A:11, in which the matrix *Y* was defined. Below the definition of *ROWCOL*, this function is executed. If we had called *ROWCOL Y*, the material of Figure A:11 would have been repeated. Instead, we have used as argument the common logarithms of the original *Y*-values (multiplied by 100). Remember that the *Y* appearing in the definition of *ROWCOL* is only a local variable, and does not affect, nor is affected by, the global variable *Y* resulting from execution of *LOBLOLLYDATA*. On the other hand, the variables mentioned in the output (*GM*, *TSS*, *RE*, . . . , *B*) are global variables, and this execution of *ROWCOL* has obliterated what was previously stored under those names in Figure A:11. A system command given just before Figure A:12 was begun reduced the maximum number of digits in output from the usual 10 to 5,

*See the note on this subject at the end of the chapter.

```
    ∇ ROWCOL Y;N;A
[1]   →2+(6≤×/ρY)∧(∧/2≤ρY)∧2=ρρY
[2]   →0,ρ⎕←'NO GO.'
[3]   'GRAND MEAN (GM) IS ';GM←(+/,Y)÷N←×/ρY;⎕←''
[4]   'TOTAL SUM OF SQUARES ABOUT MEAN (TSS), DEGREES OF FREEDOM AND MEA
      N SQUARE ARE ';TSS,(N-1),(÷N-1)×TSS←+/,RY×RY←Y-GM
[5]   '   *ROWS*',⎕←''
[6]   'EFFECTS (RE) ARE ';RE←(+/RY)÷1↓ρY
[7]   'SUM OF SQUARES (SSR), D.F. AND M.S. ARE ';SSR,(¯1+1↑ρY),(÷
      ¯1+1↑ρY)×SSR←(1↓ρY)×RE+.×RE
[8]   '   *COLUMNS*',⎕←''
[9]   'EFFECTS (CE) ARE ';CE←(+⌿RY)÷1↑ρY
[10]  'SUM OF SQUARES (SSC), D.F. AND M.S. ARE ';SSC,(¯1+1↓ρY),(÷
      ¯1+1↓ρY)×SSC←(1↑ρY)×CE+.×CE
[11]  '   *RESIDUALS*',⎕←''
[12]  'SUM OF SQUARES (RSS), D.F. (NU) AND M.S. ARE ';RSS,NU,(÷NU←×/
      ¯1+ρY)×RSS←+/,RY×RY←RY-RE∘.+CE
[13]  'THE MATRIX OF RESIDUALS IS NAMED RY.'
[14]  '   *TUKEY''S TEST*',⎕←''
[15]  'REGRESSION COEFFT. (B) OF RY ON RE∘.×CE IS ';B←(A←+/,RY×RE∘.×CE)×
      N÷SSR×SSC
[16]  'TUKEY''S 1-D.F. TERM IS ';A←A×B
[17]  'REMAINING SUM OF SQUARES, D.F. AND M.S. ARE ';A,(NU-1),(÷NU-1)×A←
      RSS-A
[18]  ⍝ THIS PROGRAM PERFORMS A STANDARD ANALYSIS OF VARIANCE ON A ROW-
      COLUMN CROSSCLASSIFICATION.
    ∇


      ROWCOL 100×10⍟Y

GRAND MEAN (GM) IS 144.69
TOTAL SUM OF SQUARES ABOUT MEAN (TSS), DEGREES OF FREEDOM AND MEAN
      SQUARE ARE 479.99  23  20.869

   *ROWS*
EFFECTS (RE) ARE 5.8535  1.5971  ¯2.9906  ¯4.46
SUM OF SQUARES (SSR), D.F. AND M.S. ARE 393.9  3  131.3

   *COLUMNS*
EFFECTS (CE) ARE ¯0.096113  ¯2.4055  ¯0.13998  0.12287  1.8329  0.68583
SUM OF SQUARES (SSC), D.F. AND M.S. ARE 38.64  5  7.728

   *RESIDUALS*
SUM OF SQUARES (RSS), D.F. (NU) AND M.S. ARE 47.45  15  3.1633
THE MATRIX OF RESIDUALS IS NAMED RY.

   *TUKEY'S TEST*
REGRESSION COEFFT. (B) OF RY ON RE∘.×CE IS 0.092868
TUKEY'S 1-D.F. TERM IS 5.4694
REMAINING SUM OF SQUARES, D.F. AND M.S. ARE 41.981  14  2.9986

      100×10*0.01×RE-RE[1]
100  90.664  81.575  78.862
```

Figure A:12

which are certainly sufficient here and easier on the eye. The last two lines in the figure express the row effect *RE* as percentages of the height for the first row. Thus we may describe the effect of seed source, as measured in this logarithmic analysis, by saying that sources no. 2, 3, 4 gave tree heights that were 91%, 82%, 79% respectively of those of source no. 1.

Tukey's test, and other examination of residuals not reproduced here, indicate very satisfactory agreement with the implied theoretical structure. The ratio of *RSS* to *TSS*, the proportion of the total variation not explained as row and column effects, is slightly lower in logarithms than in the original scale (0.099 instead of 0.104). If one feels a priori that the logarithmic scale is just as natural as the original scale, the logarithmic analysis must be judged the better—though one could hardly claim that it was "significantly" better.

Finally, we should consider the design of the loblolly-pine observations. They are a set of data that came first to hand when a modest example was sought of a one-factor experiment arranged in randomized blocks. The observations are so described by Bliss, and both he and the original author (Wakeley, 1944) made the analysis of variance shown above. Closer examination of Wakeley's paper shows, however, that the randomization of the lay-out was counterfactual-volitional rather than actual. Planting was done in 1926, before randomization had become popular, the very year, indeed, that Fisher's first paper on the arrangement of field experiments appeared. So the design was systematic, as follows. Two distinct half-acre pieces of land were planted each with 12 rows of seedlings. Each row constituted what we have termed a plot. The four seed sources were used in turn in successive rows from one side of the half-acre to the other. The observations obtained from the first half-acre, in their geographical sequence, are found from lines 13–16 of Figure A:11 by reading down the first column, then down the second column, then down the third. Similarly reading down the remaining columns in turn gives the sequence of observations in the second half-acre. Thus the six so-called blocks are two sets of three side by side, and within each block the seed-source arrangement is always the same, not randomized or otherwise varied.

That being so, the type of analysis given above may be expected to be reasonably good, but there is the possibility that a trend across the rows of a half-acre has perceptibly affected the estimated seed-source effect. Some study of this question is now summarized, without worksheets or details of calculation.

If one estimates the one-degree-of-freedom difference between the means of the two half-acres (blocks 1, 2, 3, versus 4, 5, 6) and the treatment effects (*RE*) as above, and then plots residuals against position within each half-acre, it appears that there is indeed some trend or serial dependence—the error process seems not to be one of independent identically distributed random variables, or "white noise". A theoretical structure that would fit would be a quadratic trend plus white noise—a separate trend for each half-acre. Another plausible theoretical structure would be some kind of stationary process. However, the averaged correlogram for the two sets of twelve residuals strongly suggests that the simplest kind, namely a Markov process, will not do; and fitting anything more complicated to so few readings seems over fanciful. So only the quadratic trend option has been taken. Fitting the two trends calls for estimating as many parameters as the previous analysis in terms of six blocks; and the trends analysis should be the fairer of the two. We assume that each observation is equal to a quadratic function of its position in the half-acre, plus a treatment (seed source) constant, plus error, and we fit by least squares. Trends and treatment constants are not orthogonal. We obtain the following analyses of variance, where (a) refers to the original Y-values and (b) to 100 times their common logarithms. The sum of squares for trends is the reduction in the residual sum of squares due to fitting the trends (without fitting treatment constants), and the sum of squares for seed sources is the further reduction in the residual sum of squares due to fitting treatment constants in addition to trends. The residual sum of squares after these fittings has been decomposed into Tukey's term and remainder.

	Sums of squares			Mean squares	
	(a)	(b)	D.f.	(a)	(b)
Half-acres	8.40	18.61	1	8.40	18.61
Trends	30.59	69.45	4	7.65	17.36
Seed sources	156.23	359.78	3	52.08	119.93
{ Tukey's term	3.48	4.46	1	3.48	4.46 }
{ Remainder	11.81	27.69	14	0.84	1.98 }
Residual	15.29	32.15	15	1.02	2.14
Total about mean	210.51	479.99	23	9.15	20.87

The most striking difference between this table and what we had before is that the residual mean squares are reduced to roughly two-thirds of their previous values. The ratio of residual sum of squares to total sum of squares about the mean is 0.073 for the (a) analysis and 0.067 for (b). Again the (b) analysis might be considered slightly preferable, though for both the residuals are acceptably well behaved. Tukey's test for the (a) analysis yields a ratio of mean squares of 4.13, coming at roughly the upper 6% point of the F distribution with 1 and 14 degrees of freedom—nothing to get very excited about. There is no perceptible difference in understanding to be had from the two analyses, so let us think of the (a) analysis, in the original scale.

The estimated mean tree heights in feet for the four seed sources are now

32.083 29.088 26.136 25.227 (s.e. 0.421, 15 d.f.).

These means differ only trivially in the second decimal place from $GM + RE$ in Figure A:11, but the estimated standard error of 0.421 is a bit smaller than the previous 0.494. (The estimated standard error is defined as the square root of one half the average estimated variance of the six possible differences between the four estimated treatment means. The estimated variance matrix of the treatment means is the residual mean square having 15 degrees of freedom multiplied by the inverse of the matrix of the normal equations.)

Thus our trends analysis agrees closely with the previous randomized-blocks analysis. The effect of seed source is slightly sharpened by reduction in standard error, and that is about all the change. Two morals can be drawn from the story: (i) even the simplest job of statistical analysis can provoke discussion; (ii) first impressions are not necessarily misleading.

Example 4–*Simulation of random processes.* Some illustrations of random phenomena will now be shown. Because such work invites repetition, defined functions are used, both "finished" functions that can be stored in the library, and also temporary, to be used now and scrapped.

The first function in Figure A:13 will generate a random sample of size N from a distribution over a finite set of consecutive nonnegative integers 0, 1, 2, ... N should of course be positive-integer-valued; it may be scalar, or it may be a vector, in which case the program generates the sample in the form of an array having N as its dimension vector. The second argument CP lists the cumulative probabilities associated with the value set 0, 1, 2, . . . , that is, the probabilities that an observation will be equal to 0, not greater than 1, not greater than 2, etc. The highest of these probabilities is presumably 1, and that value may be omitted from CP; the other probabilities may be listed in CP in any order. The "business" part of the program is just the single statement [3], in which a random sample of size N from the uniform distribution over the unit interval is generated, and then for each element of this array a count is made of how many elements of CP it exceeds. Statements [1], [2] and [4] of the program serve as documentation —if the user understands the function, these lines can be omitted. For the sum reduction in [3] to work correctly CP must be a vector, and not (for example) a scalar. So at the extreme right in [1] monadic comma is used to ensure that CP will be a vector in [3], even if it was not initially.

Just below the next function, *POISSON*, we see a sample of size 20 drawn from a distribution over the two values, 0 and 1, with the probability of 0 equal to $1/\sqrt{3}$. Here the right argument of *SAMPLE* is scalar.

The function *SAMPLE* can be used to sample a Poisson distribution by the device of truncating the distribution. The function *POISSON* yields the cumulative probabilities CP associated with the value set 0, 1, 2, . . . , K

```
     ∇ Z←N SAMPLE CP
[1]    →2+(∧/,(N=⌊N)∧1≤N)∧(1≥ρρN)∧∧/(1≥CP)∧0≤CP←,CP
[2]    →ρ⎕←'NO GO.',Z←''
[3]    Z←+⌿CP∘.<(÷2147483647)×?Nρ2147483647
[4]    ⍝  SAMPLE OF SIZE N FROM A DISTRIBUTION OVER NONNEGATIVE INTEGERS
       HAVING CUMULATIVE PROBABILITIES CP.
     ∇

     ∇ CP←K POISSON MU;Y
[1]    CP←(Y∘.≥Y)+.×(*-MU)×(MU*Y)÷!Y←0,⍳K
     ∇

       20 SAMPLE 3*¯0.5
0  1  0  0  0  0  1  1  1  0  0  1  0  0  0  1  0  0  0  0

       20 SAMPLE ⎕←10 POISSON 4
0.018316  0.091578  0.2381  0.43347  0.62884  0.78513  0.88933  0.94887
       0.97864  0.99187  0.99716
5  4  7  6  4  2  5  3  5  7  5  3  1  5  3  5  5  9  3  3

     ∇ Z←RNORMAL N;V
[1]    →2+∨/,(N≥1)∧(N=⌊N)∧0=ρρN
[2]    →0,ρ⎕←'NO GO.',Z←''
[3]    Z←(÷2147483647)×?(2,⌈N÷2)ρ2147483647
[4]    Z←N↑(, 1 2 ∘.○Z[2;]×○2)×V,V←(¯2×⍟Z[1;])*0.5
[5]    ⍝  N RANDOM NORMAL DEVIATES BY BOX-MULLER METHOD.
     ∇

     ∇ NORMALTEST Z;N;ZB;W;S;K
[1]    'MEAN = ';ZB←(+/Z)÷N←ρZ
[2]    'EXTREMA = ';(⌊/Z),⌈/Z
[3]    'FREQUENCIES = ';+/(¯4+⍳6)∘.=2⌊¯3⌈⌊Z;', WITH N(0,1) EXPECTATIONS '
       ;W,⌽W←0.1×⌊0.5+0.0001×N× 2275 13591 34134
[4]    'VARIANCE = ';K←(S←+/W←Z×Z←Z-ZB)÷N-1;', WITH ';N-1;' D.F.'
[5]    'G1 = ';(K*¯1.5)×÷/(+/W×Z),N+ ¯1 0 ¯2 ;', WITH S.E. ';(÷/
       6,N+ ¯2 0 1 ¯1 3)*0.5
[6]    'G2 = ';(K*¯2)×(÷/(+/W×W),N+ ¯1 0 ¯2 1 ¯3)-3×S×S÷(N-2)×N-
       3;', WITH S.E. ';(÷/24,N+ ¯3 0 ¯2 ¯1 3 ¯1 5)*÷2
[7]    ⍝  INTENDED ONLY TO TEST OUTPUT OF RNORMAL.
     ∇

     ∇ C←N AUTOCOV V;J
[1]    →2+(N≤¯1++/ρV)∧(J←2≤+/ρV)∧(1=ρρV)∧0=ρρN
[2]    →0,ρ⎕←'NO GO.',C←''
[3]    C←V+.×V
[4]    →4×N≥J←ρC←C,(J↓V)+.×(-J)↓V
[5]    ⍝  FIRST N TERMS OF CORRELOGRAM OF V ARE  1↓C÷C[1]
     ∇

     ∇ R F Y;YB;N;C
[1]    'MEAN = ';YB←(+/Y)÷N←ρY
[2]    'VARIANCE = ';(÷N-1)×1↑C←R AUTOCOV Y-YB
[3]    'AUTOCORRELATIONS:  ';0.001×⌊0.5+1000×1↓C÷C[1]
     ∇
```

Figure A:13

in a Poisson distribution with mean *MU*. Effectively the distribution is truncated so that values greater than $K + 1$ appear as equal to $K + 1$. We see this function executed with *MU* equal to 4 and *K* to 10. The cumulative probabilities are listed (with the maximum number of digits in output set to 5 rather than 10), and a sample of size 20 is generated. Since all members of this sample are less than 11, there has in fact been no truncation.

If we wish to sample without replacement a finite population whose members fall in several classes, counting how many sample members fall in each class, a procedure similar to *SAMPLE* can be followed. We think in terms of frequencies in the population rather than probabilities and use the dyadic query function instead of the monadic. Successive applications will yield a random contingency table having given marginal totals. The reader may enjoy composing and testing such a function.

Let us turn now to the normal distribution. The next function in Figure A:13, *RNORMAL*, generates random normal deviates. If z_1 and z_2 are independent $N(0,1)$ variables and if $z_1 = r\cos\theta$, $z_2 = r\sin\theta$, then r and θ are independent, $\frac{1}{2}r^2$ has an exponential distribution (so that $r = \sqrt{-2\ln u}$, where u is uniform over the unit interval), and θ is uniform over the interval $(0, 2\pi)$ (Box and Muller, 1958). An even number, N or $N + 1$, of deviates are generated; then the first N are selected by the "take" function.

The following function, *NORMALTEST*, is intended as a preliminary check on the output of *RNORMAL*. The frequency distribution obtained at statement [3] refers to the intervals $(-\infty, {}^{-}2)$, $({}^{-}2, {}^{-}1)$, $({}^{-}1, 0)$, $(0, 1)$, $(1, 2)$, $(2, \infty)$. The statistics obtained at statements [5] and [6] are Fisher's g_1 and g_2 (Fisher, 1932, appendix to chapter 3).

AUTOCOV is a function to find the sum of squares and first *N* sums of lagged products of its right argument. By dividing the first element of the explicit result into the others we obtain the first *N* terms in the correlogram, defined as

$$\left(\sum_{i=1}^{m-k} v_i v_{i+k}\right) \Big/ \left(\sum_{i=1}^{m} v_i^2\right) \quad (k = 1, 2, \ldots, N),$$

where m is the length of the vector $V = \{v_i\}$.

F is a function that applies *AUTOCOV* to a vector argument *Y* after its mean has been subtracted. This is merely a temporary function.

Use of these functions is illustrated in Figure A:14. In the first line a vector *Z* of 500 random normal deviates is called for. *NORMALTEST* and *F* are applied, with satisfactory results. This normal sample is now used for two purposes. First we see the beginning of an empirical study of the behavior of the correlogram of short segments of an autocorrelated series. The series *Y* used here, of length 500, is obtained from *Z* by a

```
      Z←RNORMAL 500

      NORMALTEST Z
MEAN = ¯0.02075943147
EXTREMA = ¯3.110112802  3.038384966
FREQUENCIES = 11  77  173  165  64  10, WITH N(0,1) EXPECTATIONS 11.4
      68  170.7  170.7  68  11.4
VARIANCE = 0.9566979401, WITH 499 D.F.
G1 = 0.06124954241, WITH S.E. 0.1092177872
G2 = 0.06936727422, WITH S.E. 0.2180060878

      7 F Z
MEAN = ¯0.02075943147
VARIANCE = 0.9566979401
AUTOCORRELATIONS:  0.071  ¯0.013  ¯0.003  ¯0.047  ¯0.006  ¯0.025  0.057

      Y←(5*¯0.5)×+⌿(⍳5)⌽ 5 500 ⍴Z

      12 F Y
MEAN = ¯0.04641949993
VARIANCE = 1.039121981
AUTOCORRELATIONS:  0.814  0.586  0.373  0.147  ¯0.04  ¯0.045  ¯0.028
      ¯0.03  ¯0.02  ¯0.023  ¯0.024  ¯0.032

      12 F 25↑Y
MEAN = ¯0.08127585001
VARIANCE = 1.347828962
AUTOCORRELATIONS:  0.775  0.55  0.379  0.225  ¯0.052  ¯0.214  ¯0.351
      ¯0.342  ¯0.402  ¯0.418  ¯0.342  ¯0.208

      12 F ¯50↑Y
MEAN = ¯0.03371413686
VARIANCE = 0.791266045
AUTOCORRELATIONS:  0.726  0.43  0.205  ¯0.077  ¯0.356  ¯0.267  ¯0.185
      ¯0.112  0.032  0.188  0.194  0.241


      ⍉+/(¯10+⍳18)∘.= 10 25 ⍴8⌊¯9⌈⌊T←÷/ 250 2 ⍴Z

  0   0   0   1   0   1   2   3  10   4   2   1   0   1   0   0   0   0
  0   0   0   1   0   1   2   1   6   7   2   0   1   0   0   0   0   4
  0   1   0   0   0   0   0   2  10   5   4   1   0   0   0   1   0   1
  1   0   0   1   1   0   0   0   7   9   1   2   0   2   0   1   0   0
  2   0   1   1   1   0   0   1   6   5   4   0   0   0   0   0   1   3
  0   0   0   0   0   1   0   5   4   6   4   0   0   0   1   0   0   4
  2   0   0   0   0   1   0   0   6   4   4   1   1   0   1   0   0   5
  2   0   0   0   1   0   1   4   3   9   1   2   1   0   0   0   0   1
  1   0   0   0   1   0   1   1   8   9   2   1   1   0   0   0   0   0
  0   0   0   0   1   1   1   3   6   8   1   0   1   0   0   0   0   3

      (⌊/T),⌈/T
¯8117.84  335.3638921
      0.1×⌊0.5+0.4×+/ 10 25 ⍴T
¯0.6  3.8  2.4  ¯0.1  ¯323.9  5  3.1  3.3  ¯0.7  13.9
```

Figure A:14

simple moving average, as follows:

$$y_i = \frac{z_{i+1} + z_{i+2} + z_{i+3} + z_{i+4} + z_{i+5}}{\sqrt{5}} .$$

The divisor on the right side has been chosen so that the y's have unit variance. The successive autocorrelations of the y-process are:

$$0.8,\ 0.6,\ 0.4,\ 0.2,\ 0,\ 0,\ 0, \ldots,$$

all being zero after the fourth. (In fact, in deriving Y from Z we have departed slightly from the above scheme by treating the z's as periodic, $z_{i+500} = z_i$ for all i, so Y has a circular pattern of correlation. The circularity can be removed by dropping the last four elements of Y.)

F is used to display the first twelve autocorrelations of Y and also of two short segments of Y, namely the first 25 elements and the last 50. By examining in this way successive segments of length 25, and then segments of length 50, and segments of length 100, each time making a little graph of the correlogram, a vivid idea can be had of typical behavior of such correlograms. The investigation can be repeated for Y's having other correlation patterns, generated by various kinds of moving-average or autoregressive schemes.

The lower part of Figure A:14 is concerned with sampling a Cauchy distribution, as a follow-up to lines 30–34 of Figure A:7. Since we already have on hand the sample of 500 normal deviates Y, a Cauchy sample T of size 250 is conveniently obtained by taking quotients of elements of Z in pairs. The 250 values are now thought of as ten samples each of size 25. The 10-by-18 table shows the frequency distributions of the ten samples over the eighteen intervals $(-\infty, {}^{-}8)$, $({}^{-}8, {}^{-}7)$, $({}^{-}7, {}^{-}6), \ldots, (7, 8), (8, \infty)$. To obtain this display, the elements of T are arranged as a 10-by-25 matrix, rounded down to the next lower integer, but values below ¯9 are raised to that value and values above 8 lowered to that value. Then a three-dimensional array is formed, the outer product (using =) of the value set (¯9, ¯8, . . . , 8) with the matrix of modified T-values. A sum reduction over the last coordinate yields an 18-by-10 matrix of frequencies, which is transposed and displayed.

The least and greatest of the 250 T-values is found, and finally the means of the ten samples of size 25 are obtained, rounded to one decimal place. As is well known, the mean of a sample from a Cauchy distribution has just the same distribution.

Example 5–*Wet and dry days.* Returning to analysis of data, let us examine some records of the incidence of wet and dry days. This material invites thinking in terms of arrays of rank greater than two.

The data refer to the three spring months, March, April and May, of the five years from 1965 to 1969—thus five periods each of 92 days' duration.

For each day (midnight to midnight) the record shows whether or not there was a measurable amount of precipitation, 0.01 inch or more of rain. For brevity we shall refer to days having measurable precipitation as wet days, and the others as dry days.

The data have been stored in a function, *WETDAYSDATA*. When the function is executed, at the top of Figure A:15, a matrix X is formed having 5 rows (for the years) and 92 columns (for the days in the three-month period), consisting of 1's for the wet days and 0's for the dry. Two lines of information are printed.

X itself is not directly shown in Figure A:15, but equivalent information in a more readable form is displayed in lines 5–19. The left portion of the display shows the sequence of wet and dry days in each month, with a "1" for each wet day and a small circle for each dry day. There is a space after every tenth day, for ease of reading. On the right the months and years are noted.

This display was formed in three steps. First the display of circles and ones was formed out of X by the various operations indicated in the left part of the preceding line (line 4)—a 5-by-92 character matrix mimicking X is set up with a circle for each zero in X and a character 1 for each numeral one in X; then a blank column is inserted corresponding to the missing date, April 31; the matrix is restructured as a three-dimensional array; finally a blank column is inserted after every tenth column. When this command had been executed, the paper was rolled back, the left-margin stop was moved 40 places to the right, and the command appearing on the right of line 4 was given, causing the names of the months to be displayed. Finally, the paper was rolled back again and the year numbers were typed after lamp signs, so that the computer would ignore them. (To get the two commands into line 4, the automatic indentation by six spaces on the left was partly undone by three backspaces.)

We might first wish to inquire whether there is any progressive change during the three-month period in the frequency of wet days, and whether the frequency varies from year to year. We see that about 37% of all the days are wet. The numerical 5-by-3 matrix shows the number of wet days in each of the three months, for each of the five years. There is no evidence of month-to-month or year-to-year variation.*

*This matter may be checked as follows. The complementary 5-by-3 matrix of dry days is formed, and hence the contingency tables: (i) a 15-by-2 table of wet and dry days in the fifteen month-year combinations, (ii) a 5-by-2 table of wet and dry days in the five years, totaling over months, (iii) a 3-by-2 table of wet and dry days in the three months, totaling over years. No deviation from expectation is large, and in all three cases the conventional χ^2 statistic is below expectation for purely random ordering of wet and dry days. Now as we shall see, there is evidence of a little positive serial correlation in the data, which would tend to inflate slightly the variability in these contingency tables. Thus all the observed variability can very comfortably be ascribed to chance.

```
      WETDAYSDATA
DAILY PRECIPITATION AT NEW HAVEN AIRPORT, MARCH-MAY, 1965-69.
X IS A LOGICAL 5 92 MATRIX SHOWING MEASURABLE PRECIP'N (1) OR NONE (0).

   (34⍴(10⍴1),0)\ 5 3 31 ⍴((61⍴1),0,31⍴1)\'∘1'[1+X]    5 3 3 ⍴'MARAPRMAY'

∘∘∘∘1∘∘∘∘1 ∘∘∘∘111111 ∘∘11∘1∘∘1∘ ∘      MAR   ⍝  1965
∘1∘∘∘∘1∘1∘ ∘1∘∘11∘1∘∘ 111∘111∘∘∘        APR
∘∘∘11∘1∘∘∘ 1∘∘∘∘∘1∘∘∘ ∘∘∘∘∘11∘1∘ ∘      MAY

1∘∘111∘∘∘∘ 111∘∘∘∘∘1∘ ∘1111∘∘∘∘∘ ∘      MAR   ⍝  1966
∘∘∘1∘∘∘1∘∘ ∘∘∘∘∘∘∘∘∘1 11111∘11∘1        APR
1∘1∘∘1∘11∘ ∘11∘∘1∘∘11 ∘1∘∘∘∘11∘∘ ∘      MAY

∘∘∘∘111∘∘∘ ∘∘11111∘∘∘ ∘1∘∘∘∘∘∘1∘ ∘      MAR   ⍝  1967
∘∘∘∘111∘11 ∘∘∘∘1∘11∘∘ ∘1∘11∘1∘∘∘        APR
∘11∘11111∘ 1∘∘11∘∘∘∘∘ ∘∘∘111∘∘1∘ ∘      MAY

1∘∘∘∘∘∘∘11 111∘∘∘11∘∘ 111∘∘∘1∘1∘ ∘      MAR   ⍝  1968
1∘∘11∘∘1∘∘ ∘∘∘∘∘∘∘∘∘∘ ∘1∘11∘1∘∘∘        APR
∘∘11∘∘∘∘∘∘ 11∘∘∘11111 1111∘∘∘111 1      MAY

∘∘1∘∘∘1∘∘∘ ∘∘∘∘∘∘∘∘∘∘ 1∘∘11∘∘∘11 ∘      MAR   ⍝  1969
∘1∘111∘∘∘1 ∘∘∘∘∘1111∘ ∘1∘1∘∘∘∘∘∘        APR
∘∘∘∘∘∘∘11∘ 11∘∘∘∘∘∘11 ∘∘∘∘11∘∘11 ∘      MAY

      (+/,X)÷460
0.37174

      +/ 5 3 31 ⍴((61⍴1),0,31⍴1)\X

 12  13   8
 12  11  13
 10  12  14
 13   8  17
  7  11  10

      (+/( 0 ¯1 ↓X)≠ 0 1 ↓X)-E←(÷46)×(92-N1)×N1←+/X
¯2.3261  ¯8.8261  ¯9.8261  ¯14.609  ¯6.9565
      ((E-1)×E÷91)*0.5
4.3843  4.5415  4.5415  4.6236  4.031

    ∇ R FF X;J
[1]   M←(R, 2 2 ,1↑⍴X)⍴J←1
[2]   M[J;;;]←+/(2 2 2 ⍴ 0 0 0 1 1 0 1 1)∧.=(2,(⍴X)-0,J)⍴(,(0,-J)↓X),,(0
      ,J)↓X
[3]   →2×⍳R≥J←J+1
[4]   'DETERMINANTS OF ESTIMATED TRANSITION PROBABILITY MATRICES (M)'
[5]   --/[2] M[;;2;]÷+/[3] M
[6]   'WITH S.E. ROUGHLY ';(¯1+¯1↑⍴X)*¯0.5;⎕←''
[7]   'DETERMINANTS OF POOLED MATRICES (SM)',⎕←''
[8]   --/SM[;;2]÷+/SM←+/M
[9]   'WITH S.E. ROUGHLY ';(+/,SM[1;;])*¯0.5
    ∇
```

Figure A:15

We should expect the weather in successive days to be correlated. A storm or other weather pattern will not always come and go entirely within a twenty-four-hour period, midnight to midnight. As one glances over the given sequence of wet and dry days, there seems to be a tendency towards runs of consecutive wet days and of consecutive dry days—though single wet days and single dry days also occur. The last bit of computing in Figure A:15 (lines 28–31) is concerned with counting numbers of runs. For each of the five sequences of 92 observations, the difference is obtained between the actual total number of runs (of both sorts of day) and the expected number under random ordering, given the total observed numbers of wet and of dry days, according to a formula due to Stevens (1939), reproduced in several textbooks. The number of pairs of unlike neighbors as one reads a row of X is one less than the number of runs, and the quantity labeled E is one less than the expectation of the number of runs. Below, the standard error of this difference is given, for each of the five sequences. It will be seen that each of the differences is negative, suggesting positive serial correlation; the largest difference in magnitude is more than three times its standard error, and the next largest more than twice its standard error. So the tendency towards runs seems real enough. (Output has been set to a maximum of five digits again.)

The simplest theoretical description of a stationary autocorrelated sequence of 0's and 1's is a Markov chain. We proceed now to consider whether the data behave like segments of a single Markov chain. For any such segment, and for a given r (where $r = 1, 2, 3, \ldots$), we may form the matrix of r-step transition frequencies

$$\begin{pmatrix} n_{r00} & n_{r01} \\ n_{r10} & n_{r11} \end{pmatrix},$$

where n_{rij} is the number of times that the value i is followed r steps later by the value j. The row totals of this matrix are nearly equal to the corresponding column totals, differing from them by at most r. If each element in the matrix is divided by its row total, we have a matrix of estimated r-step transition probabilities, say

$$\begin{pmatrix} \hat{p}_{r00} & \hat{p}_{r01} \\ \hat{p}_{r10} & \hat{p}_{r11} \end{pmatrix}.$$

Since each row of this matrix sums to one, the determinant is just

$$\hat{p}_{r11} - \hat{p}_{r01}.$$

Now for a Markov chain, as one considers $r = 1, 2, 3, \ldots$ in turn, the matrices of true (rather than estimated) transition probabilities are successive powers of the first one, and therefore so are their determinants. Thus

$$p_{r11} - p_{r01} = (p_{111} - p_{101})^r.$$

The same relation should hold approximately for the determinants of estimated transition probabilities, but will be obscured by sampling variation. In the special case that the true determinant is zero, so that $p_{r11} = p_{r01} = p$ (say), the conditional variance of the estimated determinant, $\hat{p}_{r11} - \hat{p}_{r01}$, given the row totals of the transition frequencies, that is given $n_{r00} + n_{r01} = n_0$ (say) and $n_{r10} + n_{r11} = n_1$ (say), is

$$p(1-p)\left(\frac{1}{n_0} + \frac{1}{n_1}\right),$$

and the estimate of this obtained by inserting $n_1/(n_0 + n_1)$ as an estimate of p (almost if not quite the best estimate of p, because of the near equality of row totals and column totals of the transition frequencies) is easily seen to reduce to

$$\frac{1}{n_0 + n_1},$$

the reciprocal of the sum of the elements of the transition frequency matrix. This may be taken as an approximate variance even if the true determinant is not quite zero.

The function *FF* defined at the foot of Figure A:15 is designed to find the matrices of r-step transition frequencies, for $r = 1, 2, \ldots, R$, for each row of a matrix argument X; and hence the determinants of the estimated transition probability matrices. The transition frequency matrices are stacked in a four-dimensional array M, defined by the first three statements of the program. Statement [1] forms an array M of the desired size consisting merely of ones, and then for each successive value of the first indexing variable the three-dimensional section of M is replaced (statement [2]) by a sum reduction over the last coordinate of the four-dimensional inner product of two three-dimensional arrays. Then in statement [5] the determinants of estimated transition probabilities are found. In statement [6] the above-mentioned approximate standard error for the 1-step determinants is obtained; for our given X this will be $1/\sqrt{91}$. For the r-step determinants ($r > 1$) the standard error is slightly greater—for our X it will be $1/\sqrt{92 - r}$—but the change is not worth noticing. Finally, in the last three statements, pooled results from all the rows of X are obtained, the sum reduction of M over its last coordinate being called *SM*.

In Figure A:16 we see *FF* executed, with R equal to 6 and the given X as right argument. The program does not cause M to be displayed. When execution of *FF* was complete, the three-dimensional sections of M corresponding to successive values of the last indexing variable (years) were displayed side by side in the lower half of the figure, by repeatedly rolling back the paper and moving the left-margin stop to the right. *SM* is also displayed. Thus reading down the lower part of the figure we see the 1-step, 2-step, . . . , 6-step transition frequency matrices for each year

```
      6 FF X
DETERMINANTS OF ESTIMATED TRANSITION PROBABILITY MATRICES (M)

 0.049112    0.19091     0.21869     0.31682     0.1746
¯0.0047847  ¯0.046296    0.027778    0.055556   ¯0.12698
 0.10417     0.1326     ¯0.0074074   0.021164    0.006105
¯0.049107   ¯0.072237   ¯0.20485    ¯0.014161    0.11042
 0.029954   ¯0.043285   ¯0.14725     0.026082   ¯0.0037831
 0.042857    0.11425    ¯0.16742     0.0181      0.064918

WITH S.E. ROUGHLY 0.10483

DETERMINANTS OF POOLED MATRICES (SM)
0.19587 ¯0.010385  0.058762 ¯0.043832 ¯0.022955  0.018572
WITH S.E. ROUGHLY 0.046881


      M[;;;1]      M[;;;2]      M[;;;3]      M[;;;4]      M[;;;5]          SM

  38   20      38   17      38   17      39   15      47   16      200   85
  20   13      18   18      17   19      15   22      16   12       86   84

  36   21      32   22      33   21      33   21      41   22      175  107
  21   12      23   13      21   15      20   16      21    6      106   62

  38   19      35   18      32   22      32   22      44   19      181  100
  18   14      19   17      21   14      20   15      18    8       96   68

  34   22      31   22      27   26      31   23      45   17      168  110
  21   11      23   12      25   10      20   14      16   10      105   57

  36   20      32   21      28   24      31   22      42   19      169  106
  19   12      22   12      24   11      19   15      18    8      102   58

  36   20      35   17      28   24      30   22      43   18      172  101
  18   12      19   15      24   10      19   15      16    9       96   61
```

Figure A:16

separately and (on the right) for all years together. Looking back now to the output of *FF*, we see that the determinant of pooled 1-step estimated probabilities is about 0.2, of which the successive powers 0.04, 0.008, . . . do not differ significantly from the determinants of pooled 2-step, 3-step, . . . estimated probabilities. The determinants obtained from single years, listed at the top of the figure, do not differ significantly from these same values.

Not shown in Figure A:16 are applications of *FF*, not to the whole of *X*, but to the first 46 columns of *X*, and to the remaining 46 columns, separately. The pooled estimates of the transition probabilities are nearly the same as for all of *X* together, and again the individual year sequences agree with the pooled results.

To sum up, the data are consistent with the following simple theoretical description. The sequence of wet and dry days during the three spring months constitute a stationary Markov chain, the same from year to year, with this matrix of transition probabilities:

Tomorrow:	dry	wet
Today: dry	0.7	0.3
wet	0.5	0.5

Knowledge of whether it is wet or dry today, without the other kinds of information available to meteorologists, has some predictive value for tomorrow, but scarcely any for the day after.

Example 6—*Multiple crossclassification.* As a final example of statistical programming, let us return to the generalization of our function *ROWCOL*. We shall consider the analysis of variance of a rectangular array in an arbitrarily large number of dimensions.

We have seen that arrays of known rank—vectors, matrices, and rectangular arrays in three and four dimensions—can be handled easily and elegantly. Most computational problems arising in statistics are naturally expressed in terms of such arrays of small fixed dimensionality.

Rectangular arrays of arbitrarily high dimensionality are sometimes presented by multiply classified data. To preserve the multidimensional structure during the analysis is attractive, rather than to bury it in some kind of design matrix, although the latter procedure may be advisable if the data array is incomplete.

Dealing with an array of arbitrary rank calls for a little special technique. As APL stands now, there are some things that cannot be expressed directly. For example, if X is known to be, say, of rank 3, we may refer to its (I, J, K)th element as $X[I; J; K]$. But if X is of arbitrary rank we cannot refer directly to its $(I, J, \ldots, K)$th element, because there is nothing like those three dots in APL; nor can we refer directly to its Vth element, where V is the name of a vector of length equal to the rank of X, because we cannot introduce semicolons between the elements of V (except by use of the execute function in those implementations of APL that have it). Fortunately, there is no need to do these things. A typical need is to carry out some operation, such as a reduction, over an arbitrary selection of the coordinates. A possible method is to use the transpose function (dyadic ⍉) and restructuring (dyadic ⍴) to arrange that the coordinates where the action is to take place are condensed into one known coordinate, usually either the first or the last, whereupon the operation may be carried out easily. Thus we find ourselves working with permutations of the coordinates, and the corresponding inverse permutations. Of course, a multidimensional array can always be dealt with by destructuring it into a vector and calculating the position of elements in desired sections, but that is to lose the advantage of the original structure—a Fortran-like thing to do, not worthy of the APL-man.

In Figure A:17 we see two functions, *ANALYZE* and *EFFECT*, that together constitute a possible generalization of *ROWCOL* to arrays of arbitrary rank. Statements [1], [2], [4], [5] of *ANALYZE* correspond to the first four statements of *ROWCOL*. Statement [3] of *ANALYZE* sets up global variables *J* and *DF* to be used in *EFFECT*; nothing is done with them in *ANALYZE*. *J* is scalar and will count how many terms in the analysis of variance have been computed, and *DF* is a vector, later to become a matrix, that will be used in calculating degrees of freedom. The last four statements of *ANALYZE* cause some information to be given to the user, and then execution stops.

The purpose of the function *EFFECT* is to estimate a designated main effect or a designated interaction. The argument *V* is a list of one or more coordinate (factor) numbers, just one for a main effect, two or more for an interaction. Our understanding of the meaning of the coordinates of *Y*—which coordinates correspond to crossed factors, which to nested classifications, and which of the latter is nested in which—will inform us what effects can meaningfully be asked for, and which interactions will serve as error estimates for which comparisons. Provided this is understood, successive applications of *EFFECT*, with arguments arranged in order of nondecreasing length (main effects first, then two-factor interactions, then three-factor interactions, etc.), will yield a correct analysis of variance. The rule concerning nested factors is this: if factor *i* is nested in factor *j*, then whenever *i* appears in *V* so must also *j*. More generally, if the first time that factor *i* (or a set of factors denoted collectively by *i*) appears in *V* factor (or a set of factors) *j* is also in *V* then on all subsequent occasions when *i* is in *V* so must *j* be also. If (for example) *EFFECT* 1 3 is executed, this will yield an interaction of the factors (coordinates) numbered 1 and 3, but just what this interaction means and how many degrees of freedom it has will depend on whether previously either or both of *EFFECT* 1 and *EFFECT* 3 have been executed—but not (say) on whether *EFFECT* 2 has been executed. If factors 1 and 3 are crossed, normally both *EFFECT* 1 and *EFFECT* 3 should have been executed. If factor 3 is nested in factor 1, normally *EFFECT* 1 should have been executed, but certainly not *EFFECT* 3. If neither of these main effects has been executed, all combinations of factors 1 and 3 are being treated as if they were a single factor, and should continue to be so treated in subsequent executions of *EFFECT*.

EFFECT goes to work, not on the original data array, but on the residual array *RY* left by *ANALYZE* or by the last application of *EFFECT*. The function has an explicit result, corresponding to the vectors *RE* and *CE* in *ROWCOL*. The user supplies a name (in place of the dummy *Z* shown) for this effect if he wants to keep it, a different name for each effect. If for example *V* is the vector 1 3, he might choose to name the effect matrix *E*13.

The first two statements in *EFFECT* check that *V* is made up of one or more members of the set $(1, 2, \ldots, R)$ with no repetitions, where *R* is the

```
    ∇ ANALYZE Y
[1]   →2+(∧/2≤ρY)∧2≤ρρY
[2]   →0,ρ⎕←'NO GO.'
[3]   DF←(1+ρρY)ρJ←1
[4]   'GRAND MEAN (GM) IS ';GM←(+/,Y)÷×/ρY
[5]   'TOTAL SUM OF SQUARES ABOUT MEAN (TSS), DEGREES OF FREEDOM AND MEA
      N SQUARE ARE ';TSS,NU,(÷NU←¯1+×/ρY)×TSS←+/,RY×RY←Y-GM
[6]   'PROCEED BY REPEATEDLY CALLING THE FUNCTION ''EFFECT''.'
[7]   'THE ARRAY OF RESIDUALS IS ALWAYS NAMED RY, WITH DEGREES OF FREEDO
      M NU.'
[8]   'RENAME SS EACH TIME TO SAVE IT.'
[9]   ⍝ BEGINS AN ANALYSIS OF VARIANCE OF A PERFECT RECTANGULAR ARRAY.
    ∇

    ∇ Z←EFFECT V;M;K;P;IND;R
[1]   →3-∧/∨/M←(,V)∘.=⍳R←+/ρρRY
[2]   →3+∧/0≤IND←1-+⌿M
[3]   →0,ρ⎕←'NO GO.',Z←''
[4]   DF←((J←J+1),1+R)ρ(,DF),IND,0
[5]   NU←NU-DF[J;1+R]←(×/(ρRY)[V])-+/DF[;1+R]×(+/DF[J;])=DF+.×DF[J;]
[6]   SS←K×+/,Z×Z←(÷K)×+⌿((K←×/IND/ρRY),(ρRY)[V])ρ(⍋P←(IND/⍳R),V)⍉RY
[7]   RY←RY-P⍉((IND/ρRY),(ρRY)[V])ρZ
[8]   'SUM OF SQUARES (SS), D.F. AND M.S. ARE ';SS,DF[J;1+R],SS÷DF[J;1+R
      ]
[9]   ⍝ 'ANALYZE' SHOULD BE CALLED FIRST.
    ∇


      EGGDATA
Y IS A 6 2 2 2 ARRAY OF PERCENT FAT CONTENT OF DRIED WHOLE EGGS
NESTED SAMPLING: 6 LABORATORIES, 2 ANALYSTS, 2 SAMPLES, 2 DETERMINATIONS
MITCHELL, 1950, QUOTED BY BLISS, VOL. I, TABLE 12.1.

      ANALYZE Y
GRAND MEAN (GM) IS 42.0875
TOTAL SUM OF SQUARES ABOUT MEAN (TSS), DEGREES OF FREEDOM AND MEAN
      SQUARE ARE 1.0231  47  0.02176808511
PROCEED BY REPEATEDLY CALLING THE FUNCTION 'EFFECT'.
THE ARRAY OF RESIDUALS IS ALWAYS NAMED RY, WITH DEGREES OF FREEDOM NU.
RENAME SS EACH TIME TO SAVE IT.

      E1←EFFECT 1
SUM OF SQUARES (SS), D.F. AND M.S. ARE 0.443025  5  0.088605
      SS1←SS

      E12←EFFECT 1 2
SUM OF SQUARES (SS), D.F. AND M.S. ARE 0.247475  6  0.04124583333
      SS12←SS

      E123←EFFECT 1 2 3
SUM OF SQUARES (SS), D.F. AND M.S. ARE 0.1599  12  0.013325

      RSS, NU, (÷NU)×RSS←+/,RY×RY
0.1727  24  0.007195833333
```

Figure A:17

rank of the data array. The "logical" vector *IND*, of length *R*, has 0's in the places numbered by the elements of *V*, and 1's elsewhere; thus if the data array has five dimensions and *V* is 1 3, *IND* is 0 1 0 1 1. Statements [4] and [5] are concerned entirely with finding the number of degrees of freedom associated with effect *V*, and the number of residual degrees of freedom. *DF* is a matrix, of which the first row was found in *ANALYZE*. Each time *EFFECT* is executed, a new row is catenated onto *DF*, consisting of the vector *IND* followed by (at first) a 0, which however is subsequently changed (in statement [5]) into the number of degrees of freedom in the effect. In the illustration lower down Figure A:17 the data array is four-dimensional. After the third execution of *EFFECT*, *DF* looks like this:

```
1   1   1   1    1
0   1   1   1    5
0   0   1   1    6
0   0   0   1   12
```

The reader may enjoy trying to decipher the somewhat tortuous calculation expressed by statement [5].

Statements [6] and [7] of *EFFECT* are the ones that directly tackle the multidimensional array *RY*. What we need to do is sum *RY* over all coordinates except those listed in *V*, divide by the appropriate divisor (labeled *K*) to form the effect array *Z*, and then from *K* repetitions of *Z* build up an array of fitted values, which is subtracted from *RY* to form the new array of residuals, named *RY* again. We also need to calculate the sum of squares for the effect, named *SS*.

The way this is done is first to permute the coordinates of *RY* so that the coordinates numbered *V* come last. Then the array is restructured so that all the other coordinates are elided into one coordinate, and a single summation is performed over that coordinate to yield *Z*. This, together with the finding of *SS*, is specified in statement [6]. In statement [7], the array *Z* is repeatedly stacked on itself, and then the inverse of the previous permutation is performed on the coordinates to yield the array of fitted values, to be subtracted from *RY*. Three details of notation call for explanation. The dyadic transposition symbol (⍉) takes as its left argument a permutation vector, say *T*, and its right argument is an array, say *A*, of rank ρT. The *I*th coordinate of *A* appears as the $T[I]$th coordinate of the result. We see two uses of the solidus, the reduction that we have already met, where the solidus is preceded by a primitive dyadic scalar function, and another use called "compression", in which the solidus is preceded by a logical vector. The elements in the right argument in positions corresponding to 1's in the left argument are selected, the rest being omitted. The "grade up" symbol (⍋) before a permutation vector yields the inverse permutation vector.

As a brief illustration in the amount of space available in the lower part of Figure A:17, these functions are applied to some data stored under the name *EGGDATA*, taken from Bliss's book and not reproduced here. The four-dimensional array is a simple nested arrangement with no crossed factors other than the highest-level classification (laboratories). In each of six laboratories there are two analysts (12 analysts in all), each of whom is given two samples of dried egg (24 samples in all) on each of which he makes duplicate determinations.

EGGDATA is executed, then *ANALYZE*, then *EFFECT* three times, and finally the residual sum of squares is found. The standard analysis-of-variance table looks like this:

	Sum of squares	D.f.	Mean square
Between laboratories	0.4430	5	0.0886
Between analysts within laboratories	0.2475	6	0.0412
Between samples within analysts	0.1599	12	0.0133
Between duplicate determinations	0.1727	24	0.0072
Total about mean	1.0231	47	0.0218

The effects *E*1, *E*12, *E*123 are stored and can be displayed. The sums of squares *SS*1, *SS*12, *SS* (we did not rename this last) and *RSS* are all stored, and we can check that their sum is equal to *TSS*, as it should be. The final residual array *RY* is stored and available for further study.

Of course, one never *has* to use a general program to analyze a particular data array. For this array the analysis could be built up easily out of the simple differences

```
-/Y
-/+/Y
-/+/+/Y
```

and the laboratory totals

```
+/+/+/Y.
```

Notes for Chapter 7

In Figures A:9–11 and elsewhere, numerical output has up to ten digits, as happens when printing precision is not set differently (an easy thing to do) and format is not directly controlled (another easy thing to do). The reader may perhaps find the unnecessary precision irritating. The purpose of this chapter is to show how an APL terminal can be persuaded to perform some typical statistical calculations, and the work is shown as "raw" as possible, in case the reader wishes to duplicate any of the figures. In duplicating them he may like to improve them! Each of the examples shows common procedures in a particular context. The reader should, of course, skip any examples whose subject-matter is not interesting. In Part B below, there is little consideration of how to do it; attention is on what to do.

In regard to the storing of data, the writer does not now use a defined function, as at the top of Figure A:11, but prefers a descriptively named variable, together usually with a character variable whose name is the same except prefixed by *WHAT*, that gives a brief description of the material. Thus he would define a 4-by-6 integer matrix called *LOBLOLLYDATA* and have explanations, including an instruction to divide by 10, in *WHATLOBLOLLYDATA*.

In the function *ROWCOL* in Figure A:12 and in other functions, what is termed mixed output is called for with use of semicolon separators. This method of eliciting mixed output is the only feature of *APL*\360 illustrated here that is not supported in some later implementations of APL. The later method involves translating a numerical vector (or scalar) into a character vector by means of the format function and then catenating it onto any other character vector by dyadic comma. The reader may like to compare *ROWCOL* in Figure A:12 with its namesake in Appendix 3. The latter is much longer because further tests of residuals, in addition to Tukey's test, have been included. The first 19 lines correspond to the earlier program. The later handling of mixed output is shown in lines [3], [4], [6], [7], etc. In line [1] the more stringent condition that *Y* should have at least 3 rows and 3 columns is imposed, to prevent breakdown of the tests of distribution shape and heteroscedasticity. In lines [14] and [15] checks are made that both *RSS* and $SSR \times SSC$ are positive, to prevent breakdown of Tukey's test. Such checks should have been included in the program in Figure A:12. The later version of *ROWCOL* is discussed in Appendix 2.

The writer is pleased with *ANALYZE* and *EFFECT* in Figure A:17. Smillie included a version of them in his *STATPACK*2 (1969). There has been a considerable literature on computer procedures for analysis of variance of rectangular arrays with "crossed" and "nested" factors. The procedure given here is simple to apply and renders more complicated machinery superfluous. The versions of these functions in Appendix 3 are very similar to those in Figure A:17. There are slight notation changes in *EFFECT*. The principal difference is that the later version of *EFFECT* computes names for the effect and for the sum of squares, with the aid of the execute function. That was not possible in *APL*\360.

Chapter 8

Comments

The foregoing account of APL is far from complete. Many things have not been said. The following have not been explained:

(i) how to sign on and off,
(ii) how to move material in and out of storage,
(iii) how to correct an error in typing,
(iv) how to add to, emend or "edit" a function definition,
(v) how to obtain items of information such as: (a) the amount of unused space remaining in the active workspace, (b) the number of users of the system currently signed on, (c) the amount of central-processor time used since sign-on, (d) the names of stored workspaces.

Such matters as these, for which proper provision has been made, as explained in the manual, are of immediate concern to anyone working at an APL terminal, but are beside our main purpose, which is to convey the flavor, and suggest the suitability, of APL in expressing statistical calculations. Even for that purpose, much else might be said. The programs listed in Appendix 3 and various displays of output in Part B will supplement the detailed illustrations already given, and help to show concretely what statistical analysis with APL is like.

The illustrations given above present a feature of APL terminal use that can be lost sight of. By all means let the user be uninhibited in defining functions (storing programs), and then changing and dropping them. But he should not forget that the terminal can also be used like the old desk calculator, and he can always revert to the impromptu dialog style. Often a very few APL symbols will express a useful block of computation. In a less concise language such a block would be worth storing as a subroutine. To store it as an APL function will only be profitable if the user can remember the function's name and syntax the next time he wants it—remember these sooner than he can reinvent the program. He should think of stored programs always as aids to free dialog, not as controls to his thinking (as program packages in other languages have so often been). The whole of a complex calculation need not necessarily be enshrined in a stored program.

Let us now consider the question: what is the slant of APL?

We are conscious of a great change when we lay aside paper-and-pencil calculations, assisted possibly by slide rule or desk calculator, and go courting a computer. We have to think more explicitly about our procedures, in matters that formerly were left to judgment or common sense. We must learn new terminology, and—much more important—we find ourselves thinking about the calculations in a somewhat different way. Any method of communicating with the computer (language, program package, interactive system) has a slant; it will not just transmit our thinking, it will modify our thinking. Some things are easier to do than others, our attention is directed in particular ways, unexpected questions must be answered. What slant does APL have for the statistician?

None whatever of a direct statistical kind! There is nothing in APL that can affect a statistician's philosophy, ethics, basic concepts or theoretical ideas—except what is inherent in computing rather than in APL. (The more effortless computation becomes, the more the statistician will compute, and some of his basic ideas will be changed by the experience.) By contrast, any computing system or program package designed specially to aid the statistician will also to some extent bias his statistical thinking. It will come natural to him to think of some problems in one way rather than another, because that is how the designer thought of them. The statistician will be slanted or constrained in his thinking if he relies on a library of statistical programs written in APL when he himself is too little familiar with APL to indulge effectively in free dialog. Such slanting is not necessarily bad. For some users it is a positive help.

Though APL has no substantive statistical slant, it certainly has a distinctive character among programming languages. The conciseness, the generality and the affinity with mathematical thinking lead to transparency —or (to change the metaphor) the machinery of APL creaks less than that of other languages. Apart from the sheer ease of expression, APL has at least one noticeable algorithmic slant: it encourages us to replace loops whenever possible by array operations, and therefore to think in terms of arrays—arrays in various numbers of dimensions.

Compare the mathematical concept of a finite sum, say $\sum_{i=1}^{n} a_i$, with an APL sum reduction, say $+/A$. In the mathematical expression, no particular order of summation is implied, because addition is associative. In computation, addition is associative in integer arithmetic but not necessarily in floating-point arithmetic, and there are occasions when lack of associativity matters. The sum reduction in APL is carried out in a known order (right to left). If we desire a different order of summation, we may permute the array before reducing it. Thus we can control the order of summation as we wish, and yet use a notation $(+/A)$ that views the array as a whole and does not bring the order of summation into the limelight by making us write a loop. Very often in practice that order of summation is unimportant, and we ought not to have it obtruded on us. In the element-by-element application of primitive scalar functions to arrays, as in a sum, $A + B$, the order of execution is completely irrelevant—if indeed execution is done in sequence and not simultaneously by a multiple processor. Similarly for outer products, $A \circ . \times B$.

Thus we become sharply aware of the difference between repetitive operations that can be expressed as operations on arrays defined at the outset, on the one hand, and successive approximations and recursive definitions wherein some kind of looping is essential, on the other hand. Our understanding of algorithms is increased.

Some computer scientists disapprove of the right arrow as the sole means of controlling the flow of execution in a defined function—every branch in APL is a "computed go to". They note that flow control in Algol60 is easier to read and (they feel) more elegant and natural. Any language that must specify array operations by loops needs more flow control than APL does. Among the many defined functions listed in Appendix 3 there are few loops, and in all cases the flow, as it could be represented in a flowchart, is quite simple. That is why no flowcharts are presented. Equivalent programs in Fortran or Algol would usually require much more flow control. The matter, therefore, does not seem very serious.

The general-purpose computer language seen by statisticians as the principal competitor to APL is undoubtedly Fortran, because of its very wide availability and effectiveness for numerical computation. Apart from the algorithmic differences just noted, in flow and handling of arrays, Fortran's afterthought treatment of characters, so different from APL's equal rights for characters and numbers, has had important consequences. Many persons skilled in use of Fortran for numerical computation have been disinclined to struggle with subroutines for scatterplots or other graphical displays, preferring to take whatever was in the program library. An omnibus plotting program, composed by a professional programmer, may be clever and vastly better than nothing, yet may not answer the need of the individual case very well. Size, scales, symbols and labeling are usually not controlled, or only partly controlled, by the user. If he regards such programs as out of bounds, he is discouraged from thinking how the job might be done better. Graphical techniques have therefore not developed as fast as they should have.

Obviously computers are built and operated by experts. It is easy to suppose that things are inevitably as they are; the experts know best. The more the user can control what the computer does for him, and in particular control the visual appearance of the output, the more he can see what is not inevitable. Suppose that one variable in a scatterplot is a date, and the axis needs to be labeled with the numbers 1950, 1960, 1970. The omnibus library program may possibly label the axis 1950.00, 1960.00, 1970.00, an absurdity that would not be tolerated if tampering with the program were easier. The experts who first designed and implemented Fortran, whatever their other skills, were incompetent typographers, and supposed that upper-case Roman O was necessarily hard to distinguish from numeral 0, and therefore were disposed to slash one of them. Hence the labeling of the scatterplot may come out 195Ø.ØØ, 196Ø.ØØ, 197Ø.ØØ, an even worse absurdity. Because Fortran used only upper-case letters, no need was perceived to distinguish lower-case Roman l from numeral 1, and we have been spared the insult of slashed ones also. Yes, computer languages do have an effect.

Despite the evident attractiveness of working with APL, the fact is that much statistical computing is done in other ways. There is much reliance on program packages and statistical systems, and some of those who innovate do so via Fortran and job control language. One hears three reasons for not using APL:

(i) the implementation of APL is asserted to be inefficient, or (to put it another way) computing with APL is expensive;
(ii) the implementation of APL is asserted to be unsuitable for large-scale work, such as analysis of large data sets, or extensive simulations;
(iii) it is not easy to combine work in APL with use of existing good programs written in other languages.

These are interesting issues that merit discussion.

Efficiency. When anyone first programs a computer, he becomes aware that the same job can be done in different ways, and naturally he is interested in which way is best. If a program is to be run many times, speed in execution may perhaps be economically important. The matter can easily become an obsession—hours or days may be devoted to saving some small fraction of a second of central-processor time. Indeed speed of execution of algorithms is a fascinating topic in its own right, but a topic that, in many circumstances, has negligible economic impact.

This book is about the computing that is done, or could be done, by persons engaged in statistical work. Among these persons are a small proportion who are computing experts, with wide knowledge and experience of computing. This book is addressed to all the others, who do not compute at all, or who do not wish computing to become a primary interest. For such people, there is one efficiency in a computing system that far outweights all other efficiencies: efficiency in use of personal time and effort. This is what makes the difference between doing and not doing.

This alone is vital. The first requirement is that small computing jobs can be carried out quickly, with little effort, little planning, little screwing up of courage. In this regard, APL is without rival.

What are some small computing jobs that a statistician may, at short notice, wish to do? Consider three examples.

(i) Mathematical analysis yields a function of a positive argument x whose behavior should be understood:

$$\frac{\sin(3\pi x) + \sin(\pi x)}{4\pi x(1 - 4x^2)}.$$

In fact, the behavior of this function should be compared with that of a second function,

$$\frac{\cos(3\pi x)}{1 - 36x^2}.$$

(Such functions are met with in Chapter 10 below. We suppose these functions defined by continuity where the denominator vanishes.) Armchair mathematical contemplation, without arithmetical effort, soon reveals that these functions are infinitely differentiable, and have numerous zeroes, of which the lowest is at $x = \frac{1}{2}$, and that as $x \to \infty$ the first function is $O(x^{-3})$, whereas the second is $O(x^{-2})$, in magnitude. The region where behavior of the functions is of most interest is x less than (say) 2, since that is where the functions differ perceptibly from zero. Elegant mathematical consideration leaves one vague about behavior in this region. At an APL terminal almost no time, perhaps a minute, is needed to produce a short table of each function, and only a little longer to produce a graph.

(ii) One has a small rectangular table of data; one wishes to fit an additive structure, representing the table as nearly as possible by $A \circ .+ B$, where A is a vector of row constants and B is a vector of column constants; and the fitting is to be done twice, first by least squares and then by Tukey's procedure (1977, chapters 10 and 11) of repeatedly calculating row medians and column medians. Both operations can be done easily at an APL terminal even if no general program has previously been stored—the first was illustrated in Figure A:11, and see Exercise 8 below for the second. (In fact, most of the procedures described by Tukey (1977) are quite easy and rather entertaining to program in APL.)

(iii) One has fifty observations on a pair of variables, and wishes to get an idea of their joint distribution. If the data are entered at an APL terminal, and if a suitable plotting function is available, such as *SCATTERPLOT* in Appendix 3, calling for a scatterplot of the data takes only a few seconds.

Each of these computing tasks can be carried out by hand without assistance from any machine. But each is rather tedious to do that way, and rather easy to make mistakes over, and there is a temptation not to do it, with risk of misunderstanding.

Not doing it is not the resort only of those who cannot compute. Some people who are capable of using a computer to perform substantial calculations would be reluctant to do the little jobs just mentioned, perhaps because the language for communication is so clumsy or because the turn-around time of batch processing is so long.

Granted that little computing tasks can be performed quickly and easily through an APL terminal, there are a few considerations concerning efficiency in use of the computing resource that are worth bearing in mind. Economy in storage of functions and data may be as important as economy in execution—a function that is only executed rarely may cost more to hold in storage than to execute. The most generally useful economy device for storage has already been mentioned, to hold sets of numbers as integers rather than in floating-point form, whenever that is convenient. Occasionally special codings may be resorted to. For example, if a vector of n integers contained no more than the eight different values, 0, 1, 2, . . . , 7, it could be represented as a matrix with 3 rows and n columns of which the jth column was the binary representation of the jth given number. The matrix would be stored in $3n$ bits (plus overhead for labeling and specification), as compared with n integers usually stored in $4n$ bytes, about ten times as much if n were large. (See Exercise 10 below.)

As for economy in execution, a disclaimer is advisable. There have been several implementations of APL, and no doubt there will be others. The successive manuals issued by IBM have always described APL abstractly, giving almost no information about any implementation. It is possible to discover a good deal about a particular implementation by experimenting with it, but the information obtained will not necessarily apply to other implementations. The remarks made below (like those made just above on storage) refer to the implementations with which this writer has had experience, versions of *APL*\360 and *VS APL*.

Programs written in Fortran and many other languages go through the preliminary step of being compiled, that is, translated into instructions that can be directly carried out by the computer, and then the compiled program is executed. The better compilers analyze the source program more, at higher cost of compiling, and produce a compiled program that will execute faster. When APL commands are executed, there is no preliminary step of compiling, the translation into machine instructions being done by what is called an interpreter. Each line in a defined function is interpreted as it is encountered. If a line inside a loop is encountered repeatedly, it is interpreted afresh each time. For simple repetitive steps, such as occur in Fortran array-negotiating loops, repeated interpreting is much more costly than single compiling. This leads to the one general rule for economy of execution of APL programs, that every APL user should be sharply aware of, namely that loops should usually, if possible, be avoided and replaced by array operations. There are some special situations in which loops that could be avoided are economical; but to translate a typical Fortran program directly into APL, with similar loops, will usually be disastrous. Those accustomed to Fortran must discipline themselves to think in terms of array operations, if they are to work with APL.

Apart from this major precept to think in terms of array operations, the beginner in APL is well advised to pay no heed to efficiency in execution. Attention is better directed to the objectives of the computing than to computing style. Perhaps one second piece of advice on computing style may be offered, however: not to have very many named objects in the active workspace, and in particular not to give distinct names to intermediate results that are of no interest in themselves. An example of care in this respect is shown in Figure B:8 below in Chapter 10. There two vectors of data (named *ENROLMENT* and *IMPORTS*) are each subjected to a sequence of simple operations, in preparation for Fourier analysis: logarithms are taken, then a linear trend is subtracted, then a differencing is performed, and finally a tapering. The logarithms are not named, and the result of each of the other three operations is given the same name, which therefore stands successively for different things. If each intermediate result had been named differently and so preserved, there would have been two unfortunate consequences: (i) the active workspace would have become unnecessarily full, (ii) the user would forget what most of the names meant, and when later he felt an urge to clean house he would be in danger of erasing things he wanted to keep while throwing away the junk. Such intermediate results can easily be recomputed if they are needed again for some other purpose; it's all very different from paper-and-pencil calculation, where each arithmetic step is slow and painful, and preservation of worksheets is easy.

Different ways of expressing the same computation in APL can indeed differ in speed and in the amount of space needed for execution, and one may therefore be said to be better than another. We return to this topic below.

Size. An APL system is very effective for small computations. For sufficiently large computations there are problems in execution that call for special thought. A striking example is a topic not obviously relevant to statistics, likely to interest statisticians only on their day off, namely arithmetic with large integers.

The unlabeled frontispiece to Smillie's book (1974) seems to be a printout of 500 factorial (or !500), all 1135 decimal digits of it. This number is far outside the range of numbers representable in an ordinary APL implementation as either integers or floating-point numbers. The definition can be clearly stated in an APL expression

$$\times/\iota 500$$

but that instruction cannot be executed at an APL terminal—an error message results. We are reminded that APL grew out of a notation for informing human readers, not computers. It can still do that.

The logarithm of the above expression can be calculated readily (to limited precision), and one may easily determine that 500 factorial is approximately $1.220136826E1134$, which for most purposes is good enough. But what if the exact value were required? Exact computation with indefinitely large integers is possible in APL, and presumably Smillie generated his frontispiece that way, but steps must be taken to keep within range of the implementation. Numbers can be expressed as vectors with respect to any suitable base, and defined functions can be composed to perform arithmetic operations such as multiplication on numbers so expressed. Rather than the base 2 normally used by the computer, or 10 normally used by humans, the base 10^8 (or $10*8$) is convenient for representing large integers. Then 500 factorial appears as a vector with 142 integer elements, each expressible in 8 decimal digits.

Part B of this book is concerned with statistical analysis of data. Let us consider how the size of the data set affects what is done. The size of data sets may be roughly classified in three categories:

(a) Small data sets that can be written legibly on a single sheet of paper, and which the eye can run over effectively. A typical number of numbers might be (say) a hundred. All the data sets in this book are small.

(b) Data sets of intermediate size that are too large to go on one sheet of paper and that the eye cannot run over effectively, but that are not too large to be held in the core of the computer. A typical number of numbers might be (say) ten thousand.

(c) Large data sets, too large to be held in core. A typical number of numbers might be (say) a million.

Almost anything that anyone wants to do in analyzing small data sets can be done readily in APL—certainly everything shown below in Part B.

To be able to handle intermediate-sized data sets in APL, the first requirement is a fairly large workspace. The earliest APL systems generally had an available working area of about 32000 bytes. Workspaces of six or more times that size have since become common; in some installations there is no definite limit to workspace size. Granted that the workspace is several times as large as the data, most things that would be done on a small data set can be done just the same for a larger data set. Sometimes it may be necessary to modify APL commands to reduce the amount of space needed for execution—an example is discussed below.

There are differences between analyzing small data sets and analyzing intermediate-sized data sets, but the differences are not so much in carrying out the computations as in what computations are appropriate. There is the initial problem of checking the data for accuracy or consistency. This does not usually seem difficult for a small data set, but merits careful attention for a larger data set. The remarkable book by Naus (1975) indicates some of the possibilities. Then graphical presentation of data poses different problems for different amounts of data—a scatterplot having ten thousand points may be less effective than one with a hundred points. Suggestions such as those made by Cleveland and Kleiner (1975) may prove useful.

Large data sets, too large to be held in core, take one outside a simple APL system into file management and other features of the real computer world from which the naive APL user is shielded. A few remarks are made below on APL in relation to the computing environment.

Text processing is a subject that interests most computer users sooner or later. Since so little has been said about it above, Figure A:18 may be helpful, and it provides a concrete example for some consideration of size and efficiency. At the top of the figure appears a print-out of a piece of literary text. It has been stored as a character vector named *GA*2 having 1457 elements, starting with the letter *F* and finishing with a period, the carrier returns being single elements of the vector. (The blank lines above and below the print-out are not part of *GA*2, but have been inserted for clarity by manually turning the platten.) The alphabet has also been stored under the name $\underline{A}$, and is displayed. The number of occurrences of each letter of the alphabet in the text is counted: there are 100 *A*'s, 165 *E*'s, 124 *T*'s, no *J*, *X*, *Z*. A variable *IB* is set up that indicates blanks and other non-alphabetic characters in *GA*2, and the number of times a non-alphabetic character is immediately followed by an alphabetic character is counted—this is the number of words in the text. This method of counting words works because indeed all pairs of words are separated by at least one non-alphabetic character, a blank or carrier return or dash or whatever. The first word *FOUR* is separated from the last word *EARTH* by the final period—the rotation operation treats the beginning and end as adjacent. Then two executions of a function *IN* are shown that count the number of occurrences of its left argument in its right argument. (Other executions of *IN*, not reproduced, show that *GA*2 has 11 occurrences of *THE*, 7 of *A*, 8 of *TO*, 6 of *AND*, 5 of *THIS*, 4 of *DEDICATED*, 2 of *DEDICATE*.) Finally, three different possible versions of *IN* are shown, distinguished as *IN*1, *IN*2, *IN*3; any one of these could be used as the *IN* just referred to.

Usually a function definition should contain a test that the arguments are suitable, but such tests have been omitted from these *IN* functions for ease of comparison. In each case, the first argument *W* should be a scalar or

```
      GA2

FOUR SCORE AND SEVEN YEARS AGO OUR FATHERS BROUGHT FORTH, UPON THIS
CONTINENT, A NEW NATION, CONCEIVED IN LIBERTY, AND DEDICATED TO THE
PROPOSITION THAT ALL MEN ARE CREATED EQUAL.
      NOW WE ARE ENGAGED IN A GREAT CIVIL WAR, TESTING WHETHER THAT
NATION, OR ANY NATION, SO CONCEIVED, AND SO DEDICATED, CAN LONG ENDURE.
WE ARE MET HERE ON A GREAT BATTLE-FIELD OF THAT WAR.  WE HAVE COME TO
DEDICATE A PORTION OF IT AS A FINAL RESTING PLACE FOR THOSE WHO HERE
GAVE THEIR LIVES THAT THAT NATION MIGHT LIVE.  IT IS ALTOGETHER FITTING
AND PROPER THAT WE SHOULD DO THIS.
      BUT IN A LARGER SENSE WE CAN NOT DEDICATE--WE CAN NOT CONSECRATE--
WE CAN NOT HALLOW THIS GROUND.  THE BRAVE MEN, LIVING AND DEAD, WHO
STRUGGLED HERE, HAVE CONSECRATED IT FAR ABOVE OUR POOR POWER TO ADD OR
DETRACT.  THE WORLD WILL LITTLE NOTE, NOR LONG REMEMBER, WHAT WE SAY
HERE, BUT CAN NEVER FORGET WHAT THEY DID HERE.  IT IS FOR US, THE
LIVING, RATHER TO BE DEDICATED HERE TO THE UNFINISHED WORK WHICH THEY
HAVE, THUS FAR, SO NOBLY CARRIED ON.  IT IS RATHER FOR US TO BE HERE
DEDICATED TO THE GREAT TASK REMAINING BEFORE US--THAT FROM THESE HONORED
DEAD WE TAKE INCREASED DEVOTION TO THAT CAUSE FOR WHICH THEY HERE GAVE
THE LAST FULL MEASURE OF DEVOTION--THAT WE HERE HIGHLY RESOLVE THAT
THESE DEAD SHALL NOT HAVE DIED IN VAIN; THAT THIS NATION SHALL HAVE A
NEW BIRTH OF FREEDOM; AND THAT THIS GOVERNMENT OF THE PEOPLE, BY THE
PEOPLE, FOR THE PEOPLE, SHALL NOT PERISH FROM THE EARTH.

      ⍴GA2
1457
      A̲
ABCDEFGHIJKLMNOPQRSTUVWXYZ
      +/A̲∘.=GA2
100 14 31 54 165 25 26 79 69 0 3 41 13 76 91 16 1 81 45 124 20 23 27 0
      10 0
      IB+.∧1⌽~IB←~GA2∊A̲
269
      'THAT' IN GA2
12
      'HERE' IN GA2
9

      ∇ Z←W IN1 T;M
[1]   M←⍴W←,W
[2]   Z←+/((M+1)⌽IB)∧IB∧W∧.=(⍳M)⌽(M,⍴T)⍴T
      ∇

      ∇ Z←W IN2 T;M
[1]   M←⍴W←,W
[2]   Z←+/((M+1)⌽IB)∧IB∧∧⌿(⍳M)⌽W∘.=T
      ∇

      ∇ Z←W IN3 T;J;M
[1]   J←1+Z←M-M←⍴W←,W
[2]   Z←Z+(⍱/T[J+0,M+1]∊A̲)∧W∧.=T[J+⍳M]
[3]   →2×((⍴T)-M+1)≥J←J+1
      ∇
```

Figure A:18

vector of alphabetic characters, and the second argument T should be a character vector having a non-alphabetic character at one end. For $IN3$, T should have a non-alphabetic character at each end, because rotation does not occur, and so if $IN3$ were applied to $GA2$, the latter should have a carrier return (or something) catenated onto the front of it. $IN1$ and $IN2$ refer to a global variable IB that is supposed to have been derived from T in just the way that, higher up the page, IB was in fact derived from $GA2$. Of course IB could have been a local variable defined inside the function, but if the function is to be executed several times with the same right argument, a single global specification of IB is presumably more efficient than repeated local specifications.

$IN1$ is the sort of program that an APL user might first think of for making the desired word count. W is compared with every possible set of m consecutive elements of T, where m is the number of letters in W; and wherever there is a match, the character next before and the character next after must both be non-alphabetic, to ensure that the letters of W are not embedded in a larger word (like *THE* in *THEY* or *RATHER*). How much room is needed in the workspace to execute this function? Suppose the function is executed with $GA2$ as its right argument. Then T will be set equal to $GA2$. In the body of the function no assignment into T occurs, that is, T is never altered. With the implementation of APL that this writer currently uses, such a function argument is treated as a pointer to the global variable to which it is set equal; a copy of $GA2$ is not made for use inside the function. So almost no space is taken up by the argument T. Let the number of characters in $GA2$ (or whatever T is equal to when $IN1$ is executed) be denoted by n. Then on the right side of line [2] a character matrix is defined with m rows and n columns, and then the rows are rotated, which requires a second matrix of the same size to be formed; the room needed is essentially $2mn$ bytes. Thereafter the first matrix is discarded and the second matrix enters into the inner product with W, the result of which is a logical vector stored in n bits (approximately), like IB. Thus the greatest amount of room needed in the execution of $IN1$ is about $2mn$ bytes, or $2m$ times the space required to hold the piece of text.

Now $GA2$ is so short that there can be little problem about holding $2m$ times that amount in the workspace, at least when m is small, say 4. But we may wish to apply $IN1$ to a much longer piece of text and also have a first argument W with many more than 4 letters. Then we may regret $IN1$'s lordly expansiveness.

$IN2$ is very similar to $IN1$, but the two m-by-n matrices are logical, stored in bits rather than bytes. Thus $IN2$ only needs about one-eighth of the space needed by $IN1$. Its execution time is roughly the same (when both can be executed)—in fact, it seems to execute a little less fast than $IN1$. Therefore we can judge $IN2$ to be on the whole the better program.

Even *IN*2 requires space of order *n*. What could be done if the text occupied most of the active workspace? *IN*3 does the same calculation as *IN*1, but in a different order, and requires space depending on *m* but not at all on *n*. The dramatic saving of space is accompanied by a dramatic increase in execution time. When $m = 4$ and the text is *GA*2, *IN*3 takes about 70 times as long to execute (central-processor time) as *IN*1, or about 50 times as long as *IN*2. This sort of loop should never be written in APL —or rather, if written, never executed.

Although *IN*3 as it stands is a disaster, the essential idea is good. A version of *IN* can be composed that breaks the text up into pieces of convenient size, perhaps about 1000 characters, finishing always with a carrier return, and then applies *IN*1 or *IN*2 to each piece. A long text can be handled effectively in APL that way.

The generalization that one might make from this example seems to be usually correct. APL programs often need a good deal of room for execution. With ingenuity they can sometimes be rearranged to use less space, without noticeable loss in speed. Usually an array cannot be processed "in place"—except by use of Fortran-like loops whose execution is unacceptably slow. For example, the "fast Fourier transform" used in analysis of time series can be programmed in Fortran so that the series is transformed in place and only a tiny amount of additional space is needed for execution. The method can be programmed effectively in APL (a possible version is given in Appendix 3) but needs space several times as large as that occupied by the series to be transformed.

Counting the frequency of occurrence of a given word in a piece of text may seem a rather unimportant thing to do, but is a useful preliminary to the more ambitious task of constructing a concordance. Suppose we really needed to execute *IN* so many times on such long text that costs mounted and became a major consideration. How can such a function as *IN* be executed most economically? A professional programmer would arrange that full use was made of whatever facilities for that kind of task were available in the particular computer. He might perhaps do this by writing the program in a list-processing language and using a good compiler. The resulting machine code would be likely to execute much faster than any equivalent program in APL, as generally implemented. If the whole job to be done would cost, in APL, only two or three dollars and not much thought, clearly consulting a professional programmer would be unwise; the cost in money and personal effort would be substantial, even if execution eventually cost only a few cents. But when a well-defined computing job is to be done on such a scale that it would cost, in APL, thousands of dollars, the assistance of the professional programmer is likely to be much appreciated. Large-scale production runs do not show APL to best advantage. Experimental or exploratory computing, even on a sizeable scale, does.

Interaction with the environment. The earliest implementations of APL were closed systems. Input and output went only through typewriter terminals. There was no interaction with other computing activities or access to other peripheral equipment.

Users were not slow to demand a more flexible way of handling files of data than what came with the APL system. Suppliers of APL service added file management to the standard APL system. Eventually IBM augmented APL by the device of shared variables. The shared variable acts as an interface between the APL user and any facility of the computer that he desires to interact with: input and output devices, programs in other languages. Naturally the user needs to know how to address these facilities if he were approaching them directly rather than through APL. A description of shared variables will be found in Polivka and Pakin (1975), together with descriptions of the file systems of *APL*PLUS* (Scientific Time Sharing Corporation) and of *APL*/700 (Burroughs Corporation).

Since these features are not required for the purposes of this book, they are not discussed further.

We have seen that APL developed out of Iverson's original "programming language", which was an adaptation of ordinary mathematical notation intended for unambiguous description of computing procedures for human readers. Computer implementation of the language was apparently an afterthought. In this book, APL is considered primarily as a means of addressing a computer. For the most part, mathematics addressed to the reader is expressed in ordinary mathematical notation, not in APL. To have done otherwise would have impeded readers, not already familiar with APL, who might be disposed to dip into the book without reading consecutively; and it would perhaps have distracted attention from the main theme, which is the role of computing in statistical science.

However, since APL was first put forward, there have been developments not only in the computer implementations but also in the language as a means of human communication, as a style of mathematics that can be pursued without reference to a computer. Mathematical work is often more algorithmic in character than is generally recognized. In some contexts APL has been found to be a better medium of exposition than traditional notation. APL can be adapted to expression of infinite sets and limit operations not directly realizable in a finite computer. And APL can be applied to computer manipulation of mathematical expressions, which is a different matter from numerical evaluation of mathematical expressions.*

All the mathematics referred to in this book comes under the heading of calculus methods, mathematical analysis, or elementary algebra. All expressions given in ordinary mathematical notation could be given, equally

*See, for example, Berry et al. (1978), Iverson (1976, 1979, 1980), Orth (1976).

precisely and sometimes a little more concisely, in APL. These are the kinds of expressions for which ordinary mathematical notation was developed. APL recognizes that the basic operations of algebra, addition, subtraction, multiplication, division, and raising to a power, are syntactically alike, being scalar dyadic functions, and APL makes them look alike, a symbol between its arguments. But these basic functions do not behave alike; multiplication distributes over addition and subtraction, not the other way round, and raising to a power distributes over multiplication and division. The peculiarities of ordinary notation seem to conform rather well to typical needs, and often permit some qualitative features of an expression to be seen at a glance, more easily than if it were written in APL. For example, the first gain function mentioned above,

$$\frac{\sin(3\pi x) + \sin(\pi x)}{4\pi x(1 - 4x^2)},$$

can be expressed in APL thus:

```
(+/1○X∘.×○ 3 1)÷○4 × X × 1 − 4 × X*2
```

Because in APL the division sign does not catch the eye as fast as the horizontal line in the traditional notation, the fact that the function is $O(x^{-3})$ as $x \to \infty$ is seen more easily in the first version. Consider Fisher's formula for the variance of his kurtosis statistic g_2, for a sample of size n from a normal population, namely

$$\frac{24n(n-1)^2}{(n-3)(n-2)(n+3)(n+5)}.$$

This can be expressed in APL in several ways, including:

```
(24 × N × (N − 1)*2) ÷ (N − 3) × (N − 2) × (N + 3) × N + 5
÷/24,N + ¯3 0 ¯2 ¯1 3 ¯1 5
(24 × N⊥ 1 ¯2 1 0) ÷ N⊥ 1 3 ¯19 ¯27 90
```

The first of these is a straightforward translation, valid if N is any numerical array, not necessarily scalar. The other two assume that N is a single number. The second expression is the briefest, and is used in the function *SUMMARIZE* shown in Appendix 3. The third expression involves representation of polynomials by the "decode" function. Such a representation elegantly brings the vector of coefficients into view, and the argument is written only once. This third expression may therefore be judged the most insightful. Anyone equally conversant with APL and with ordinary notation will perceive equally fast from the third APL expression and from the ordinary representation that the function behaves like 24 ÷ N (or $24n^{-1}$) when the argument is large—but probably less fast with either of the other two APL expressions.

Thus our use of ordinary notation can be defended on grounds other than familiarity or laziness. (Ordinary mathematical notation must be thoroughly familiar to any possible reader of this book, whether or not APL is also familiar.) Admittedly, there is a pleasure in using the same style of notation for thought as for computing, encouraging immediate transitions from the one to the other. In exposition of statistical methods for a mathematically underprivileged audience, if computing procedures are expressed in APL, the same medium should surely be used for precise thought.

Exercises for Part A

1. Become familiar with the primitive dyadic and monadic scalar functions. If you are not sure of the definition, a little experimenting soon settles the matter. Try commands such as these:

```
(⍳3) ∘.* ⍳5
(0,⍳5)!5
!¯3.5 + ⍳6
1 2 ∘.○(⍳12) ÷ 6
2.5 ¯2.5 ∘.|⍳6
```

(The last of these has been affected by one of the very few changes that have taken place in the meaning of APL expressions—when the left argument of the residue function is negative, later versions of APL yield a result that is negative or zero, earlier versions one that is positive or zero.)

2. Get some error messages. Try to understand what each message means. Let *Q* not have been defined.

```
2 × Q + 7
4 ÷ 0
!100
×/⍳100
(⍳5) + ⍳6
(⍳4),'ABCDEF'
(⍳6) +'ABCDEF'
(⍳3)[4]
(⍳3) + 3 3 ⍴0
((⍳5) + 2
```

3. (a) It is essential to become familiar with the concepts of active workspace and stored workspaces. The use of various system commands must be understood, particularly these:

```
)SAVE
)LOAD
)COPY
)CLEAR
)ERASE
)WSID
)FNS
)VARS
)SI
)LIB
)OFF
```

The importance of the state-indicator command may not be perceived at first. As soon as defined functions are used, there will be suspensions and consequent blocking of global variables by local variables.

(b) If your active workspace has identification *A*, you have defined a variable *X* in it, and you wish to put a copy of *X* in another workspace named *DATA*, give this sequence of system commands.

```
)SAVE
)LOAD DATA
)COPY A X
)SAVE
```

(c) Find out how to determine how much room (working area) is left in the active workspace. Change your password.

4. (a) Enter the population of the New England states (ME, NH, VT, MA, RI, CT) in 1960 in thousands:

```
P60←975 609 389 5160 855 2544
```

The populations in 1970 and the areas in square miles:

```
P70←997 742 446 5706 951 3041
A←33215 9304 9609 8257 1214 5009
```

The 1970 population densities (thousands per square mile), and the 1970 population density for the whole of New England, are:

```
P70 ÷ A
(+/P70) ÷ +/A
```

The percentage population increases from 1960 to 1970 are:

```
100 × ¯1 + P70 ÷ P60
```

The permutation that arranges the 1970 populations in ascending order is:

```
S←⍋P70
```

The mean and the median of the 1970 populations are:

```
(+/P70) ÷ 6
0.5 × +/P70[S[3 4]]
```

(b) Now include information about the Middle Atlantic states (NY, NJ, PA):

```
P60←P60, 16838 6103 11329
P70←P70, 18268 7193 11813
A←A, 49576 7836 45333
```

Repeat the above calculations for the nine states of the Northeast region. Which state has the greatest 1970 population density, which the least, which the median density? Which state has the greatest percentage population increase from 1960 to 1970, which the least, which the median increase?

(c) For the 9-element vector of 1970 populations defined above, enter a vector of possible location measures and a matrix of deviations, say

$M \leftarrow 2\ 3\ 4\ 5\ 6\ 7\ 8\ 9 \times 1000$
$D \leftarrow P70 \circ . - M$

The sum of squares of deviations of the populations from each location value, the sum of absolute deviations, and the greatest absolute deviation, are

```
+ ⌿ D × D
+ ⌿ | D
⌈ ⌿ | D
```

Which of the location values is closest to the mean population, which to the median population, which to the mid-range population?

5. If *OF* is a vector of observed frequencies, corresponding to equal intervals of an observed variable, such as in Figure A:9, a simple unlabeled histogram can be generated with the instruction

```
' | ',' ⎕'[1+OF∘.≥⍳⌈/OF]
```

If you would like the columns to be broader, *OF* could first be triplicated with the redefinition

```
OF←,OF∘.+3⍴0
```

Make sure you understand how these instructions work—each outer product can be executed separately. Can you include a labeling of the values of the variable?

Note on printing APL. Whereas in the numbered figures, and in the program listings in Appendix 3, APL expressions are reproduced from output of a typewriter terminal fitted with the original APL typeball, brief APL expressions occurring in the text of this book are sometimes set in the font used for mathematics in ordinary notation. The two styles look rather different, not only because the characters are shaped differently (a matter of little importance provided they are recognized), but also because of the spacing. The typewriter terminal spaces its output equally, that is, the centers of consecutive non-blank characters are equally spaced across the page, and most members of the APL character set appear to be nearly equally broad. The numeral 1, with large serifs, seems nearly as wide as the letter *M* (which it is not in the font used in setting this sentence). The period and the vertical stroke (residue function) are narrow, but they are placed in the center of the same space allocation as all other printed characters, which is also the allocation for the blank character. Consecutive printed characters are therefore clearly either adjacent, with no space inserted between, or else separated by one or more blanks; and there can be no uncertainty about whether two consecutive characters do or do not form part of a name or of the decimal representation of a number. Moreover, since blank spaces come in integer multiples of the basic space allocation for individual characters, the number of blanks separating two printed characters in a line of output can if necessary be unambiguously determined (provided it is not very large). By contrast, the characters in a traditional printer's font are not all of the same width, there is no common space allocation for all characters, and when space is inserted between consecutive printed characters it is finely adjustable in extent. Unlike APL, traditional mathematical notation does not usually require spaces as separators (separations are made with parentheses, commas, subscripts and superscripts, and changes of font); appropriate small spaces can improve legibility, but are usually not absolutely necessary. In reading ordinary mathematical notation, one need not be conscious of the spaces—most readers are not. In APL, wherever spacing is optional, the reader may again ignore it; but he must be able to tell whether a pair of consecutive characters that might be part of a name are indeed separated or not separated by a space. Provided that names with more than one character are perceived as names, some kinds of APL expressions can be satisfactorily set in a font used for setting ordinary mathematics.

A grave difficulty arises with such a font, if an APL expression contains characters in quotes and is an instruction to generate a display on a typewriter terminal (or similar output device). The difficulty has no parallel in ordinary mathematical notation, which never refers to its own characters. The first displayed line in Exercise 5 contains a character vector, which must be recognized as having precisely two elements, a blank and a quad. Recognition cannot be counted on if there is possible uncertainty about the width or space allocation of characters. Anyone writing such an expression by hand is likely to indicate the blank character positively with a caret or other special mark. In Exercise 5, the displayed lines are taken from a typewriter terminal.

6. Make a short table of values of the two trigonometric functions mentioned in Chapter 8. Try to produce it within one minute. Take care to avoid x-values for which either denominator vanishes. A convenient choice would be

$$X \leftarrow {}^{-}0.025 + 0.05 \times \iota 40$$

7. Information on participants in a conference is entered in a character matrix, one row for each participant, arranged in order of registration. The first twenty characters in each row give the participant's name, surname followed by other names or initials, separated by single blanks and with trailing blanks to make 20 characters. Then starting in the 21st column other information is entered, such as where the participant is staying, his or her business affiliation, or whatever.

Consider how to print the matrix with rows permuted so that the names are (almost) in alphabetical order. A fast and easy method, good enough for most practical purposes, is to order the rows by the first five characters only, leaving names that agree in their first five characters in the order of original entry. The five characters are transformed to numbers by dyadic iota (a blank becomes 0, A becomes 1, . . . , Z becomes 26), then the 5-element numerical vectors are interpreted as numbers in the scale of 27 and evaluated by the decode function, and finally the grade-up function is used to order the rows of the matrix.

At a well-run conference, an updated list of participants is posted each day.

8. Tukey (1977) suggests fitting an additive structure to a two-way table by repeatedly subtracting medians of rows and medians of columns. That is, at step 1, the median of each row is found, and the corresponding median is subtracted from every entry in each row. Then (step 2) the median of each column is found, and the corresponding median is subtracted from every entry in each column. Then back to rows (step 3), and back to columns (step 4), and so on repeatedly. If, at any step after the first, every one of a set of medians (whether of rows or of columns) turns out to be zero, the procedure stops; the matrix of residuals has zero median for every row and for every column. Tukey has called the procedure *median polish*.

We may suppose without loss of generality that the data are a matrix of integers —observations are rational numbers and become integers upon multiplying by a suitable factor. Suppose first that the number of rows and the number of columns are both odd. Then the median of the entries in any row, or in any column, is unique and an integer, and the residuals are integers.

(a) Define a function to carry out this procedure. The medians of the rows, or of the columns, may be determined with the primitive "grade-up" function. Since grade-up takes a vector argument but not a matrix argument, the permutations needed to order each row (or column) of the matrix cannot be found by simply applying grade-up to it. An easy method is to find the median of each row successively with a loop, and then the medians of columns with another loop. Try that. It is possible to avoid such loops and grade the rows simultaneously by the trick of adding to all the entries in each row some constant number multiplied by the row number, and then grading at once all the entries in the matrix, destructured into a vector by monadic comma; and similarly for columns. The constant number must be so chosen that all the entries in any row become, after the multiples of the constant have been added, certainly greater than all entries in preceding rows and certainly less than all entries in following rows; and similarly for columns. It seems that an adequate value for such a constant is eight times the maximum magnitude of entries in the data matrix, or four times the difference between the greatest entry and the least. McNeil (1977, p. 116) gives an APL program of this kind, but his choice of the constant is not always satisfactory. The wrong answer is obtained for this data matrix:

```
7 3 ρ 1 3 1 1 3 3 3 1 3 3 1 1 3 1 2 3 1 2 3 1 2
```

(b) Prove that the procedure described above, when the number of rows and number of columns are both odd, certainly terminates after a finite number of steps, and give an upper bound to that number. (At each step, if the medians do not all vanish, at least one is not less than 1 in magnitude, and the sum of magnitudes of the residuals is reduced by at least 1 when the medians are subtracted.)

(c) Show by example that interchanging the sequence of operations, first removing medians of columns, then medians of rows, and so on (or, in other words, applying the above procedure to the transpose of the data matrix and at the end transposing the residual matrix), may lead to different results, with a different sum of magnitudes of residuals. Thus median polish does not necessarily minimize the sum of magnitudes of residuals. Small matrices of not-too-small random numbers are easily found that have this property. An elegant example with very small entries, due to J. A. Hartigan, is

```
3 3 ρ 0 2 0 0 0 2 2 1 1
```

(d) The median of an even number of values is usually defined to be the average of the two middle values when the set of values is ordered (graded). Any value in the interval bounded by the two middle values serves to minimize the sum of magnitudes of residuals, and could be called a median. With the usual definition, show by example that, when the number of rows and number of columns of the data matrix are not both odd, median polish, carried out with unlimited precision, does not necessarily terminate in a finite number of steps. However, when the data matrix has only 2 rows and 2 columns, median polish certainly terminates in at most 2 steps; the result is the same whether row medians or column medians are subtracted first; and the sum of magnitudes of residuals is minimized.

(e) To prevent the number of steps from tending to infinity when there are more than 2 rows and 2 columns and one or both of these numbers are even, precautions may be taken such as: (i) restrict all medians to integer values, say the integer of lesser magnitude adjacent to the median as usually defined, whenever the latter is not an integer; then all fitted values and residuals are integers; (ii) with the usual definition of median, terminate the procedure as soon as every one of a set of medians does not exceed some tolerance (less than 1) in magnitude; (iii) identify the type of convergence and infer the limiting residuals. Of these, (i) and (ii) are easy to program, and (ii) has the advantage that computation can be resumed with a smaller tolerance, by applying the function to the previous residuals—a possible program named *MP* is shown in Appendix 3. There are plenty of questions to investigate. Does precaution (i) or precaution (ii) ensure that the procedure will terminate? What is the relation between median polish and strictly minimizing the sum of magnitudes of residuals?

9. (a) Find out how to use a plotting function, such as *SCATTERPLOT* in Appendix 3, or a plotting function in the public library to which you have access; or compose your own plotting function. Use the function to graph the U.S. population recorded in the decennial censuses as ordinate against the date as abscissa. Choose the scales so that the points representing 1790 and 1970 are separated horizontally by roughly the same amount as vertically, and the plot looks roughly square.

Year	Population (millions)	Year	Population (millions)
1790	3.929	1890	62.9
1800	5.308	1900	76.0
1810	7.24	1910	92.0
1820	9.64	1920	105.7
1830	12.87	1930	122.8
1840	17.07	1940	131.7
1850	23.19	1950	151.3
1860	31.44	1960	179.3
1870	39.82	1970	203.2
1880	50.16		

(b) Fit a straight line to the data, by least squares or by any other method that you prefer, and calculate the residuals (a residual is the difference between an actual population value and the ordinate of the fitted line for the same date). Plot the residuals against the date, using the same abscissa scale as before, and using an ordinate scale four times as large as the previous ordinate scale.

(c) Repeat (a), replacing the population values by a transform, such as square root or cube root or fourth root or logarithm—indeed try each in turn. Use always the same abscissa scale, and each time let the ordinate scale by chosen so that the vertical separation between the points representing 1790 and 1970 is the same.

(d) Now repeat (b) for each set of transformed data—fit a straight line, calculate residuals, plot with four times the previous ordinate scale and always the same abscissa scale.

(e) Which transform most nearly shows linear growth? Predict the size of the U.S. population in 1980, 2000, 2500.

(f) Similarly examine population growth for another country.

10. This question and the next relate to computation with APL in a workspace of limited size. Results will depend on the implementation. Comments about what the writer finds refer to the implementation to which he currently has access.

Determine by how many bytes the working area is reduced by each of the following commands.

$$A \leftarrow 1000\rho 1.5$$
$$B \leftarrow 1000\rho\,^{-}1 + \iota 8$$
$$C \leftarrow 1 = 2\ 2\ 2\ \top B$$
$$D \leftarrow 500 + \iota 1000$$

Note that if C were stored, B could be recovered by the command

$$B \leftarrow 2 \perp C$$

[The first two characters in the definition of C, namely "1 =", do not alter the values of the elements but ensure that C is recognized as having only the "logical" values 0 and 1. The writer finds that D occupies much less space than B.]

11. (a) Let M be a matrix with a large number of rows, in a nearly full workspace. The following three commands, if executed, will each yield the number of rows of M, a one-element vector for the first and third commands and a scalar for the second. Determine how much room (working area) is needed for each command to be executable.

$$R \leftarrow 1\uparrow\rho M$$
$$R \leftarrow (\rho M)[1]$$
$$R \leftarrow \rho M[;1]$$

[The writer finds that the third command requires more room than the other two.]

(b) Let V be a vector with many elements (such as one of the longer explanatory variables shown in Appendix 3), in a nearly full workspace. Let M and N be rather large integers (say 1000), with $M + N$ less than ρV. The following commands, if executed, will each yield a print-out of N elements of V, starting just after the *M*th. Determine how much room is needed.

```
N↑M↓V
V[M + ⍳N]
```

[The writer finds that neither command needs much room.]

(c) Let *U* and *V* be long integer-valued vectors of equal length, in a nearly full workspace. Determine how much room is needed to execute each of the first pair of equivalent expressions and each of the second pair.

```
U +.× V
+/U × V
V +.× V
+/V × V
```

[The writer finds that the first command requires less room than the other three.]

12. A possible way of arranging some text, such as *GA*2 in Figure A:18, is by two steps. First the material of each paragraph is entered without carrier returns as though it were to be displayed in one long line. Then carrier returns are inserted and other adjustments made so that the material will print in the desired form.

(a) Write a function to facilitate entering the material of a paragraph as if it were to be displayed on one line. The quote-quad can be used to call for pieces of the text of convenient size, to be catenated together. If instead of a piece of text a special symbol (not otherwise needed) is entered, such as a right-arrow, the program will take this as a signal that entry of material is complete.

(b) Write a second function to arrange the material assembled by the first function so that it will print in the desired form. This function may be tailored to your stylistic preferences. The function that was used to form *GA*2 refers to two global variables, *PW* (meaning page width), a numerical scalar giving the maximum number of characters to a line, and *PI* (paragraph indentation), a numerical scalar giving the number of blanks to be inserted at the outset. The three paragraphs of *GA*2 were formed with *PW* set equal to 72 and *PI* set to 0 for the first paragraph and to 5 for the other two. Each line should print with the greatest number of characters not exceeding *PW*, so that the line ends at a permissible break-point. Permissible break-points are just before a blank or just before a double hyphen (representing a dash) or just after a double hyphen or just after a single hyphen.

[Writing functions for text editing, that work satisfactorily as intended, is harder than one might expect, because there are many details to think about. Congratulate yourself if you do a good job within a week. Text material does not have to be printed in Italic capitals, just because the editing is done in APL. Any other typeball (or character set) may be used to which you have access. The second function suggested above needs to be adjusted to the particular typeball or character set, since the hyphen (and some other characters that may interest you, such as the underscore) is placed differently on different typeballs.]

13. (a) The text *GA*2 shown in Figure A:18 is unusual in that no apostrophe is required for the possessive case of a noun. What effect would such an apostrophe have on the method shown for counting the number of words, that led to the answer 269? *GA*2 has one instance of two words joined by a hyphen. Note that they have been counted as two words. (As it happens, in the first draft of the speech the words are separate without a hyphen, and in none of the six extant drafts are they fused together as a single word; counting the hyphenated expression as two words therefore seems reasonable.)

(b) Suppose that some English text has possessive-case apostrophes, and that it contains numerals (as it might be the date *NOVEMBER* 19, 1863). Consider how to count the number of words, if we do not wish an apostrophe to break a word in two and if we wish to count numbers like 19 and 1863 as a single word each. Observe that all words, under these conditions, begin with a letter of the alphabet or a numeral and are preceded (except perhaps for the first word) by a blank or hyphen or carrier return.

14. (a) Write a function that will display a vector of text such as *GA*2 with numbering of the lines, either on the left or on the right, as you prefer.

(b) Modify one of the functions *IN*1 or *IN*2 shown in Figure A:18 so that, if there are any occurrences of the first argument in the second argument, the line number of each occurrence is given.

15. Find out how to use a format function, either the primitive format function if your version of APL has it, or a function in the public library to which you have access; or compose your own. Use it to reproduce the display in Figure A:19, or something similar.

In constructing Figure A:19, the probabilities were first computed. The probability of a count equal to r in a binomial distribution with mean μ and exponent n is (in ordinary mathematical notation)

$$\binom{n}{r}\left(\frac{\mu}{n}\right)^r\left(1-\frac{\mu}{n}\right)^{n-r} \quad (0 \leqslant r \leqslant n);$$

in a Poisson distribution with mean μ it is

$$\mu^r e^{-\mu}/r! \quad (r \geqslant 0);$$

in a negative binomial distribution with mean μ and exponent k it is

$$\frac{k(k+1)\ldots(k+r-1)}{r!}\left(\frac{\mu}{\mu+k}\right)^r\left(\frac{k}{\mu+k}\right)^k \quad (r \geqslant 0).$$

The table in the upper part of the figure (all the material printed with an APL typeball, lying between the two single lines printed with a standard typeball having upper-case and lower-case Roman letters) was formed as a single character matrix, the columns listing the probabilities being obtained with the format function, the various labelings catenated onto the left side and above, and blank rows and columns inserted for ease of reading.

In the lower part of the figure, the random samples were obtained with the function *SAMPLE* shown in Figure A:13.

To get the whole figure printed, changes of typeball must be made. One possibility is to put the printing instructions into a single defined function, having the following line wherever a change of typeball is needed:

```
[ ]      M←⍞
```

Probabilities in some distributions with mean 4

DISTRIBUTION:	*BINOMIAL*		*POISSON*	*NEGATIVE BINOMIAL*				
MEAN:	4	4	4	4	4	4	4	4
EXPONENT:	8	16		8	4	2	1	0.5
COUNT = 0:	0.0039	0.0100	0.0183	0.0390	0.0625	0.1111	0.2000	0.3333
1:	0.0313	0.0535	0.0733	0.1040	0.1250	0.1481	0.1600	0.1481
2:	0.1094	0.1336	0.1465	0.1561	0.1563	0.1481	0.1280	0.0988
3:	0.2188	0.2079	0.1954	0.1734	0.1563	0.1317	0.1024	0.0732
4:	0.2734	0.2252	0.1954	0.1590	0.1367	0.1097	0.0819	0.0569
5:	0.2188	0.1802	0.1563	0.1272	0.1094	0.0878	0.0655	0.0455
6:	0.1094	0.1101	0.1042	0.0918	0.0820	0.0683	0.0524	0.0371
7:	0.0313	0.0524	0.0595	0.0612	0.0586	0.0520	0.0419	0.0306
8:	0.0039	0.0197	0.0298	0.0383	0.0403	0.0390	0.0336	0.0255
9:	--	0.0058	0.0132	0.0227	0.0269	0.0289	0.0268	0.0214
10:	--	0.0014	0.0053	0.0129	0.0175	0.0212	0.0215	0.0181
11:	--	0.0002	0.0019	0.0070	0.0111	0.0154	0.0172	0.0153
12:	--	0.0000	0.0006	0.0037	0.0069	0.0111	0.0137	0.0131
13:	--	0.0000	0.0002	0.0019	0.0043	0.0080	0.0110	0.0112
14:	--	0.0000	0.0001	0.0009	0.0026	0.0057	0.0088	0.0096
15:	--	0.0000	0.0000	0.0005	0.0016	0.0041	0.0070	0.0082
16:	--	0.0000	0.0000	0.0002	0.0009	0.0029	0.0056	0.0071
17:	--	--	0.0000	0.0001	0.0005	0.0020	0.0045	0.0061
18:	--	--	0.0000	0.0000	0.0003	0.0014	0.0036	0.0053
19:	--	--	0.0000	0.0000	0.0002	0.0010	0.0029	0.0046
≥20:	--	--	0.0000	0.0000	0.0002	0.0023	0.0115	0.0310

A random sample of size 25 from each of the above distributions

BINOMIAL, *N* = 8
3 2 5 4 3 4 5 6 5 3 5 6 3 6 4 4 4 5 5 4 6 4 4 5 5

BINOMIAL, *N* = 16
5 8 7 6 3 4 4 2 4 4 6 4 5 6 5 5 5 5 1 6 4 4 2 5 4

POISSON
8 2 2 3 5 2 5 4 6 3 4 3 2 1 7 3 2 7 3 2 6 2 2 1 3

NEGATIVE BINOMIAL, *K* = 8
1 5 3 4 2 1 5 5 8 3 4 6 0 1 4 5 0 3 1 3 5 4 8 6 4

NEGATIVE BINOMIAL, *K* = 4
1 5 3 5 8 6 2 0 5 2 4 6 13 3 2 12 5 6 5 1 4 7 2 3 6

NEGATIVE BINOMIAL, *K* = 2
3 1 2 2 1 3 9 9 0 9 3 3 2 15 3 2 0 11 0 3 2 2 9 3 3

NEGATIVE BINOMIAL, *K* = 1
12 0 6 6 7 0 0 5 9 4 5 5 31 9 1 1 1 3 4 8 2 8 1 2 3

NEGATIVE BINOMIAL, *K* = 0.5
15 0 0 0 10 3 0 4 1 1 2 0 3 8 3 17 2 0 24 1 0 2 0 2 1

Figure A:19

This instruction causes the keyboard to unlock during execution to receive material to be entered into M (a local variable not used for any other purpose). The typeball is changed and an empty vector is returned harmlessly to M.

16. The computer can be used for algebraic manipulation. As a first step, consider how to multiply polynomials.

(a) Polynomials in a single variable can be represented simply by a vector of coefficients, provided that zeroes are inserted for missing terms and that there is a fixed understanding that the terms are arranged in ascending order (or in descending order) of powers of the variable, with the lowest power always the zeroth. For example, the polynomial

$$1 + 5x - 2x^3$$

can be represented by the vector 1 5 0 ¯2 if powers are ascending, or by the same in reverse order if powers are descending. To multiply two polynomials we have to form the outer product (multiplication) of the vectors of coefficients and sum along diagonals. Summing by diagonals is easily achieved by initially catenating some zeroes onto one of the vectors, doing a suitable rotation over one coordinate of the outer product, and then summing over the other coordinate. A suitable APL function is described below under the name *PP* (for polynomial product). Multiple use of such a function makes light work of simplifying a clumsy expression like the following (encountered in obtaining the third moment of a kurtosis statistic):

$$\begin{aligned} &n^2(n+2)^2(512n^3 + 13440n^2 + 105856n - 226560) \\ &\quad -24(n-1)n(n+2)(n+8)(n+10)(64n^2 + 496n - 864) \\ &\quad +1024(n-1)^3(n+4)(n+6)(n+8)(n+10). \end{aligned}$$

(b) Another representation of a polynomial, in either one or several variables, is by a matrix with one more rows than there are variables. Each column represents a term, the first row giving the coefficient and the other rows the powers of the variables, each row referring to a particular variable. All powers are supposed to be nonnegative integers. Thus with two variables, x and y, the polynomial

$$1 + 5xy^2 - 2x^3y$$

can be represented by the matrix

```
1   5   ¯2
0   1    3
0   2    1.
```

There is no need to include terms with zero coefficients, and the terms may appear in any order. A polynomial in only one variable may be more conveniently represented in this style than in the preceding vector style if the greatest power is large but the number of nonzero coefficients is small.

The sum of two polynomials (in the same variables) can be represented by catenating the two matrices. The product of two polynomials can be represented by a matrix specifying each of the possible products of a term in one polynomial and a term in the other, the number of columns being the product of the numbers of terms in the polynomials. Sometimes the result of adding or multiplying polynomials in this way can be simplified by collecting like terms. The operation of collection may conveniently be separated from addition and multiplication.

The functions *PP* and *COLLECT* in Appendix 3 show possible ways of performing multiplication and collection. If two polynomials in one variable are represented by vectors, as in (a) above, line [4] of *PP* finds the product vector. If the first argument of *PP* is scalar and the second argument is the matrix representation of a polynomial in one or more variables, line [5] finds the product matrix. If both arguments of *PP* are matrices with the same number of rows, representing two polynomials in the same variables, and if all elements in rows other than the first row of each matrix (representing the powers of the variables) are nonnegative integers, line [7] finds the product matrix. The remaining lines in *PP* are concerned with testing the arguments to see which of [4], [5] or [7], if any, does the appropriate job. Line [7] involves a trick. The product matrix could be found straightforwardly row by row. The first row is the content of the outer product (using multiplication) of the first rows of the given matrices; each remaining row is the content of the outer product (using addition) of the corresponding rows of the given matrices. The rows could therefore be found easily one by one in a loop. In line [7] all rows below the first are handled simultaneously by a suitable use of the decode function. If the meaning of the line is not clear to the reader, intermediate steps can be examined by inserting ⎕← at one or more suitable places.

The same sort of trick is used in *COLLECT*, to avoid a loop. An elegant feature of line [5] is a "scan" with the function $<$, which has the effect, in each column of the following matrix, of replacing every 1 after the first by 0. The last operation in *COLLECT* is rearrangement of the columns of the resulting matrix, after collection, into a "dictionary" order, so that of two terms in the polynomial one precedes the other if the power of the first variable is lower or, in the event that the powers of the first variable are equal, if the power of the second variable is lower, and so on. The function *COLLECT* is applied only to a matrix argument. When polynomials in one variable are represented by vectors, collection happens automatically in addition and multiplication, and no separate operation of collection is needed.

Many interesting programs, to manipulate polynomials, differentiate functions, and perform other feats of computerized mathematics, have been given by Orth (1976).

PART B
EXPERIMENTS IN STATISTICAL ANALYSIS

Chapter 9

Changing Attitudes

Statistical analysis of data is the oldest part of statistical science as we have it today—going back more than three hundred years to John Graunt's study of the London bills of mortality. The purpose of statistical analysis may be stated vaguely as obtaining a right understanding of the data. Some kinds of data do not pose any special problem of understanding, and then nothing called statistical analysis is done. Other kinds of data seem to defy complete detailed explanation and understanding. They invite being thought of in terms of variability, randomness or noise, masking the relations or properties that we should like to examine.

Two extreme and opposed views concerning the nature and due process of statistical analysis have been maintained and acted on by different persons, and in some cases by the same person at different times. These two views may be labeled the *arithmetic* view and the *theory-directed* view.

According to the arithmetic view, statistical analysis consists of arithmetic operations and manipulations intended to summarize and reveal interesting features of the data. The operations need not be motivated by any theoretical belief concerning the statistical character of the data. The analyst tries to let the data speak for themselves, to hear what they are saying. John Graunt belonged to this school—there was no statistical theory in his day to help or confuse, nor did he invent any. At one time R. A. Fisher maintained that the analysis of variance was a purely arithmetic procedure, not dependent on probabilistic assumptions about normal distributions or anything like that. There are proponents of the arithmetic view today.

The theory-directed view of statistical analysis is based on a definite assumption concerning the statistical character of the data. An incompletely specified theoretical structure or "model" is postulated, and the job of statistical analysis is seen as assessing the bearing of the data on such features of the structure as are not fully specified. The data are viewed with blinkers on; answers are sought to preconceived questions in relation to a preconceived structure. Almost all of statistical theory concerning analysis of data is addressed to this kind of situation. R. A. Fisher was much concerned with exact sampling distributions of statistics on well-defined hypotheses, he originated the ideas of sufficiency and exhaustive estimation of parameters, he fiduciated. Not surprisingly, he usually seemed to approach statistical analysis from a theory-directed point of view—as have many others since.

One may conjecture that in fact most statistical analysis is done unselfconsciously, without declaration of aims and principles, without much thought for consistency, not according perfectly with either of the above views. Theory and empiricism are blended, discretion is exercised. The result is not just a mixture of the arithmetic and theory-directed views, for it is not true to either.

John Tukey has suggested that both the above views of statistical analysis are correct and needed. They complement each other, and resemble the two sides of the legal process. Proper administration of justice requires: (i) discovery, detection, the testimony of witnesses; and also (ii) judgment, reckoning, the verdict of the jury. Neither of these is effective without the other. The two functions are sharply separated in legal practice today; though in twelfth-century England when trial by a jury of twelve men was instituted the functions of testimony and judgment were combined. If it is indeed true that there are in principle two sorts of statistical analysis, exploratory and judgmental, responsibility for them is not separated in present-day statistical practice. The legal analogy invites us to consider whether separation is possible and desirable.

A couple of small examples may help to fix ideas. Consider first the problem of obtaining a right understanding concerning the following set of eighteen numbers:

3, 1, 1, 0, 3, 7, 5, 5, 2, 4, 2, 1, 0, 2, 6, 4, 3, 0.

Presented thus nakedly, without explanation of their origin, those numbers might perhaps stimulate a statistician of the arithmetic school to make some summaries.

(a) He might make a frequency count. There are three each of 0's, 1's, 2's, 3's; two each of 4's, 5's; one 6 and one 7. He might present these frequencies in a histogram.

(b) He might calculate some moments. The mean is 2.72, the variance 4.45.

(c) He might examine the serial relation of the numbers. The number of local maxima and minima within the sequence is 6.

Are such statements useful? Do they help us to understand the given numbers?

Suppose I say that during a recent period I made a weekly trip to New

York City and would find myself trying to cross Fifth Avenue at about 10 a.m. Each time I made a count of the number of buses I could see within two blocks of my crossing point, and obtained the above numbers.

We know a good deal about road traffic. There is little prospect of an interesting explanation of the individual numbers, and we resort naturally to a description in terms of a distribution of frequencies. In some circumstances numbers of road vehicles are known to have something close to a Poisson distribution, and it is natural to ask whether these numbers of buses follow such a distribution. One may easily think of reasons why the numbers should *not* follow a Poisson distribution. If the buses could adhere exactly to their schedules they would no doubt show less variability than a Poisson distribution. Traffic congestion and lights, on the other hand, tend to cause an uneven flow having large variance. A Poisson distribution is a natural reference point; we should not be surprised if the data conformed with one, but should not strongly expect this.

Granted that we shall think in terms of frequencies, the above arithmetic summaries are relevant and good. If we are contemplating a Poisson distribution, the sample mean is a sufficient statistic for the one parameter, and the variance provides a test of conformity. So we might finally summarize the data by saying that they are reasonably close to what might be expected of a random sample from a Poisson distribution with mean somewhere in the neighborhood of 3, but there is some suggestion of overdispersion.

Now in fact the story about buses on Fifth Avenue is partly invention—I did not make any systematic count of buses. Instead, the numbers quoted above have an entirely different origin. They are the first eighteen digits of the expression in the octal scale of the number π, familiar to us in the decimal scale as 3.14159 approximately. To say that the given numbers are the first eighteen octal digits of π is to give a complete description of them and convey a right understanding. The arithmetic statistical summaries contribute nothing. Of course, if someone had the quixotic idea of using digits of π as pseudorandom numbers he would be interested in their statistical properties, but that is another matter. This rather crude example is intended to suggest the moral that there is no such thing as a good statistical analysis of a set of numbers about which we know absolutely nothing. Any statistical analysis points towards an explanation or description of a certain type. It cannot be a good analysis unless that type is appropriate.

For a second example, let us turn to R. A. Fisher's *Statistical Methods for Research Workers*. This book has had an immense influence on statistical science, through its particular methods and examples, through its general attitude, and through its sense of identity and scope of the subject. The book was conditioned not only by the author's achievements in statistical theory and his experience in statistical practice, but also by the desk-calculator technology of the era. Like other great landmarks in the history of statistics from Graunt onwards, it is worth careful reappraisal in the light of present-day understanding and technique.

Fisher made a point of illustrating the procedures described in his book by application to real data. His appointment as statistician at Rothamsted had been for the purpose of studying the records of the long-term experiments. To illustrate linear least-squares regression he used some records from the Broadbalk wheat experiment. He quoted the yields, expressed in bushels of dressed grain per acre, of two plots during the thirty years from 1855 to 1884. The plots had received identical amounts and kinds of artificial manures each year, except that one of the plots, labeled "9a", had received nitrogen in the form of nitrate of soda, whereas the other, "7b", had received ammonium salts.* Fisher remarked that the individual yields varied greatly from year to year, but the difference in yield of the two plots each year was much less variable. The differences are mostly negative in the earlier years and positive in the later years; they are reproduced here in Figure B:1. Fisher commented: "In the course of the experiment plot '9a' appears to be gaining in yield on plot '7b'. Is this apparent gain significant?"

He answered this question by calculating the least-squares linear regression of the yield differences on the year numbers. He carefully showed how to obtain the residual sum of squares and associated number of degrees of freedom, and he made a Student t test of the null hypothesis that the true regression coefficient was zero ($t = 2.2814$, 28 d.f.). "The result must be judged significant, though barely so; in view of the data we cannot ignore the possibility that on this field, and in conjunction with the other manures used, nitrate of soda has conserved the fertility better than sulphate of ammonia; these data do not, however, demonstrate the point beyond possibility of doubt."

The average yield difference, denoted by $\bar{y}$, was 4.47. Fisher commented: "The standard error of $\bar{y}$, calculated from the above data, is 1.012, so that there can be no doubt that the difference in mean yields is significant; if we had tested the significance of the mean, without regard to the order of the values, that is by calculating s^2 by dividing 1020.56 by 29, the standard error would have been 1.083. The value of b was therefore high enough to have reduced the standard error. This suggests the possibility that if we had fitted a more complex regression line to the data the probable errors would be further reduced to an extent which would put the significance of b beyond doubt. We shall deal later with the fitting of curved regression lines to this type of data."

* Fisher described the nitrogen fertilizer applied to 7b as sulphate of ammonia, but that seems to be not quite accurate, according to the Rothamsted report cited below.

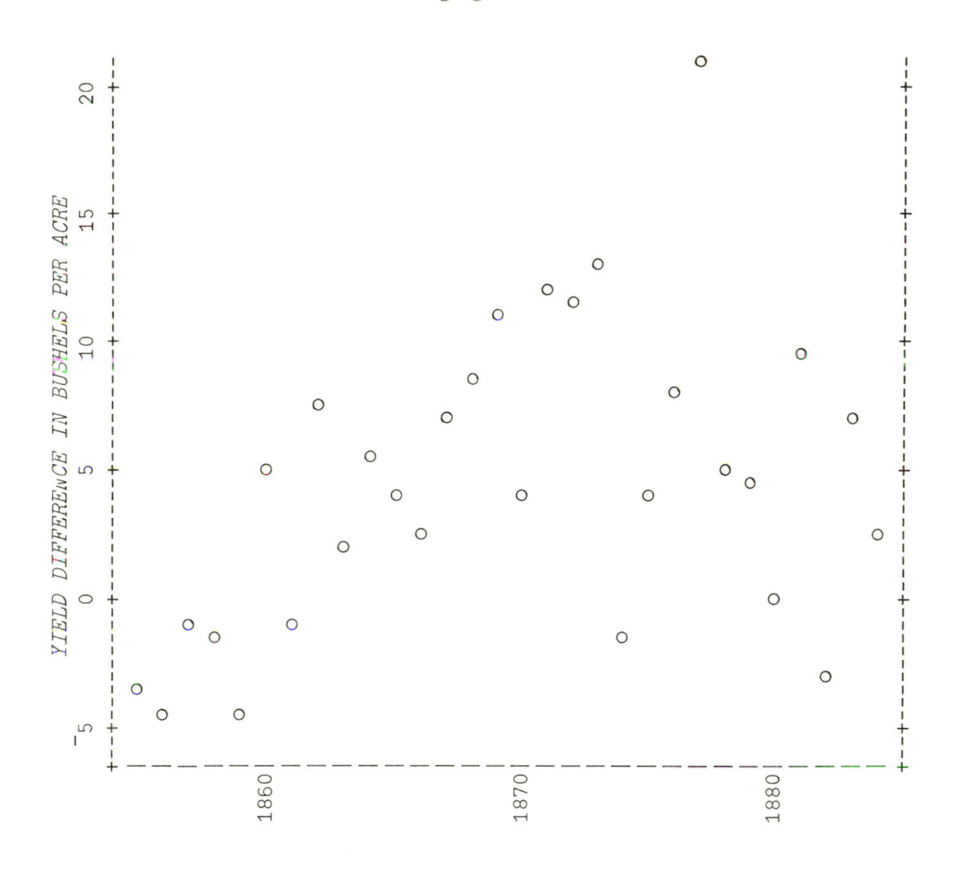

YIELD DIFFERENCE OF TWO PLOTS IN BROADBALK WHEAT FIELD, BUSHELS PER ACRE
(FISHER, STATISTICAL METHODS FOR RESEARCH WORKERS, TABLE 29)
THE THIRD COLUMN SHOWS WINTER RAINFALL IN INCHES, NOVEMBER TO FEBRUARY

1855	¯3.38	5.1	1870	4.13	9.5
1856	¯4.53	8.1	1871	12.13	7.1
1857	¯1.09	7.9	1872	11.63	8.2
1858	¯1.38	5.2	1873	13.06	13.3
1859	¯4.66	6.2	1874	¯1.37	6.4
1860	4.90	9.7	1875	3.87	9.3
1861	¯1.19	7.2	1876	7.81	10.5
1862	7.56	7.9	1877	21.00	17.3
1863	1.90	7.9	1878	5.00	11.0
1864	5.28	6.0	1879	4.69	12.8
1865	3.84	8.9	1880	¯0.25	5.1
1866	2.59	11.3	1881	9.31	11.2
1867	6.97	9.4	1882	¯2.94	11.4
1868	8.62	7.8	1883	7.07	14.4
1869	10.75	10.9	1884	2.69	8.7

Figure B:1

Undoubtedly Fisher's prime purpose in considering this example was to exhibit the numerical process of calculating a linear regression and making associated tests. This was a proper purpose. However, since he used real data, it is not unfair to ask whether his treatment was appropriate. Moreover, since here and elsewhere in the book he emphasized accurate calculations, efficient estimation and exact significance tests, it is not unreasonable to ask whether all his thinking was as precise as some evidently was.

Let us first of all examine the yield differences as Fisher quoted them. Afterwards we shall look for further evidence that might corroborate or refute the findings.

Applying a Student t test to the mean yield difference $\bar{y}$ or to the regression coefficient b, as Fisher did, is valid if we may suppose that the observed yield difference in year j, say y_j, can be expressed in the form

$$y_j = \alpha + \beta j + \epsilon_j,$$

where the "errors" $\{\epsilon_j\}$ are generated by an independent normal process, i.e. each ϵ is independently distributed in a common normal (Gauss-Laplace) distribution. Some departure from precise normality and precise identity in the distributions will scarcely affect the behavior and appropriateness of the t statistic, but the independence is important. In some kinds of experimental work independence of the errors is ensured, after a fashion, by randomizing the treatment allocation; but nothing of that sort has happened here. There seems to be no theoretical reason to expect that the $\{y_j\}$ consist of linear trend plus white noise.

Fisher often seemed indifferent to whether an implied theoretical structure fitted well. The reason was perhaps partly technological. The methods he advocated involved about as much effort with a desk calculator as flesh and blood could stand, and for that amount of effort the return was good (if not the best possible). In the second chapter of his book Fisher had spoken of the usefulness of graphs, but he omitted the elementary precaution of making a scatterplot of these yield differences. Such a plot can be made by hand, but is more time-consuming than one might expect; and making a plot by hand neatly enough for reproduction in a book is far harder than persuading a computer to generate the plot in Figure B:1. When we examine the plot, the data seem ill chosen to illustrate linear regression. If we are to think of $\{y_j\}$ as trend plus white noise, should not the trend be curved?

Another scatterplot is advisable at this point, of the yield differences against the yield sums (for the two plots each year). That plot, not reproduced here, suggests that differences and sums vary independently, and confirms Fisher's decision to confine his attention to differences. We shall do likewise.

How can the trend in the yield differences be represented? A quadratic trend fits notably better than a linear one, as shown in this analysis of variance.

	Sum of squares	D.f.	Mean square
Linear trend	160.0	1	160.0
Quadratic trend	268.3	1	268.3
Residual	592.3	27	21.9
Sum of last two terms	860.6	28	30.7
Total about mean	1020.6	29	35.2

Taking out the quadratic term reduces the residual mean square from 31 to 22, clearly a significant reduction. The residuals now appear to be nothing but white noise.

Although low-degree polynomials are popular as expressions for smooth trends, we could consider other functions. In a periodogram analysis, regression is performed on sine and cosine waves whose period is the length of the sequence of observations, here 30 years, and on sine and cosine waves for simple harmonics of that period, here 15, 10, 7.5, 6, 5, . . . , 2 years; each period accounts for two degrees of freedom, except the last having one degree of freedom. The cumulative sum of the terms in this analysis of variance can be plotted against the cumulative degrees of freedom and the straightness of the plot examined—something similar was shown in Figure A:7. All the terms for harmonics of the fundamental period seem to be consistent with white noise, but the fundamental period itself gives a larger sum of squares. So we might propose the following analysis of variance.

	Sum of squares	D.f.	Mean square
Thirty-year period	394.2	2	197.1
Residual	626.4	27	23.2
Total about mean	1020.6	29	35.2

The residual mean square is almost equal to that from fitting a quadratic trend.

Thus we have two equally good representations of the given yield differences, (a) a quadratic trend plus white noise, (b) a sinusoidal wave of period thirty years plus white noise. Though these two structures fit almost equally well, they, and also Fisher's linear trend, would imply very different extrapolations if we were to speculate concerning a continuation of the observations beyond 1884 under similar conditions. Fisher's trend would imply that the yield differences would ever increase, the quadratic trend that they would ever decrease (algebraically), and the cyclic trend that they would slowly oscillate. Although there is something ludicrous about the cyclic trend, its period having been determined by the length of the given series, one might guess that it would be a better forecaster than the other two trends.

But what about right understanding? Is it reasonable to be speculating about trend plus white noise? Why should the noise be white? The yields of a single plot treated always the same in successive years are likely to show serial cohesion. Annual rainfall does; and the incidence of weeds, diseases and insect pests in any year will be affected by their incidence the previous year. We should therefore not be surprised if the yields had a spectrum that was not flat but had much power on slower oscillations or longer periods—what might be termed red noise. We are, however, concerned here with the differences in yield of two plots in the same wheat field. The factors likely to cause redness in the spectrum of yields of one plot may be almost the same for both plots, and perhaps will not redden the spectrum of the yield differences. Yet, unless experience strongly suggests otherwise, we ought at least to entertain the possibility that the yield differences have the structure: constant plus red noise. The periodogram suggests that very slow oscillations are prominent in the spectrum, if there is no trend. Against a background of sufficiently red noise, even Fisher's assertion that $\bar{y}$ is significantly different from 0 is unjustified.

Without further information or experience to predispose us to view the data in a certain light, the best we can do by way of a summary is to point to the scatterplot in Figure B:1 and remark that in the middle of the thirty-year period the yield differences seem to have been somewhat higher than at the beginning (and perhaps higher than at the end).

But indeed there is much further information available that could throw light on this particular set of observations—information that was presumably also accessible to Fisher, writing his book some forty years after 1884. There have been many small changes in the treatments applied to the plots of Broadbalk field since the wheat experiment began in the fall of 1843, but individual plots have been treated in the same way for many successive years. After the harvest of 1925 a system of fallowing was introduced. A more radical change was made after the harvest of 1967; whereas until then wheat had been the only crop grown, now there is a crop rotation. A special report reviewing the whole experiment up to 1967 has been published (1969). From this several facts can be gleaned that bear on Fisher's data.

The particular comparison of types of nitrogen dressing on plots 9a and 7b could only be made during the thirty years quoted by Fisher. But various comparisons of nitrogen dressings can be made with other plots during the period of continuous wheat up to 1925, and these are summarized in the report. The effectiveness of artificial nitrogenous manures depended on the time of year of their application. Application in the autumn was substantially less effective than in the following spring, the difference in mean grain yield being roughly 2 cwt per acre on the average (or 4 bushels per acre); the difference was greater in the wetter years. A comparison of the two types of nitrogen fertilizer, nitrate of soda and ammonium sulphate (or other ammonium salts), at the same levels as for Fisher's data and in the presence of the same "complete mineral fertilizers", can be made on a pair of plots during 1885–1925. The authors of this part of the report, H. V. Garner and G. V. Dyke, state that nitrate of soda averaged 0.9 cwt per acre higher in grain yield than ammonium sulphate (or 1.6 bushels per acre); and they indicate that there was scarcely any long-term trend in the yield differences. But throughout this period there was a difference in time of application for the two types of fertilizer—nitrate of soda was applied in the spring, whereas ammonium sulphate was applied one quarter in the autumn and three quarters in the spring. (Artificial fertilizers other than nitrogenous fertilizers were always applied in the autumn.)

We are therefore led to inquire when the nitrogenous fertilizers were applied on plots 9a and 7b during 1855–1884. The report states (p. 23) that ammonium salts, as on plot 7b, were applied all in autumn up to 1877, all in spring for 1878–1883, and for 1884 one quarter in autumn and three quarters in spring. Sodium nitrate, as on plot 9a, was always applied all in spring from 1867 onwards; the report does not state when it was applied before 1867. Mr. Dyke has explained in response to an inquiry that the time of application of nitrate of soda is not known for certain before 1867, "except for the years 1853, 1858, 1859, 1865 when it was stated to have been applied in spring. As far as I know Lawes and Gilbert never mentioned putting nitrate of soda on in autumn and I think we must presume that it was applied in spring throughout."

If we ignore the minor change in the time of application of ammonium salts in 1884, we see that the whole 30 years consist of 23 years (1855–1877) during which there was a sharp inequality in time of application of the two types of nitrogenous fertilizer, and then 7 years (1878–1884) in which the time was more or less the same. During the 23 years we should expect the yield differences to be positively associated with the amount of winter rainfall. Information about rainfall at Rothamsted has been supplied by Mr. A. T. Day of the Physics Department, and total rainfall in the four months, November to February, of the winter preceding harvest is shown in Figure B:1 above. For 1855–1877, the mean value of the yield difference is 4.71 bushels per acre, and the regression coefficient on winter rainfall is 1.69 bushels per acre per inch. For 1878–1884, the mean yield difference is 3.65 bushels per acre, and there is little association with winter rainfall (none is expected). The residual sum of squares when these means, and also the regression in the earlier period, are subtracted is 541.9, with 27 degrees of freedom; the mean square is 20.1. The residuals show little evidence of trend and are reasonably consistent with white noise.

Thus the trend in yield differences noticed by Fisher should apparently not be taken as an indication that nitrate of soda conserved fertility better than ammonium salts, but should rather be attributed to a striking trend in winter rainfall during the 23 years in which winter rainfall would be expected to be influential.

We have tried to suggest that statistical analysis should not generally be regarded as a purely arithmetic process independent of probabilistic or inferential theory, nor should it be directed by theoretical suppositions treated as though they were given when they are no such thing. What then is a tenable view of the nature of statistical analysis? We suggest that the aim is to find a theoretical description of the data having three qualities that can be summed up as: (a) agreement with the data, (b) simplicity, (c) accordance with theory and other experience. All three qualities are ill defined, though in particular instances we may perceive them, or their lack, very clearly. A few words about (b) and (c) first, and then let us turn to (a).

Simplicity is in the eye of the beholder. Opinions can differ about the relative simplicity of different types of theoretical description. Roughly speaking, the fewer parameters needed the better; but it is unwise to try to measure simplicity by counting parameters.

Accordance with theory can be both retrospective and prospective. A theoretical description should not, without good reason, violate existing received theoretical ideas, and it should if possible help stimulate further theoretical development. A theoretical description that suggests a causal explanation is likely to be more valuable than one that does not.

By quality (a), agreement with the data, we mean that the theoretical description should relate to the whole of the data, or to the whole of some portion or aspect of the data, and be reasonably consistent with the data. If the data are voluminous we are usually content if the theoretical description reflects only the broader features of the data. The theoretical description should not imply properties of the data that can be seen to violate the facts, unless the violation, duly considered, is judged unimportant for our purposes.

Statistical analysis is, of course, generally done with purpose; some parameters or features of a theoretical description are of primary interest, others of no interest at all for their own sake, though perhaps of some secondary interest because they bear on our understanding of the features of primary interest. For example, the sole reason for making a statistical analysis may be to measure the linear dependence (regression) of a variable y on another variable x; the regression coefficient is the only parameter of primary interest. Yet other features of the data, the variance of the x's, and the amount and kind of residual variation of the y's about the regression line, vitally affect our understanding of the regression coefficient, its precision and significance. The theoretical description should therefore reflect these features of secondary interest reasonably faithfully.

In some contexts it is held that the purpose of a statistical analysis is to estimate one or more well defined quantities, and that attention should be directed to ensuring that these quantities are indeed estimated; then secondary features can be largely ignored. This is the tradition in the analysis of surveys. A survey is conducted to measure, say, the productive capacity of a country or region in some commodity. It is decided that the mean productive capacity per unit of the population of producers (or equivalently the total capacity of the whole population) is to be determined, and great care is exercised that the measurement shall be unbiased and adequately precise. If anyone were to suggest that the logarithms of productive capacities were nearly normally distributed in the population and therefore the mean and variance of the logarithms should first be calculated and the mean capacity deduced from these rather than estimated directly as the mean of the untransformed observations, the objection would probably be raised that it was exactly the mean capacity that was needed and only the direct calculation could be relied on to estimate this. If indeed the mean capacity in the population is the only matter of interest, this objection has force. But it is doubtful whether the purpose of the study should be so narrowly prescribed. Whatever the reason for seeking the information about productive capacity, the way the capacity is distributed over the population cannot be entirely a matter of indifference. If the ultimate objective were to stimulate growth in productivity, the actions to be taken would be different if (a) production were almost all in the hands of a single large producer, or (b) production were distributed evenly over many small producers. Thus information about the population in addition to the mean productivity is really needed, and the statistical analysis should yield a description of the whole distribution, not only an estimate of its mean.

Just because we have no direct interest in some aspect of the data, we are not justified in making erroneous assumptions about it. Erroneous assumptions are not justified by meticulous optimization of the procedure of analysis in the light of those assumptions. Statistical analysis has far too often been performed as a rite or mystery, a votive offering to Fortuna, a sop to the editor of a scientific journal who, unable to ensure that statistical analyses are relevant, insists that they shall have a semblance of propriety.

If statistical analysis is to be properly sensitive to the facts, and also sensitive to theory and experience and common sense, it will generally have to be done in several steps. It should not be completely automated. There is no unique correct method. We cannot say at the outset what approach will prove successful. Possibly more than one satisfactory theoretical description will be found.

The following three long chapters aim to illustrate statistical analysis of data, and the development of statistical methods, when we have ready access both to mathematical reasoning and to computing. The topics are ones with which the author happens to have been concerned during recent years—quite other topics might have been chosen for the same purpose. The sets of data presented will, it is hoped, be found interesting in their own right, and they are considered for their own sake—statistical methods should, of course, serve the data, rather than the reverse. Our purpose is, however, methodological, and no claim can be made that the treatments contribute much to the fields of study to which the data are obviously relevant.

Chapter 10
Time Series: Yale Enrolment

A. Elementary Treatment

As an example of examination of a time series, let us consider the enrolment of students at Yale University. Information concerning the numbers of students enrolled in the various schools and programs can be obtained for each academic year from 1796/97 onwards. The University was founded in 1701. For most years before 1796 the enrolment is not known directly, though Welch and Camp (1899) made estimates from the numbers of degrees awarded. For present purposes we shall start at 1796, and use the counts of total enrolment listed in Figure B:2, a time series of length 180.

Accuracy. Checking these figures for accuracy is less easy than one might hope it would be. The nature of the educational activities at Yale has undergone many changes. New programs have been introduced, and old ones reorganized. In addition to students enrolled in regular degree courses, there have been various special students, part-time students, summer students. Nearly every year from 1814 onwards there has been a summary page in the University Catalogue showing the numbers of students in the various schools or programs, with a grand total. Just which kinds of students are included in the total has varied—sometimes there is a note mentioning that so many students of some special kind have not been included. For consistency with other years when apparently all kinds of students were counted, any such excluded number has here been added to the given grand total. Enrolments up to 1898/99 were also listed by Welch and Camp, in good agreement with the Catalogue.

Three times during this century there seems to have been a muddle over the annual report of enrolment in the Catalogue. Nothing was reported for 1927/28, and the figure used here has been supplied from other sources by Professor George W. Pierson, the historian of the University. Doubt attaches to the figure for 1938/39, since a footnote in the Catalogue says the

TOTAL STUDENT ENROLMENT AT YALE UNIVERSITY

+	0	1	2	3	4	5	6	7	8	9	
1790							115	123	168	195	*A*
1800	217	217	242	233	200	222	204	196	183	228	*B*
1810	255	305	313	328	350	352	298	333	349	376	*C*
1820	412	407	481	473	459	470	454	501	474	496	*D*
1830	502	469	485	536	514	572	570	564	561	608	*E*
1840	574	550	537	559	542	588	584	522	517	531	*F*
1850	555	558	604	594	605	619	598	565	578	641	*G*
1860	649	599	617	632	644	682	709	699	724	736	*H*
1870	755	809	904	955	1031	1051	1021	1039	1022	1003	*I*
1880	1037	1042	1096	1092	1086	1076	1134	1245	1365	1477	*J*
1890	1645	1784	1969	2202	2350	2415	2615	2645	2674	2684	*K*
1900	2542	2712	2816	3142	3138	3806	3605	3433	3450	3312	*L*
1910	3282	3229	3288	3272	3310	3267	3262	2006	2554	3306	*M*
1920	3820	3930	4534	4461	5155	5316	5626	5457	5788	6184	*N*
1930	5914	5815	5631	5475	5362	5493	5483	5637	5747	5744	*O*
1940	5694	5454	5036	5080	4056	3363	8733	8991	9017	8519	*P*
1950	7745	7688	7567	7555	7369	7353	7664	7488	7665	7793	*Q*
1960	8129	8221	8404	8333	8614	8539	8654	8666	8665	9385	*R*
1970	9214	9231	9219	9427	9661	9721					*S*

Figure B:2

summary is an estimate for the following year, but there are grounds to doubt that assertion, and with no other information available the figure has been accepted—it is plausible, anyway. Nothing was reported for 1972/73; the figure given was supplied by the Secretary's office.

The date of the Catalogue's census is usually early in the academic year, typically November 1. (For 1945/46 it was March, 1945.) For this reason, in Figure B:2 the names of the academic years have been abbreviated to the calendar year of beginning, so that 1796 means 1796/97 and 1975 means 1975/76.

In sum, each of the entries in Figure B:2 can be claimed to have some authority behind it, if not to be actually correct. No doubt there have been errors and inconsistencies in the Catalogue's reckoning. Professor Pierson is bringing out a revised tabulation showing some small differences from the Catalogue. None of them would be perceptible for present purposes.

Graphs. Clearly the first thing to do with these enrolment figures is to graph them, and that has been done in three different ways in Figures B:3, B:4, B:5. Figure B:3 shows the counts directly plotted against date. Because such a graph, produced on a typewriter terminal rather than on a cathode-ray tube, has rather low precision, the earlier years are poorly presented, and attention is focused on the big changes occurring since about

1890. In particular, the disturbances associated with the two world wars are prominent.

Figure B:4 shows the common logarithms of the counts plotted against date. Now one notices that the main trend is linear, and that the plot wobbles nearly as much in the earlier years as in the later. Figure B:5 shows residuals of the logarithms of counts from a fitted straight line, on a larger scale. These three diagrams illustrate a general principle of graphical analysis emphasized by Tukey (1977), that relations between variables should if possible be linear, for ready comprehension, and that when some simple large effect is seen it should be removed and the residuals examined more closely.

The most striking feature of the residuals is their lack of independence. They may be red noise, perhaps, but are certainly not white noise. Complete independence of the residuals is not to be expected, since most students are enrolled for several consecutive years. But Figure B:5 shows pronounced local trends or movements lasting ten, twenty, or even fifty, years. One might consider representing these movements by a polynomial function of time. There is no suggestion of any quadratic relation with time. A fourth or fifth degree polynomial can indeed mimic some of the movements, but others would require a polynomial of much higher degree.

Choice of theoretical description. If we fit a trend to these enrolment figures, we are presumably groping towards a "law of growth". Can any broad generalization be made about the growth, that is possibly inherent in the nature of the institution and its environment, and not merely accidental? Is there a main course that manifests itself in spite of particular events? Enrolment has certainly been affected by external events, obviously by the two world wars, perhaps by other economic or political occurrences. Enrolment has also certainly been affected by internal decisions and policies, by development in physical facilities and in programs. Such internal and external events can be pointed to in connection with the last few entries in our table of enrolments. The rise from 1968 to 1969 was caused mainly by the internal decision, taken early in the 1968/69 academic year, that Yale College, hitherto all-male, should henceforth admit women students also. That 1970 does not show a further rise above 1969 but instead a slight fall is due mainly to the external event of reduction in federal support for graduate study and a consequent internal decision to reduce enrolment of new students in the Graduate School. In the following five years there has been some further growth in Yale College, and smaller increases in the Graduate School and the School of Medicine.

Similarly detailed and particular explanations could be found for all the preceding changes in enrolment. We are observing a kind of random walk, one that manages, despite its irregularity, to show (in logarithms) a striking linear component of trend, and no discernible quadratic component. This

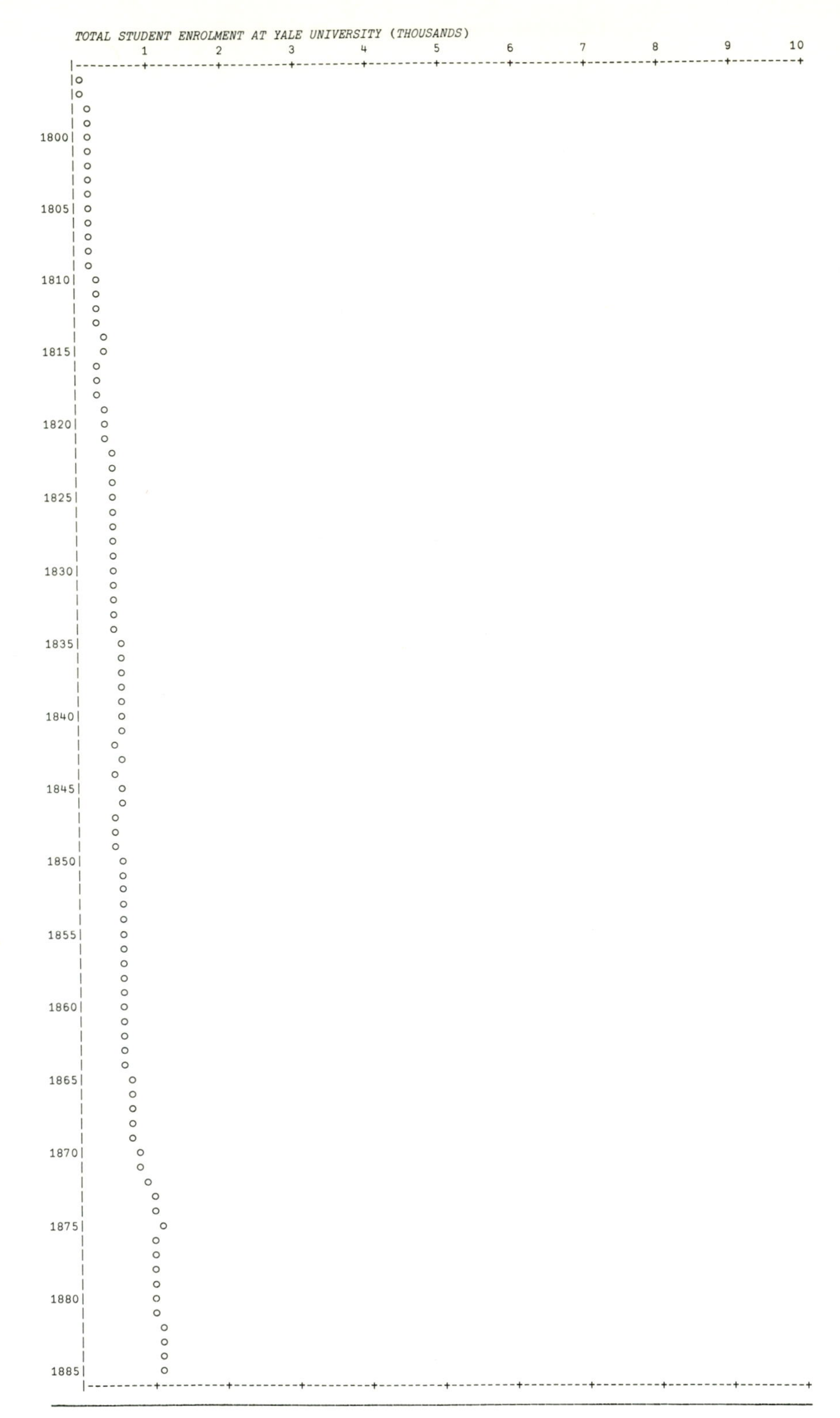

Figure B:3 (Part 1)

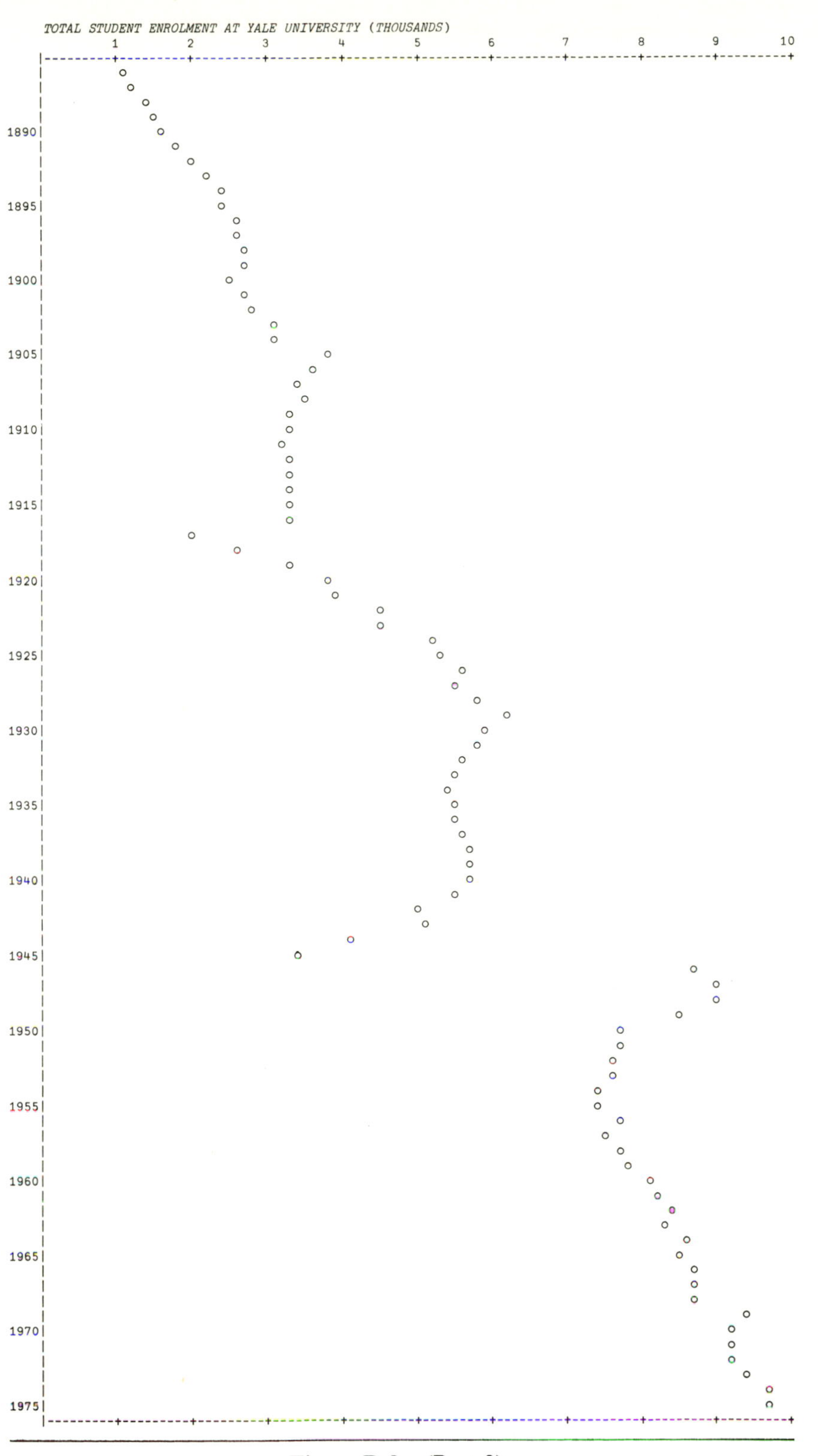

Figure B:3 (Part 2)

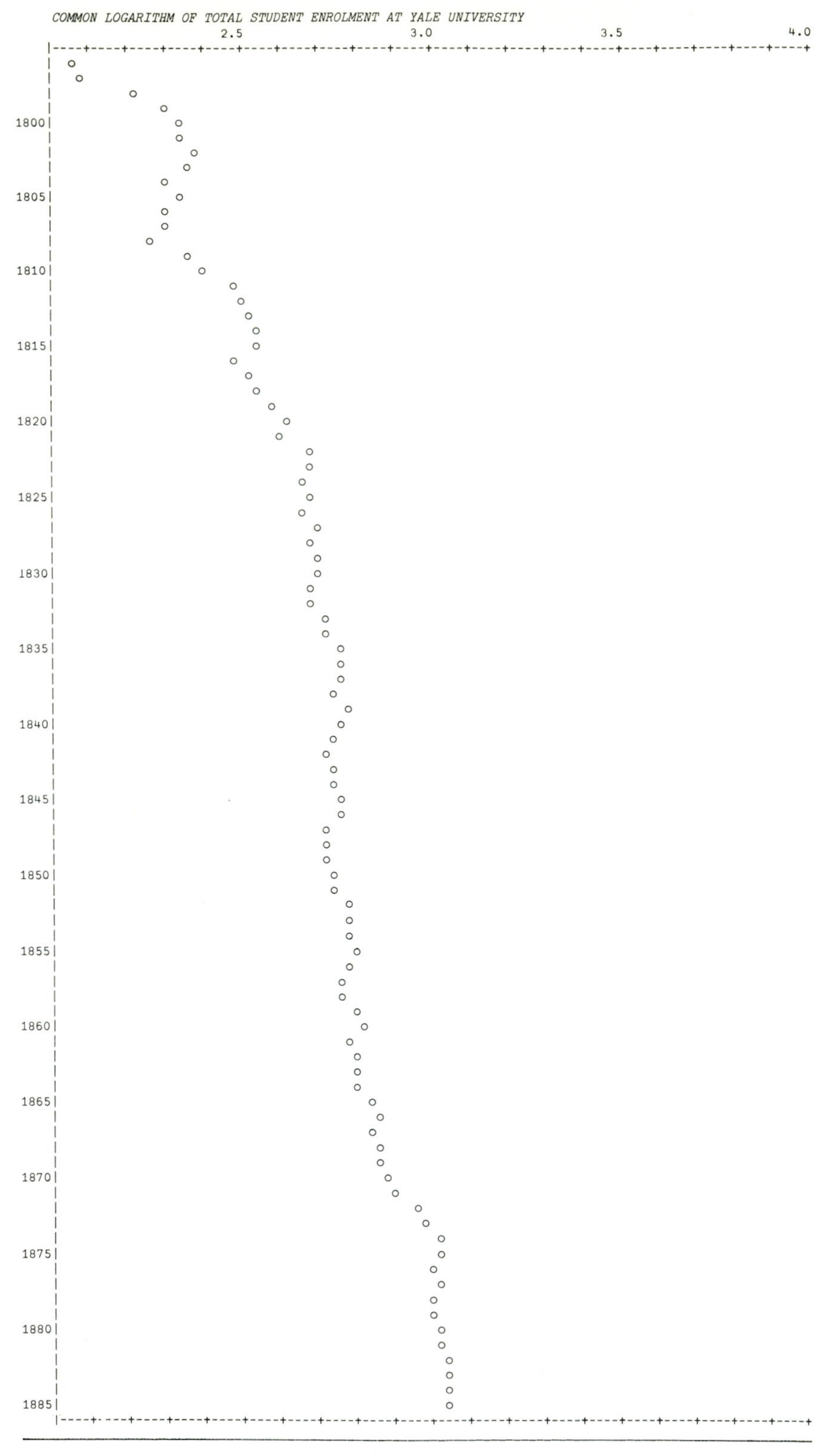

Figure B:4 (Part 1)

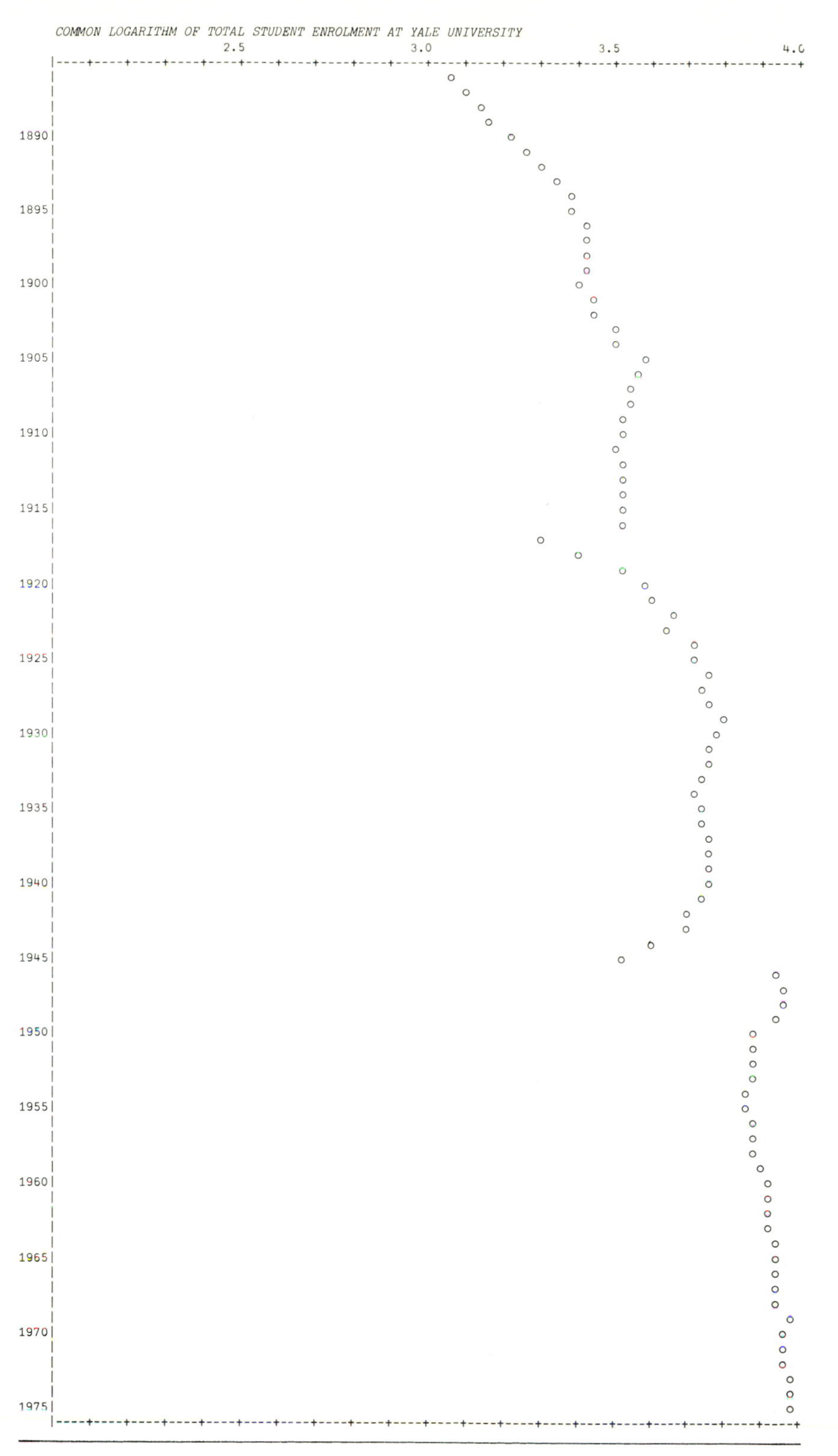

Figure B:4 (Part 2)

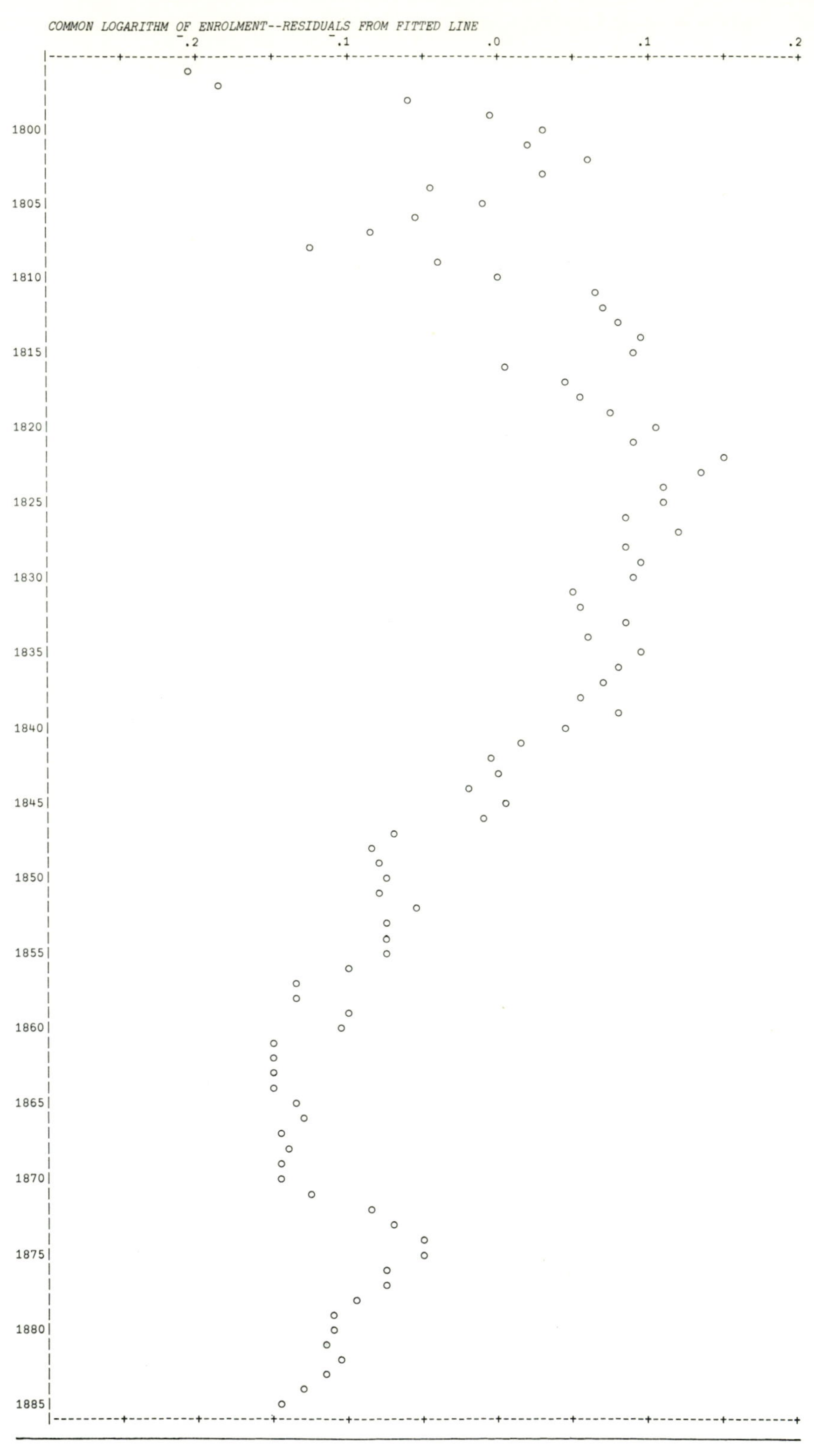

Figure B:5 (Part 1)

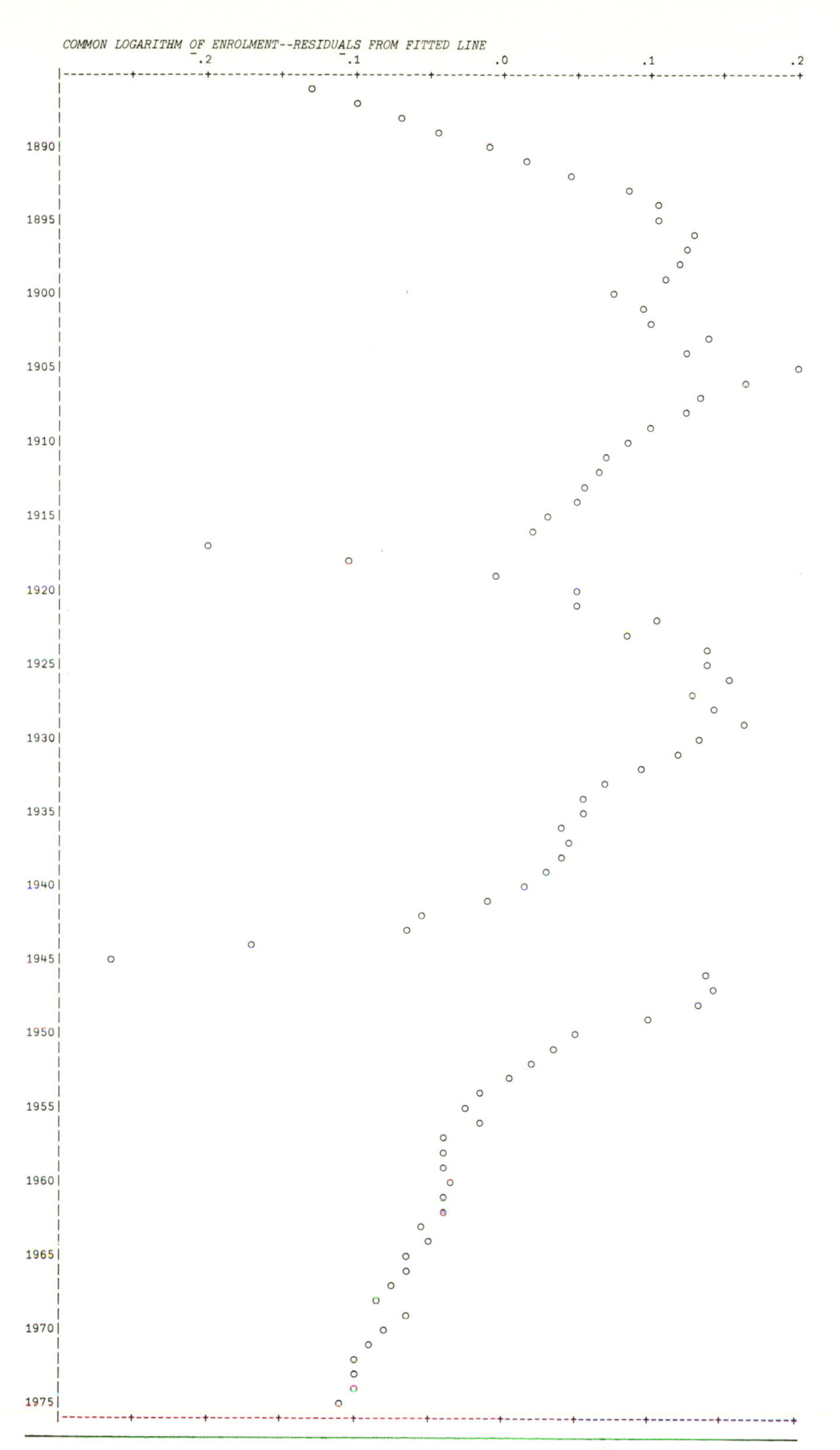

Figure B:5 (Part 2)

trend contrasts with that of the decennial census figures for the whole population of the United States, during roughly the same period of years. There, on the logarithmic scale, the rate of growth has been decreasing; growth appears roughly linear in the cube root of the population size (see Exercise 9 at the end of Part A).

An attractive theoretical description of the Yale enrolment figures is therefore that their logarithms consist of a linear trend plus stationary red noise. The linear trend can be entertained as a possible long-term course, and the red noise represents the details of history.

Fitting the trend. In preparing Figure B:5, the linear trend was estimated by ordinary least squares. The fitted expression for common logarithm of enrolment is

$$3.180 + 0.01024\ (t - 1885.5),$$

where t is the year number, $1796 \leqslant t \leqslant 1975$. The slope implies doubling of enrolment in 30 years, or multiplication more than tenfold in a century.

Fitting a relation by least squares is theoretically good if the deviations from the supposed relation are random variables, independent and normally distributed with constant variance. A sufficiently nonnormal distribution of deviations may cause the method of least squares to work badly. Recently there has been much interest in development of robust methods of estimation having a wider range of efficacy than traditional least squares.

In the present case, it seems to matter little how the linear trend is fitted. The residuals from the least-squares line show one somewhat large negative deviation, ¯0.263 for 1945, but otherwise their distribution seems to depart from the normal shape in being short-tailed rather than long-tailed. In fact, Fisher's g_2 statistic, calculated as though the residuals were a random sample from some population, comes out at ¯0.94 with formal standard error 0.36. This standard error is too small because of the evident serial correlation of the residuals; but at least one can say there is little evidence of positive kurtosis.

If the line is fitted by Huber's method, as described in the next chapter, with one fifth of the readings given reduced weight, the result is for practical purposes identical to what we had by simple least squares, namely

$$3.182 + 0.01021(t - 1885.5)\ .$$

The residuals from this line differ from the previous least-squares residuals only in the third decimal place; the residual of largest magnitude is still ¯0.263, for 1945.

As for the possibility of fitting by least squares a polynomial trend of degree higher than one, the following analysis of variance can be obtained, showing the effect of fitting successive orthogonal polynomials.

	Sum of squares	D.f.	Mean square
Linear trend	50.924	1	50.924
Quadratic trend	0.005	1	0.005
Cubic trend	0.182	1	0.182
Quartic trend	0.310	1	0.310
Quintic trend	0.377	1	0.377
Residual	0.801	174	0.0046
Sum of last five terms (Residual from linear trend)	1.675	178	0.0094
Total about mean	52.599	179	0.2939

It will be seen that, whereas fitting a linear trend cuts the residual mean square by a factor of 30, going as high as the fifth degree cuts the residual mean square by a further factor of 2. The distribution of residuals from the fifth-degree polynomial is strikingly nonnormal, with five large negative residuals causing an appearance of both skewness and kurtosis. These residuals could be used for correlating with a possible explanatory variable, though below we shall in fact use other residuals. The fifth degree polynomial trend does not inspire confidence for extrapolation. Linear trend plus red noise seems a much preferable description.

Before 1796. Though they are not reproduced here, Welch and Camp's conjectured enrolments in the years preceding 1796 have been examined also. On the logarithmic scale, they are somewhat more variable than the post-1796 figures, but as far back as the mid 1720's they seem to show roughly the same linear trend. One could claim, therefore, that the growth behavior of doubling in about 30 years has been observed over a period of not merely 180 years but 250 years. On the other hand, during the first 20 years or so of Yale's existence, the growth rate was considerably higher than that.

Forecasts. This is as far as it seems reasonable to go in statistical analysis of the enrolment figures, considered by themselves without relation to other information. Past behavior does not imply any necessary consequence for the future, but in so far as past behavior shows some regularity and constancy a continuation of those features should at least be entertained as a possibility, if not accepted as fate. The constancy we have observed has been the linear trend in logarithms plus rather sluggish "red noise" deviations.

If we think of the far future this behavior seems unlikely to be maintained, for it suggests that in only some six centuries from now the whole United States population (9 billion) will be enrolled as Yale students and then where will Harvard be? But if we look merely to the year 2000, the linear trend implies an expected Yale enrolment of about 23000 (and the superimposed red noise implies that with about 95% probability the enrolment will lie between 15000 and 37000). These figures are not beyond credibility, unwelcome though such increase in size may appear.

One might base a forecast on more recent experience only, say on the enrolment figures for the last fifty years. Then a more modest rate of exponential growth would be suggested, with doubling in about 50 years rather than 30. A forecast of future growth made in that way in 1870 would have been wide of the mark. So may such a forecast be now.

B. Harmonic Analysis

Harvey Brenner (1973) has made some remarkable analyses of health series, showing in many cases an association between the incidence of illness and the state of the economy, as measured by an employment or unemployment index. To try to relate the Yale enrolment figures to some indicator of the United States economy is tempting. Unemployment is not a suitable measure, however, over the long period with which we are concerned. For most of the nineteenth century unemployment figures are not available, and even if they were they would have a much narrower significance in the largely rural economy than they have today. Summary information is available concerning foreign trade of the United States from 1790 onwards, and in particular the volume of imports has been held to be a good indicator of economic prosperity. Accordingly series U193, general imports of merchandise expressed in millions of dollars, in *Historical Statistics of the United States* (1975) has been adopted for the present purpose, updated by recent entries in the annual *Statistical Abstract of the United States* (up to the edition for 1976). The figures have been taken without any adjustment, as given, except for one modification. The published figures relate to years ending September 30 until 1842, then to years ending June 30 until 1915, and thereafter to calendar years. The entry for 1843 is for a nine-month period, and there is separate information for the last six months of 1915. For present purposes phase effects are of interest and could be obscured by uncorrected shifts in the fiscal year. So simple linear interpolation between adjacent entries has been used up to 1915 to estimate values for the calendar year. The results rounded to the nearest integer are given in Figure B:6. This series has no doubt been affected by various changes in definitions, and certainly by changes in the value of the dollar and by growth of population, as well as by changing rates of taxation and by changing taste and need. Some of its oscillations nevertheless seem to reflect the major historical movements between depression and prosperity. (We shall consider later some other possible indicators of the economy.)

Figure B:7 graphs residuals of the logarithms of the imports series from a fitted straight line, in the same style as Figure B:5 for enrolment. To the

TOTAL IMPORTS OF MERCHANDISE (MILLIONS OF DOLLARS)

+	0	1	2	3	4	5	6	7	8	9
1790	25	30	32	32	44	73	80	74	72	82
1800	96	102	73	70	94	123	132	119	58	66
1810	77	59	63	20	38	122	135	105	113	84
1820	69	61	78	72	77	87	76	74	78	66
1830	71	96	97	103	116	147	165	122	111	142
1840	104	116	86	80	108	116	120	136	145	158
1850	193	209	236	281	278	284	329	306	297	343
1860	322	239	216	280	278	337	416	377	388	427
1870	478	574	635	605	550	497	456	444	442	557
1880	656	684	724	696	623	607	664	708	735	767
1890	817	836	847	761	694	756	773	691	657	774
1900	837	863	965	1009	1055	1173	1331	1314	1253	1435
1910	1542	1590	1733	1854	1784	1750	2392	2952	3031	3904
1920	5278	2509	3113	3792	3610	4227	4431	4185	4091	4399
1930	3061	2091	1323	1450	1655	2047	2423	3084	1960	2318
1940	2625	3345	2756	3381	3929	4159	4942	5756	7124	6622
1950	8852	10967	10717	10873	10215	11384	12615	12982	12792	15207
1960	14654	14714	16380	17138	18684	21364	25542	26812	33226	36043
1970	39952	45563	55563	69476	100997	96940				

Figure B:6

eye there is little relation between these two graphs. They are roughly similar in character, but their movements do not seem to run consistently in parallel. The imports residuals show a suggestion of quadratic trend. They also have the peculiarity that they are less variable in the middle years than either earlier or later—the graph is smoother in the middle part. To some extent this is due to the interpolation. The entries in Figure B:6 relating to years between 1843 and 1915 have been derived by simple (equal-weighted) averaging of consecutive pairs of the source series, a procedure, as we shall see below, that suppresses high frequencies. The averaging used for years before 1843, being unequally weighted, has much less effect on high frequencies, and entries relating to years after 1915 are not averaged at all.

How close a relation could be expected between an economic series and the enrolment series? We have already remarked that enrolment is affected by many institutional happenings little or not at all dependent on the economic health of the nation. It may also be affected by national events, notably the two world wars, that are not primarily economic. One may reasonably expect, however, that a prolonged period of economic prosperity in the nation would tend to be for the university a time of increasing endowment, improved physical facilities and enlarged enrolment; and that contrariwise a prolonged economic recession would tend to be a time for

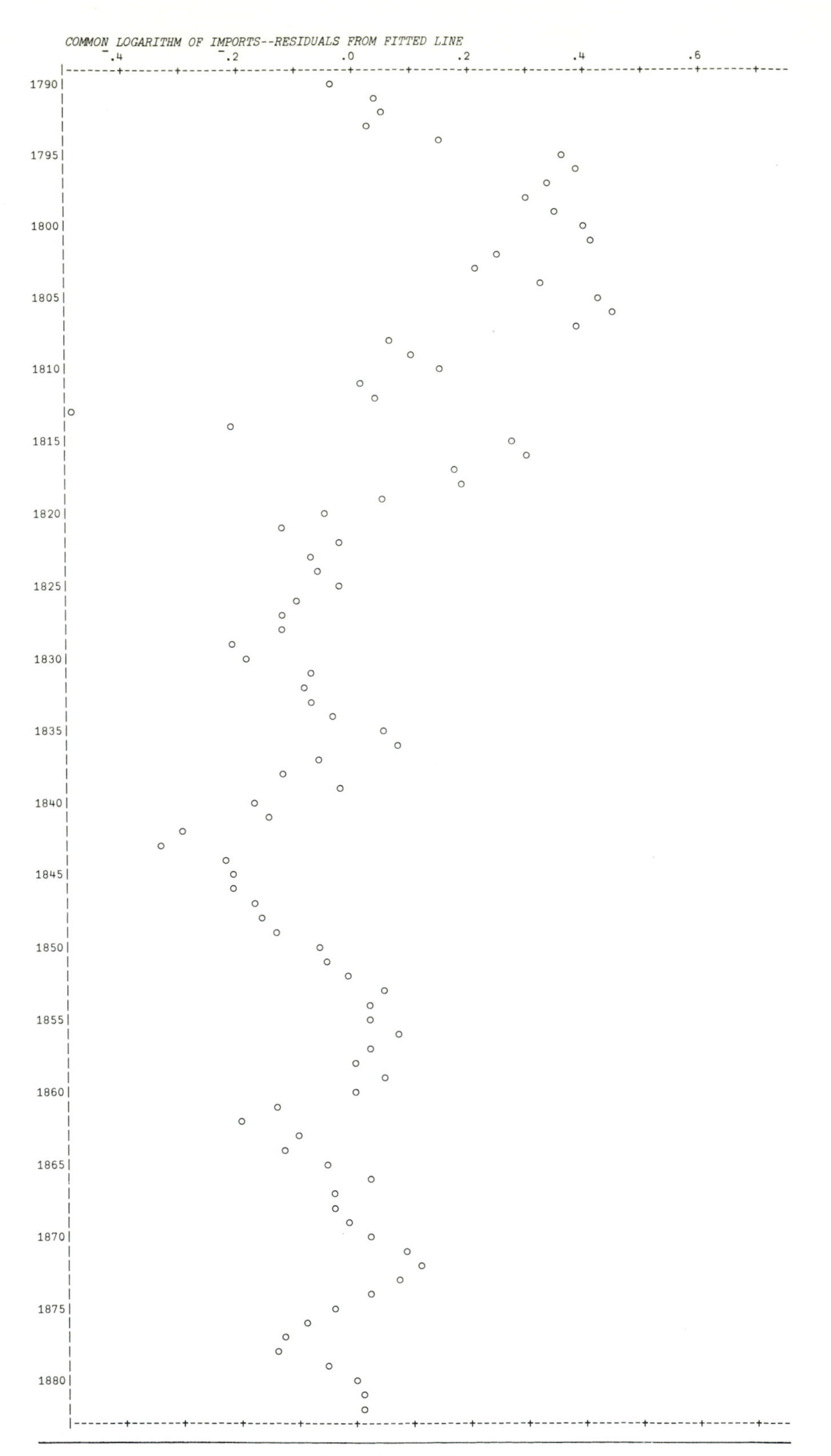

Figure B:7 (Part 1)

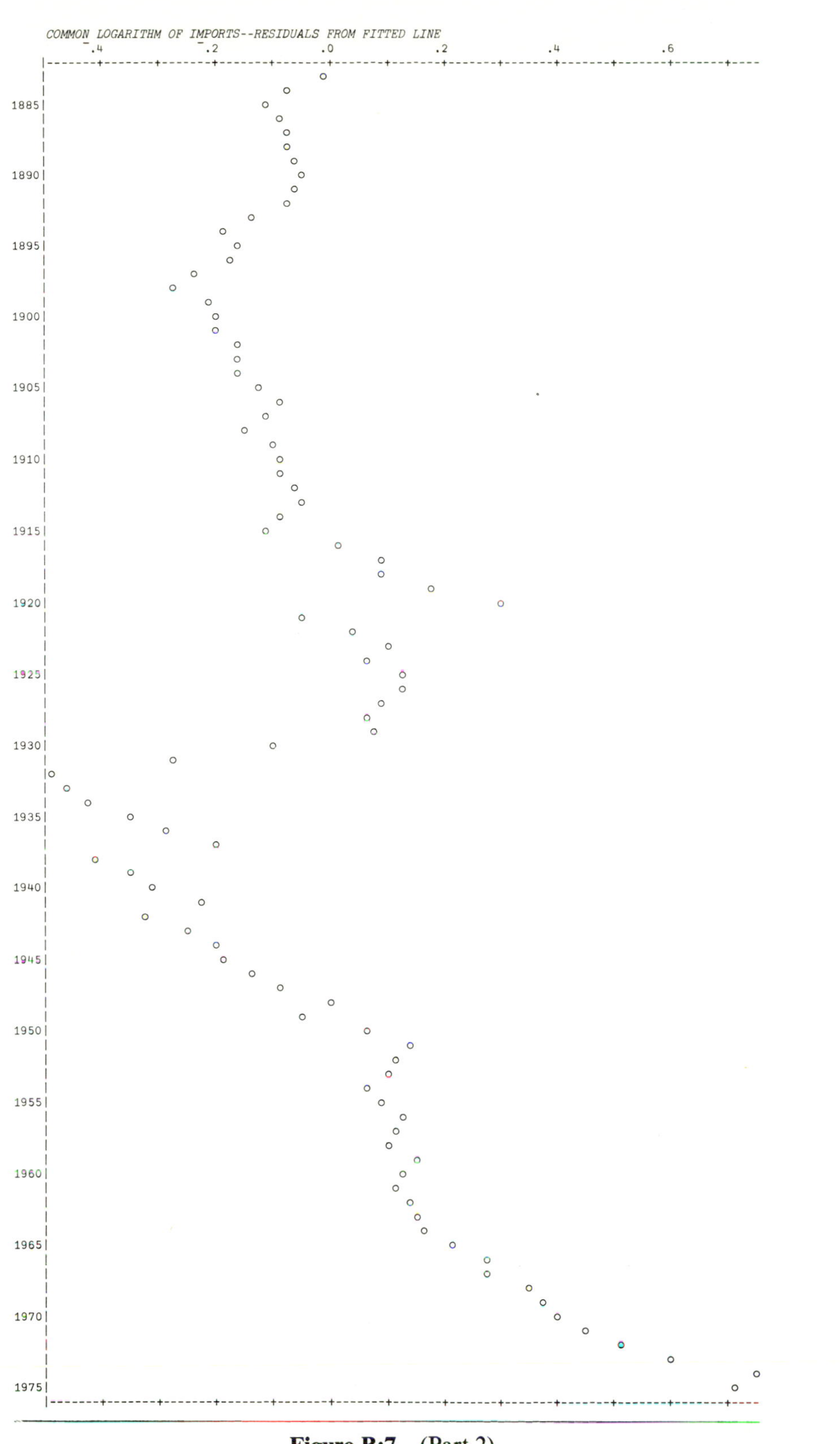

Figure B:7 (Part 2)

holding or retrenchment. That is, whereas economic movements of short duration may well have no detectable effect on enrolment, longer-term economic changes could be reflected visibly in enrolment, perhaps with several years' lag. In fact, the diverse ways in which the state of the economy might influence the university, through endowment, building, and pressure for admission of students, could well bear results on enrolment having diverse lags. That in itself would be a reason to expect only rather slow economic changes to have visible effects.

We are indeed trying to gauge the state of the economy through gross dollar value of imports of merchandise. That is almost certainly not a good measure of the characteristics of the economy affecting Yale's growth. Slow economic changes, which would be particularly interesting if they could be correctly measured, are all too likely to be concealed by the other influences already mentioned.

Thus if the economy, as indicated by imports, is reflected in Yale's enrolment, the relation will probably be frequency-dependent. The amount and perhaps also the lag will vary with the rapidity of oscillation—little relation probably at the highest and the lowest frequencies, but perhaps some relation at intermediate frequencies, as has usually been the case in Harvey Brenner's studies. Frequency-dependent relations between processes are familiar in engineering. Consider, for example, in a high-fidelity sound system, the relation between the input to the amplifier from a pick-up cartridge or tape deck or radio tuner and the output to the speakers. Over the intended audible range of frequencies the two processes should be almost perfectly correlated; but the amplifier is designed to ignore input frequencies below the audible range, arising perhaps from warp in a phonograph disc, and possibly to ignore very high frequencies constituting a hiss.

The natural way to study a frequency-dependent relation between series is to perform a harmonic analysis on each and compare corresponding components. The literature of time-series analysis is formidable in size and character, and methods are often described in arm-chair style as though the writer did not actually practise them. We therefore give below a sketch of the principal theoretical ideas, before proceeding to examination of the data at hand.

This book is no skilfully graded textbook for a course. Easy and difficult things are jumbled together, as in life. The gentle reader is invited to skip freely in the rest of this chapter.

Fourier representation of a vector

Consider an n-element vector $\{x_t\}$, where $t = 1, 2, \ldots, n$; n is a positive integer large enough for ensuing expressions to be defined. We shall think of t as measuring time in years and refer to $\{x_t\}$ as a time series. (Sometimes it is more convenient to let the n consecutive integer values assumed by t be $0, 1, \ldots, n-1$. Usually the choice is unimportant.)

The n-dimensional linear space R^n is spanned by n orthogonal vectors of unit length,

$$\left\{\frac{1}{\sqrt{n}}\right\}, \left\{\sqrt{\frac{2}{n}} \cos\frac{2\pi t}{n}\right\}, \left\{\sqrt{\frac{2}{n}} \sin\frac{2\pi t}{n}\right\}, \left\{\sqrt{\frac{2}{n}} \cos\frac{4\pi t}{n}\right\},$$
$$\left\{\sqrt{\frac{2}{n}} \sin\frac{4\pi t}{n}\right\}, \ldots$$

These may be expressed as

$$\{A_t^{(r)}\} = \left\{\sqrt{\frac{d_r}{n}} \cos\frac{2\pi rt}{n}\right\} \quad (0 \leqslant r \leqslant \tfrac{1}{2}n),$$

and

$$\{B_t^{(r)}\} = \left\{\sqrt{\frac{2}{n}} \sin\frac{2\pi rt}{n}\right\} \quad (1 \leqslant r < \tfrac{1}{2}n),$$

where r is always integer-valued, and $d_r = 2$ if $1 \leqslant r < \frac{1}{2}n$, but $d_r = 1$ if $r = 0$ or (when n is even) $\frac{1}{2}n$. We say that $\{A_t^{(r)}\}$ or $\{B_t^{(r)}\}$, for any r, represents a *simple harmonic oscillation* or *sinusoid* with *frequency* r/n cycles per year and *period* n/r years per cycle. These uses of "frequency" and "period" are special to the context of harmonic analysis, whereas elsewhere in this book the words mean "number of occurrences" and "interval of time", respectively. For given n we speak of the frequency r/n as the rth *harmonic*. If a frequency, in cycles per year, is multiplied by 2π it is sometimes referred to as an *angular frequency*, in radians per year.

The regression coefficients of $\{x_t\}$ on $\{A_t^{(r)}\}$ and on $\{B_t^{(r)}\}$ are respectively

$$a_r = \Sigma_t x_t A_t^{(r)}, \quad b_r = \Sigma_t x_t B_t^{(r)}. \tag{1}$$

If the elements of $\{x_t\}$ are realizations of independent random variables, each normally distributed with zero mean and variance σ^2, that is, if $\{x_t\}$ is Gaussian white noise, the same is true of the coefficients a_r, b_r, for all values of r specified above. No value for b_r has been defined when $r = 0$ or (if n is even) $\frac{1}{2}n$. Defining b_r to be zero when $r = 0$ or $\frac{1}{2}n$, we call the set of elements

$$\{z_r\} = \{a_r^2 + b_r^2\} \quad (0 \leqslant r \leqslant \tfrac{1}{2}n), \tag{2}$$

the *line spectrum* of $\{x_t\}$. (More properly, the spectrum is the set of frequencies r/n for which $z_r > 0$, and the z's are the corresponding powers; but to refer to $\{z_r\}$ as the spectrum is usual and convenient. Other names include the *empirical* spectrum, *discrete* spectrum, *raw* spectrum.) The sum of all the elements of the line spectrum is equal to $\Sigma_t x_t^2$. Provided $z_r > 0$ we can define an angle ϵ_r, where $0 \leqslant \epsilon_r < 2\pi$, such that

$$a_r = \sqrt{z_r} \cos \epsilon_r \ , \quad b_r = -\sqrt{z_r} \sin \epsilon_r \ . \tag{3}$$

Then we have

$$x_t = \Sigma_r \sqrt{\frac{z_r d_r}{n}} \cos\left(\frac{2\pi r t}{n} + \epsilon_r\right), \tag{4}$$

the summation being over all nonnegative integer r not exceeding $\frac{1}{2}n$. We call $\sqrt{z_r d_r / n}$ the *amplitude* and ϵ_r the *phase* of the rth harmonic in the Fourier representation of $\{x_t\}$. (Different writers use various slightly different definitions.)

The serial correlation coefficients of $\{x_t\}$ can be expressed in terms of the line spectrum $\{z_r\}$, and conversely; see Exercise 1 below. Before the discovery of the "fast Fourier transform", the line spectrum was usually computed from the serial correlations. Now the line spectrum $\{z_r\}$, together with the phases $\{\epsilon_r\}$, are computed directly.

If $\{x_t\}$ is Gaussian white noise with zero mean and variance σ^2, each element z_r in the line spectrum is equal to σ^2 multiplied by a random variable having the tabulated χ^2 distribution with 2 degrees of freedom, except that if $r = 0$ or $\frac{1}{2}n$ the χ^2 has only 1 degree of freedom. All elements in the line spectrum are independent. The line spectrum is an analysis of variance of $\{x_t\}$.

We allude in this way to Gaussian white noise because it is a natural reference point. Usually when we do harmonic analysis of a time series, Gaussian white noise is not a plausible theoretical description of the process. More likely we are willing to suppose that the given series is realized from a *stationary random sequence*, and to think of the line spectrum as estimating properties of the sequence.

Stationary random sequences

A sequence of random variables $\{\xi_t\}$, where t may have any (unlimited) integer value, is said to be *stationary* if the joint distribution of any finite set of variables, $(\xi_t, \xi_u, \xi_v, \ldots)$, is identical to that of the set $\{\xi_{t+h}, \xi_{u+h}, \xi_{v+h}, \ldots)$, in which all subscripts have been increased by an arbitrary integer h. Then ξ_t has the same marginal distribution for every t, and any pair of variables (ξ_t, ξ_u) has a joint distribution depending only on $t - u$.

It is usually supposed that $\mathcal{E}(\xi_t) = 0$ and that all second moments are finite, $\mathcal{E}(\xi_t^2) = \sigma^2$. Then there is a spectral decomposition theorem (see Doob, 1953, chapter 10, or Cox and Miller, 1965, chapter 8) to the effect that any such stationary random sequence can be expressed as the sum of (infinitely many) sinusoids whose frequencies are in the interval $(0, \frac{1}{2})$, and whose phases and amplitudes are random (but in general not independent).

For various reasons the kind of stationary random sequence that is of greatest interest, in relation to linear (harmonic) methods of time-series analysis, is a normal or Gaussian sequence, in which the joint distribution of any finite set of variables $(\xi_t, \xi_u, \ldots)$ is normal. For such a sequence the spectral decomposition can be exhibited in a direct and elementary fashion by summing independent random sinusoids, as follows.

Random sinusoids

Consider the random sinusoid

$$\xi_t = \sqrt{2}\cos(2\pi\lambda t + \epsilon) \quad (t = 0, \pm 1, \pm 2, \ldots), \tag{5}$$

where λ and ϵ are independent random variables, the phase ϵ uniformly distributed over the interval $(0, 2\pi)$ and the frequency λ distributed over the interval $(0, \frac{1}{2})$ with distribution function $F(\lambda)$; $F(\lambda)$ is nondecreasing, $F(0-) = 0$, $F(\frac{1}{2}+) = 1$. (We follow the common practice of letting the name of the random variable λ serve also in $F(\lambda)$ as a member of its value set.) It is easily seen that, because of the uniform distribution for ϵ, this random sequence is stationary, with zero expectation:

$$\mathcal{E}(\xi_t) = 0.$$

For any integer h,

$$\begin{aligned} \mathcal{E}(\xi_t \xi_{t+h}) &= 2\mathcal{E}\{\cos(2\pi\lambda t + \epsilon)\cos(2\pi\lambda(t+h) + \epsilon)\} \\ &= \mathcal{E}\{\cos(2\pi\lambda(2t+h) + 2\epsilon)\} + \mathcal{E}\{\cos(2\pi\lambda h)\}. \end{aligned}$$

The first term vanishes because of the uniform distribution for ϵ, and the second gives

$$\mathcal{E}(\xi_t \xi_{t+h}) = \int_0^{1/2} \cos(2\pi\lambda h)\, dF(\lambda).$$

In particular, $\mathcal{E}(\xi_t^2) = 1$. Thus the serial correlation coefficient of the sequence $\{\xi_t\}$ at lag h, say ρ_h, is a Fourier transform of the distribution of the random frequency λ:

$$\rho_h = \int_0^{1/2} \cos(2\pi\lambda h)\, dF(\lambda) \quad (h = 0, \pm 1, \pm 2, \ldots). \tag{6}$$

We shall usually suppose that $F(\lambda)$ is absolutely continuous, the integral of its derivative $f(\lambda)$, so that

$$\rho_h = \int_0^{1/2} \cos(2\pi\lambda h)\, f(\lambda)\, d\lambda. \tag{7}$$

Then $f(\lambda)$ can be expressed as a Fourier series in terms of the correlations

$\{\rho_h\}$, and therefore the function $f(\lambda)$ and the sequence $\{\rho_h\}$ are equivalent ways of expressing the correlation structure of $\{\xi_t\}$. To obtain the Fourier series, let the definition of $f(\lambda)$ be extended over the interval $(-\frac{1}{2},\frac{1}{2})$ by the relation $f(-\lambda) = f(\lambda.)$ The Fourier series for $f(\lambda)$ over $(-\frac{1}{2},\frac{1}{2})$ is then easily seen to be

$$f(\lambda) = 2 + 4 \sum_{h=1}^{\infty} \rho_h \cos(2\pi\lambda h). \tag{8}$$

That is, the right side is the Fourier series of the left side; if $f(\lambda)$ is sufficiently well behaved, the series converges to $f(\lambda)$. An important special case is an uncorrelated random sequence having $\rho_h = 0$ for $h \neq 0$, and $f(\lambda) = 2$ for all λ in $(0,\frac{1}{2})$.

The marginal distribution for ξ_t is continuous with U-shaped density function; the kurtosis coefficient, fourth cumulant divided by variance squared, is ¯1.5. If independent random variables distributed like ξ_t are summed, normality is approached somewhat faster than it is by the binomial distribution with $p = \frac{1}{2}$, because the elementary indicator variables that are summed to make the binomial distribution have kurtosis coefficient ¯2. Thus to obtain a close approximation to a Gaussian stationary random sequence having zero mean and serial correlations given by (6) above, we need only add a large number N of independent sinusoids as defined above. That is, we let

$$\xi_t = \sqrt{2/N} \sum_{i=1}^{N} \cos(2\pi\lambda_i t + \epsilon_i) \quad (t = 0, \pm 1, \pm 2, \ldots), \tag{9}$$

where $\{\lambda_i\}$ and $\{\epsilon_i\}$ are independent random samples of size N from the distributions specified above. The first and second moments of this sequence are identical to those of (5). It is at first sight surprising that a Gaussian sequence can be approximated in this way, since any realization of the single sinusoid (5) is utterly non-random in appearance, being in general completely determined by any three of its values, such as ξ_0, ξ_1, ξ_2. If attention is restricted to some number n of consecutive values for t, there will not likely be any detectable periodicity in a realization of (9) unless N is much smaller than n.

It can be proved (Wiener-Khintchine-Wold theorem) that the serial correlations $\{\rho_h\}$ of any stationary random sequence with finite variance can be expressed in the form (6), where $F(\lambda)$ is the distribution function of a random variable taking values in the interval $(0,\frac{1}{2})$. We have seen that, conversely, given any distribution function $F(\lambda)$ over $(0,\frac{1}{2})$, the sequence $\{\rho_h\}$ defined by (6) is the serial correlation function of a stationary random sequence. $F(\lambda)$ is called the *spectral distribution function* of the sequence, and its derivative (if there is one) is called the *spectral density function*.

We have generated a Gaussian sequence by adding many sinusoids of equal amplitude but random frequency and random phase. There are other ways to reach the same result. One slight variant will be of interest later. Let $G(\lambda)$ be a (nonrandom) function, which without loss of usefulness may be supposed continuous for $0 \leqslant \lambda \leqslant \frac{1}{2}$. In place of (5), let a single sinusoid be defined thus:

$$\xi_t = \sqrt{2}\, G(\lambda) \cos(2\pi\lambda t + \epsilon), \tag{10}$$

where λ and ϵ are independent random variables as before. Then $\mathcal{E}(\xi_t) = 0$ and

$$\mathcal{E}(\xi_t \xi_{t+h}) = \int_0^{1/2} \cos(2\pi\lambda h) \left(G(\lambda)\right)^2 dF(\lambda). \tag{11}$$

The variance is $\mathcal{E}(\xi_t^2) = \int_0^{(1/2)} (G(\lambda))^2 dF(\lambda) = \sigma^2$, say. Then if the distribution for λ is absolutely continuous with density $f(\lambda)$, the spectral density of the sequence (10) is $(G(\lambda))^2 f(\lambda)/\sigma^2$. The same is true for a sum of independent such sinusoids.

The line spectrum obtained from a realization of a stationary random sequence with variance σ^2, possessing a spectral density $f(\lambda)$, constitutes an estimate of $\sigma^2 f(\lambda)$, in a somewhat similar sense to that in which an ordered sample of independent observations of a random variable having an absolutely continuous distribution constitutes an estimate of the density function. Asymptotic results are available, valid as the length n of the realization tends to infinity, of this type: if the spectral density $f(\lambda)$ is continuous at a frequency λ_0 $(0 < \lambda_0 < \frac{1}{2})$, elements of the line spectrum whose frequencies are close to λ_0 are independent, have expectations equal to $\sigma^2 f(\lambda_0)$ and are distributed exponentially. If we consider a fixed neighborhood of λ_0, for sufficiently large n the element of the line spectrum whose frequency is closest to λ_0 is almost independent of behavior of the spectral distribution $F(\lambda)$ outside that neighborhood. Such asymptotic results are not altogether satisfactory as a guide to interpreting the line spectrum of a not-very-long realization. With the spectral decomposition of a stationary sequence in mind, let us examine the line spectrum of a single sinusoid.

Spectrum of a single sinusoid

Consider the sinusoid

$$x_t = \sqrt{2} \cos(2\pi\lambda t + \epsilon) \quad (t = 1, 2, \ldots, n), \tag{12}$$

where for the moment λ and ϵ are fixed. We can suppose that $0 \leqslant \lambda \leqslant \frac{1}{2}$, for if that were not so λ could be replaced by such a value without effect on the n values of x_t.

It is easy to see that if $n\lambda$ is an integer there is just one element of the line spectrum, namely $z_{n\lambda}$, that is positive, all the rest being zero. But if $n\lambda$ is not an integer, all elements of the line spectrum are positive; z_r becomes smaller as r is taken further from $n\lambda$ in either direction, but is never zero. The dispersion is greatest when $n\lambda$ is equal to an integer plus one-half. This phenomenon is referred to as *leakage*.

The Fourier representation (4) of a given n-element vector $\{x_t\}$ may be considered to be defined for all real t, and is then a periodic function of t with period n. The phases are changed, but the amplitudes are not changed, if we perform the operation on $\{x_t\}$ termed "rotation" in APL, namely if we remove the first few elements from the beginning of $\{x_t\}$ and catenate them on the end. We can explain the presence of frequencies far removed from λ in the Fourier representation of the sinusoid (12) by noting that although the sequence varies smoothly as t goes from 1 to n, the two ends of the sequence do not fit together smoothly except when $n\lambda$ is an integer.

Thus if we are observing a realization of the stationary process (9), each element z_r of the calculated line spectrum is dependent on the behavior of the spectrum $F(\lambda)$, not only at the value r/n, but also at all other values of λ that are not integer multiples of n^{-1}. If there were any reason to think that $F(\lambda)$ would have a jump at some specific frequency λ_0, it would be desirable, if possible, that n should be an integer multiple of the corresponding period λ_0^{-1}, to avoid leakage from that discrete component. In social-science contexts, few if any discrete components are expected in annual series—conceivably some U.S. series could show a four-year cycle related to the political process. Discrete components whose frequencies are multiples of 1/12 cycle per month are certainly to be expected in monthly series; and similarly for quarterly or weekly series. Otherwise, the spectrum may be plausibly expected to have a continuous density function $f(\lambda)$ that is nearly constant over any short interval. Let us determine the magnitude of the leakage effect in this case.

Upon substituting (12) into (1), we have

$$\begin{aligned} a_r &= \sqrt{2d_r/n}\,\Sigma_t \cos(2\pi\lambda t + \epsilon)\cos 2\pi rt/n \\ &= \sqrt{d_r/2n}\,\Sigma_t\{\cos(2\pi\delta t + \epsilon) + \cos(2\pi\sigma t + \epsilon)\} \\ &= \sqrt{d_r/2n}\,\big\{\cos((n+1)\pi\delta + \epsilon)\sin(n\pi\delta)(\sin\pi\delta)^{-1} \\ &\qquad + \cos((n+1)\pi\sigma + \epsilon)\sin(n\pi\sigma)(\sin\pi\sigma)^{-1}\big\}, \end{aligned}$$

where $\delta = \lambda - r/n$, $\sigma = \lambda + r/n$ (the difference and sum frequencies of λ and r/n).[†] A similar expression can be found for b_r, and hence an expression for $z_r = a_r^2 + b_r^2$.

[†]Here briefly, and in a similar place later, σ stands for a sum frequency. Elsewhere σ is the standard deviation of a stationary process. The author apologizes.

Now let us treat λ and ϵ as independent random variables, ϵ being uniformly distributed over $(0, 2\pi)$ and λ distributed with density $f(\lambda)$ over $(0, \frac{1}{2})$. Taking expectation first with respect to ϵ and then with respect to λ, we find

$$\mathcal{E}(z_r) = \tfrac{1}{2} d_r \int_0^{1/2} \{ h^*(\lambda - r/n) + h^*(\lambda + r/n) \} f(\lambda)\, d\lambda, \tag{13}$$

where

$$\left.\begin{aligned} h^*(x) &= n\left(\frac{\sin n\pi x}{n \sin \pi x} \right)^2 \quad \text{when } x \neq 0, \quad |x| < 1, \\ &= n \qquad\qquad\qquad\; \text{when } x = 0. \end{aligned}\right\}$$

Two steps make this expression easier to study. First, extend the definition of $f(\lambda)$ to the interval $(-\frac{1}{2}, \frac{1}{2})$ by the relation $f(\lambda) = f(-\lambda)$. Then

$$\mathcal{E}(z_r) = \tfrac{1}{2} d_r \int_{-1/2}^{1/2} h^*(\lambda - r/n) f(\lambda)\, d\lambda.$$

Next, use the well-known expression for the squared cosecant in terms of its poles:

$$h^*(x) = n^{-1} (\sin n\pi x)^2 \sum_{k=-\infty}^{\infty} \left[(x - k)\pi \right]^{-2},$$

and still further extend the definition of $f(\lambda)$ to the whole real line by the relations $f(\lambda) = f(-\lambda) = f(1 + \lambda)$. Then

$$\mathcal{E}(z_r) = \tfrac{1}{2} d_r \int_{-\infty}^{\infty} h(\lambda - r/n) f(\lambda)\, d\lambda, \tag{14}$$

where

$$\left.\begin{aligned} h(x) &= n\left(\frac{\sin n\pi x}{n\pi x} \right)^2 \quad \text{when } x \neq 0, \\ &= n \qquad\qquad\quad\;\; \text{when } x = 0. \end{aligned}\right\}$$

This function $h(x)$ is continuous, and vanishes when x is an integer multiple of n^{-1}, positive or negative but not zero; otherwise $h(x) > 0$; and $\int_{-\infty}^{\infty} h(x)\,dx = 1$. Thus (14) implies that $\mathcal{E}(z_r)$ is a weighted average of the values of $f(\lambda)$, multiplied by $\frac{1}{2} d_r$ which equals 1 for $1 \leqslant r < \frac{1}{2} n$. The function $h(x)$ has the properties of a probability density function, and its tails are long, like the Cauchy distribution.

It is interesting to suppose that $f(\lambda) = n$ for λ in an interval $((r_0 - \frac{1}{2})/n, (r_0 + \frac{1}{2})/n)$, where r_0 is an integer, $0 < r_0 < \frac{1}{2} n$, and $f(\lambda) = 0$ elsewhere in $(0, \frac{1}{2})$. Then (12) is a random sinusoid whose frequency is closer to r_0/n than to any other integer multiple of $1/n$. Suppose that n is large and that r_0 is not close to 0 or $\frac{1}{2} n$. Then as an approximation we may ignore contributions to the right side of (14) from values of λ above $\frac{1}{2}$ or below 0, and we find (setting $d_r = 2$)

$$\mathcal{E}(z_r) \approx n \int_{(r_0 - r) - (1/2)}^{(r_0 - r) + (1/2)} \left(\frac{\sin \pi x}{\pi x} \right)^2 dx. \tag{15}$$

Some values for $\mathcal{E}(z_r)$, divided by n and multiplied by 100 to yield percentages of the expected total sum of squares, are shown below. The distribution is symmetric about r_0.

$r - r_0 = \cdots$	¯2	¯1	0	1	2	3	4	5	6	7	8	...
$\mathcal{E}(z_r) \propto \cdots$	1.40	7.87	77.37	7.87	1.40	0.59	0.32	0.21	0.14	0.10	0.08	...

The total expected contribution from elements of the line spectrum for which $|r - r_0| \geqslant 4$ is about 3%; from those for which $|r - r_0| \geqslant 11$ it is about 1%.

Since a Gaussian process can be approached by summing independent sinusoids, as at (9), the expression (14) for $\mathcal{E}(z_r)$ applies to any stationary Gaussian sequence with zero mean, unit variance and spectral density $f(\lambda)$. $\mathcal{E}(z_r)$ is equal to an average value of $f(\lambda)$, most weight being on values of λ close to r/n, but some weight on distant values of λ. If $f(\lambda)$ varies little over the whole frequency interval $(0, \frac{1}{2})$, $\mathcal{E}(z_r)$ will be close to $f(r/n)$ for all r $(1 \leqslant r < \frac{1}{2}n)$. But if, for example, $f(\lambda)$ has values in the interval $(0, 0.1)$ that are 100 or 1000 times greater than its values in $(0.2, 0.5)$, say, and if n is not extremely large, leakage from the high-power low frequencies may cause the line spectrum to misrepresent $f(\lambda)$ seriously at the higher frequencies. The actual distribution of the individual regression coefficients a_r and b_r defined at (1) is normal with zero mean. If n is large, the elements z_r of the line spectrum are approximately independent and are distributed approximately proportionally to χ^2 variables with 2 degrees of freedom. See for example Anderson (1971, chapter 8) and Exercise 7 below.

Two devices are commonly used, either singly or together, to reduce the disturbing effects of leakage. One is to "filter" the given series $\{x_t\}$, by differencing it, or subtracting a moving average, or performing some other such linear operation, so that the spectrum of the process is changed and becomes more uniform. This is sometimes called "prewhitening". If indeed the spectrum becomes nearly uniform, leakage will for most purposes have no noticeable consequence. Apart from the effect of leakage, the correlation between two series in the neighborhood of a particular frequency, as we shall be considering below, is not affected by identical filtering of both series—nor in some instances by different filtering.

The other device is to "taper" the series, overlapping the two ends so that they fit together more smoothly. This amounts to a redefinition of the harmonic analysis, leading to a diminution in the leakage phenomenon itself. Both filtering and tapering involve a shortening of the given series and consequent reduction in information in the line spectrum—but the information will be cleaner. Reduction in bias is achieved at the cost of some (usually little) loss in precision.

Filters

A linear filter is defined by a vector of m coefficients $\{w_j\}$, for $j = 0, 1, \ldots, m-1$ (m being an integer not less than 2). From a given time series $\{x_t\}$, for $t = 1, 2, \ldots, n$, we derive the filtered series

$$y_{t+l} = \Sigma_j w_j x_{t+j} \quad (t = 1, 2, \ldots, n+1-m). \tag{16}$$

Here l is an arbitrary constant, not necessarily an integer, representing the time lag between the time index of the first member of $\{x_t\}$ used in obtaining any y-value and the time to which the y-value may conveniently be referred. We call the series $\{x_t\}$ the *input* to the filter and $\{y_{t+l}\}$ the *output*.

As before, we begin by considering the special case of a pure sinusoid input, say

$$x_t = \cos(2\pi\lambda t + \epsilon) \quad (t = 1, 2, \ldots, n), \tag{17}$$

where $0 \leqslant \lambda \leqslant \frac{1}{2}$ and $0 \leqslant \epsilon < 2\pi$. The right side is of course defined for all real t, not just the first n integers, and we use this fact in a moment. It is easy to show that if l satisfies the equation

$$\Sigma_j w_j \sin[2\pi\lambda(j - l)] = 0, \tag{18}$$

then

$$y_{t+l} = G(\lambda)\cos[2\pi\lambda(t + l) + \epsilon],$$

where

$$G(\lambda) = \Sigma_j w_j \cos[2\pi\lambda(j - l)]. \tag{19}$$

Thus the output, indexed at time $t + l$, is equal to the sinusoid input at time $t + l$ multiplied by the factor $G(\lambda)$, which is called the *gain* of the filter.

Some interesting filters have the symmetry property that for all j

$$w_j = w_{(m-1)-j} \quad (j = 0, 1, \ldots, m-1).$$

Then it is easy to see that (18) has just one solution that does not depend on λ, namely

$$l = \tfrac{1}{2}(m - 1). \tag{20}$$

For an unsymmetric filter, the l determined by (18) generally depends on λ, and we may write it $l(\lambda)$. (18) and (19) do not determine $l(\lambda)$ and $G(\lambda)$ uniquely. Given any solution, we may obtain other solutions by adding $M/(2\lambda)$ to $l(\lambda)$ and multiplying $G(\lambda)$ by $(-1)^M$, for any integer M. For convenience, one can define a "principal" solution as one for which $l(\lambda)$ is continuous for $0 < \lambda \leqslant \frac{1}{2}$ and, if possible, $l(\lambda)$ has a finite limit as $\lambda \to 0$, or, if that is not possible, $|l(\lambda)|$ is asymptotically minimal as $\lambda \to 0$; see Exercise 10 below. We call $l(\lambda)$ the *lag* of the filter; the product $2\pi\lambda\, l(\lambda)$ may be referred to as the *phase shift* of the filter.

If the input to the filter is realized from a stationary process with spectral density function $f(\lambda)$, the output is a realization of a process with spectral density proportional to $f(\lambda)[G(\lambda)]^2$. If the filter is symmetric, the l in (16) can be chosen to be that defined at (20), and then the relation between input and output is especially simple, since it involves amplitude changes but not phase changes; for each sinusoid component of the input, its value at time $t + l$ is multiplied by the gain at that frequency to form the corresponding sinusoid component of y_{t+l}. But if the filter is unsymmetric, there is generally no such l that can be used in (16): sinusoid components of the input suffer different lags as well as amplitude changes. For the purpose of studying phase and amplitude relations between two given series, defined over a common set of consecutive years, we may, without harm to phase differences, apply different symmetric filters to the two series. (Use of different symmetric filters does not affect correlation of amplitudes, though regression coefficients will be changed.) But if an unsymmetric filter is to be applied to either series, then preferably exactly the same filter should be applied to both, so that the lag effects will be the same.

Let us consider now some specific filters and their effect on the sinusoid input (17). The most common kind of unsymmetric filter is this:

$$m = 2, \quad w_0 = -a, \quad w_1 = 1. \tag{21}$$

Here a is a constant, and we may suppose that $|a| \leqslant 1$; for if $|a| > 1$ we may readily infer the behavior of the filter from that of the filter in which a is replaced by its reciprocal. The case $a = 0$ is trivial, and the case $a = ^{-}1$ is special because the filter is then symmetric. We find from (19)

$$G(\lambda) = \sqrt{1 + a^2 - 2a\cos 2\pi\lambda}\ ,$$

and $l(\lambda)$ may be defined, from (18), to be the least positive number l satisfying

$$\tan 2\pi\lambda l = \frac{\sin 2\pi\lambda}{(\cos 2\pi\lambda) - a}\ .$$

If $a = 1$, the filter is a simple first difference, and the above formulas simplify to

$$G(\lambda) = 2\sin \pi\lambda, \quad l(\lambda) = \tfrac{1}{2} + (4\lambda)^{-1}.$$

The following table indicates the behavior of $G(\lambda)$ and $l(\lambda)$ for various values of a. The two values for $l(\lambda)$ shown in parentheses are limits as $\lambda \to 0$ or $\frac{1}{2}$, as the case may be.

The simple first-difference filter ($a = 1$) has been used frequently in analysis of red-noise series. As the gain vanishes when $\lambda = 0$, slow-moving trends are almost eliminated. Some writers have advocated choosing a equal to the first serial correlation coefficient of the given series. If the process were in fact a stationary Markov sequence with first serial correlation coefficient equal to a, the filter (21) would precisely yield a sequence of uncorrelated variables (not necessarily independent, not necessarily Gaussian) with constant spectral density. If the input process had a spectral density that decreased as λ increased, even though not exactly like a Markov density, the spectral density of the output might well be near enough uniform to be handled easily.

Gain and lag for the filter (21)

	$a = 1$		$a = 0.75$		$a = 0.5$		$a = 0$		$a = {}^{-}0.5$		$a = {}^{-}1$	
λ	$G(\lambda)$	$l(\lambda)$	$G(\lambda)$	$l(\lambda)$	$G(\lambda)$	$l(\lambda)$	$G(\lambda)$	$l(\lambda)$	$G(\lambda)$	$l(\lambda)$	$G(\lambda)$	$l(\lambda)$
0.0	0.00	(∞)	0.25	4.00	0.50	2.00	1.00	1.00	1.50	0.67	2.00	0.50
0.1	0.62	3.00	0.59	2.34	0.66	1.73	1.00	1.00	1.43	0.67	1.90	0.50
0.2	1.18	1.75	1.05	1.60	0.97	1.41	1.00	1.00	1.25	0.69	1.62	0.50
0.3	1.62	1.33	1.42	1.28	1.25	1.21	1.00	1.00	0.97	0.73	1.18	0.50
0.4	1.90	1.13	1.67	1.11	1.43	1.08	1.00	1.00	0.66	0.82	0.62	0.50
0.5	2.00	1.00	1.75	1.00	1.50	1.00	1.00	1.00	0.50	1.00	0.00	(0.50)

Let us turn now to symmetric filters, choosing l always by (20). If each coefficient w_j is positive and $\Sigma_j w_j = 1$, the filter is called a *moving average* of length m. From (19), the gain is equal to 1 when $\lambda = 0$, and never exceeds 1 in magnitude for any λ. Usually the gain is small when λ is large, in which case the moving average passes low-frequency input but suppresses high-frequency input. If its coefficients are divided by 2, the filter (21) just considered, with $a = {}^{-}1$, is an instance of a moving average.

Other symmetric filters subtract a symmetric moving average (whose length m is odd) from the input to form a "detrended" series. More generally, a symmetric filter may yield the difference between two symmetric moving averages of the input; the lengths of these moving averages must be either both odd or both even, since otherwise the filter could not be symmetric. The gain function for the difference of two such moving averages is just the difference of the gains of each separately, and the gain for a detrended series is just one minus the gain for the moving-average trend. For understanding of these various kinds of filters, therefore, it suffices to study moving averages.

The following four types of symmetric moving average are of particular interest. The first is the *uniform* filter with equal weights. The second has weights proportional to equal-spaced ordinates of $\sin x$ for x in $(0, \pi)$. The third has weights similarly determined by the function $\sin^2 x$ or $1 - \cos 2x$; they are sometimes called *cosine weights*. The second filter is intermediate in character and effect between the first and the third. There are other possibilities for intermediate filters, and a fourth filter is defined below whose weights are a mixture of uniform and cosine weights. All four filters are identical if $m = 2$. When $m = 5$, for example, the weights are approximately

(i)	0.20	0.20	0.20	0.20	0.20
(ii)	0.13	0.23	0.27	0.23	0.13
(iii)	0.08	0.25	0.33	0.25	0.08
(iv)	0.13	0.25	0.25	0.25	0.13

(i) *Uniform filter.*

$$w_j = m^{-1} \quad (j = 0, 1, \ldots, m - 1).$$

The gain is given by (19), with l determined by (20). If m is large the sum is well approximated by an integral, yielding

$$G(\lambda) \sim \frac{\sin \pi\lambda m}{\pi\lambda m}.$$

(ii) *Sine-weighted moving average.*

$$w_j = \left(\tan \frac{\pi}{2(m + 1)}\right) \sin \frac{\pi(j + 1)}{m + 1} \quad (j = 0, 1, \ldots, m - 1).$$

Here the first factor serves to make the weights sum to 1. We may if we like consider w_j to be defined also for $j = {}^{-}1$ and m, since the above expression then vanishes. The gain is found by substitution into (19). If m is large we may use an integral approximation yielding

$$G(\lambda) \sim \frac{\cos[\pi\lambda(m + 1)]}{1 - 4\lambda^2(m + 1)^2}.$$

(iii) *Cosine-weighted moving average.*

$$w_j = \frac{2}{m + 1} \sin^2 \frac{\pi(j + 1)}{m + 1} = \frac{1}{m + 1}\left(1 - \cos \frac{2\pi(j + 1)}{m + 1}\right)$$
$$(j = 0, 1, \ldots, m - 1).$$

Again we may think of w_j as defined also for $j = {}^{-}1$ and m. The gain is found by substitution into (19). Again there is an integral approximation, yielding

$$G(\lambda) \sim \frac{\sin[\pi\lambda(m + 1)]}{\pi\lambda(m + 1)[1 - \lambda^2(m + 1)^2]}.$$

This is a better-quality approximation than the preceding ones, because not only the integrand but also its first derivative vanish at both ends of the range of integration, and so the first Euler-McLaurin correction term vanishes.

(iv) *Mixed uniform/cosine-weighted moving average.* For $m = 3k - 1$, where k is an integer,

$$\left.\begin{aligned} w_j &= \frac{1}{4k}\left(1 - \cos\frac{\pi(j+1)}{k}\right) && (j = 0, 1, \ldots, k-1), \\ &= \frac{1}{2k} && (j = k, k+1, \ldots, 2k-1), \\ &= \frac{1}{4k}\left(1 - \cos\frac{\pi(j+1-k)}{k}\right) && (j = 2k, 2k+1, \ldots, 3k-2). \end{aligned}\right\}$$

Again w_j may be supposed to vanish for $j = {}^-1$ and m. The integral approximation, of better quality like that for (iii), is

$$G(\lambda) \sim \frac{\sin(3\pi\lambda k) + \sin(\pi\lambda k)}{4\pi\lambda k(1 - 4\lambda^2 k^2)} .$$

The table shows behavior of the gain function for these four types of filter, when $2 \leqslant m \leqslant 5$ (just $m = 5$ for the fourth filter) and also when m is large. To show the asymptotic approximation, λ values are given as multiples of m^{-1} for the uniform weighting and as multiples of $(m+1)^{-1}$ for the other three types of weighting. For the small values of m, a line is drawn below the entry for $\lambda = \frac{1}{2}$. Usually attention should be restricted to frequencies in the interval $(0, \frac{1}{2})$, but if for some purpose frequencies outside that interval need to be considered the table may be extended indefinitely by the relations

$$G(\lambda) = G(-\lambda) = (-1)^{m-1} G(1 - \lambda).$$

As λ increases from 0, zero gain is first attained by the uniform filter at m^{-1}, by filters (ii) and (iv) at $1.5(m+1)^{-1}$, and by filter (iii), if $m \geqslant 3$, at $2(m+1)^{-1}$. For filter (iii) the gain remains less than 0.03 in magnitude when $2(m+1)^{-1} \leqslant \lambda \leqslant \frac{1}{2}$; for most practical purposes the gain is effectively zero throughout that range. Filter (iii), with a somewhat large value for m, makes an attractive trend remover, when its output is subtracted from corresponding elements of the input series. For input frequencies below $0.2(m+1)^{-1}$ the gain is less than 0.03; when $\lambda = (m+1)^{-1}$ the gain is 0.5; when $\lambda \geqslant 2(m+1)^{-1}$ the gain is effectively 1. In order to suppress both low-frequency and high-frequency components of input, corresponding elements in the output of two filters of type (iii), with different m's, may be subtracted.

Filters (ii) and (iv) have quite similar gain functions. In the interval $1.5(m+1)^{-1} \leqslant \lambda \leqslant \frac{1}{2}$, the maximum value of $|G(\lambda)|$ is nearly twice as large

Gains $G(\lambda)$ for symmetric moving average filters

	(i) Uniform						(ii) Sine-weighted					(iii) Cosine-weighted					(iv) Mixed	
	Value of m						Value of m					Value of m					Value of m	
λm	2	3	4	5	∞	$\lambda(m+1)$	2	3	4	5	∞	2	3	4	5	∞	5	∞
0.25	0.92	0.91	0.91	0.90	0.90	0.25	0.97	0.96	0.95	0.95	0.94	0.97	0.96	0.96	0.96	0.96	0.95	0.95
0.5	0.71	0.67	0.65	0.65	0.64	0.5	0.87	0.83	0.81	0.80	0.79	0.87	0.85	0.85	0.85	0.85	0.81	0.81
0.75	0.38	0.33	0.32	0.31	0.30	0.75	0.71	0.64	0.61	0.60	0.57	0.71	0.69	0.69	0.69	0.69	0.60	0.60
1	0.00	0.00	0.00	0.00	0.00	1	0.50	0.41	0.38	0.37	0.33	0.50	0.50	0.50	0.50	0.50	0.38	0.37
1.25		¯0.24	¯0.21	¯0.20	¯0.18	1.25	0.26	0.19	0.17	0.16	0.13	0.26	0.31	0.32	0.32	0.32	0.16	0.16
1.5		¯0.33	¯0.27	¯0.25	¯0.21	1.5	0.00	0.00	0.00	0.00	0.00	0.00	0.15	0.16	0.17	0.17	0.00	0.00
1.75			¯0.18	¯0.16	¯0.13	1.75		¯0.13	¯0.10	¯0.08	¯0.06		0.04	0.06	0.06	0.06	¯0.10	¯0.10
2			0.00	0.00	0.00	2		¯0.17	¯0.12	¯0.10	¯0.07		0.00	0.00	0.00	0.00	¯0.13	¯0.13
2.25				0.14	0.10	2.25			¯0.08	¯0.06	¯0.04			¯0.01	¯0.02	¯0.02	¯0.10	¯0.12
2.5				0.20	0.13	2.5			0.00	0.00	0.00			0.00	¯0.02	¯0.02	¯0.06	¯0.08
2.75					0.08	2.75				0.05	0.02				¯0.01	¯0.01	¯0.02	¯0.04
3					0.00	3				0.07	0.03				0.00	0.00	0.00	0.00
3.25					¯0.07	3.25					0.02					0.01		0.02
3.5					¯0.09	3.5					0.00					0.01		0.02
3.75					¯0.06	3.75					¯0.01					0.00		0.02
4					0.00	4					¯0.02					0.00		0.01

for (iv) as for (ii), and usually the output of (ii) appears smoother than that of (iv). When m is large, $|G(\lambda)|$ is closer to 0 for (iv) than for (ii) at large values of λ, but this advantage of (iv) over (ii) is unimportant because $G(\lambda)$ is so close to 0 in any case for large λ. (There has been widespread reliance on asymptotic results in the literature of time-series analysis. Often, as here, such results are misleading.)

For the purpose of smoothly modifying the spectrum of a given time series, before harmonic analysis or other kind of study, type (iii) moving averages seem more attractive on the whole than types (i), (ii) or (iv). We reduce a part of the spectrum, only in order to see more clearly what remains; a smooth gain function without perceptible wobbles is therefore desirable. There is, in fact, not quite as much difference between the types as appears from the formulas and from the table of gains. To detrend a series by subtracting a moving average of length 3, of any one of the three types, is equivalent to forming second differences of the series; only with length 5 do we reach essentially different detrendings.

Moving averages and other linear filterings are used, not only to detrend and prewhiten series before harmonic analysis, but also to smooth line spectra and to exhibit relations between simultaneous series. Needs and objectives vary. When great smoothness is important, cosine weights are usually satisfactory. When the utmost damping of high frequencies for a given m is needed, uniform weights are appropriate. Filters of type (ii) or (iv) are compromises between these extremes.

We turn now to tapering. We shall see that tapering involves moving-average filtering, not of the input series, but of beats generated in the harmonic analysis. The beats are annoying, we should like to get rid of them, we do not care about modifying them smoothly, and it turns out that the uniform filter (i) is on the whole more effective than (ii) or (iii).

Tapering

A k-element tapering of an n-element series $\{x_t\}$ $(t = 1, 2, \ldots, n)$, where $n \geqslant 2k + 1$ and $k \geqslant 1$, can be expressed in terms of weights $\{W_t\}$ satisfying the conditions

$$W_t = 1 \quad \text{for } k + 1 \leqslant t \leqslant n - k,$$
$$W_t = 0 \quad \text{for } t \geqslant n + 1, \quad \text{or} \quad t \leqslant 0,$$

and for $1 \leqslant t \leqslant k$

$$0 < W_t < 1, \quad W_t < W_{t+1}, \quad W_t + W_{k+1-t} = W_t + W_{t+n-k} = 1.$$

The tapered series $\{y_t\}$, of length $n - k$, is defined by

$$y_t = W_t x_t + W_{t+n-k} x_{t+n-k} \quad (t = 1, 2, \ldots, n - k). \tag{22}$$

Thus the middle $n - 2k$ elements of $\{x_t\}$ appear unchanged in $\{y_t\}$, but the first k and the last k elements of $\{x_t\}$ are combined to form k elements of $\{y_t\}$, which are specified above to be the first k elements, though they could equally well have been the last k, or split, some at the beginning and some at the end of $\{y_t\}$. Tapering is usually thought of differently, namely as a device to make the given series $\{x_t\}$ merge smoothly into zeroes to be catenated at both ends, extending the series to length greater than n; $\{x_t\}$ is replaced by $\{W_t x_t\}$ before the catenations. Here we think of the two ends of $\{x_t\}$ as overlapped or spliced to form $\{y_t\}$, so that the series is reduced in length and circularized. "Circularizing" would have been a better name than "tapering", but we retain the usual term. (Hannan, 1970, calls $\{W_t\}$ a "fader".)

We shall consider the line spectrum of $\{y_t\}$, namely $\{z_r\}$, for $0 \leqslant r \leqslant \frac{1}{2}(n - k)$, at the frequencies $\{r/(n - k)\}$; the definitions are as above, (1)–(4), with $\{y_t\}$ replacing $\{x_t\}$ and $(n - k)$ replacing n.

As usual, we proceed by seeing what happens to a pure sinusoid input, which we shall take to be as at (12) above, with λ and ϵ at first thought of as fixed. If $(n - k)\lambda$ is an integer, $\{y_t\}$ is identical to the first $n - k$ elements of $\{x_t\}$, and only one element of its line spectrum, namely $z_{(n-k)\lambda}$, is positive. If $(n - k)\lambda$ is not an integer, $\{y_t\}$ is not a pure sinusoid, and all elements of the line spectrum are positive. We have

$$a_r = \sqrt{\frac{2d_r}{n - k}} \sum_{t=1}^{n} W_t \cos(2\pi\lambda t + \epsilon) \cos \frac{2\pi r t}{n - k}$$

$$= \sqrt{\frac{d_r}{2(n - k)}} \sum_{t=1}^{n} W_t \left[\cos(2\pi\delta t + \epsilon) + \cos(2\pi\sigma t + \epsilon) \right],$$

where δ is the difference (beat) frequency $\lambda - r/(n - k)$ and σ is the sum frequency $\lambda + r/(n - k)$. Let the differences of $\{W_t\}$ be defined thus:

$$w_j = W_{j+1} - W_j \quad (j = 0, 1, \ldots, k).$$

From the conditions imposed above on $\{W_t\}$, we see that these $k + 1$ differences constitute the weights of a symmetric moving-average filter. Upon performing the operation of "summing by parts", analogous to the more familiar integration by parts, we find

$$a_r = \sqrt{\frac{d_r}{2(n - k)}} \left\{ w_0 \sum_{t=1}^{n} + w_1 \sum_{t=2}^{n-1} + \cdots + w_k \sum_{t=k+1}^{n-k} \right\}$$

$$\left[\cos(2\pi\delta t + \epsilon) + \cos(2\pi\sigma t + \epsilon) \right].$$

Each of these $k + 1$ sums can be evaluated, using the result:

$$\sum_{t=p}^{q} \cos(2\pi\alpha t + \epsilon) = \cos[\pi\alpha(p + q) + \epsilon]\sin[\pi\alpha(q + 1 - p)]/\sin(\pi\alpha), \tag{23}$$

for integers p and q $(q > p)$ and $0 < |\alpha| < 1$. A similar expression can be found for b_r, and hence an expression for $z_r = a_r^2 + b_r^2$.

Now let λ and ϵ be independent random variables, ϵ uniform over $(0, 2\pi)$ and λ distributed over $(0, \frac{1}{2})$ with density $f(\lambda)$. Corresponding to (13) we find

$$\mathcal{E}(z_r) = \tfrac{1}{2}d_r \int_0^{1/2} \{h^*(\lambda - r/(n-k)) + h^*(\lambda + r/(n-k))\} f(\lambda)\, d\lambda, \tag{24}$$

where

$$h^*(x) = (n-k)\left(\frac{w_0 \sin n\pi x + w_1 \sin(n-2)\pi x + \cdots + w_k \sin(n-2k)\pi x}{(n-k)\sin \pi x}\right)^2 \quad \text{when } x \neq 0, |x| < 1,$$

$$= n - k \qquad \text{when } x = 0.$$

The summation in the numerator of $h^*(x)$ is a linear filtering with weights $\{w_j\}$ of the function $\sin[2\pi x(\frac{1}{2}n - t)]$. If $G(\cdot)$ is the gain function of this filter, the summation may be written

$$\sum_{j=0}^{k} w_j \sin(n-2j)\pi x = G(x)\sin(n-k)\pi x.$$

Finally, extending the definition of $f(\lambda)$ as before, we obtain an expression corresponding to (14),

$$\mathcal{E}(z_r) = \tfrac{1}{2}d_r \int_{-\infty}^{\infty} h(\lambda - r/(n-k)) f(\lambda)\, d\lambda, \tag{25}$$

where

$$\left.\begin{aligned} h(x) &= (n-k)\left(\frac{G(x)\sin(n-k)\pi x}{(n-k)\pi x}\right)^2 && \text{when } x \neq 0, \\ &= n - k && \text{when } x = 0. \end{aligned}\right\}$$

This $h(x)$ differs from the $h(x)$ in (14) by, first, substitution of $n - k$ for n and, second, multiplication by the square of the gain function $G(x)$ of the filtering of beats between the input frequencies λ and the Fourier frequencies $r/(n-k)$.

For the special case that $f(\lambda) = n - k$ for λ in the interval $((r_0 - \frac{1}{2})/(n-k), (r_0 + \frac{1}{2})/(n-k))$, where r_0 is an integer, $0 < r_0 < \frac{1}{2}(n-k)$, and

$f(\lambda) = 0$ elsewhere in $(0, \frac{1}{2})$, we find, corresponding to (15),

$$\mathcal{E}(z_r) \approx (n-k) \int_{(r_0-r)-1/2}^{(r_0-r)+1/2} [G(x/(n-k))]^2 \left(\frac{\sin \pi x}{\pi x}\right)^2 dx, \tag{26}$$

for $n-k$ large and $r_0/(n-k)$ not close to 0 or $\frac{1}{2}$. For example, if the taper weights $\{W_t\}$ are the linear weights defined below, so that the $\{w_j\}$ are equal, we have

$$\mathcal{E}(z_r) \approx (n-k) \int_{(r_0-r)-1/2}^{(r_0-r)+1/2} \left(\frac{\sin \pi x \xi}{\pi x \xi}\right)^2 \left(\frac{\sin \pi x}{\pi x}\right)^2 dx, \tag{27}$$

where $\xi = (k+1)/(n-k)$. So for 5% tapering ($\xi = 1/19$) and 10% tapering ($\xi = 1/9$) we find, corresponding to the values previously given for no tapering,

$r - r_0 = \cdots$	¯2	¯1	0	1	2	3	4	5	6	7	8	$\cdots$
For $\xi = 1/19$,												
$\mathcal{E}(z_r) \propto \cdots$	1.38	7.96	78.70	7.96	1.38	0.55	0.29	0.17	0.10	0.07	0.04	$\cdots$
For $\xi = 1/9$,												
$\mathcal{E}(z_r) \propto \cdots$	1.26	7.96	80.13	7.96	1.26	0.43	0.17	0.07	0.03	0.01	0.00	$\cdots$

As before, the values of $\mathcal{E}(z_r)$ have been scaled to sum to 100. The total expected contribution from elements of the line spectrum for which $|r - r_0| \geqslant 4$ is about 1.5% for 5% tapering and about 0.6% for 10% tapering. The tails are much shorter than they were without tapering.

Let us consider now some particular choices for the taper weights $\{W_t\}$, for $1 \leqslant t \leqslant k$. We suppose always that $W_t = 1$ for $k+1 \leqslant t \leqslant n-k$, and $W_t = 1 - W_{k+t-n}$ for $n+1-k \leqslant t \leqslant n$.

(i) *Linear weight function*:

$$W_t = \frac{t}{k+1} \quad (t = 1, 2, \ldots, k).$$

The first differences are

$$w_j = W_{j+1} - W_j = \frac{1}{k+1} \quad (j = 0, 1, \ldots, k).$$

A filter with these weights $\{w_j\}$ is the symmetric filter of type (i) above with $m = k+1$.

(ii) *Cosine weight function* whose first derivative with respect to t vanishes at 0 and $k+1$:

$$W_t = \tfrac{1}{2}\left(1 - \cos \frac{\pi t}{k+1}\right) \quad (t = 1, 2, \ldots, k).$$

The first differences are

$$w_j = \left(\sin \frac{\pi}{2(k+1)}\right) \sin \frac{\pi(j+\frac{1}{2})}{k+1} \qquad (j = 0, 1, \ldots, k).$$

A $(k+1)$-element filter with these weights $\{w_j\}$ has almost (asymptotically for large k) the same effect as the k-element symmetric filter of type (ii) above. (Tukey, 1967, section 20.)

(iii) *Integrated cosine weight function* whose first and second derivatives with respect to t vanish at 0 and $k+1$:

$$W_t = \frac{t}{k+1} - \frac{1}{2\pi} \sin \frac{2\pi t}{k+1} \qquad (t = 1, 2, \ldots, k).$$

The first differences are

$$w_j = \frac{1}{k+1}\left[1 - \left(\frac{k+1}{\pi} \sin \frac{\pi}{k+1}\right) \cos \frac{2\pi(j+\frac{1}{2})}{k+1}\right] \qquad (j = 0, 1, \ldots, k).$$

A $(k+1)$-element filter with these weights $\{w_j\}$ has almost (asymptotically for large k) the same effect as the k-element symmetric filter of type (iii) above.

Applying our knowledge of these filters, we see that the factor $(G(x))^2$ in $h(x)$ in (25) first vanishes, as x moves away from 0 (i.e. as λ moves away from $r/(n-k)$) when $|x| = 1/(k+1)$ if the taper weights are (i), when $|x| = 1.5/(k+1)$ if the taper weights are (ii), when $|x| = 2/(k+1)$ for taper weights (iii). For greater values of $|x|$, taper weights (iii) are the most successful in keeping $h(x)$ almost constantly zero; but even for taper weights (i), $(G(x))^2$ is never greater than 0.06 for $|x| > 1/(k+1)$, if $k \geqslant 5$.

The table below shows some specimen calculations, not with the approximate formulas (15) and (26), but with exact expressions like (13) and (24), for the special case of the random sinusoid whose frequency is uniformly distributed between $(r_0 \pm \frac{1}{2})/(n-k)$. The length n of the series before tapering has been taken to be 65, which is perhaps somewhere near the lower limit at which harmonic analysis is worth while. The expected frequency of the input is always close to 0.25; with no tapering it is 16/65, and with k-element tapering it is

$$\frac{16 - (k/4)}{65 - k};$$

that is, r_0 has been chosen equal to $(n-k-1)/4$. The expected line spectrum is almost perfectly symmetrical about the r_0th harmonic. The first column shows the ordinate at that harmonic, the next five columns refer to the next five higher harmonics, and the seventh column gives the sum of ordinates for all remaining higher harmonics. The same values repeat themselves in reverse order for harmonics below the r_0th, which are therefore not shown in the table. The (constant) amplitude of the random sinusoid input has been adjusted so that its expected sum of squares after tapering, equal to the sum of the expected spectral ordinates, should be equal to 100.

The first row of the table, labeled "$k = 0$, $r_0 = 16$", refers to absence of tapering; the entries differ only slightly from the approximate values given previously just below (15). Then we see results for tapering with $k = 4, 8, 16, 32$, in turn, each time with each of the three choices of taper weights just considered. It will be seen that the greater k is the more the spectrum is concentrated close to the r_0th harmonic. The harmonics themselves become further separated in frequency as k increases; the frequency difference between adjacent harmonics is always $(n - k)^{-1}$, and so about twice as large when $k = 32$ as when there is no tapering.

Expected line spectrum of a random sinusoid

	Ordinate at harmonic no.							Leakage		Equivalent
	r_0	$r_0 + 1$	$r_0 + 2$	$r_0 + 3$	$r_0 + 4$	$r_0 + 5$	$\geqslant r_0 + 6$	var	m.d.	d.f.
$k = 0, r_0 = 16$										
	77.39	7.89	1.42	0.60	0.34	0.22	0.84	2.31	0.46	65
$k = 4, r_0 = 15$										
(i)	79.35	7.98	1.34	0.51	0.24	0.12	0.14	0.74	0.31	60.3
(ii)	78.92	7.97	1.37	0.54	0.27	0.15	0.23	0.91	0.34	60.5
(iii)	78.66	7.96	1.38	0.56	0.29	0.17	0.31	1.10	0.36	60.6
$k = 8, r_0 = 14$										
(i)	81.19	7.89	1.10	0.30	0.08	0.02	0.02	0.47	0.26	55.8
(ii)	80.30	7.95	1.23	0.41	0.16	0.06	0.03	0.57	0.29	56.2
(iii)	79.81	7.97	1.29	0.46	0.20	0.09	0.08	0.68	0.31	56.3
$k = 16, r_0 = 12$										
(i)	85.24	6.99	0.35	0.01	0.01	0.01	0.00	0.32	0.21	46.7
(ii)	83.49	7.50	0.68	0.08	0.00	0.00	0.00	0.39	0.24	47.4
(iii)	82.51	7.69	0.87	0.16	0.02	0.00	0.00	0.46	0.27	47.7
$k = 32, r_0 = 8$										
(i)	94.97	2.45	0.05	0.01	0.00	0.00	0.00	0.22	0.10	28.6
(ii)	91.80	4.10	0.00	0.00	0.00	0.00	0.00	0.32	0.16	29.9
(iii)	89.89	5.04	0.02	0.00	0.00	0.00	0.00	0.40	0.20	30.5

The eighth and ninth columns give summary measures of spread of the spectrum—that is, they measure the leakage. If we think of the ordinates (divided by 100) as probabilities associated with the values of the harmonic number r, the variance and mean deviation of $nr/(n - k)$, that is of n times the frequency, have been found. This random variable r has moments because its value set is finite; it will be recalled that the approximation (15) had long tails and no moments. The variances and mean deviations of leakage confirm the impression formed by examining the spectral ordinates, that for any given k the linear taper weights (i) are more effective in reducing leakage than (ii), which in turn are more effective than (iii).

The last column headed "Equivalent d.f." presents a slight attempt to measure the price we pay for reduced leakage, in reduced ability to estimate the spectral density of the process. How to measure this loss of precision is not obvious. A crude argument is that a tapered series consists of $n - 2k$ members of the original series, together with a further k elements that are averages of pairs of members of the original series, having therefore smaller variance. Both sorts of elements, the unaveraged and the averaged, are treated alike in calculating the line spectrum, and that is presumably not the optimum treatment. A plausible conjecture is that the tapered series will yield spectral estimates that are less precise than would be obtained from an untapered series of length $n - k$, but more precise than what would be obtained from an untapered series of length $n - 2k$.

Consider now a very special situation that may perhaps serve as a guide in more general contexts. Suppose the given series $\{x_t\}$ $(t = 1, 2, \ldots, n)$ is realized from a stationary sequence of independent $N(0, \sigma^2)$ variables. The spectral density is constant, and there is nothing to estimate but σ^2. Suppose the series is tapered, yielding $\{y_t\}$ $(t = 1, 2, \ldots, n - k)$, and the line spectrum $\{z_r\}$ is calculated and averaged—because we know the spectral density is constant, there is nothing else to do with the line spectrum. Then our statistic for estimating σ^2 is

$$\Sigma_r z_r = \Sigma_t y_t^2.$$

This is not the best statistic for σ^2, given $\{y_t\}$, because in it the y's have equal weight, but they are not equally distributed. It is easy to show that

$$\mathcal{E}(\Sigma_t y_t^2) = \left[(n - k) - 2\Sigma_t W_t(1 - W_t)\right]\sigma^2,$$

$$\operatorname{var}(\Sigma_t y_t^2) = 2\left[(n - k) - 4\Sigma_t W_t(1 - W_t) + 4\Sigma_t W_t^2(1 - W_t)^2\right]\sigma^4,$$

where on the left sides the summations for t are from 1 to $n - k$, and on the right sides from 1 to k. If we approximate the distribution of $\Sigma_t y_t^2$ by a constant multiplied by a χ^2 variable with ν degrees of freedom, we may estimate ν as follows:

$$\nu = \frac{2\left[\mathcal{E}(\Sigma_t y_t^2)\right]^2}{\operatorname{var}(\Sigma_t y_t^2)} \approx (n - k) - 4\Sigma_t W_t^2(1 - W_t)^2. \tag{28}$$

Since the maximum possible value for $4W_t^2(1 - W_t)^2$ is $1/4$, this approximate equivalent number of degrees of freedom lies between $n - k$ and $n - (5k/4)$. The right side of (28) is what is shown in the last column of the table above. It will be seen that taper weights (ii) and (iii), which are less effective (for given k) than (i) in reducing leakage, do a little better than (i) in "equivalent d.f.".

Time series of real data are not ordinarily known to have a zero mean, and our series of enrolment and imports residuals have had a linear trend subtracted. The effect of such detrending on estimation of a common σ^2 would presumably be roughly to reduce the right side of (28) by 2.

The above has been a theoretical study of the effect of tapering on the line spectrum of a random sinusoid. Since aggregation of independent random sinusoids yields a stationary Gaussian random sequence, we see the effect of tapering on the closeness with which an empirical line spectrum reflects the spectral density.

Below we shall be interested, not merely in the spectrum of a single time series, but in the relation of two series. One may obtain some idea of the effect of tapering in such a study, by using the random number generator to construct a pair of artificial series having a known relation to each other, and then trying to detect that relation through harmonic analysis, after various degrees of tapering. Even when the spectral densities are flat and tapering is unnecessary to prevent serious leakage in the line spectra, tapering sometimes helps to clean up the phase-difference plot described below, when one series is equal to a multiple of the other series *with a lag*, together with superimposed independent error. The cleaning-up occurs when the length k of the taper is roughly equal to the lag.

To sum up, the following suggestions seem reasonable for tapering to be applied to given series before harmonic analysis:

(a) Let $n - k$ be divisible by any period at which a discrete component of the spectrum of the process is suspected.

(b) If k is small compared with n, the linear weight function (i) should be used.

(c) In comparing two series, k should be as large as (or not much less than) the largest lag to be sought.

(d) Even after prewhitening, some tapering may be advisable to guard against residual unevenness in the spectral density. If n is very small, of the order of 50 or 100, perhaps k should be about one-tenth of n. The larger n is, the larger k should probably be, but perhaps the ratio of k to n should decrease.

(e) If k is chosen larger than as just suggested, in order to reduce the effective length of the series and so the computing cost, the integrated cosine weight function (iii) may be preferred to (i) for its slight improvement in "equivalent d.f.".

The above suggestions have been made with the series presented in this chapter in mind. Those series are all short, ranging in length from about 90 to about 200 terms. The total set of harmonics is small. Tapering, to be effective, must substantially reduce leakage over just a few adjacent harmonics. How perfectly leakage would be prevented at a frequency distance of many harmonics is irrelevant, because the whole width of the spectrum, from frequency 0 to frequency $\frac{1}{2}$, is not a great many harmonics. In other contexts, needs may be different. Some physical processes can be observed at equally spaced times for many thousands of readings. Minute examination of spectra then becomes possible, and there may be primary interest in a scheme of tapering that very strongly reduces leakage over frequency distances of many harmonics. Taper weights (iii), with k a substantial fraction (10% or more) of n, may be preferred to (i) if the spectral density seems to be far from flat and has not been effectively prewhitened. An extreme type of tapering for long series, the "4π prolate spheroidal wave function", has been suggested by Thomson (1977, section 3.1). This weight function is unusual in that it does not satisfy the conditions imposed above on $\{W_t\}$ and so cannot be regarded as a device for circularizing the series.

Spectral analysis of a pair of series

Let $\{x_t^{(0)}\}$ and $\{x_t^{(1)}\}$ be two time series, each defined for some set of consecutive integer values for t, possibly not identical sets. Before doing a harmonic analysis we shall usually filter the series in order to prewhiten them somewhat, observing the precaution that if an unsymmetric filter is applied to either series the same filter is applied to the other series. Possibly also one series may be lagged relative to the other, an integer constant being added onto all the t-values for one series. Extreme members of either series are dropped, for which there is no corresponding member of the other series with the same t-value; both series are then of the same length and have the same set of t-values. Finally each series is tapered in the same way. Let the series so adjusted be denoted by $\{y_t^{(0)}\}$ and $\{y_t^{(1)}\}$, where (say) $t = 1, 2, \ldots, n$. Let $\{z_r^{(i)}\}$ be the line spectrum and $\{\epsilon_r^{(i)}\}$ the phases, and $\{a_r^{(i)}\}$ and $\{b_r^{(i)}\}$ the cosine and sine coefficients, in the harmonic analysis of $\{y_t^{(i)}\}$ ($i = 0$ or 1, $0 \leqslant r \leqslant \frac{1}{2}n$), as defined above in equations (1)–(3).

We shall entertain the possibility that there is some sort of linear regression relation between the two series, possibly of the form

$$y_t^{(0)} = \beta y_{t-l}^{(1)} + \eta_t, \tag{29}$$

where the regression coefficient β and the integer lag l are constants to be estimated and $\{\eta_t\}$ is a stationary process independent of $\{y_t^{(1)}\}$. A more general possibility is that

$$y_t^{(0)} = \Sigma_j \beta_j y_{t-j}^{(1)} + \eta_t, \tag{30}$$

where the β's are coefficients of a linear filter, to be estimated, j ranging over some set of integers, and $\{\eta_t\}$ is a stationary process independent of $\{y_t^{(1)}\}$. If the process $\{y_t^{(1)}\}$ is driving $\{y_t^{(0)}\}$, one might expect that j would range over a set of nonnegative values, say $0 \leqslant j \leqslant m - 1$; but in principle a regression of $\{y_t^{(0)}\}$ on $\{y_t^{(1)}\}$ could involve time shifts in either direction. Equation (30) seems to be the proper formulation of a linear regression relation between two time series, under an assumption that the errors are stationary, corresponding to the usual formulation of regression between two variables, where the errors are treated as independent.

Since, as we have seen, the gain and lag of a filter generally depend on frequency, relation (30) will seem to be frequency dependent. Consider a frequency band, a subinterval of $(0, \frac{1}{2})$, within which the gain G and lag l of the filter with weights $\{\beta_j\}$ are nearly constant; and let $\{y_t^{(0)*}\}$ denote the sum of the harmonic components of $\{y_t^{(0)}\}$ in this frequency band, and $\{y_t^{(1)*}\}$ denote the sum of harmonic components of $\{y_t^{(1)}\}$ in the same band, when $\{y_t^{(0)}\}$ and $\{y_t^{(1)}\}$ are expressed as sums of random sinusoids. Then a relation like (29) will hold, namely

$$y_t^{(0)*} = G y_{t-l}^{(1)*} + \eta_t^*, \tag{31}$$

where $\{\eta_t^*\}$ is independent of $\{y_t^{(1)*}\}$; but the lag l and the gain G will in general depend on the frequency band, and l will not necessarily be an integer.

For an r such that r/n is within the frequency band under consideration, what does (31) imply about the relation between $a_r^{(0)}$, $b_r^{(0)}$ and $a_r^{(1)}$, $b_r^{(1)}$? Ignoring leakage from outside the band, we have (approximately)

$$\begin{aligned} a_r^{(0)} &= \sqrt{d_r/n}\,\Sigma_t y_t^{(0)*}\cos(2\pi rt/n) \\ &= G\sqrt{d_r/n}\,\Sigma_t y_{t-l}^{(1)*}\cos(2\pi rt/n) + \text{error} \\ &= G\sqrt{d_r/n}\,\Sigma_t y_t^{(1)*}\cos(2\pi r(t+l)/n) + \text{error}, \end{aligned}$$

where "error" stands for the corresponding transform of $\{\eta_t^*\}$, which we suppose to be independent of $\{y_t^{(1)}\}$. In the last summation above, t ranges over values $1-l, 2-l, \ldots, n-l$. If l is small compared with n, we should obtain nearly the same result if instead t ranged over $1, 2, \ldots, n$. Thus we have (approximately)

$$a_r^{(0)} = G\{a_r^{(1)}\cos(2\pi rl/n) - b_r^{(1)}\sin(2\pi rl/n)\} + \text{error}. \tag{32}$$

Similarly

$$b_r^{(0)} = G\{b_r^{(1)}\cos(2\pi rl/n) + a_r^{(1)}\sin(2\pi rl/n)\} + \text{error}. \tag{33}$$

If the error terms could be ignored, these relations would imply

$$\left.\begin{aligned} \epsilon_r^{(0)} - \epsilon_r^{(1)} &= -2\pi rl/n \quad \text{if } G > 0, \\ &= \pi - 2\pi rl/n \quad \text{if } G < 0, \end{aligned}\right\} \tag{34}$$

$$z_r^{(0)} = G^2 z_r^{(1)}. \tag{35}$$

(An arbitrary multiple of 2π may be added to the right side of (34).) Thus if a relation of the form (30) exists, with the variance of the first member on the right side (filtering of $\{y_t^{(1)}\}$) not drowned by the variance of the second member (the error process $\{\eta_t\}$), we might hope to detect the relation by plotting $\epsilon_r^{(0)} - \epsilon_r^{(1)}$ against r, and also by scatterplotting $z_r^{(0)}$ against $z_r^{(1)}$—the first plot should show a progressive drift, the second should show correlation. It turns out that the first of these kinds of plots (phase differences against harmonic number) is indeed effective in uncovering a relation of the form (30) between the given series. Moreover, since different r-values do not fall on top of each other, there is no need to make separate plots for different frequency bands; one plot for all frequencies suffices. The other kind of plot, of the line spectra, $z_r^{(0)}$ against $z_r^{(1)}$, turns out to have low sensitivity, because signs are lost when the coefficients $a_r^{(i)}$ and $b_r^{(i)}$ are squared and summed to form $z_r^{(i)}$. In any case this plot should be restricted to r-values in a not-too-broad band, such that the correlation is nearly constant; several plots may be required to show the correlation in different frequency bands. A much more sensitive procedure (but more troublesome) for revealing the correlation within a frequency band is to estimate l in that band from the plot of phase differences, and then, for each value of r in the band, plot the individual harmonic components $a_r^{(0)}$ and $b_r^{(0)}$ as ordinates against the corresponding lag-corrected harmonic components of the other series as abscissas, namely

$$a_r^{(1)}\cos(2\pi rl/n) - b_r^{(1)}\sin(2\pi rl/n)$$

and

$$b_r^{(1)}\cos(2\pi rl/n) + a_r^{(1)}\sin(2\pi rl/n).$$

What we shall actually show in examples below will be (i) the phase-difference plot and (ii) tabulated correlation coefficients between the individual harmonic coefficients of one series and the corresponding lag-corrected coefficients of the other series, for various frequency bands.

It is useful to have a systematic calculation procedure for estimating G and l in (31), for any given frequency band. If we suppose that, over the band, the error process $\{\eta_t\}$ is close to Gaussian white noise, the terms on the right side of (32) and (33) labeled "error" are approximately independent normal variables with zero mean and equal variance; and therefore it would be reasonable to estimate G and l by least squares, minimizing the sum (over the frequencies of the band) of the squared differences between the left sides of (32) and (33) and the first members of their respective right sides. But first a small change in the framing of the question turns out to be helpful. We have supposed that the lag l of the filter $\{\beta_j\}$ appearing in (30) was constant over the frequency band being considered. But in general l will depend on the frequency λ. Let φ be the phase shift of the filter, defined thus:

$$\varphi(\lambda) = 2\pi\lambda l(\lambda). \tag{36}$$

In place of $2\pi rl/n$ as argument of the cosines and sines in (32) and (33), let us put $\varphi(r/n)$, and let us suppose that over the band $\varphi(\lambda)$ is linear in λ. Let r_0/n be a central frequency in the band. We shall set

$$\varphi(\lambda) = A + (\lambda - r_0/n)B, \tag{37}$$

where A and B are constants. Previously we assumed that $\varphi(r/n)$ was exactly proportional to r; now we make the weaker assumption that locally $\varphi(r/n)$ is linear in r, and we now have two parameters, A and B, to describe φ instead of the one before, l. The change makes the estimation problem not only more plausible but more tractable (assuming that l is constant is equivalent to constraining the zero-intercept value $A - (r_0/n)B$ to be a multiple of π, a more troublesome parameter to estimate than an unconstrained zero-intercept). The parameter B is the derivative of φ at r_0/n, and is known in some circles as the "group delay" or "envelope delay" of the filter $\{\beta_j\}$. If $l(\lambda)$ is constant, $B = 2\pi l$, but if $l(\lambda)$ is not constant, B is in general not equal to $2\pi\, l(r_0/n)$.

We therefore frame the least-squares estimation problem as follows. Let the frequency band correspond to $2s + 1$ consecutive integer values for r, centered at r_0, namely $r_0 - s, r_0 + 1 - s, \ldots, r_0 + s$. Let $\{w_j\}$ be a given set of positive weights, for $j = -s, 1 - s, \ldots, s$, satisfying $\Sigma_j w_j = 1$. The w's could be all equal to $1/(2s + 1)$, but other choices may be interesting. We propose that G, A and B be chosen to minimize

$$S = \sum_{r=r_0-s}^{r_0+s} w_{r-r_0}\Big\{\big(a_r^{(0)} - G\big[a_r^{(1)}\cos\varphi(r/n) - b_r^{(1)}\sin\varphi(r/n)\big]\big)^2 + \big(b_r^{(0)} - G\big[b_r^{(1)}\cos\varphi(r/n) + a_r^{(1)}\sin\varphi(r/n)\big]\big)^2\Big\}, \tag{38}$$

where $\varphi(r/n)$ is given in terms of A and B by (37). We find easily that

$$S = C_{00} - 2C_{01}G + C_{11}G^2, \tag{39}$$

where

$$C_{00} = \Sigma_r w_{r-r_0} z_r^{(0)}, \quad C_{01} = \Sigma_r w_{r-r_0}\sqrt{z_r^{(0)} z_r^{(1)}}\cos\big[(\epsilon_r^{(0)} - \epsilon_r^{(1)}) + \varphi(r/n)\big],$$
$$C_{11} = \Sigma_r w_{r-r_0} z_r^{(1)}.$$

S is quadratic in G, and the parameters A and B appear only in C_{01}. Minimization of S can therefore be done in two distinct steps. First the magnitude of C_{01} is maximized with respect to A and B; call the result $\hat{C}_{01}$. Then S is minimized with respect to G at the value $\hat{G} = \hat{C}_{01}/C_{11}$. The minimized S is

$$S_{\min} = C_{00} - \hat{C}_{01}\hat{G} = (1 - R^2)C_{00}, \tag{40}$$

where $R^2 = \hat{C}_{01}^2/[C_{00}C_{11}]$. Clearly $\hat{G}$ has the same sign as $\hat{C}_{01}$, and therefore $\hat{C}_{01}\hat{G}$ and R^2 are necessarily nonnegative. The maximization of C_{01}

may be done by Newton-Raphson iteration using starting values suggested by the phase difference plot of $\epsilon_r^{(0)} - \epsilon_r^{(1)}$ against r. The maximizing A and B are not unique: any multiple of π may be added to A, the sign of $\hat{C}_{01}$ and $\hat{G}$ being changed if the multiple is odd; any multiple of $2n\pi$ may be added to B, and if attention is restricted to values of B less than $n\pi$ in magnitude there may still be more than one value of B yielding a local extremum of C_{01}.

The various terms calculated have individual interpretations. C_{00} is an averaged value of the line spectrum $\{z_r^{(0)}\}$ in the neighborhood of the r_0th harmonic. If $\{y_t^{(0)}\}$ is realized from a stationary random sequence having variance σ_0^2 and spectral density $f_0(\lambda)$ that does not change rapidly, C_{00} may be regarded as an estimate of $\sigma_0^2 f_0(r_0/n)$. Making the approximation that each element of the line spectrum entering C_{00} is independently exponentially distributed with mean $\sigma_0^2 f_0(r_0/n)$, we see that C_{00} is distributed nearly proportionally to a χ^2 variable with number of degrees of freedom $\nu = 2/\Sigma_j w_j^2$. That is, this is the number of degrees of freedom for which a χ^2 variable has the correct ratio of variance to squared mean. So if $\{w_j\}$ is uniform, $\nu = 2(2s + 1)$; and if $\{w_j\}$ is a set of sine weights (filter (ii) above with $m = 2s + 1$),

$$\nu = \frac{2}{s+1}\cot^2\frac{\pi}{4(s+1)} \sim \frac{32(s+1)}{\pi^2}$$

when s is large. Similarly C_{11} is an estimate, with ν equivalent degrees of freedom (approximately), of the variance multiplied by the spectral density at r_0/n of the process generating $\{y_t^{(1)}\}$. For a suitable fixed set of weights $\{w_j\}$, tabulation of C_{00} and C_{11} for successive values of r_0 therefore gives an indication of the behavior of these two spectral densities.

Like C_{00} and C_{11}, $S_{\min}$ is a smoothed estimate of a product of variance and spectral density, this time of the error process $\{\eta_t\}$. We have not, however, been able to work directly with $\{\eta_t\}$, because its values have not been observed; instead we have used residuals from our "harmonic regression" of $\{y_t^{(0)}\}$ on $\{y_t^{(1)}\}$, in which three parameters have been estimated. Had the estimation been of the familiar linear kind, three degrees of freedom would have been used in the estimation, and $S_{\min}$ would have $\nu - 3$ equivalent degrees of freedom. Now if the regression relation between $\{y_t^{(0)}\}$ and $\{y_t^{(1)}\}$ is strong, S can be well approximated by a quadratic function of all three parameters, and the nonlinearity of the estimation can then be ignored in determining standard errors. Some interest, however, attaches to testing significance of the regression effect, that is, to testing the null hypothesis that $G = 0$ and A and B are indeterminate. Here the nonlinearity of the estimation may be more important. Estimation is nonlinear because of the parameter B—were B given or set arbitrarily, S could be expressed as a quadratic function of $G\cos A$ and

$G \sin A$, and then if we had $\{w_j\}$ uniform and if $\{\eta_t\}$ could be assumed to be Gaussian white noise, minimization of S with respect to the two parameters would yield exactly a standard analysis of variance:

$$C_{00} = \hat{C}_{01}\hat{G} + S_{\min},$$

where the two terms on the right were independent sums of squares with 2 and $\nu - 2$ degrees of freedom; $\nu = 2(2s + 1)$. If B is also estimated, it is tempting to conjecture that this result remains nearly true with the degrees of freedom changed to 3 and $\nu - 3$. How correct the conjecture is, the writer does not know.

R^2 is the squared correlation coefficient, within the frequency band, with the weights $\{w_j\}$, between the harmonic components of $\{y_t^{(0)}\}$ and the lag-corrected harmonic components of $\{y_t^{(1)}\}$. To assess significance of the difference of R^2 from 0, the above conjecture suggests treating

$$\frac{(\nu - 3)R^2}{3(1 - R^2)}$$

as an F statistic with 3 and $\nu - 3$ degrees of freedom, or $R^2 = 1/\{1 + (\nu - 3)/3F\}$.

Now let us consider C_{01} further. Two vectors of data enter C_{01}, namely $\left\{\sqrt{z_r^{(0)} z_r^{(1)}}\right\}$ and $\{\epsilon_r^{(0)} - \epsilon_r^{(1)}\}$. The first is evidently a set of weights indicating the informativeness of the second (the phase differences) in estimating the function $\varphi(\lambda)$. The two vectors together constitute what is generally called the (raw) cross spectrum of the two series. When below we plot the phase differences against r, the associated weights will be indicated roughly by the plotting symbols. Six symbols are used,

· ∘ ○ ⊖ ⎕ ⌹

In looking for trends among the plotted points, the viewer lets his eye be guided by the larger symbols.

In a full plot, phase differences are plotted against r for $1 \leqslant r < \frac{1}{2}n$. The phase differences are determined only to withhin a multiple of 2π. To plot them in a fixed interval of width 2π, such as $(0, 2\pi)$, is a little confusing, for we wish to regard values just above 0 as close to values just below 2π, but they are widely separated visually. Representing each phase difference by two values, spaced 2π apart, in the interval $(0, 4\pi)$ permits us always to see values regarded as close represented by plotted points that are close. Representation of phase differences by (say) three points in $(0, 6\pi)$ would sometimes make trends still easier to see, but the two-point representation in $(0, 4\pi)$ is always tolerable.

In the procedure just outlined, no attempt is made to estimate directly the coefficients $\{\beta_j\}$ appearing in (30); what is estimated is their Fourier transform, the gain and lag (or phase shift) functions, from which (if they were precisely known) the β's could be inferred. Although the procedure has been presented unsymmetrically as regression of one series on the other, much of the treatment is symmetric. The phase-difference plot (or cross spectrum) is unaffected by interchange of the two series, except that the signs of phase differences are changed; and therefore R^2 is unaffected by interchange.

One detail of notation should be commented on. In the next chapter we consider regression of Y on X, and regularly in the computer programs the predictor variables X are mentioned before the predicend Y, as the letter X precedes Y in the alphabet and abscissas are customarily mentioned before ordinates. Here it is inconvenient to use different letters of the alphabet for the different series, because there are many derived quantities. Instead the variables are distinguished by numerical superscripts: we consider regression of $\{y_t^{(0)}\}$ on $\{y_t^{(1)}\}$ and later on $\{y_t^{(2)}\}, \ldots$. For better or worse, the predicend series has been numbered before the predictor series, and this order is followed consistently. The difference in convention between the two chapters is regretted.

Harmonic regression of Yale enrolment on U.S. imports

Figures B:8 and B:9 show the first steps to study the interdependence of the Yale enrolment series and the U.S. imports series, which were given in Figures B:2 and B:6. Figure B:8 shows some preliminary computations, and Figure B:9 the phase-difference plot. The latter is of substantive interest; the former is given for two reasons. First, Figure B:8 and the following short explanation indicate what calculations on the data have been carried out, and this any reader may wish to know. Second, a reader interested in doing such calculations in APL, possibly using the programs listed in Appendix 3, may like to see an example.

Before the segment of terminal session reproduced in Figure B:8, the Yale enrolment series was copied into the active workspace as a vector named *ENROLMENT*, and the imports series as a vector named *IMPORTS*. The functions *AUTOCOV* and *FIT* (shown in Appendix 3) were also copied. At the top of the page the length of each vector is checked and the last ten elements are displayed. Then each series is transformed to 100 times the natural logarithm of the data, and residuals are calculated from simple linear regression on the year number, using the function *FIT*. *FIT* causes information to be displayed about the means of the variables and the regression coefficient, together with a conventional estimated standard error for the latter, calculated as though the errors were independent. As we have already noted, Figures B:5 and B:7 show very clearly that the residual variability is not independent, and therefore the calculated standard error is inappropriate and should be ignored. The regression coefficients can be

```
      ⍴ENROLMENT
180
      ¯10↑ENROLMENT
8654 8666 8665 9385 9214 9231 9219 9427 9661 9721

      ⍴IMPORTS
186
      ¯10↑IMPORTS
25542 26812 33226 36043 39952 45563 55563 69476 100997 96940

      Y0←(1795+⍳180) FIT 100×⍟ENROLMENT
MEANS ARE  1885.5 732.28
REGRESSION COEFFICIENT IS 2.357
   WITH ESTIMATED STANDARD ERROR  0.032045   (178 D.F.)

      Y1←(1795+⍳180) FIT 100×⍟6↓IMPORTS
MEANS ARE  1885.5 666.43
REGRESSION COEFFICIENT IS 3.5652
   WITH ESTIMATED STANDARD ERROR  0.070782   (178 D.F.)

      1↓C÷1↑C←6 AUTOCOV Y0
0.86893 0.74754 0.65881 0.58261 0.52415 0.48344
      1↓C÷1↑C←6 AUTOCOV Y1
0.86419 0.72155 0.63222 0.55189 0.48896 0.42609
      0.86*⍳6
0.86 0.7396 0.63606 0.54701 0.47043 0.40457

      ⍴Y0←(1↓Y0)-0.86×¯1↓Y0
179
      ⍴Y1←(1↓Y1)-0.86×¯1↓Y1
179

      )ERASE ENROLMENT IMPORTS FIT AUTOCOV C
      )COPY 1234 ASP2 HARGROUP
SAVED 15:43:32 05/21/77

      ⍴Y0←19 TAPER Y0
19 LINEAR TAPER WEIGHTS WITH INCREMENT 0.05
160
      ⍴Y1←19 TAPER Y1
19 LINEAR TAPER WEIGHTS WITH INCREMENT 0.05
160

      ⍴FT←1 FFT 10 8 2 ⍴Y0
81 2
      ⍴FT←FT,[2.5] 1 FFT 10 8 2 ⍴Y1
81 2 2

      )ERASE Y0 Y1

      PREHAR
PHASE DIFFERENCE PLOT FOR  Y0  AND  Y1 ?  Y OR N
YES
DISTRIBUTION OF SYMBOLS (.∘○⊖⎕⌹):  3 13 24 26 10 3
```

Figure B:8

interpreted as average percentage rates of annual increase, thus about 2.4% for Yale enrolment and 3.6% for imports. When these operations are performed on *IMPORTS*, the first six elements of *IMPORTS* are dropped, so that both series are of the same length, referring to the 180 years from 1796 to 1975. The two residual series are the same (except for a trivial change of scale, and except for the shortening of the import series) as the residuals plotted in Figures B:5 and B:7.

Next, the first six serial correlation coefficients of each series are calculated, using the function *AUTOCOV*. Conveniently, the two sets of serial correlations are not very different from each other, and not very different from the first six powers of an approximate value of 0.86 for the first serial correlation. Therefore to complete the prewhitening process both series are filtered with the unsymmetric two-point filter having weights ($^{-}0.86, 1$). The series are now of length 179. At this point the original series and the two functions just used are no longer needed, and they are erased.

A simpler procedure for prewhitening that might have been followed, after the logarithmic transformation of the data, would have consisted of taking the simple first difference, that is, filtering with weights ($^{-}1, 1$).

Functions to perform the remaining operations of harmonic regression of one "dependent" series on one or more "independent" series have been grouped under the name *HARGROUP*, and these are now copied. (Printouts are given in Appendix 3.) The two series are tapered, using the function *TAPER*. We have suggested that linear tapering by something like 10% is advisable, and that the reduced length of the series should perhaps be divisible by 4 in case the four-year U.S. political cycle should have a perceptible effect. (In fact in this study virtually no evidence has appeared of any such 4-year cycle.) Thus with $n = 179$, values for k of 15 or 19 are suggested, in the notation of (22). We have chosen 19, and the tapered series therefore have length 160. The two series, after these various adjustments, are named $Y0$ and $Y1$, corresponding to our earlier notation $\{y_t^{(0)}\}$ and $\{y_t^{(1)}\}$; they are subsequently referred to by these names.

Next the series are Fourier transformed by a function *FFT* implementing a version of the "fast Fourier transform" algorithm. (*FFT* is based directly on Bloomfield, 1976, chapter 4; instructions for use are given in Appendix 3.) *FFT* can be used to transform either a single series or simultaneously two series of equal length. For no compelling reason, each series is here transformed separately. When *FFT* is applied to $Y0$, the result, called *FT*, is a matrix having 81 rows and 2 columns, the columns being the vectors of cosine and sine coefficients previously denoted by $\{a_r^{(0)}\}$ and $\{b_r^{(0)}\}$. Then *FFT* is applied to $Y1$, and the result is stacked against the transform of $Y0$ to make a three-dimensional array to which the name *FT* is reassigned. Since $Y0$ and $Y1$ are no longer needed, they are erased.

Finally a function *PREHAR* is invoked, to perform some preliminary

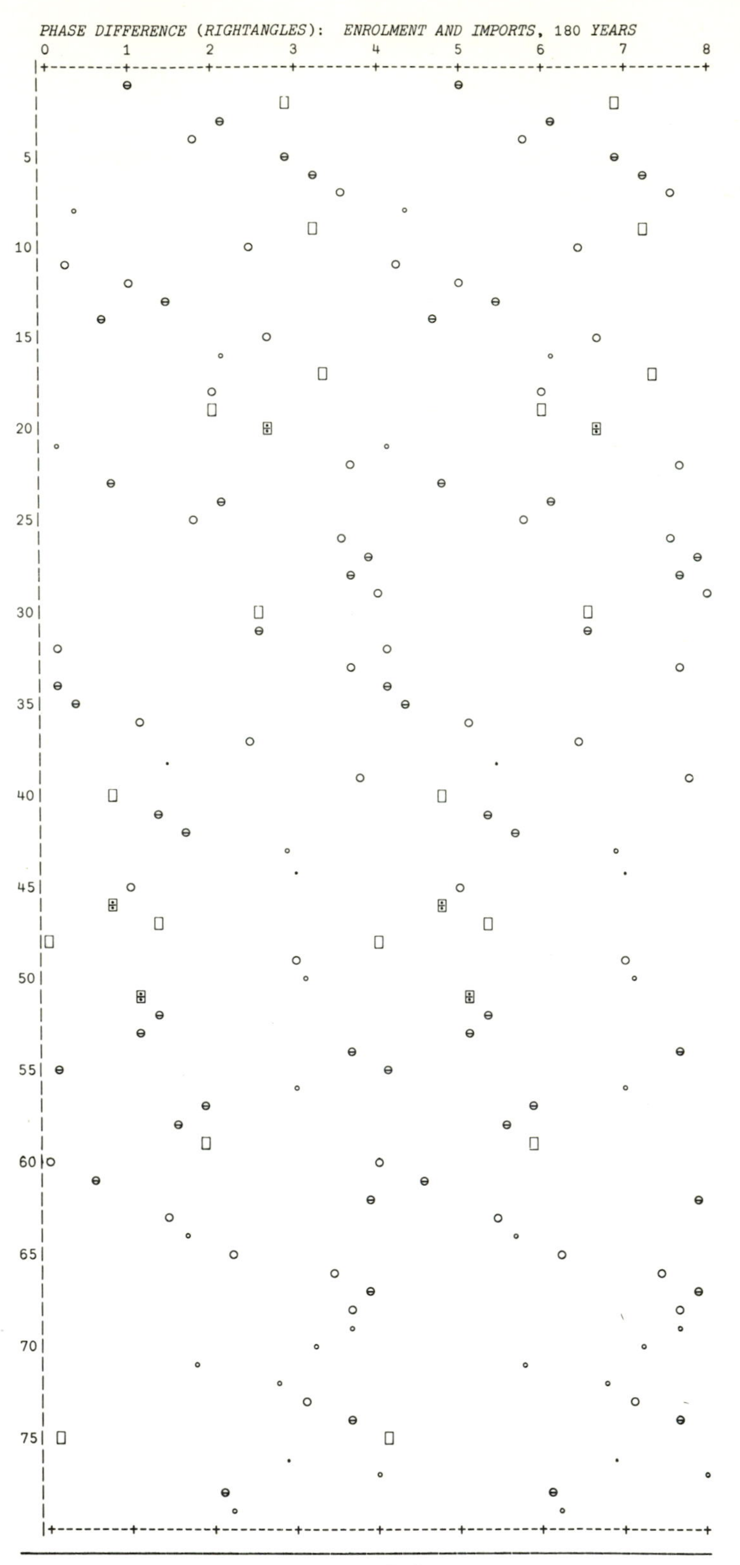
PHASE DIFFERENCE (RIGHTANGLES): ENROLMENT AND IMPORTS, 180 YEARS
0
1
2
3
4
5
6
7
8
5
10
15
20
25
30
35
40
45
50
55
60
65
70
75

Figure B:9

steps of harmonic regression. The function has no explicit argument, but goes to work on the global variable *FT*, which should consist of the Fourier transforms of however many series are to be studied, stacked in a three-dimensional array. *PREHAR* translates each Fourier transform to polar form, that is, to squared amplitudes and phases, and arranges these in a manner convenient for the intended later computations. Then the user is asked if he wants to see the phase-difference plot for each possible pair of series. In the present instance there are only two series, *Y*0 and *Y*1, under consideration, and therefore only one possible pair of series. Upon the answer *YES* by the user, a frequency distribution of symbols in the following plot is shown—the symbols represent size categories of $\{\sqrt{z_r^{(0)} z_r^{(1)}}\}$, as mentioned previously.

The phase-difference plot yielded by *PREHAR*, of $\epsilon_r^{(0)} - \epsilon_r^{(1)}$ against r, based on 180 years' contemporaneous data of Yale enrolment and U.S. imports, is shown separately in Figure B:9. What one looks for in such a plot is trends. At first glance the points seem utterly random and trend-free. The plot could hardly be more unlike the plots of artificial data suggested below in Exercises 15 and 16.

Let us look more carefully at Figure B:9. If the enrolment and imports series imitated each other without any lag in some frequency band, the phase differences should seem to be nearly constant in that band. Since each phase difference is plotted twice in the interval $(0, 4\pi)$, or 0 to 8 rightangles, a relation with zero lag in some frequency band would lead to two vertical arrangements of points in that band with two blank areas between—the sort of effect shown in Figure B:21 at all frequencies. There are only fleeting suggestions of such an effect in Figure B:9. Between the 35th and 55th harmonics, many of the phase differences seem near to 1 (or 5) rightangles. Beyond the 65th harmonic most phase differences are a little less than 4 (or 8) rightangles.

But if the economy, reflected (imperfectly) in the import series, is influencing the enrolment series, there may well be a lag. If the enrolment reflects the imports of several years earlier, the trend in phase differences will be a line (or curve) sloping upwards to the right in Figure B:9. We should therefore look for such sloping trends. Indeed conceivably imports might lag after enrolment and the trend would be upwards to the left, but a priori such an effect is implausible.

Consider a line joining the label "15" (harmonic number) in the left margin to the label "8" (rightangles) in the top margin—that is, a line passing through the centers of these two labels. It passes within one rightangle (of the horizontal scale) of most of the first fifteen plotted points (not numbers 1, 5 and 11, which are missed by nearly two rightangles, and numbers 6 and 15 are barely missed also by one rightangle). Since phases are defined only modulo 2π, the line may be displaced 2π (four rightangles) to the right without changing its meaning. Let it be then continued downward with a slight bend so that it passes through the label "25" in the left margin. Let it again be displaced four rightangles to the right and continued downward to pass through the label "35". This nearly straight line passes within one rightangle of most of the first 35 points. If the line is continued, descending by 10 harmonic numbers per 4 rightangles, it will reach the foot of the diagram (the 80th harmonic) at a total horizontal displacement from where it crossed the top of the diagram (the 0th harmonic) of about 34 rightangles or 17π. The curve seems to pass fairly close to most of the plotted points down to about the 55th harmonic. The suggestion is therefore that at frequencies between the 1st and the 55th harmonics enrolment lags some 17 years after imports—or if account is taken of the slight bend in the curve, the suggested lag is about 20 years at the very lowest frequencies and about 16 years at not-so-low frequencies.

If one looks for a trend curve sloping upwards to the right, among the lower frequencies, the one just described is the most visible—though perhaps visible only to the credulous. If one entertains a trend curve sloping upwards to the left, downwards to the right, a case can be made for a line passing through the label "1" in the top margin and reaching the right side of the diagram (8 rightangles) at about the 24th harmonic. That would imply that imports followed enrolment about 12 years later, at low frequencies. When the points are as scattered as these are, very different trend curves may seem to fit equally well, and none really well.

To clinch and assess the suggestions of the phase-difference plot, the calculation procedure outlined above may be resorted to. The computer implementation is through functions *HARINIT* and *HAR*1. First the set of weights $\{w_j\}$ must be specified (as a global variable named W). There are two decisions to be made: (a) the pattern of the weights—should they be uniform, or sine weights as in the type (ii) moving average described above, or what? (b) the number of weights—how large should s be? The equivalent degrees of freedom ($\nu = 2/\Sigma_j w_j^2$) are increased if (a) we change from sine weights to uniform weights, with s constant, and if (b) we increase s, with the pattern of weights constant. As a general rule, the estimated coefficients $\hat{G}$, $\hat{A}$, $\hat{B}$ and the multiple R^2 depend noticeably on the weights. The greater ν is, the smaller R^2 usually turns out to be. However, for testing significance of R^2, the critical values for R^2 decrease as ν increases, and experimentation suggests that judgments of significance do not depend so much on the choice of the weights. Smoothness in the estimates, as r_0 changes, is desirable, both for plausibility and for the sake of convergence of the computational procedure. All the computations shown here have been made with sine weights. As for how many weights, one does not want s to be so large that interesting detail is blurred away, but neither does one want the smoothed spectral estimates to be subject to high sampling error. A possible rule of thumb is that ν ought not to be much less than 40, and should be higher if n is large enough. The computation shown in Figure B:10 below is based on a 29-point sine-weighted moving average, for which $\nu = 48.5$. We shall also try a 23-point moving average.

HARINIT calculates the smoothed spectral estimates for the two given series, denoted by C_{00} and C_{11} in (39). A global matrix *SSE* (smoothed spectral estimates) is formed, of which the first column lists the possible values for r_0 and the second and third columns the corresponding values for C_{00} and C_{11}. The remaining four columns are at first filled with dummy values, but later the function *HAR*1 inserts in these columns estimated values for (respectively), G, A, B and R^2.

To run *HAR*1, we specify what value for r_0 is to be tried first, and whether subsequent values for r_0 are to be stepped forwards or backwards; and a trial value for B is specified. The iterative procedure involves calculating A for the given B, then performing one step of Newton-Raphson iteration to improve B, then back to A again, and so on until either a preselected number of iterations has been performed or the change in B is less than a preselected tolerance. Finally, G is calculated; and then the value for r_0 is stepped and the procedure is repeated with the last value for B as the new trial value. What happens depends on where one begins. If one starts at the minimum value for r_0, namely 14, with a trial value for B corresponding to a lag of $^{-}12$ (B is 2π times this, say $^{-}75$), and if r_0 is stepped forward by 1, the procedure converges reasonably well (the poorest convergence is for r_0 between 55 and 58) all the way to $r_0 = 66$. But if one starts at $r_0 = 14$ with a trial value for B corresponding to the more plausible lag of 20 (say $B = 120$), and r_0 is stepped forward by one, the computation proceeds smoothly as far as $r_0 = 30$, but thereafter becomes unstable until it locks into the negative values for B previously found.

These two attempts to run *HAR*1 are summarized in Figure B:10. The first seven columns, going all the way down the page, are a print-out of *SSE* when one looks for a negative lag at low frequency (imports following Yale enrolment). As already mentioned, the successive columns show r_0, C_{00}, C_{11}, $\hat{G}$, $\hat{A}$, $\hat{B}$ and R^2, in the notation of (37)–(40) above. The four columns on the right, going only part way down the page, show $\hat{G}, \hat{A}, \hat{B}$ and R^2, respectively, obtained from a run of *HAR*1 looking for positive lag at low frequency (Yale enrolment following imports). The minimized sum of squares, $S_{\min}$, is not tabulated, but could be calculated from (40) as $(1 - R^2)C_{00}$.

At the three lowest values for r_0 ($14 \leqslant r_0 \leqslant 16$), the larger R^2 occurs at the negative values for B (the apparent lag, namely $B/2\pi$, is about $^{-}10$ years). Then for $17 \leqslant r_0 \leqslant 25$ the larger R^2 occurs when B is positive (corresponding to a lag of about 15 years). Thereafter negative lags win again. The greatest R^2 is for $r_0 = 46$ (apparent lag about $^{-}1$ year). At the

Print-out of *SSE* for 29-point sine-weighted moving average

14	210	795	.201	4.49	¯65.7	.153	.179	5.17	115.4	.121
15	210	803	.191	4.14	¯63.2	.139	.174	5.84	107.9	.116
16	210	808	.178	3.78	¯61.6	.122	.178	.19	100.2	.121
17	211	811	.164	3.39	¯61.7	.103	.188	.82	94.9	.137
18	213	824	.148	3.08	¯57.5	.084	.192	1.43	92.8	.143
19	213	842	.134	2.85	¯50.2	.071	.190	2.00	91.7	.143
20	213	863	.128	2.65	¯39.5	.066	.184	2.59	91.9	.137
21	215	879	.133	2.38	¯32.5	.072	.178	3.20	91.8	.130
22	220	888	.141	2.14	¯30.1	.080	.176	3.80	90.7	.125
23	223	893	.146	1.93	¯29.2	.086	.172	4.39	90.0	.119
24	224	895	.151	1.74	¯28.8	.091	.168	4.98	89.2	.112
25	224	899	.150	1.55	¯28.1	.090	.161	5.55	89.7	.104
26	224	906	.152	1.36	¯27.4	.093	.149	6.09	88.9	.090
27	223	907	.154	1.15	¯27.4	.096	.134	.36	88.8	.073
28	223	903	.157	.95	¯28.0	.100	.123	1.18	97.4	.061
29	222	894	.161	.76	¯28.8	.105	.117	1.94	105.7	.055
30	220	880	.165	.58	¯30.4	.108	.118	2.64	112.2	.056
31	217	859	.169	.41	¯31.7	.113				
32	216	849	.173	.26	¯31.0	.118				
33	214	857	.175	.09	¯29.0	.123				
34	214	862	.175	6.19	¯26.1	.123				
35	215	868	.177	6.04	¯24.6	.126				
36	216	885	.180	5.90	¯23.6	.133				
37	218	900	.187	5.77	¯21.7	.144				
38	221	909	.196	5.65	¯19.8	.158				
39	222	919	.204	5.52	¯17.8	.172				
40	225	927	.207	5.40	¯16.2	.177				
41	226	934	.207	5.29	¯14.2	.176				
42	226	932	.208	5.18	¯11.4	.178				
43	225	935	.212	5.06	¯8.1	.187				
44	226	936	.220	4.98	¯6.5	.200				
45	227	941	.227	4.91	¯6.8	.214				
46	229	943	.229	4.85	¯6.6	.216				
47	230	947	.227	4.80	¯5.7	.211				
48	230	959	.220	4.77	¯4.1	.202				
49	231	965	.214	4.75	¯2.9	.192				
50	231	967	.209	4.73	¯2.9	.182				
51	234	962	.204	4.69	¯4.0	.171				
52	237	950	.198	4.65	¯4.9	.157				
53	238	934	.188	4.64	¯4.6	.139				
54	238	915	.176	4.67	¯3.2	.119				
55	236	891	.162	4.75	¯.7	.099				
56	234	867	.150	4.89	2.7	.084				
57	232	838	.141	5.03	5.7	.072				
58	229	803	.133	5.18	8.8	.063				
59	224	763	.126	5.35	12.6	.054				
60	218	720	.123	5.48	16.5	.050				
61	214	675	.129	5.49	19.3	.052				
62	211	637	.137	5.54	20.7	.057				
63	207	612	.149	5.61	21.2	.066				
64	206	589	.155	5.73	22.1	.068				
65	204	566	.159	5.84	24.3	.070				
66	200	544	.164	5.91	28.5	.073				

Figure B:10

highest frequencies R^2 is very small and B goes positive (up to a lag of about 4 years).

Should any of these values for R^2 be judged significantly different from zero? They are very highly correlated and will therefore seem to be significant or non-significant in consecutive bunches. 5%, 1% and 0.1% values for R^2 calculated from tabulated F values (with 3 and 45.5 degrees of freedom), as tentatively suggested above, are 0.16, 0.22 and 0.30. Perhaps, therefore, the relatively high correlation when r_0 is in the 40's should be rated mildly significant, but neither the negative nor the positive lag at low frequencies.

The reader may like to see the effect of shortening the extent of the moving average $\{w_j\}$, and may also like to see the running of *HARINIT* and *HAR*1. The top of Figure B:11 is a brief segment of terminal session. In the first line, a 23-point sine-weighted moving-average filter is defined, and the sum of the weights is checked. Then *HARINIT* is executed. Its argument 0 1 signifies that we are interested in regression of $Y0$ on $Y1$, a matter about which there is little choice, since no other series are available. In its output, *HN* stands for $\frac{1}{2}n$, and S stands for what we have previously called s. The number of equivalent degrees of freedom is stated. The last line concerning *CC* should be ignored unless the programming is of interest. The user then specifies the maximum number of iterations and the tolerance for change in B, needed in *HAR*1. *HAR*1 is executed, with 120 as first trial value for B, and r_0 first set equal to 11, stepping by 1 up to 13. The output comes in pairs of lines, the first line giving the number of iterations performed and the last change in B, the second line giving the row of *SSE* that has just been completed. The rest of this terminal session, exploring the whole of *SSE*, is not reproduced, but the lower part of Figure B:11 shows a partial print-out of the result, in the same style as Figure B:10. To save space, harmonics 31–39 and 51–69 have been omitted. The calculation with positive lag at low frequency becomes unstable after the 28th harmonic, and so the print-out stops there.

The values for A and B are mostly quite close to those in Figure B:10; we have gained three extra values at the beginning. The values for G and R^2 are mostly rather higher. 5%, 1% and 0.1% significance values for R^2, calculated as before, are 0.19, 0.27 and 0.36, and so the appearance of significance is much the same as before.

The estimated values for G yielded by *HAR*1 are always nonnegative, and one cannot tell at a glance whether the two series, in any frequency band, are positively or negatively correlated. Sometimes there is no single answer. Sometimes one can say, about equally well, that in a frequency band the two series are positively correlated at one lag, and negatively correlated at another lag. We saw in (34) that if throughout a frequency band one series were exactly proportional to the other series at some lag, the

```
      +/W←(3○○÷48)×1○(⍳23)×○÷24
1

      HARINIT 0 1
HN = 80
S = 11
EQUIVALENT D.F. = 38.796
COLUMNS (CC) OF  ZZ  AND  DE  = 1

      ITS←5
      TOL←0.1

      120 HAR1 11 1 13
3 ¯0.015022
 11   207   765  .210  2.98 122.8  .163
3 0.010233
 12   207   782  .216  3.75 120.2  .175
2 0.097042
 13   207   790  .217  4.53 117.8  .180
```

```
  Print-out of parts of SSE for 23-point sine-weighted moving average
11   207   765 | .244   5.73  ¯66.8   .220 | .210   2.98  122.8   .163
12   207   782 | .238   5.30  ¯67.8   .214 | .216   3.75  120.2   .175
13   207   790 | .229   4.87  ¯68.4   .200 | .217   4.53  117.8   .180
14   207   792 | .217   4.44  ¯70.8   .181 | .213   5.31  113.9   .173
15   211   788 | .204   3.99  ¯71.1   .155 | .211   6.03  111.3   .167
16   212   805 | .187   3.58  ¯68.9   .133 | .204    .40  107.3   .158
17   212   827 | .169   3.26  ¯62.9   .111 | .195    .97  102.1   .149
18   213   842 | .160   2.96  ¯54.5   .101 | .187   1.53   95.9   .138
19   218   851 | .154   2.65  ¯47.2   .092 | .190   2.12   89.9   .142
20   221   858 | .149   2.41  ¯38.5   .087 | .200   2.70   86.0   .156
21   222   876 | .150   2.22  ¯31.8   .089 | .204   3.27   83.1   .163
22   222   900 | .145   2.07  ¯28.0   .086 | .198   3.80   81.9   .159
23   220   922 | .144   1.90  ¯25.1   .088 | .187   4.33   82.1   .146
24   221   932 | .146   1.70  ¯24.0   .090 | .176   4.90   83.2   .130
25   223   930 | .148   1.51  ¯23.9   .091 | .168   5.47   84.2   .118
26   225   920 | .150   1.34  ¯24.1   .092 | .159   6.05   87.1   .103
27   225   911 | .147   1.19  ¯25.8   .088 | .152    .36   92.1   .094
28   224   892 | .146   1.04  ¯26.8   .085 | .147    .97   96.7   .086
29   222   885 | .150    .88  ¯28.6   .089
30   219   889 | .150    .68  ¯29.6   .091

40   219   892 | .208   5.27  ¯15.6   .176
41   220   911 | .222   5.13  ¯11.4   .203
42   222   920 | .240   5.02   ¯8.5   .240
43   226   928 | .248   4.95   ¯5.9   .253
44   229   939 | .250   4.89   ¯2.0   .257
45   228   960 | .248   4.86    1.9   .260
46   226   975 | .245   4.84    4.6   .259
47   226   986 | .240   4.84    3.5   .251
48   231  1000 | .232   4.80   ¯1.9   .233
49   236  1007 | .226   4.74   ¯5.0   .219
50   238  1012 | .219   4.70   ¯6.1   .204
```

Figure B:11

phase differences would vary linearly with frequency, and the zero-intercept of this linear relation would be 0 (mod 2π) if the correlation were positive and π (mod 2π) if the correlation were negative. Thus one looks for a linear approximation to the trend in phase differences having zero-intercept of 0 or π (mod 2π), and the zero-intercept determines the sign of the correlation. If one wants to exhibit the correlation by lagging one series relative to the other, then filtering the series to favor the frequency band, and scatterplotting, the slope of the linear approximation should correspond to an integer lag.

In each frequency band, centered at r_0, the estimated A and B specify a linear approximation to the phase-shift curve, a chord, nearly a tangent. It will be noticed that B, multiplied by the frequency difference between consecutive harmonics, namely $1/n$ (here 1/160), is nearly equal to the change in A as r_0 increases by 1. Suppose that one plots the value of A given in column 5, or in column 9, of the print-out of *SSE*, either in Figure B:10 or in Figure B:11, against r_0 (given in column 1), with judicious increments of 2π wherever there appears to be a discontinuity. The points lie on a fairly smooth curve whose tangent at each point has slope nearly equal to the corresponding B. Where the curve is nearly straight, the A-values given in Figure B:10 are almost the same as the corresponding values in Figure B:11; but where the curve is less straight, the values in Figure B:11 tend to have the greater curvature, or more wobble.

For the frequencies where R^2 is largest, with r_0 between (say) 41 and 50, the trend in phase differences can be approximated, though not very closely, by a line whose zero-intercept is 0 (mod 2π) and slope corresponds to lag $^{-}1$. The total range of frequencies comprised in the frequency bands centered at values of r_0 between 40 and 50 is something like 25/160 to 65/160. This part of the phase difference plot therefore suggests that if the original series of enrolment and imports are trimmed to the same length, so that the imports refer to one year later than the enrolment—that is, enrolment from 1796 to 1974 and imports from 1797 to 1975—and then if the logarithms of both series are filtered to favor frequencies in the range 25/160 to 65/160, a scatterplot of the resulting series against each other should show positive correlation. This is tried below.

We may similarly consider the low frequencies, with r_0 below (say) 22, where R^2 is also substantial, though perhaps not large enough to be judged significant. For the more plausible positive lag, the trend in phase differences can be well approximated by a line with zero-intercept π (mod 2π) and slope corresponding to lag 15, for r_0 above 15. But there is a suggestion of curvature in the trend at the lowest values for r_0, especially when we use the A and B values from Figure B:11 (columns 9 and 10). If A and B could be observed for r_0 below 11, a reasonable conjecture is that the trend would bend a little further and have zero-intercept 0 (mod 2π), in which case in the neighborhood of the 15th harmonic the correlation between the series would be positive at lag about 20. The suggestion is therefore that at low frequencies we could look for positive correlation if enrolment were lagged 20 years after imports, or negative correlation if enrolment were lagged 15 years after imports. Similar considerations apply to negative lags at low frequency—negative correlation at lag ¯9 or positive correlation at lag ¯15 or so.

The plausibility of these various lags deserves thought. It is not unreasonable that there should be a detectable simultaneous association between "the state of the economy" and the Yale enrolment. If the state of the economy means how businessmen perceive and assess the general economic condition and the economic prospect for the near future, such perception and assessment might well be indicated by actual financial transactions at a later date. The value of imports in any year is probably strongly influenced by the economic climate just before that year, since orders take time to be issued, filled and shipped. A correlation between Yale enrolment and U.S. imports of the following year is therefore not implausible, and does not require us to suppose that Yale enrolment directly influences future U.S. imports. That Yale enrolment should follow U.S. imports by as much as 15 or 20 years at low frequencies does seem surprising, though not utterly incredible. Increases in endowment or availability of capital funds may lead to planning and construction of new buildings, permitting increases in the student body—this chain of events will sometimes be quite slow to complete. That U.S. imports should follow Yale enrolment by 9 or 15 years seems at first glance even more surprising, though again not utterly incredible. Yale enrolment is no doubt correlated with enrolment in other institutions of higher learning, and fluctuations in the rate at which graduates are released from these institutions could conceivably have some detectable effect on the national economy at a later date—or is that just a teacher's fantasy?

Let us now look at plots "in the time domain" to verify the results of the harmonic analysis. First we consider middle-frequency filtering of enrolment (1796–1974) and imports of the following year (1797–1975). The original series of enrolment (with the last reading dropped) and imports (with the first seven readings dropped) have been prewhitened as before by the successive steps of (i) taking logarithms, (ii) subtracting a linear trend, (iii) filtering by the two-point filter (¯0.86, 1). Then middle frequencies have been emphasized by filtering both series with a 12-point filter whose weights are proportional to the coefficients in the polynomial $(1 - u)^7(1 + u)^4$. This filter produces the effect of seven first-differencings, to suppress low frequencies, together with four symmetric two-point averagings, to suppress high frequencies; the gain function is greatest near frequency 0.29, and falls to about one-tenth of its maximum value at frequencies 0.15 and

0.42. The filtered series have correlation coefficient 0.40, which may be compared with the maximum value of 0.156 for R^2 for 39-point sine weights in the harmonic regression analysis, achieved when the central frequency r_0 is the 44th harmonic. A higher correlation coefficient could be obtained by more drastic filtering, corresponding to the higher values for R^2 obtained with shorter sets of sine weights, as shown above.

The filtered series are plotted against each other in Figure B:12. The decade of the enrolment is indicated by the letters given on the right side of Figure B:2—*B* for 1800–09, *C* for 1810–19, etc. Stars indicate coincident points. The scales are not marked, because the values of the filtered series have little interest; but zero is shown on each axis.

Much of the appearance of correlation is caused by outlying points, which are all labeled *C* or *M* or *P*. These were the decades of the war of 1812, the first world war and the second world war. The imports series reacted strongly to the first of these wars, the enrolment series reacted strongly to the other two. If all points in the three decades are omitted, the remaining points have correlation coefficient 0.20, which is surely not significantly different from zero.

An effect that may be looked for in Figure B:12 is a progressive change

0
0
Enrolment vs. imports of the following year: middle-frequency filtering.

Figure B:12

with time in the relation between the two series. Indeed the regression relation suggested by the *C*-points is quite different from that suggested by the *M*-points or the *P*-points, but what about the more tranquil decades? Unfortunately many of the remaining points appear as stars, and further plotting (not reproduced) is needed to separate them—the same sort of plot can be made on a larger scale, and partial plots on the same scale can be made for the earlier decades and for the later decades. It appears that there has been little association between the series (as filtered) in most decades, and no very clear progressive change.

The middle-frequency association between the series therefore inheres mainly in their reactions to three particular times of war. Even if we accept the association as statistically significant, it is unsatisfactory as evidence that the state of the national economy affects Yale enrolment; rather it indicates that both are affected by wars.

As for low-frequency association between the series, our harmonic analysis suggested either a large positive lag of 15 or 20 years (enrolment following imports) or a large negative lag of 9 or 15 years (imports following enrolment). We explore only the first possibility. One may repeat the plot and calculations shown in Figures B:9–11 with the series suitably displaced at the outset, so that the estimated lag will appear smaller. In our treatment is an assumption that lags are small, and the values of R^2 obtained are presumably more reliable when the phase-shift derivative B is near zero. So the calculation has been repeated for the enrolment series with the first 9 members dropped (representing 1805 to 1975) and the imports series with the last 15 members dropped (1790 to 1960); enrolment is therefore initially lagged 15 years after imports, and the series are of length 171. They have been prewhitened as before (to length 170), and then tapered by 14 to final length 156. The phase-difference plot suggests a positive lag of about 6 years at low frequencies (between the first and twentieth harmonics), but thereafter the lag seems close to zero. The values of *SSE* corresponding to columns 8–11 in Figures B:10–11 look rather similar, except that the values of B are reduced by roughly 2π times the initial lag of 15. All the values of R^2 are now a little smaller than before. For 29-point sine weights, the maximum R^2 is now 0.113 (for $r_0 = 17$), instead of the value 0.143 shown in Figure B:10; and for 23-point sine weights, the maximum R^2 is now 0.151 (for $r_0 = 13$), instead of the value 0.180 shown in Figure B:11. Realigning the series has not helped the appearance of association. The estimated phase-shift curve suggests that we may look for positive association at the lowest frequencies at a lag of about six years, or negative association at slightly higher frequencies at zero lag—these lags being additional to the fifteen-year lag already introduced.

In any of these plots and calculations, how strong the correlation is depends on just what filtering is done. It turns out that at very low frequencies we get greater correlation in the time domain with less prewhitening. Figure B:13 is one possible low-frequency analog of Figure B:12. Enrolment (1811 to 1975) has been lagged 21 years after imports (1790 to 1954); logarithms have been taken and the linear trend subtracted, but the usual prewhitening step of filtering with weights (¯0.86, 1) has been omitted. High frequencies have then been suppressed with a 13-point cosine-weighted moving average. The resulting series are plotted against each other in Figure B:13; their correlation coefficient is 0.44. There was no particular reason for choosing the length 13 for the cosine-weighted moving average; the correlation coefficient between the filtered series increases as the length of the moving average increases, but the frequency band narrows and if one were trying to assess significance the critical levels for the correlation coefficient would increase. With very low frequencies strongly present in the filtered series, the plot has evident dependence between temporally adjacent points. Starting with the three points marked *C*, one can easily trace (with occasional small uncertainties) the whole sequence of 153 points in order, finishing with the ten *R*'s—this sort of effect is unusual in scatterplots.

At the bottom left are the *H*'s—enrolment in the 1860's was low (see Figure B:5), as were imports 21 years earlier (see Figure B:7). At the top right are *C*'s and *D*'s—enrolment in the 1820's was high, as also imports 21 years earlier. The *N*'s at the top left spoil the appearance of association—enrolment in the 1920's was high, but imports 21 years earlier were low. And so on; one sees which decades conform with the positive correlation, and which do not. Figure B:13 shows (the author believes) about as interesting an association at low frequencies as can be obtained by lagging and filtering the series. It is very dubious whether any claim of statistical significance should be made.

To sum up, we have tried to detect influence of the state of the national economy on enrolment at Yale University, by relating the enrolment series to the value of general imports of merchandise. We have systematically correlated the series throughout the entire spectrum of frequencies. We have found some suggestions of association, but nothing compelling.

Other economic series

Perhaps our failure to achieve decisive positive results is due to an unfortunate choice of economic indicator. Perhaps the value of imports does not well reflect the features of the economy affecting Yale enrolment.

Economic series going back as far as 1800 and still continued today, that might plausibly reflect the state of the economy, are not abundant. One that has been considered is a series of total revenue from federal taxation—

Enrolment vs. imports of 21 years before: low-frequency filtering.

Figure B:13

mostly customs duties in the earlier years, mostly income tax in recent times. The series is no doubt responsive to the state of the economy—it shows a proper dip in the early 1930's—but its most dramatic movements are very large permanent increases in three times of war, the American civil war and the two world wars, reflecting altered responsibilities of the federal government. Neither the Yale enrolment series nor the imports series reacted in that way to those wars. The revenue series is not promising material for present purposes.

If one is content with series going back less far, the field of choice is wider. Two such series are presented in Figure B:14. The first gives the wholesale price of butter in cents per pound at New York. It is remarkable among price series for its length; it runs (with, admittedly, some changes in definition) from 1830 to 1970 as series K 608 in *Historical Statistics of the U.S.*, and is updated to 1975 from the *Statistical Abstract.* Apparently no value is available for 1973; that shown has been inserted by linear interpolation between 1972 and 1974. The other series is N 156 in *Historical Statistics*, updated. It gives the total number of new housing units started each year, in thousands, from 1889 onwards. This ought, on the whole, to be a good indicator of businessmen's perception of immediate economic

WHOLESALE PRICE OF BUTTER AT NEW YORK (CENTS PER POUND)

+	0	1	2	3	4	5	6	7	8	9	
1830	13.9	14.9	15.2	15.8	14.4	19.2	23.9	21.6	23.4	22.9	*E*
1840	17.4	18.6	16.5	13.3	15.2	17.7	16.7	20.7	20.1	18.9	*F*
1850	19.6	18.4	23.6	23.0	23.0	26.4	25.8	25.7	23.8	23.9	*G*
1860	21.9	19.4	20.9	28.2	43.7	39.8	42.7	34.8	44.7	43.3	*H*
1870	38.1	33.6	32.0	35.4	36.2	32.8	31.3	28.5	27.3	24.2	*I*
1880	30.5	31.8	35.6	31.2	30.3	26.6	26.8	26.7	27.5	24.4	*J*
1890	23.7	26.2	26.3	27.1	23.0	21.2	18.5	19.0	19.6	21.3	*K*
1900	22.2	21.4	24.7	23.4	21.7	24.6	24.6	28.1	27.6	29.9	*L*
1910	31.1	27.9	31.6	32.2	29.8	29.8	34.0	42.7	51.5	60.7	*M*
1920	61.4	43.3	40.6	46.9	42.6	45.3	44.4	47.3	47.4	45.0	*N*
1930	36.5	28.3	21.0	21.6	25.7	29.8	33.1	34.4	28.0	26.0	*O*
1940	29.5	34.3	40.1	44.8	42.2	42.8	62.8	71.3	75.8	61.5	*P*
1950	62.2	69.9	73.0	66.6	60.5	58.2	59.9	60.7	59.7	60.6	*Q*
1960	59.9	61.2	59.4	59.0	59.9	61.0	67.2	67.5	67.8	68.5	*R*
1970	70.4	69.3	69.6	[67.0]	64.4	81.8					*S*

TOTAL NUMBER OF NEW HOUSING UNITS STARTED (THOUSANDS)

+	0	1	2	3	4	5	6	7	8	9	
1880										342	*J*
1890	328	298	381	267	265	309	257	292	262	282	*K*
1900	189	275	240	253	315	507	487	432	416	492	*L*
1910	387	395	426	421	421	433	437	240	118	315	*M*
1920	247	449	716	871	893	937	849	810	753	509	*N*
1930	330	254	134	93	126	221	319	336	406	515	*O*
1940	603	706	356	191	142	326	1023	1268	1362	1466	*P*
1950	1952	1491	1504	1438	1551	1646	1349	1224	1382	1536	*Q*
1960	1283	1354	1487	1635	1561	1510	1196	1322	1545	1500	*R*
1970	1469	2085	2379	2057	1352	1171					*S*

Figure B:14

prospects. Series N 156 comes in two parts: (a) from 1889 to 1962 with farm housing excluded, (b) from 1959 onwards with farm housing included. The entries for each of the four years of overlap, 1959 to 1962, are nearly equal. The two versions have been amalgamated by linear splicing. (For years $1958 + t$, where $1 \leqslant t \leqslant 4$, the (a) value has been multiplied by $1 - t/5$ and added to the (b) value multiplied by $t/5$, and the result rounded to the nearest integer.) The series are referred to as *BUTTER* and *NEWHOUSES*.

Figures B:15 and B:16 show logarithms of these series with linear trend subtracted. Figure B:15 (butter) resembles the corresponding part of Figure B:7 (imports) fairly closely, especially between about 1875 and about 1965. Since what the imports and butter series measure are so different, one is tempted to conclude from the resemblance that each series is a broad

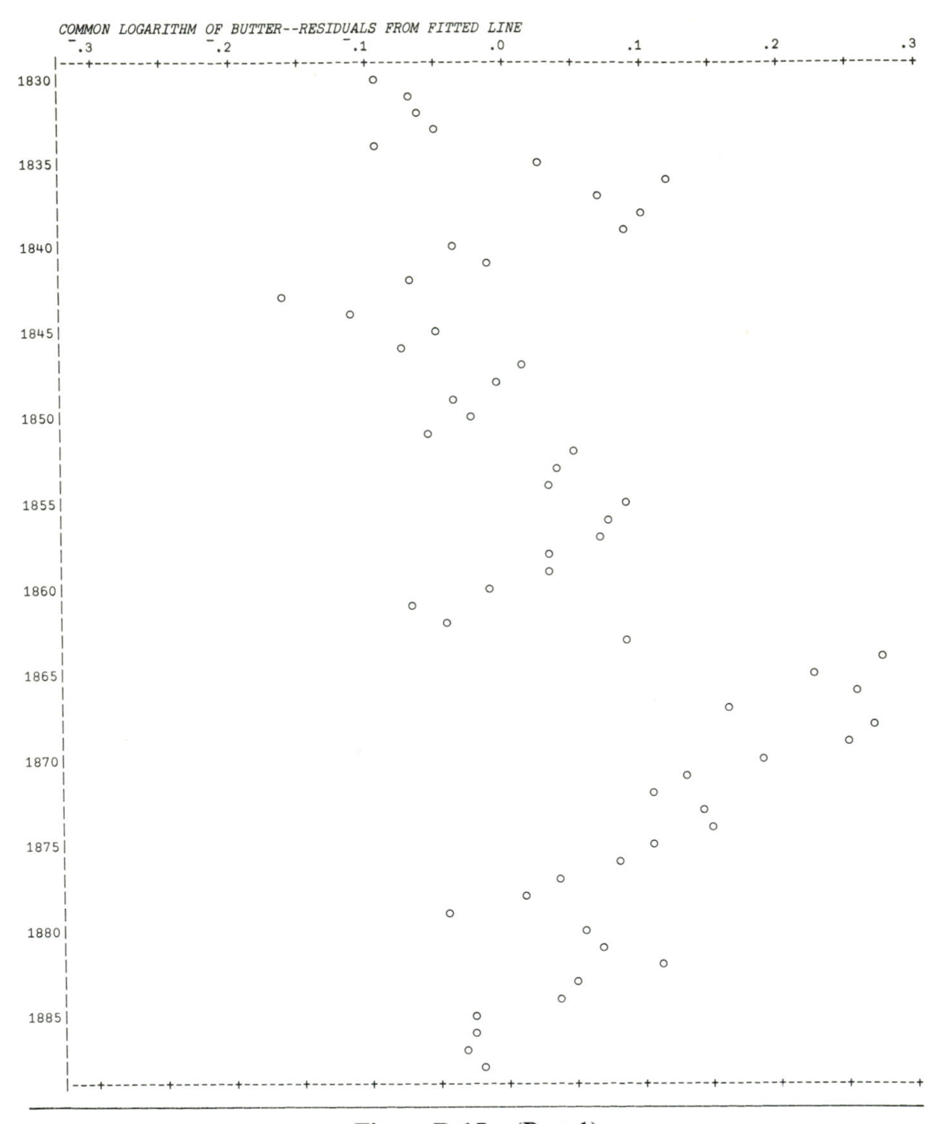

Figure B:15 (Part 1)

indicator of the economy. Our first choice of imports is supported. On the other hand, if we are looking for indicators of other features of the economy than those measured by imports, for possible stronger correlation with the enrolment series, then because of the resemblance butter is not very promising. Figure B:16 (housing) shows somewhat more different behavior, and in particular the reactions to the two world wars are much like those of enrolment. It is a pity that the housing series, though longer than most available economic series, is so much shorter than the enrolment series.

We proceed to examine the four series in their 87 common years, 1889–1975. Figures B:17–23 correspond to Figures B:8–9 in our study of

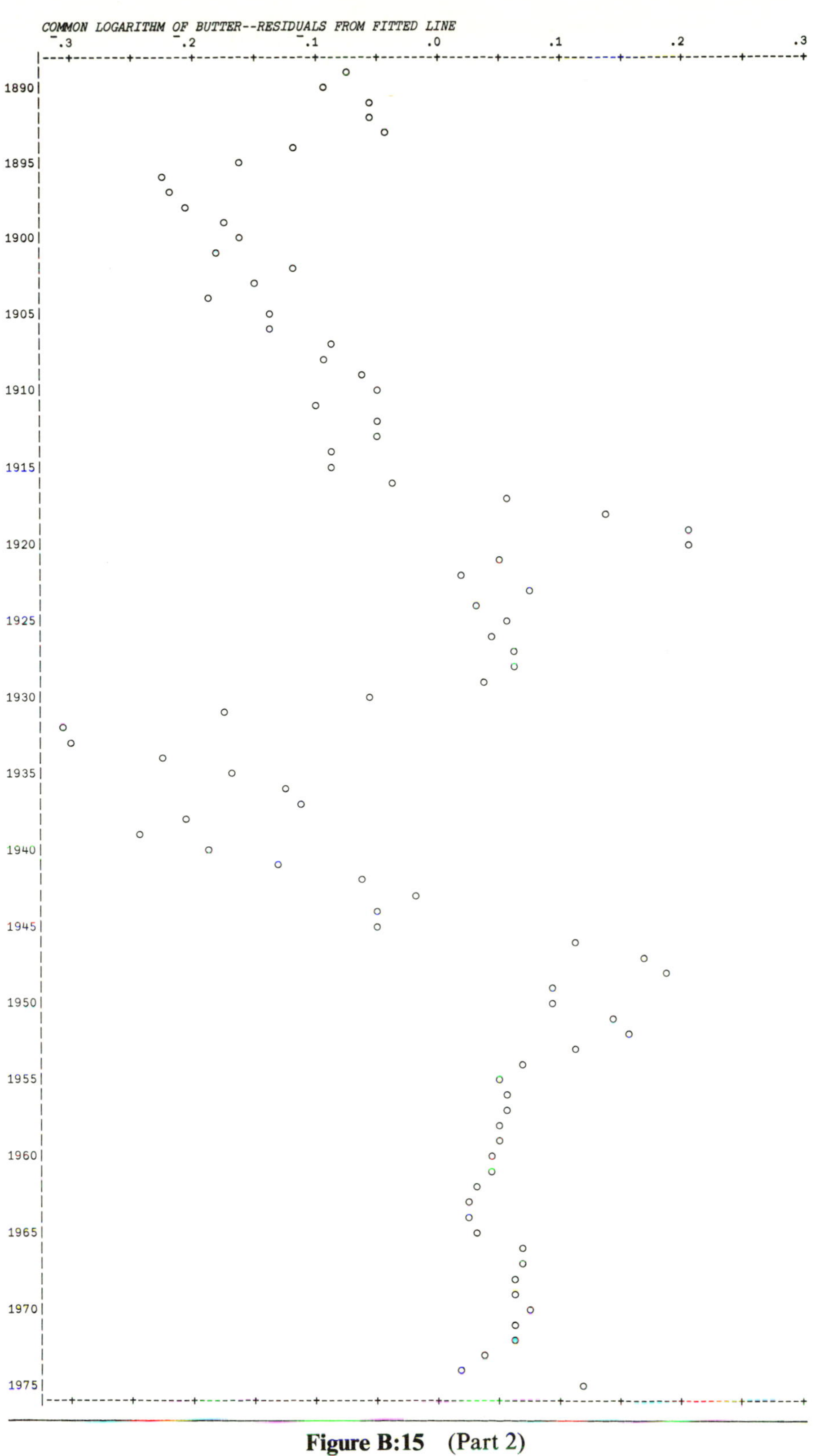

Figure B:15 (Part 2)

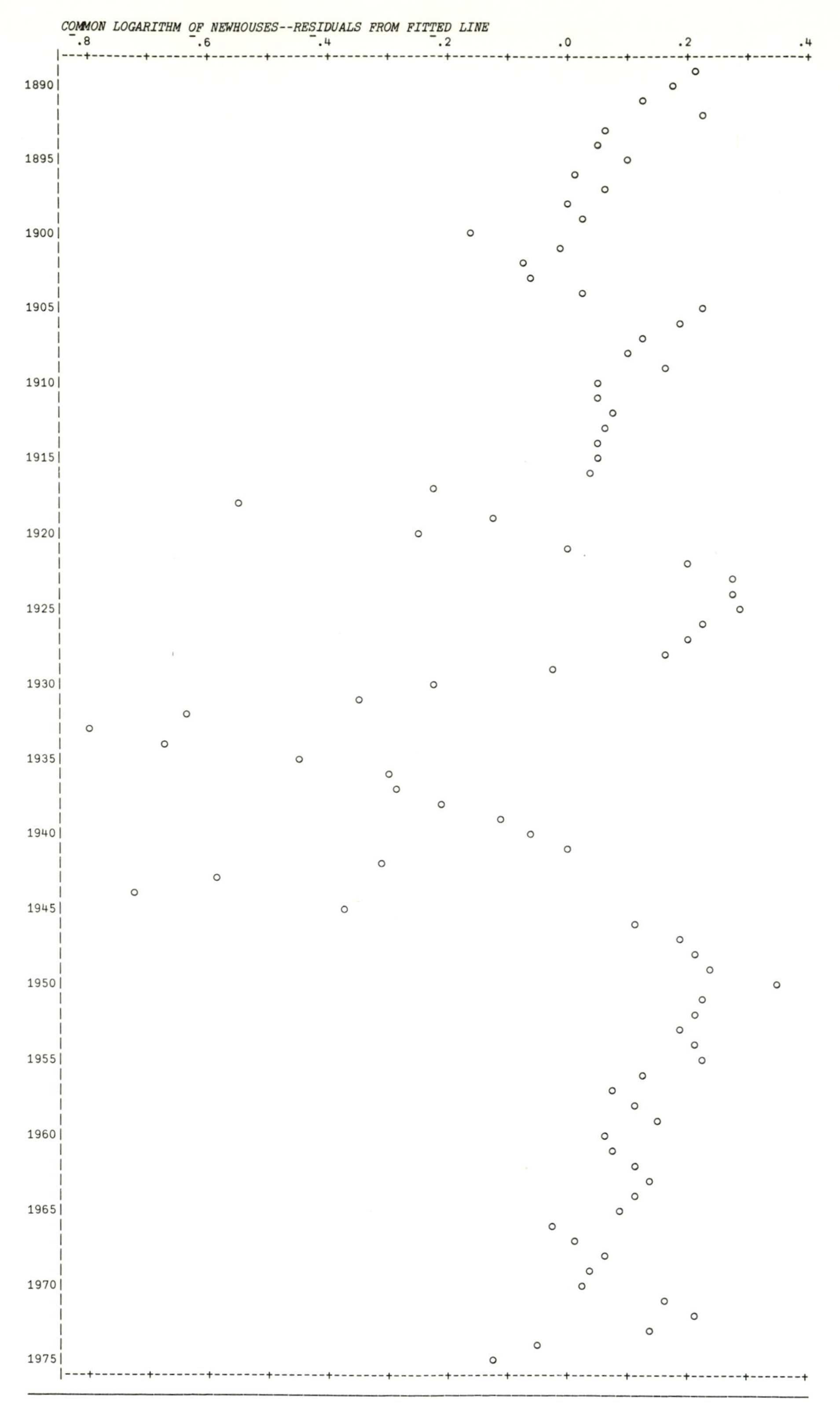
COMMON LOGARITHM OF NEWHOUSES--RESIDUALS FROM FITTED LINE
-.8
-.6
-.4
-.2
.0
.2
.4
1890
1895
1900
1905
1910
1915
1920
1925
1930
1935
1940
1945
1950
1955
1960
1965
1970
1975

Figure B:16

the 180 common years of enrolment and imports. In Figure B:17 we see the four series prewhitened in the usual way and tapered by 6 to final length 80. When their Fourier transform *FT* has been found, *PREHAR* is invoked. This causes phase difference plots to be produced for each pair of series. Figures B:18–20 show phase differences for enrolment with each of the three economic series in turn, and Figures B:21–23 show phase differences between pairs of the economic series. Clearly Figure B:18 is similar in character to Figure B:9, Figure B:19 is not much different, but Figure B:20 indicates a more definite relation between enrolment and the housing series, with almost no lag. Figure B:21 shows that imports and butter are quite strongly related at most frequencies, least strongly in the middle near frequency 1/4 (the 20th harmonic). Imports and housing seem strongly related at high frequency; butter and housing are not so strongly related.

At this point, if we were doing ordinary linear regression, it would be natural to perform a step of Gram-Schmidt orthogonalization of the "independent" variables. Residuals of the other three series after linear regression on the housing series would be calculated; and the enrolment residuals would be plotted against the imports residuals and also against the butter residuals, to see whether either or both pairs of variables were noticeably related. The residuals would be uniquely defined, and (in the absence of singularity) a complete regression relation could be determined by steps, the "independent" variables being introduced one at a time, as illustrated in the next chapter.

Unfortunately, our harmonic regression involves a sequence of overlapping regression calculations, depending on a set of moving-average weights; and each regression involves estimation of a partly nonlinear relation. We have on our hands many nonlinear regressions rather than one linear one. Whether harmonic regression of one "dependent" series on more than one "independent" series can be done correctly by steps, the "independent" series being introduced and worked with one at a time, may be doubted. However, phase-difference plots for residual series, after harmonic regression on one "independent" series, may perhaps be a useful guide to the introduction of a second "independent" series, and seem to be worth trying.

We accordingly use the functions *HARINIT* and *HAR*1 to perform harmonic regression of *Y*0 (as defined in Figure B:17, the filtered and tapered enrolment series) on *Y*3 (filtered and tapered housing), and then harmonic regression of *Y*1 (imports) on *Y*3, and finally harmonic regression of *Y*2 (butter) on *Y*3. The resulting matrices of smoothed spectral estimates are denoted by *SSE*03, *SSE*13 and *SSE*23, respectively, and are reproduced in Figure B:24. Before making such calculations, the length of the sine-weighted moving average to be used must be decided on. The rather low value of 17 has been chosen, for which the equivalent degrees of freedom $\nu = 29.0$, and approximate 5%, 1% and 0.1% values for R^2, calcu-

```
      Y0←(⍳87) FIT 100×⍟¯87↑ENROLMENT
MEANS ARE  44 849.08
REGRESSION COEFFICIENT IS 1.8716
   WITH ESTIMATED STANDARD ERROR  0.075222   (85 D.F.)
      Y1←(⍳87) FIT 100×⍟¯87↑IMPORTS
MEANS ARE  44 829.71
REGRESSION COEFFICIENT IS 4.8237
   WITH ESTIMATED STANDARD ERROR  0.19884   (85 D.F.)
      Y2←(⍳87) FIT 100×⍟¯87↑BUTTER
MEANS ARE  44 599.14
REGRESSION COEFFICIENT IS 1.4487
   WITH ESTIMATED STANDARD ERROR  0.10113   (85 D.F.)
      Y3←(⍳87) FIT 100×⍟NEWHOUSES
MEANS ARE  44 635.3
REGRESSION COEFFICIENT IS 2.327
   WITH ESTIMATED STANDARD ERROR  0.24131   (85 D.F.)

      1↓C÷1↑C←6 AUTOCOV Y0
0.67439 0.40581 0.2255 0.082183 ¯0.0048858 ¯0.050721
      1↓C÷1↑C←6 AUTOCOV Y1
0.89105 0.77173 0.67553 0.5783 0.48932 0.43427
      1↓C÷1↑C←6 AUTOCOV Y2
0.87456 0.6819 0.50476 0.38276 0.30749 0.23036
      1↓C÷1↑C←6 AUTOCOV Y3
0.83053 0.59036 0.36306 0.18497 0.052379 ¯0.031766

      ⍴Y0←(1↓Y0)-0.86×¯1↓Y0
86
      ⍴Y1←(1↓Y1)-0.86×¯1↓Y1
86
      ⍴Y2←(1↓Y2)-0.86×¯1↓Y2
86
      ⍴Y3←(1↓Y3)-0.86×¯1↓Y3
86

      ⍴Y0←6 TAPER Y0
6 LINEAR TAPER WEIGHTS WITH INCREMENT 0.14286
80
      Y1←6 TAPER Y1
6 LINEAR TAPER WEIGHTS WITH INCREMENT 0.14286
      Y2←6 TAPER Y2
6 LINEAR TAPER WEIGHTS WITH INCREMENT 0.14286
      Y3←6 TAPER Y3
6 LINEAR TAPER WEIGHTS WITH INCREMENT 0.14286

      ⍴FT←2 FFT (10 8 ⍴Y0),[2.5] 10 8 ⍴Y1
41 2 2
      ⍴FT←FT,2 FFT (10 8 ⍴Y2),[2.5] 10 8 ⍴Y3
41 2 4

      PREHAR
PHASE DIFFERENCE PLOT FOR  Y0  AND  Y1 ?  Y OR N
YES
DISTRIBUTION OF SYMBOLS (.∘○⊖⎕⍠):  0 9 13 7 8 2
```

Figure B:17

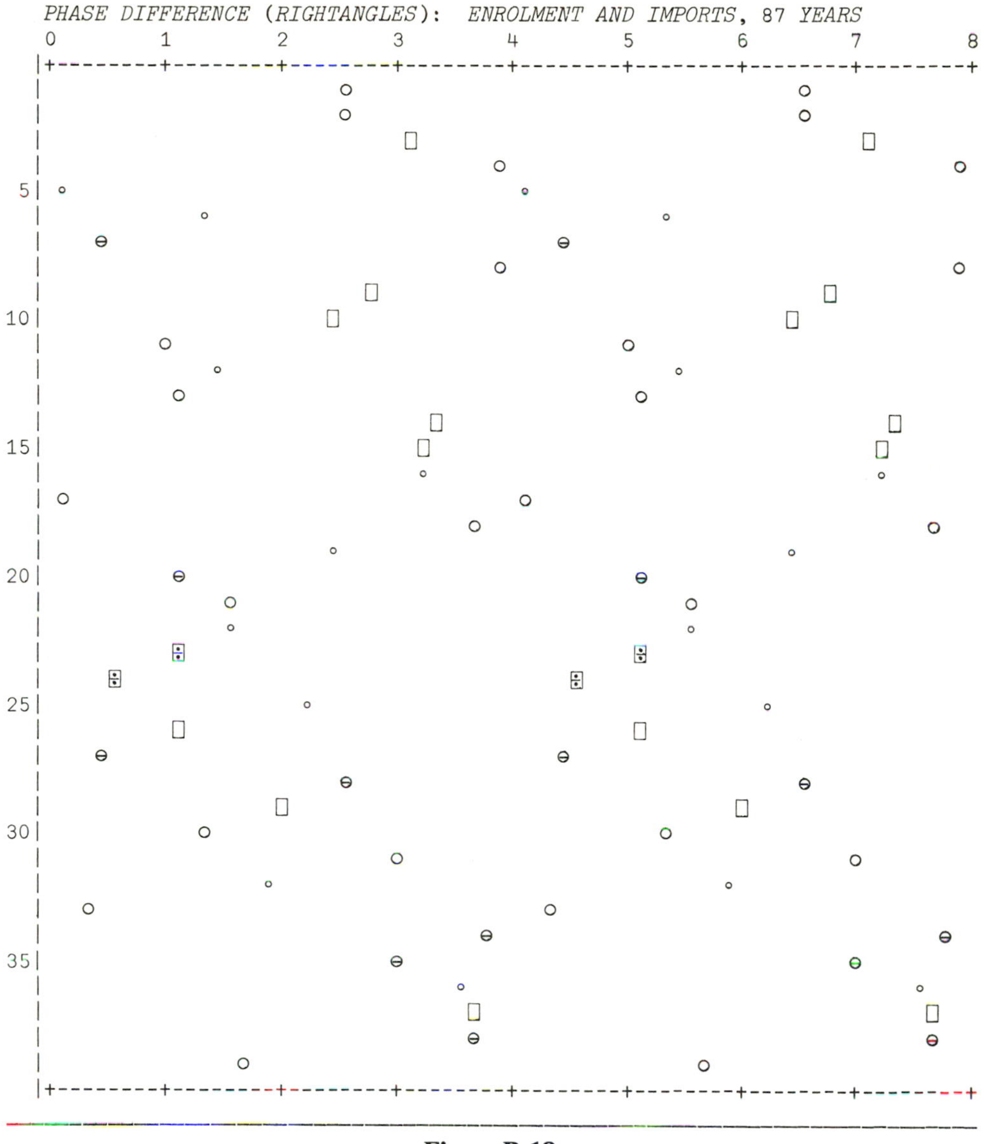

Figure B:18

lated as before, are 0.26, 0.35 and 0.46. With this value, *SSE*03 and *SSE*13 can both be obtained directly in one execution of *HAR*1, r_0 going from its lowest value, 8, by steps of 1 to its greatest value, 32. The coefficient $\hat{B}$ changes smoothly in *SSE*03, and the procedure converges easily. $\hat{B}$ makes a substantial jump in *SSE*13 as r_0 goes from 22 to 23, but with enough iterations the procedure accomplishes this jump. On the other hand, *SSE*23 as shown cannot be obtained in one execution of *HAR*1, because the very large jump in $\hat{B}$ as r_0 goes from 17 to 18 is not yielded by the iterative procedure. *SSE*23 has been pieced together from two execu-

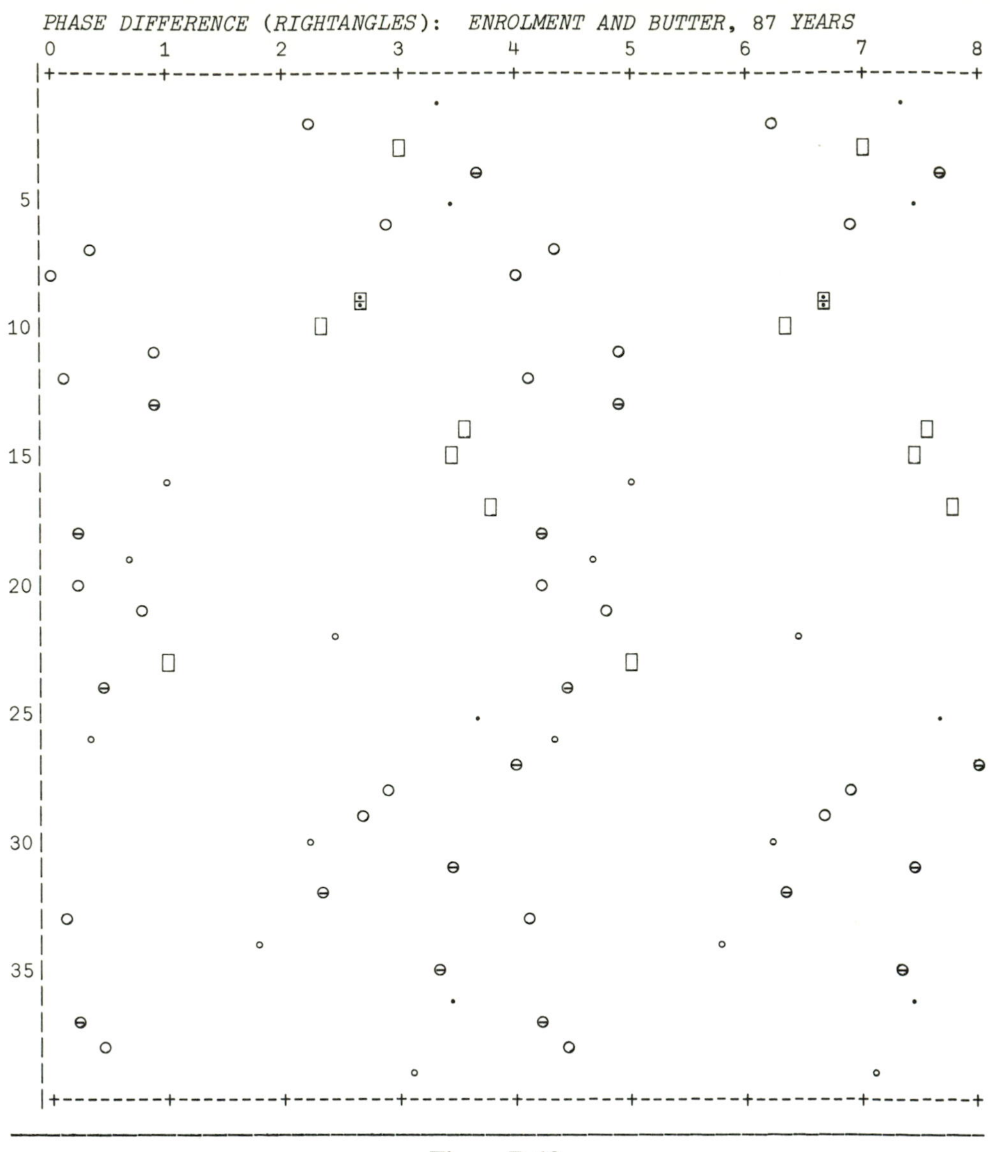

Figure B:19

tions of *HAR*1, in one of which r_0 stepped up and in the other r_0 stepped down, the solution with the greater R^2 having been chosen.

The upper part of Figure B:24 presents *SSE*03 in the same style as Figure B:10. The successive columns give r_0, C_{00}, C_{33}, $\hat{G}_{03}$, $\hat{A}_{03}$, $\hat{B}_{03}$ and R_{03}^2, to use our previous notation modified by addition of suffixes making clear that we are concerned with regression of *Y*0 on *Y*3. Whereas in Figure B:10 the frequency of the r_0th harmonic was $r_0/160$, here the frequency of the r_0th harmonic is $r_0/80$. The lower part of Figure B:24 shows the additional information contained in *SSE*13 and *SSE*23. The third column of each of

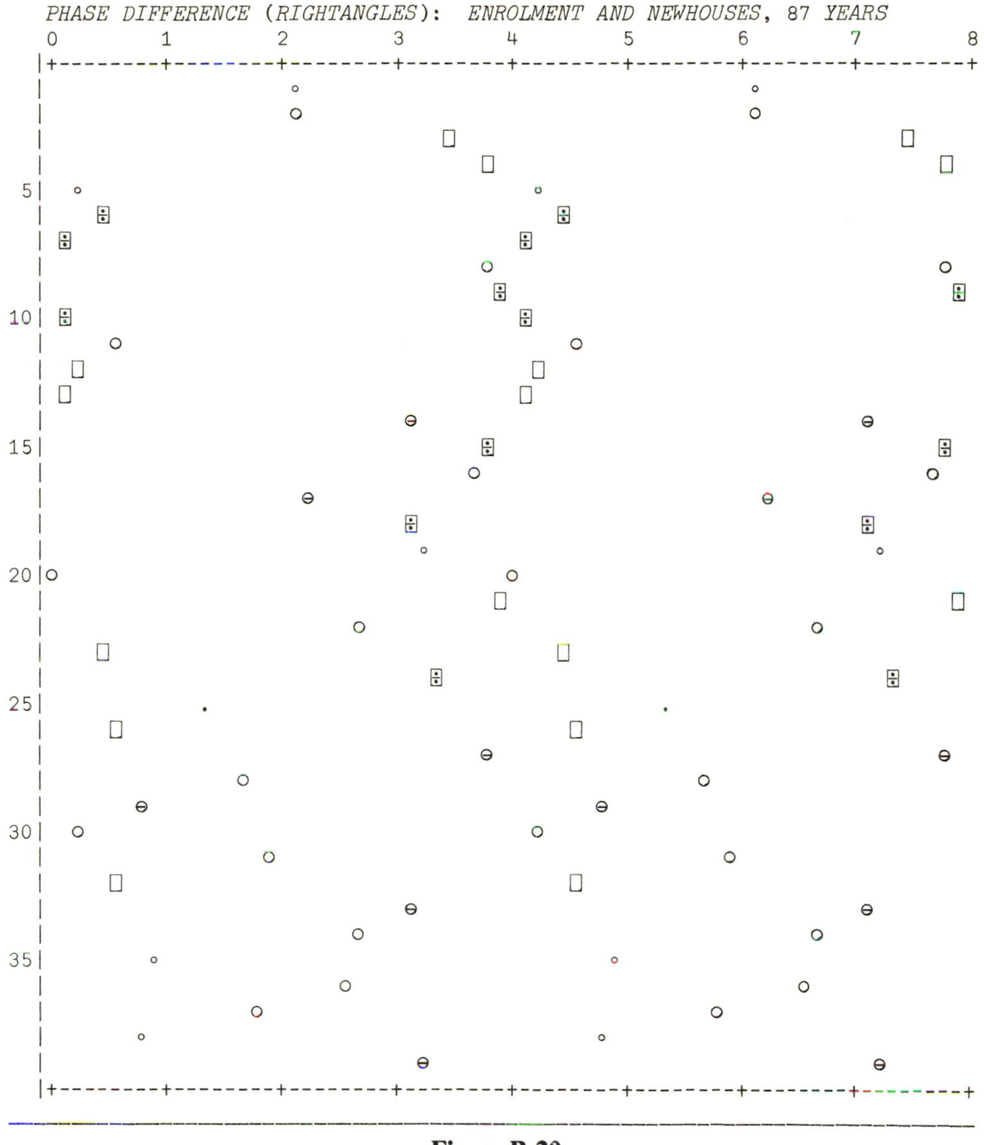

Figure B:20

these matrices contains C_{33}, which has already been shown in the print-out of *SSE*03. So the columns in the lower part of Figure B:24 present r_0, C_{11}, C_{22}, $\hat{G}_{13}$, $\hat{A}_{13}$, $\hat{B}_{13}$, R^2_{13}, $\hat{G}_{23}$, $\hat{A}_{23}$, $\hat{B}_{23}$ and R^2_{23}. The first, second, fourth, fifth, sixth and seventh of these columns are precisely *SSE*13 with its third column omitted; the first, third, eighth, ninth, tenth and eleventh of these columns are *SSE*23 with its third column omitted.

*SSE*03 confirms the indication of Figure B:20 that there is a strong relation between $Y0$ and $Y3$ at low frequencies, with little lag. *SSE*13 shows relations between $Y1$ and $Y3$ both at low and at high frequencies—at

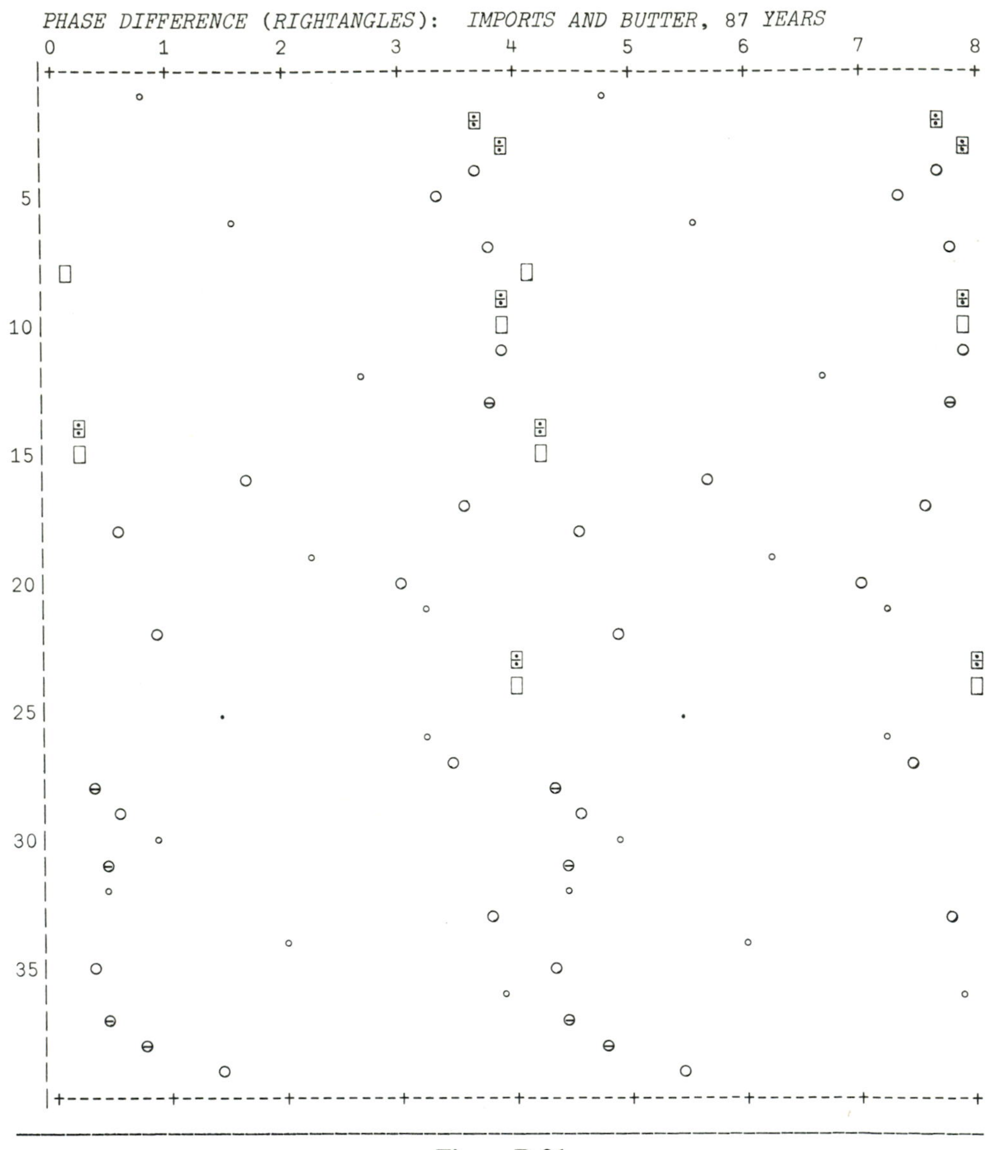

Figure B:21

low frequencies $Y3$ follows some eleven years after $Y1$, at high frequencies there is little lag between the series. None of the R^2 values in $SSE23$ is impressive.

We may now find the residuals of the Fourier transforms of $Y0$, $Y1$ and $Y2$ after harmonic regression on $Y3$. This has been done by a function $HAR1R$ that is executed as soon as SSE has been determined by $HAR1$.

Let us first turn back to equation (38), where we were considering regression of $Y0$ on $Y1$. The residual of the Fourier transform of $Y0$ after regression on $Y1$ may be defined to be the coefficients $\{a_r^{(01)}, b_r^{(01)}\}$, for

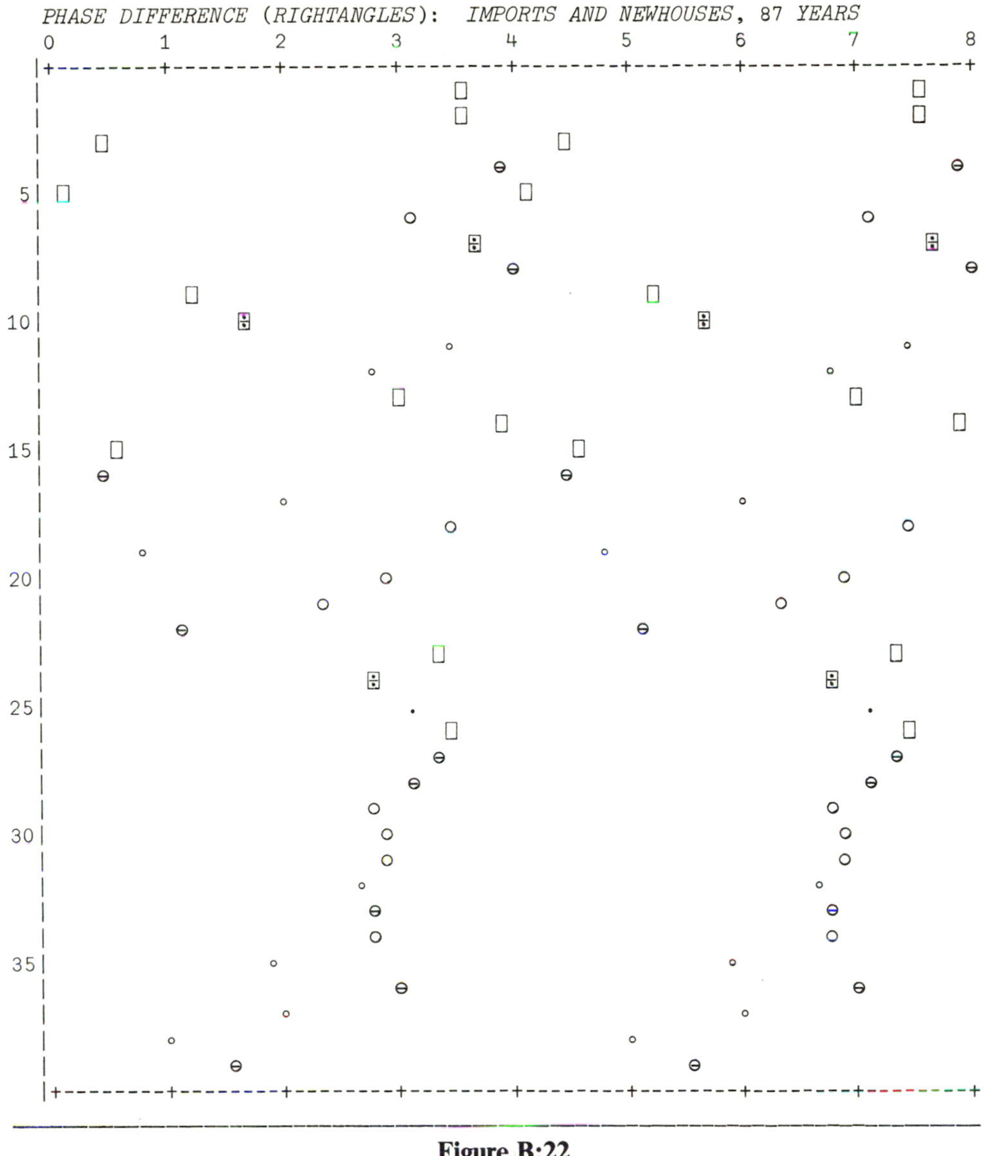

Figure B:22

$0 \leqslant r \leqslant \frac{1}{2}n,$

$$a_r^{(01)} = a_r^{(0)} - G\left[a_r^{(1)}\cos\varphi - b_r^{(1)}\sin\varphi\right],$$

$$b_r^{(01)} = b_r^{(0)} - G\left[b_r^{(1)}\cos\varphi + a_r^{(1)}\sin\varphi\right],$$

where the gain G and phase shift φ, both depending on r, are to be estimated. In *HAR1R* the following simple and crude procedure is used to estimate G and φ for each r. For any r that falls among the values of r_0 in the first column of *SSE*, G and φ are set equal to the values of $\hat{G}$ and $\hat{A}$ in

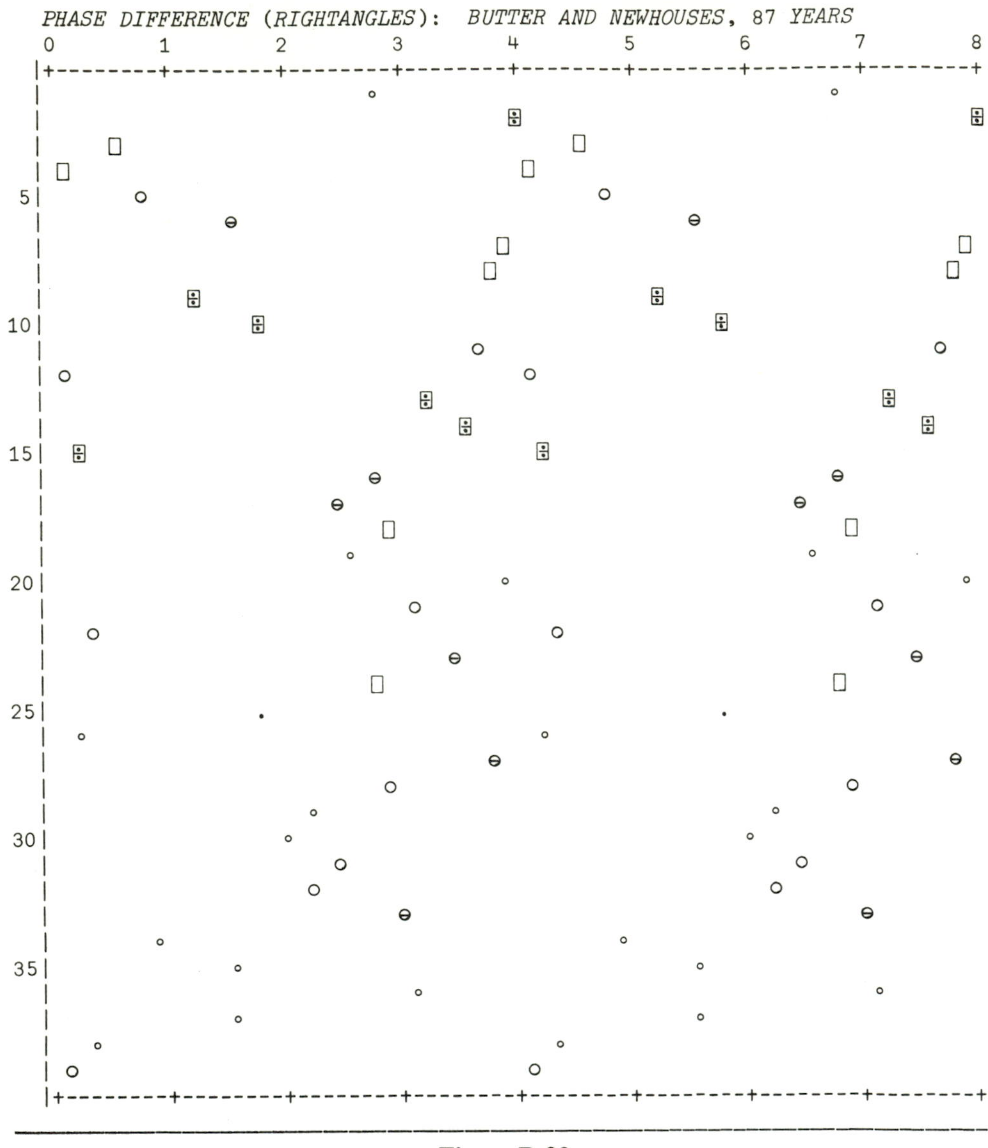

Figure B:23

the fourth and fifth columns of SSE. For values of r below the lowest r_0, the $\hat{G}$ and $\hat{A}$ values are extrapolated, $\hat{G}$ being constantly equal to $SSE[1; 4]$ and $\hat{A}$ changing linearly in accordance with the value of $\hat{B}$ in $SSE[1; 6]$; and similarly for r above the highest r_0. (See the definition of $HAR1R$ in Appendix 3.)

Thus after each SSE matrix has been found (which we have named $SSE03$, $SSE13$ and $SSE23$), $HAR1R$ has been executed, yielding residual Fourier transforms, say $FT03$, $FT13$ and $FT23$. These have been stacked in a three-dimensional array, renamed FT; and $PREHAR$ has been exe-

Print-out of *SSE*03 for 17-point sine-weighted moving average

8	314	3183	.235	6.19	2.0	.560
9	334	3208	.240	6.22	3.8	.553
10	355	3192	.247	.03	7.0	.550
11	367	3101	.257	.14	9.3	.556
12	373	2949	.266	.26	11.0	.558
13	380	2760	.272	.37	11.1	.535
14	382	2551	.275	.49	10.8	.506
15	383	2334	.268	.57	9.4	.439
16	388	2187	.264	.61	7.3	.393
17	388	2096	.256	.65	5.7	.355
18	385	1992	.244	.65	2.2	.308
19	390	1887	.250	.63	¯2.2	.301
20	391	1813	.257	.60	¯5.1	.306
21	395	1700	.271	.52	¯9.1	.316
22	399	1573	.290	.41	¯11.7	.330
23	398	1472	.297	.25	¯14.0	.327
24	398	1363	.316	.10	¯15.8	.343
25	400	1297	.315	6.19	¯15.3	.322
26	395	1248	.300	5.99	¯14.2	.284
27	390	1176	.282	5.82	¯13.4	.240
28	385	1173	.258	5.58	¯16.0	.203
29	380	1142	.240	5.26	¯20.4	.173
30	372	1082	.223	4.92	¯23.9	.145
31	362	1045	.207	4.47	¯27.1	.124
32	346	1034	.182	4.21	¯26.8	.099

Print-out of selected columns of *SSE*13 and *SSE*23 for 17-point s.w.m.a.

8	675	435	.252	6.12	¯85.0	.299	.174	6.21	¯88.1	.221
9	639	449	.254	5.05	¯81.5	.323	.176	5.11	¯86.6	.221
10	586	453	.258	3.98	¯76.7	.363	.180	4.01	¯84.7	.228
11	545	443	.274	2.99	¯71.9	.427	.185	2.92	¯81.6	.240
12	511	435	.289	2.09	¯70.3	.481	.198	1.87	¯78.4	.266
13	489	426	.295	1.23	¯70.4	.491	.213	.90	¯77.2	.293
14	461	410	.293	.30	¯72.0	.475	.221	6.24	¯76.3	.305
15	463	388	.270	5.84	¯68.0	.366	.218	5.30	¯75.7	.286
16	474	360	.235	5.62	¯53.1	.255	.201	4.23	¯78.6	.246
17	485	322	.226	5.13	¯47.5	.220	.180	3.12	¯82.0	.210
18	502	284	.225	4.61	¯43.2	.200	.189	1.21	4.6	.251
19	523	257	.234	4.07	¯39.8	.198	.200	1.22	3.7	.293
20	562	233	.244	3.61	¯38.2	.192	.192	1.23	2.2	.287
21	595	204	.252	3.14	¯35.7	.182	.185	1.26	.8	.284
22	614	172	.264	2.65	¯31.3	.179	.174	1.29	¯.7	.276
23	624	148	.323	1.16	8.2	.246	.172	1.34	¯2.6	.295
24	635	139	.375	1.42	1.2	.302	.176	1.43	¯.7	.304
25	638	138	.423	1.47	¯.3	.364	.178	1.47	4.3	.300
26	634	134	.454	1.48	1.2	.406	.185	1.50	7.6	.319
27	619	136	.483	1.51	2.1	.443	.193	1.57	9.9	.323
28	590	136	.489	1.54	2.8	.476	.189	1.68	9.7	.308
29	559	134	.493	1.58	3.9	.497	.184	1.80	9.3	.289
30	527	131	.498	1.64	4.7	.509	.178	1.91	9.1	.264
31	490	125	.478	1.77	6.6	.487	.162	2.01	7.8	.220
32	444	117	.419	1.91	9.0	.409	.136	2.02	4.5	.163

Figure B:24

cuted again, yielding phase difference plots of the residual series in pairs. The two of these plots that are of immediate interest are given in Figures B:25–26.

Figure B:25, showing phase differences for enrolment residuals and imports residuals, after regression on housing, suggests that there is perhaps some association at high frequencies with little lag, and that if there is any association at low frequencies it is with a sizeable lag; but no strong association of any kind strikes the eye. Figure B:26, showing phase differences for enrolment residuals and butter residuals, after regression on

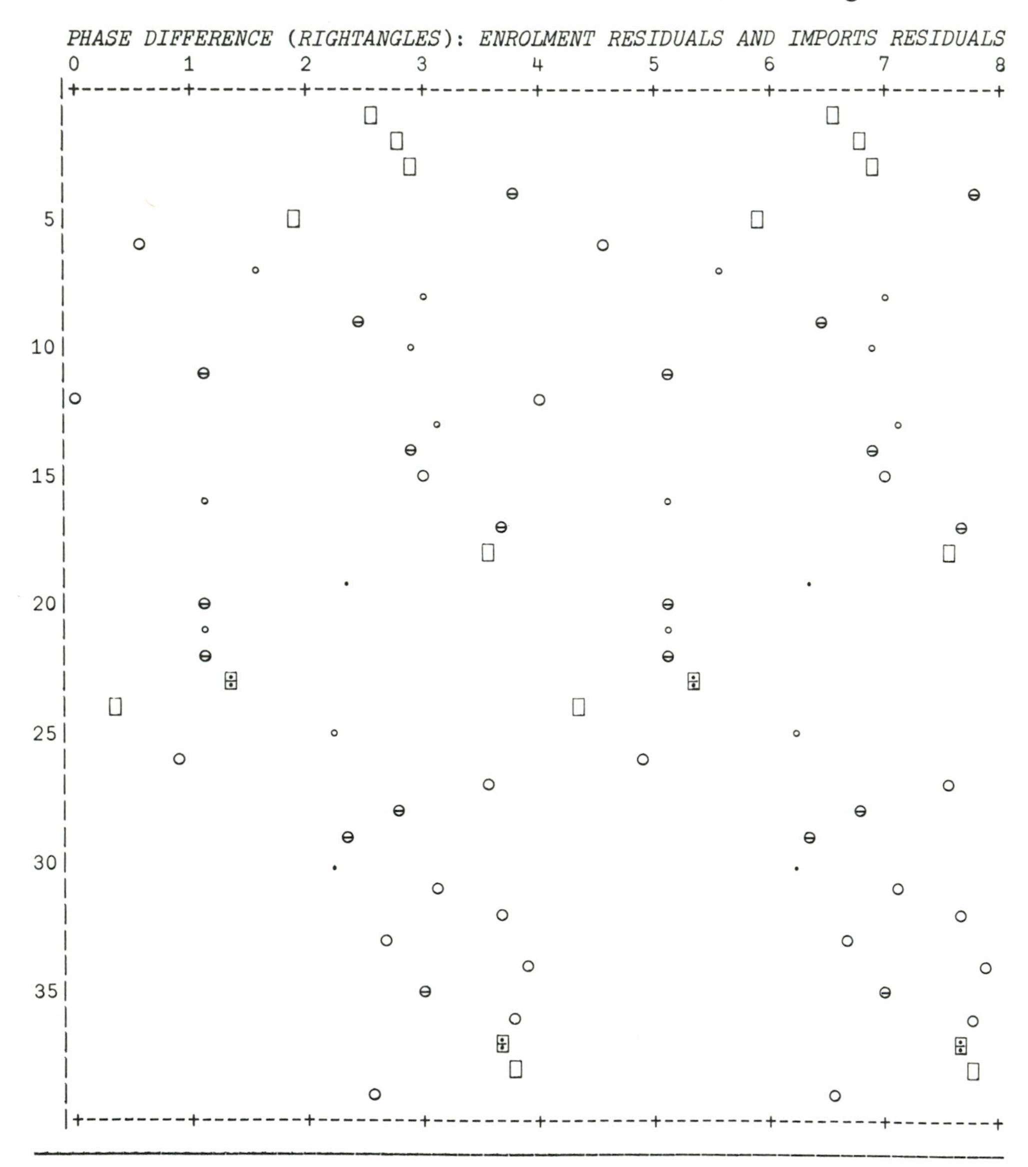

Figure B:25

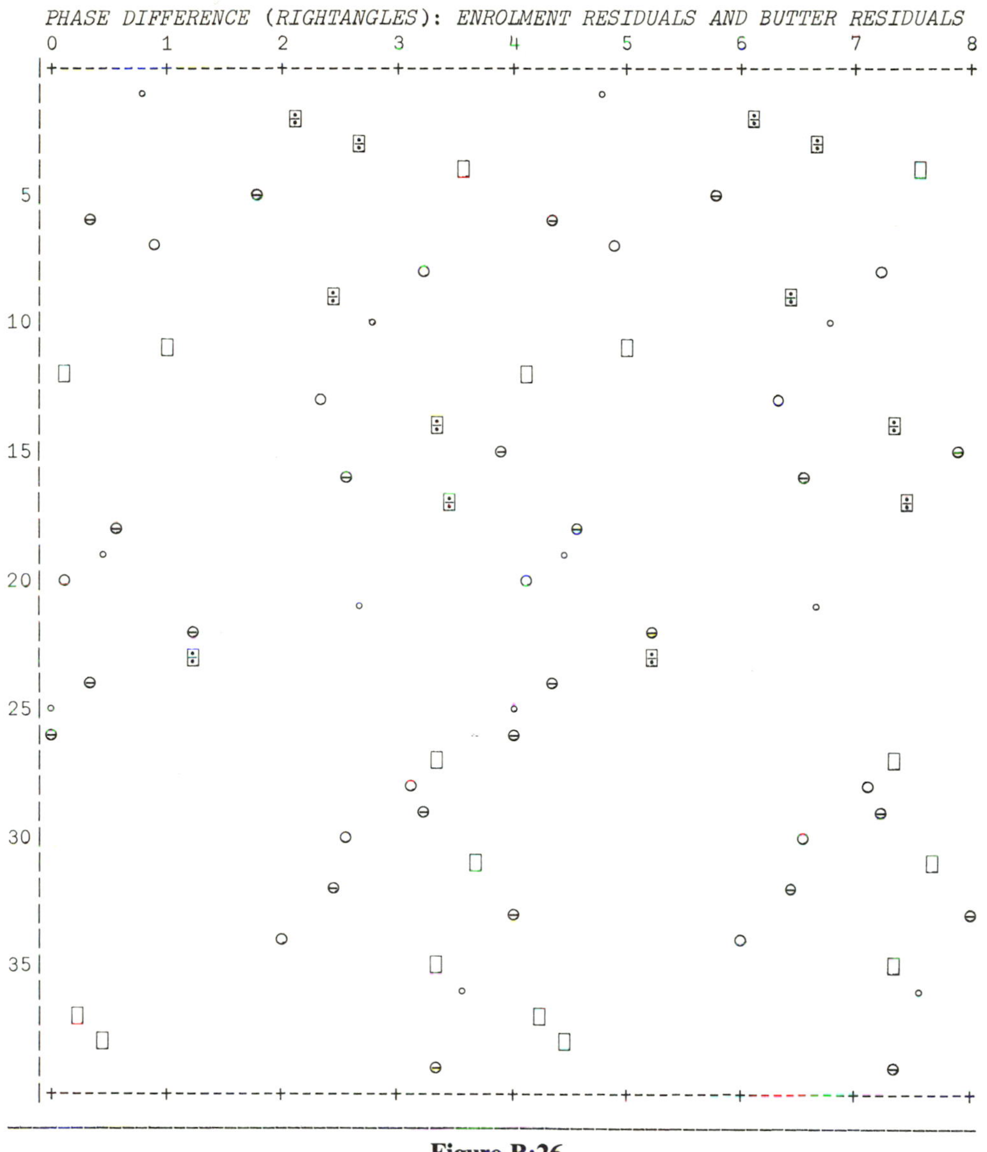

Figure B:26

housing, strikes the eye even less. Not reproduced is a plot of phase differences for imports residuals and butter residuals, which resembles Figure B:21 fairly closely.

Because Figures B:25–26 are so similar to each other, they afford no firm ground for choice between imports and butter as a second series to be introduced into the regression relation, in addition to housing—and indeed little encouragement to introduce either one. The imports series is a priori more plausible as an indicator of the economy than the butter series, and it is therefore preferred now.

To perform harmonic regression of $Y0$ (enrolment) on $Y1$ (imports) and $Y3$ (housing) jointly, we use the functions $HARINIT$ (with argument 0 1 3) and $HAR2$. The latter is analogous to $HAR1$, but accommodates two "independent" series instead of one; see Exercise 17 below. The resulting matrix of smoothed spectral estimates has columns for

$$r_0, C_{00}, C_{11}, C_{33}, \hat{G}_{01}, \hat{A}_{01}, \hat{B}_{01}, \hat{G}_{03}, \hat{A}_{03}, \hat{B}_{03}, R^2.$$

In each minimization of the weighted sum of squares, six parameters are estimated, namely the two G's, two A's and two B's. To judge significance of the difference of R^2 from zero, our previous tentative reasoning yields

$$\frac{(\nu - 6)R^2}{6(1 - R^2)}$$

as an F statistic with 6 and $\nu - 6$ degrees of freedom. For the 17-point sine weights used above in Figure B:24, and now used again, we have $\nu = 29.0$, and approximate 5%, 1% and 0.1% values for R^2 are now 0.40, 0.49 and 0.60.

The SSE matrix obtained is not reproduced here. The values of R^2 are highest when r_0 is small; $R^2 = 0.670$ at $r_0 = 8$, and $R^2 > 0.60$ for $8 \leqslant r_0 \leqslant 14$. R^2 also has a local maximum value of 0.508 when $r_0 = 22$. At low frequencies, enrolment seems to follow imports by about 13 years, and at medium frequencies by about 8 years; there is little lag between enrolment and housing. These lags of 13 or 8 years between enrolment and imports are greater than the amount of tapering that has been done (6 years). They would be more reliably estimated if some such lag were introduced into the data at the outset. Accordingly the calculation has been repeated with $Y0$ and $Y3$ defined as in Figure B:17, using the data for 1889–1975, but $Y1$ defined similarly except for a ten-year displacement, using the imports series for 1879–1965. The SSE matrix yielded by $HAR2$ is nearly the same as before except that values of $\hat{B}_{01}$ are reduced by about 20π and $\hat{A}_{01}$ is correspondingly changed. This second version of SSE, referred to as $SSE013$ to indicate which variables are considered, is reproduced in Figure B:27. At low frequencies there is positive association of enrolment with (i) imports of $10 + 3 = 13$ years earlier and (ii) new housing of the same year. At medium frequencies there is negative association of enrolment with (i) imports of $10 - 2 = 8$ years earlier and (ii) new housing of two years later. (These lags have been estimated, as usual, by plotting the estimated phase-shift curve and fitting to the relevant part of it a straight line having zero-intercept equal to 0 or π (mod 2π) and slope corresponding to an integer lag.)

Necessarily, every entry in the last column of Figure B:27 is not less than the corresponding entry in the upper part of Figure B:24. Distinguishing the entries by the notation R^2_{013} and R^2_{03}, we have $R^2_{013} \geqslant R^2_{03}$ for each r_0.

Print-out of *SSE*013 for 17-point s.w.m.a., *Y*1 displaced by ten years

8	314	629	3183	.261	1.25	19.4	.218	5.92	1.9	.655
9	334	610	3208	.293	1.47	21.9	.218	5.91	3.0	.657
10	355	574	3192	.301	1.75	26.0	.221	6.00	6.3	.644
11	367	525	3101	.306	2.14	29.7	.231	6.13	8.7	.642
12	373	510	2949	.300	2.79	30.3	.263	6.25	11.8	.632
13	380	492	2760	.371	3.33	28.4	.298	.06	13.7	.631
14	382	471	2551	.403	3.76	29.8	.314	.20	14.4	.613
15	383	473	2334	.351	3.77	24.6	.290	.35	13.8	.528
16	388	487	2187	.304	3.83	21.1	.271	.45	11.2	.474
17	388	500	2096	.285	4.05	19.8	.257	.54	9.5	.436
18	385	518	1992	.239	4.40	21.7	.236	.50	4.0	.374
19	390	543	1887	.221	4.88	20.5	.249	.48	¯1.2	.361
20	391	579	1813	.370	6.18	¯18.2	.206	.63	¯14.7	.447
21	395	611	1700	.370	6.00	¯16.9	.240	.45	¯16.2	.491
22	399	628	1573	.356	5.82	¯15.1	.267	.28	¯15.8	.513
23	398	635	1472	.345	5.61	¯12.4	.274	.10	¯16.1	.507
24	398	641	1363	.327	5.45	¯10.7	.287	6.24	¯16.6	.505
25	400	641	1297	.322	5.35	¯11.6	.289	6.02	¯15.3	.476
26	395	635	1248	.325	5.28	¯10.0	.283	5.81	¯13.8	.444
27	390	618	1176	.332	5.25	¯6.8	.273	5.65	¯11.8	.405
28	385	586	1173	.339	5.15	¯5.5	.250	5.50	¯12.0	.367
29	380	552	1142	.333	4.95	¯6.5	.229	5.27	¯14.9	.324
30	372	519	1082	.309	4.71	¯9.0	.217	5.02	¯17.4	.269
31	362	481	1045	.294	4.50	¯10.9	.207	4.61	¯21.9	.233
32	346	435	1034	.299	4.43	¯15.0	.199	4.36	¯22.3	.205

Figure B:27

To assess whether the increment, $R_{013}^2 - R_{03}^2$, is significantly greater than 0, we may proceed as before, as though linear regression coefficients were being estimated by ordinary least squares, and conjecture that

$$\frac{(\nu - 6)(R_{013}^2 - R_{03}^2)}{3(1 - R_{013}^2)}$$

behaves approximately like an F statistic with 3 and $\nu - 6$ degrees of freedom. For $r_0 = 9$, our calculations give $R_{013}^2 = 0.657$ and $R_{03}^2 = 0.553$, and with $\nu = 29.0$ the F statistic is about 2.33, with 3 and 23.0 degrees of freedom; this is close to the upper 10% point of the distribution. Thus the inclusion of imports as an explanatory variable in addition to housing has had a positive effect at low frequencies hardly large enough to be reckoned significant. For $r_0 = 22$ (the other value at which R_{013}^2 has a local maximum), we have $R_{013}^2 = 0.513$, $R_{03}^2 = 0.330$, whence $F = 2.88$, which is about at the upper 6% point. Again, at middle frequencies, the incremental effect of imports is not very impressive.

In sum, we have found a rather strong (highly significant) positive association between enrolment and housing, without lag, at low frequencies; and there is a suggestion of negative association between enrolment and housing of two years later at middle frequencies. When we bring in imports as a second explanatory variable together with housing, both associations with enrolment, at low frequencies and at middle frequencies, are strengthened, but by amounts that cannot be judged very significant.

As before, we may attempt to exhibit the associations in the time domain by scatterplotting the series after appropriate lagging and filtering. We shall in fact only consider the association at low frequencies. To achieve a correlation in the time domain similar to that observed in the frequency domain, we try to find a filter for the series whose gain function behaves like the sine weights used in the harmonic analysis, in determining *SSE*. That is, our values for R^2 when $r_0 = 9$ (in Figures B:24, 27) were obtained with sine weights having positive values for the harmonics numbered from 1 to 17 (frequencies 1/80 to 17/80), the greatest value being for the ninth harmonic (frequency 9/80); zero weights were given to all other harmonics (numbered 0, 18, 19, . . . , 40). If the gain function of the filter used in the time domain has values at these frequencies in substantially different proportion from those frequency-domain weights, we cannot be surprised if the correlation turns out differently—the correlation could be higher or (of course, more likely) lower. We need to suppress high frequencies above about 0.23, which can conveniently be done by a filter with cosine weights of extent 7 or 8. It turns out to be important also to suppress the very lowest frequencies, as by a simple first differencing. A filter representing the combination of a 7-point cosine-weighted moving average and a first differencing has weights in the proportion

$$\bar{}0.41,\ \bar{}1,\ \bar{}1,\ \bar{}0.41,\ 0.41,\ 1,\ 1,\ 0.41$$

approximately. This has maximum gain near frequency 0.10, and one-tenth of the maximum gain near frequencies 0.01 and 0.23; and so the gain function does very roughly approximate the shape of our 17-point sine weights when $r_0 = 9$.

What has been done, then, is to take the enrolment and housing series for 1889 to 1975, and imports of 13 years earlier, 1876 to 1962. They have been prewhitened by the usual steps of taking logarithms, subtracting a linear trend, and filtering by (¯0.86, 1). Finally the above further 8-point filter has been applied. Figure B:28 shows the filtered enrolment series plotted against the filtered housing series. The correlation coefficient is 0.77 (of which the square is 0.59). A similar plot, not reproduced, has been made of the filtered enrolment series against a linear combination of the filtered housing and imports series. Only a slight increase in the correlation coefficient can be attained this way, with the filtering we have used, to 0.79 (with square 0.62); and the plot looks only a little tighter than Figure B:28, and therefore seems rather uninteresting.

The appearance of strong correlation in Figure B:28 is largely due to the points marked *P*—both series reacted similarly to the second world war. A lesser contribution comes from the points marked *M*, for the first world war. These similarities are obvious when Figures B:5 and B:16 are compared. Other prominent low-frequency features of those figures do not correspond.

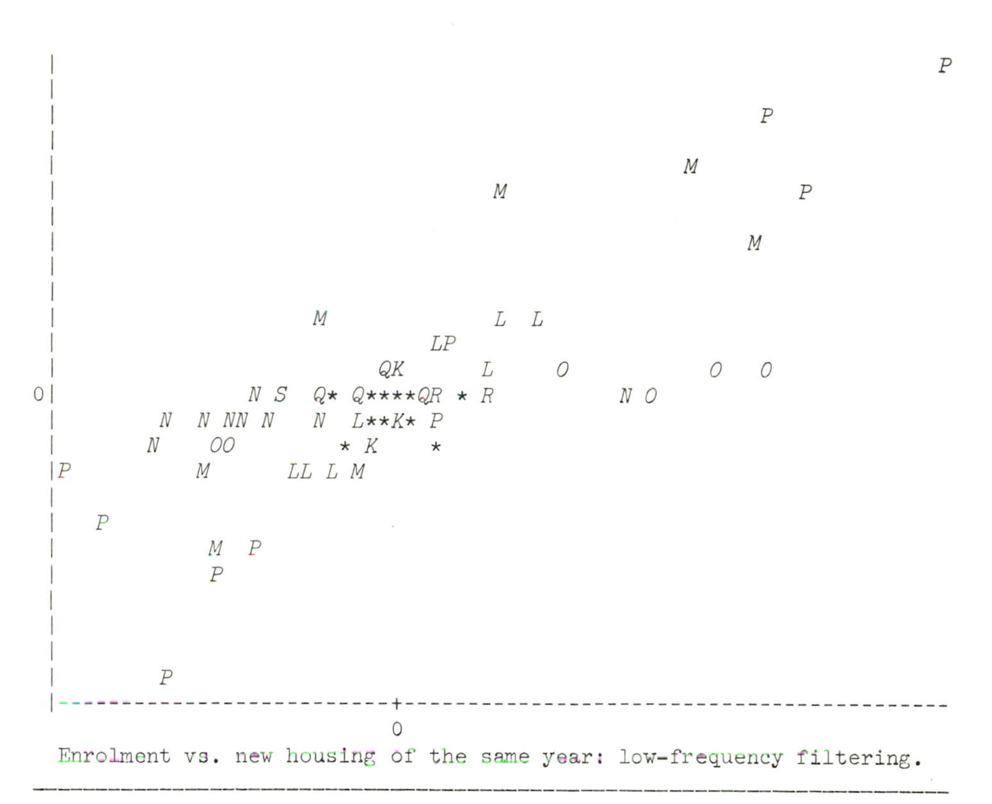
0
0
Enrolment vs. new housing of the same year: low-frequency filtering.

Figure B:28

The great rise in enrolment up to 1896 has no parallel in the housing series. The great smooth dip of the housing series from 1928 to 1941 has only a weak reflection in the enrolment series. To achieve a high correlation between the filtered series it was necessary not only to suppress the upper half of the frequency spectrum (with the cosine-weighted moving average) but also to suppress the lowest frequencies (with the first differencing).

Thus our examination of the four series in 87 years has led to a similar conclusion to that of the two series in 180 years. We have found a strong medium-low-frequency association between enrolment and housing due mostly to the strong similar reactions of both series to the two world wars. There is otherwise not much indication of similarity between the series, little to support a case that the state of the national economy has a visible effect on Yale's enrolment. We have used a technique of search for association, throughout the whole spectrum of frequencies, that appears sensitive and effective in revealing associations between artificial series related as in equation (30) above. Yale's enrolment was disturbed in an immediate and evident way by the national emergencies of the two world wars, as was also the housing series. But other national emergencies, such as the civil war of the eighteen-sixties and the economic depression of the nineteen-thirties, had no such vivid effect on the enrolment. One is tempted to conclude that Yale's development has been independent of the national economy.

Perhaps the dependence of an educational institution on its economic environment is too manifold, too complicated, or even too rapidly changing, to be detected by any sort of regression analysis of time series.

Some questions

There is a considerable difference in attitude between this chapter and the next. Ordinary regression calculations, in which a linear relation between given variables is estimated by the method of least squares, have been doable correctly for a century and a half. During the last decade somewhat different computational methods (Gram-Schmidt orthogonalization, Householder reflections) have been widely used, as alternatives to Gauss's solution of the normal equations; but the results are the same when the data are well conditioned. Interest in the methodology of least-squares regression today centers on assessing the appropriateness in retrospect of the regression calculation. We look at the diagonal elements of the matrix named Q in the next chapter, to see whether any single observations have had a dominating influence in estimation of any of the regression coefficients. We study the residuals in various ways to judge the correctness of the supposed linear structure. A great variety of alternatives to the method of least squares have been developed and studied.

By contrast, regression analysis of time series in the frequency domain is not a widely used and well understood technique. Our first concern has been to be able to perform it at all. The reader may have noticed some loose ends in the foregoing account that the author has been unable to tie satisfactorily. Critical assessment of results has not been undertaken.

How satisfactory is it to treat the series considered here as if they were realized from stationary Gaussian processes? Would it have been better to exclude the readings referring to the two world wars, and if so how?

How does the understanding obtained from analysis in the frequency domain compare with that to be had by the time-domain methods of Box and Jenkins (1970)?

One very modest critical step appears in Appendix 2, where kurtosis tests are suggested to detect departure from joint normality in a stationary process, in view of the possible importance (not widely recognized today) of joint normality for harmonic analysis.

Notes for Chapter 10

The great length of this chapter is due to interpolation, into the examination of the enrolment data, of an account of pertinent parts of the theory of harmonic analysis. Anyone familiar with the literature will be inclined to skip this material, but one detail should be noted, that tapering is defined in a slightly different way from usual, as a circularizing of the series, after which the set of frequencies used in the Fourier analysis is reduced. For general accounts of theory and methods see Bartlett (1955), Blackman and Tukey (1959), Parzen (1967), Hannan (1970), Anderson (1971), Brillinger (1975), Bloomfield (1976), Brillinger and Tiao (1980).

In our regression analysis, we try to estimate the Fourier transform of the coefficients $\{\beta_j\}$ appearing in equation (30); the transform is expressed as two real functions, the gain $G(\lambda)$ and the phase shift $\varphi(\lambda)$. We suppose that in any narrow frequency band $G(\lambda)$ is nearly constant and $\varphi(\lambda)$ is nearly linear in λ, and so in each band we estimate the three real parameters G, A, B, by minimizing the weighted sum of squares (38), a nonlinear (but otherwise simple) regression problem. Thus many small regressions are calculated. This may be compared with direct estimation of the parameters of (30) by some kind of multiple regression; the number of these parameters can easily be greatly in excess of three, and the conditioning poor; and the character of the error process, usually not white noise, should be allowed for.

Cleveland and Parzen (1975) have suggested an elegant quick (non-iterative) method of approximately minimizing (38). They estimate $G(\lambda)$ and $\varphi(\lambda)$ in separate steps. First they choose a set of weights $\{w_j\}$ that are possibly of short extent and are very smooth, such as cosine weights; and they minimize (38) with respect to G and A, setting B equal to 0. This is a linear estimation problem, as we have noted. The resulting A is taken as an estimate of $\varphi(r_0/n)$, for each possible r_0. Second, they choose another set of weights $\{w_j\}$, perhaps of greater extent, perhaps less smooth, like our sine weights, or uniform weights; and they minimize (38) with respect to G after inserting the previous estimates of $\varphi(r/n)$. If the derivative of $\varphi(\lambda)$, which we have denoted by B, is small in magnitude, this procedure does seem attractive. However, when we studied the relation of enrolment and imports at low frequency we found large-magnitude values for B, and this behavior of the phase-shift function cannot be obtained by the Cleveland-Parzen procedure. Admittedly our relation at low frequency was of dubious significance, but the discrepancy between results suggests caution in the use of their procedure.

The methods described here have been applied to another set of annual data, with rather similar results (Anscombe, 1977).

Exercises for Chapter 10

1. Let the given series $\{x_t\}$ $(t = 1, 2, \ldots, n)$ be extended circularly so that, for all integer t, $x_{t+n} = x_t$. Then from (4) we have, for any integer h and for $1 \leqslant t \leqslant n$, $0 \leqslant r \leqslant \frac{1}{2}n$, r integer,

$$\Sigma_t x_t x_{t+h} = \Sigma_r z_r \cos(2\pi rh/n).$$

Let the serial correlation coefficients of the series be defined circularly:

$$R_h = (\Sigma_t x_t x_{t+h})/(\Sigma_t x_t^2).$$

Then we have, as a discrete analog of (7),

$$R_h = \left[\Sigma_r z_r \cos(2\pi rh/n)\right]/(\Sigma_r z_r).$$

A Fourier representation of $\{z_r\}$, analogous to (8), can be found by setting $z_{-r} = z_r$ for all r $(0 \leqslant r \leqslant \frac{1}{2}n)$ and expressing z_r/d_r as a finite Fourier series over $-\frac{1}{2}n < r \leqslant \frac{1}{2}n$:

$$z_r/\Sigma_s z_s = (d_r/n)\,\Sigma_h d_h R_h \cos(2\pi rh/n),$$

where the summations are for integer values, $0 \leqslant s \leqslant \frac{1}{2}n$, $0 \leqslant h \leqslant \frac{1}{2}n$.

2. Let $\{x_t\}$ be defined for t equal to n consecutive integers:

$$\left.\begin{aligned} x_t &= 0 \quad \text{for} \quad t \neq k, \\ &= 1 \quad \text{for} \quad t = k, \end{aligned}\right\}$$

where k is some one of the integers. Determine the line spectrum $\{z_r\}$ of $\{x_t\}$, and show that the phases $\{\epsilon_r\}$ are congruent, modulo 2π, to a multiple of $\{kr\}$. What is the effect of a single greatly outlying reading in an otherwise well-behaved time series on the Fourier transform of the series?

3. Consider four stationary random sequences, each having $\rho_1 = \frac{2}{3}$ and ξ_t distributed $N(0, \sigma^2)$ for every t. The first three are jointly normal sequences.

(i) A Markov sequence $\{\xi_t\}$ satisfying (for all integer t) the relation

$$\xi_t = a\xi_{t-1} + \eta_t,$$

where $|a| < 1$ and $\{\eta_t\}$ are independent variables identically distributed $N(0, \sigma^2(1 - a^2))$. For any integer h, $\rho_h = a^{|h|}$. Consider $a = \frac{2}{3}$.

(ii) A moving-average sequence $\{\xi_t\}$ defined in terms of independent variables $\{\eta_t\}$ identically distributed $N(0, \sigma^2/3)$:

$$\xi_t = \eta_{t+1} + \eta_t + \eta_{t-1}.$$

We find $\rho_1 = \frac{2}{3}$, $\rho_2 = \frac{1}{3}$, $\rho_h = 0$ for $h \geqslant 3$.

(iii) Another (smoother) moving-average sequence $\{\xi_t\}$ defined in terms of independent variables $\{\eta_t\}$ identically distributed $N(0, \sigma^2/6)$:

$$\xi_t = \eta_{t+1} + 2\eta_t + \eta_{t-1}.$$

Now $\rho_1 = \frac{2}{3}$, $\rho_2 = \frac{1}{6}$, $\rho_h = 0$ for $h \geqslant 3$.

(iv) A Markov sequence $\{\xi_t\}$ such that, given ξ_{t-1}, the conditional distribution for ξ_t is: $\xi_t = \xi_{t-1}$ with probability a, ξ_t is distributed $N(0, \sigma^2)$ with probability $1 - a$. For this "jump" process the serial correlations are the same as for (i). Consider $a = \frac{2}{3}$.

(**a**) From (8) or otherwise, find the spectral density functions of these sequences, and graph them. For (i) and (iv), verify that the function can be written

$$f(\lambda) = \frac{2(1 - a^2)}{1 - 2a\cos(2\pi\lambda) + a^2}.$$

(**b**) Generate a sample of 50 (or so) readings from each of the sequences, and plot them on the same scale, to compare behavior. If you were shown such a sample from one of the sequences but were not told which, how might you guess?

The generating of samples may be done as follows, using *RNORMAL* (listed in Appendix 3).

```
      X1←RNORMAL 50
      X1[1+⍳49]←X1[1+⍳49]×(5÷9)*0.5
      ∇M;J
[1]   J←2
[2]   X1[J]←X1[J]+2×X1[J-1]÷3
[3]   →2×50≥J←J+1   ∇
      M
```

$X1$ is now a sample of 50 readings from (i). (But while we were about it we should have defined a function with two arguments to generate n observations from (i) for any given value of a.) For samples of size 50 from (ii), (iii) and (iv):

```
      X2←(RNORMAL 52)÷3*0.5
      X2←¯2↓+⌿ 0 1 2 ⌽ 3 52 ⍴X2
      X3←(RNORMAL 52)÷6*0.5
      X3←¯2↓ 1 2 1 +.× 0 1 2 ⌽ 3 52 ⍴X3
      X4←RNORMAL +/T←1,1=?49⍴3
      X4←X4[+\T]
```

As a check, *AUTOCOV* could be used to find the sum of squares and the first six (say) autocorrelations of each vector. Plotting could be done with *DOWNPLOT*.

4. Suppose the stationary random sequence $\{\xi_t\}$ $(t = 0, \pm 1, \pm 2, \ldots)$ has spectral density $\pi\cos\pi\lambda$. Show that

$$\rho_1 = \frac{1}{3}, \quad \rho_2 = -\frac{1}{15}, \quad \rho_3 = \frac{1}{35}, \ldots$$

These are coefficients in a Fourier-series representation of the spectral density over $(-\frac{1}{2}, \frac{1}{2})$:

$$\pi\cos\pi\lambda = 2 + \frac{4}{3}\cos 2\pi\lambda - \frac{4}{15}\cos 4\pi\lambda + \frac{4}{35}\cos 6\pi\lambda - \ldots$$

5. (a) Let ϵ be uniformly distributed over $(-\frac{1}{2}\pi, \frac{1}{2}\pi)$—or equally well over $(0, 2\pi)$. Then $Y = \sqrt{2}\sin\epsilon$ has mean 0, variance 1, and density $1/(\pi\sqrt{2 - y^2})$ for $|y| < \sqrt{2}$. Sketch the density.

(b) Consider the standardized sum of two such Y's, that is now let

$$Y = \sin\epsilon_1 + \sin\epsilon_2,$$

where ϵ_1 and ϵ_2 are independent and uniformly distributed over $(-\frac{1}{2}\pi, \frac{1}{2}\pi)$. Y has mean 0 and variance 1; let the density be $f(y)$ for $|y| < 2$. Show that if δ is a small positive number, prob$(Y > 2 - \delta) \sim \delta/2\pi$, and so $f(2-) = 1/2\pi$. A possible expression (not good for computation) for $f(y)$ when $0 < y < 2$ is

$$f(y) = f(-y) = \frac{1}{\pi^2}\int \frac{d\epsilon}{\sqrt{1 - (y - \sin\epsilon)^2}},$$

where the limits of integration are $\sin^{-1}(y - 1)$ and $\frac{1}{2}\pi$. Show that $f(0) = \infty$. Sketch the density. Clearly the distribution for Y is less platykurtic than a uniform distribution. The sum of three or more such variables (six or more independent sin ϵ's) has a density quite close to normal.

6. Take a random sample of (say) eight λ's from the uniform distribution over $(0, \frac{1}{2})$ and eight ϵ's from the uniform distribution over $(0, 2\pi)$, taking care to note the values obtained, and make a plot of

$$\xi_t = \sqrt{\frac{2}{8}}\sum_{j=1}^{8}\cos(2\pi\lambda_j t + \epsilon_j)$$

for $1 \leqslant t \leqslant 50$ (say). Can you perceive any periodicity or other sign that the ξ's are not independent variables identically distributed $N(0, 1)$? Compute the line spectrum of the first 20 ξ's, and of the first 50 or 100 ξ's, and plot against the frequency. (Ignore the 1-d.f. elements in the line spectrum.) Do the $\{z_r\}$ $(0 < r < \frac{1}{2}n)$ look like a sample from the tabulated χ^2 distribution with 2 degrees of freedom? Count how many fall between, say, the quintiles of the distribution and make a test of goodness of fit. Try also plotting the 0.3 power of $\{z_r\}$ against r, as suggested below in Exercise 9.

For comparison, similarly compute and examine the line spectrum of independent observations identically distributed $N(0, 1)$.

Possible instructions to generate 50 values of the sum of eight random sinusoids and plot the spectrum:

```
+L←(?8ρ1000000000)÷2000000000
+E←(?8ρ1000000000)×○2÷1000000000
X+.×X←0.5×+/2○(50 8 ρE)+(ι50)∘.×○2×L
+/Z←+/FT×FT←1 FFT 5 5 2 ρX
DPS← 0 0.2 ,⌈⌈/Z
DPM←'+----'
DPH←' ', 5 0 ⍕0,ιDPS[3]
DPL←4ρ5
DOWNPLOT 1↓¯1↓Z
+⌿(1↓¯1↓Z)∘.≤V←¯2×⍟ 0.8 0.6 0.4 0.2
```

7. Consider the distribution of the line spectrum $\{z_r\}$ of an n-element time series $\{x_t\}$ $(1 \leqslant t \leqslant n)$ supposed to be realized from a jointly normal stationary random sequence having zero mean, unit variance and spectral density $f(\lambda)$. Let k-element tapering be performed as specified at (22), and let $G(.)$ be the gain function of the moving average whose coefficients $\{w_j\}$ are differences of the taper weights $\{W_t\}$.

The line spectrum is not affected by a shift in the time origin, but the phases $\{\epsilon_r\}$ and the coefficients $\{a_r\}$ and $\{b_r\}$ do depend on the time origin. For present purposes it is convenient to move the origin to the middle of the series. Let us define the cosine coefficients, for integer r $(0 \leqslant r \leqslant \frac{1}{2}(n-k))$, by

$$a_r = \sqrt{\frac{d_r}{n-k}}\ \Sigma_t W_t x_t \cos \frac{2\pi r(t - \frac{1}{2}(n+1))}{n-k}$$

and b_r similarly, with sine instead of cosine.

At first let $\{x_t\}$ be a single random sinusoid,

$$x_t = \sqrt{2}\cos\left(2\pi\lambda\left(t - \tfrac{1}{2}(n+1)\right) + \epsilon\right),$$

where λ and ϵ are independent random variables, λ having density $f(\lambda)$ in $(0, \frac{1}{2})$ and ϵ uniformly distributed over $(0, 2\pi)$. Then, using (23), show that

$$a_r = \sqrt{\frac{d_r}{2(n-k)}}\left\{\frac{G(\sigma)\sin(n-k)\pi\sigma}{\sin\pi\sigma} + \frac{G(\delta)\sin(n-k)\pi\delta}{\sin\pi\delta}\right\}\cos\epsilon,$$

$$b_r = \sqrt{\frac{d_r}{2(n-k)}}\left\{\frac{G(\sigma)\sin(n-k)\pi\sigma}{\sin\pi\sigma} - \frac{G(\delta)\sin(n-k)\pi\delta}{\sin\pi\delta}\right\}\sin\epsilon,$$

where $\sigma = \lambda + r/(n-k)$, $\delta = \lambda - r/(n-k)$. Because of the uniform distribution for ϵ, we have at once

$$\mathcal{E}(a_r) = \mathcal{E}(b_r) = \mathcal{E}(a_r b_r) = \mathcal{E}(a_r b_{r'}) = 0,$$

for any r and r' (where r' may have any of the values that r may have). In terms of $\mathcal{E}(z_r)$, given at (25), we find

$$\operatorname{var}(a_r) = \mathcal{E}(a_r^2) = \tfrac{1}{2}\mathcal{E}(z_r) + C,$$

$$\operatorname{var}(b_r) = \mathcal{E}(b_r^2) = \tfrac{1}{2}\mathcal{E}(z_r) - C,$$

where

$$C = \frac{d_r}{2(n-k)}\mathcal{E}\left\{\frac{G(\sigma)\sin(n-k)\pi\sigma}{\sin\pi\sigma}\ \frac{G(\delta)\sin(n-k)\pi\delta}{\sin\pi\delta}\right\}.$$

Similar expressions can be given for $\mathcal{E}(a_r a_{r'})$ and $\mathcal{E}(b_r b_{r'})$, where $r \neq r'$; for example,

$$\mathcal{E}(a_r a_{r'}) = \frac{d_r}{4(n-k)}\mathcal{E}\left\{\frac{G(\delta)\sin(n-k)\pi\delta}{\sin\pi\delta}\ \frac{G(\delta')\sin(n-k)\pi\delta'}{\sin\pi\delta'} \right.$$

$$\left. + \text{three similar terms}\right\}.$$

When $r = 0$, $b_0 = 0$, $\sigma = \delta = \lambda$, and $\text{var}(a_0) = \mathcal{E}(z_0)$. If $n - k$ is even, when $r = \frac{1}{2}(n - k)$ either a_r or b_r vanishes, according to whether n is even or odd, and the other of them has variance equal to $\mathcal{E}(z_r)$.

By adding N independent random sinusoids as at (9) and letting $N \to \infty$ we approach a sample from the joint normal random sequence indicated at the outset. Now all the coefficients $\{a_r\}$ and $\{b_r\}$ are jointly normally distributed, and the moments just found for a single sinusoid still apply. Thus every cosine coefficient a_r is independent of every sine coefficient $b_{r'}$. If $1 \leqslant r < \frac{1}{2}(n - k)$, z_r is the sum of squares of two independent normal variables having zero means and, in general, unequal variances, but if $|C|$ is small the variances are nearly equal and z_r is nearly exponentially distributed (is nearly proportional to a variable having the tabulated χ^2 distribution with 2 d.f.). If $\mathcal{E}(a_r a_{r'})$ and $\mathcal{E}(b_r b_{r'})$ are small in magnitude, z_r is nearly independent of $z_{r'}$.

As for the size of C and of the covariances $\mathcal{E}(a_r a_{r'})$ and $\mathcal{E}(b_r b_{r'})$, the Dirichlet kernel has an orthogonality property:

$$\int_{-1/2}^{1/2} \frac{\sin(n - k)\pi\delta_1}{\sin \pi\delta_1} \, \frac{\sin(n - k)\pi\delta_2}{\sin \pi\delta_2} \, d\lambda = 0,$$

where $\delta_1 = \lambda - r_1/(n - k)$, $\delta_2 = \lambda - r_2/(n - k)$, and r_1 and r_2 are unequal integers less than $\frac{1}{2}(n - k)$ in magnitude. If there is no tapering ($k = 0, G(\cdot) = 1$) and $f(\lambda)$ is constant, one application of this result shows that $C = 0$ and two applications that either of the above covariances is zero. Otherwise, we are interested in integrals of the above type with $G(\delta_1)G(\delta_2)f(\lambda)$ inserted as an extra factor in the integrand. If $n - k$ is large, we may expect that such an integral will nearly vanish if the extra factor is nearly constant or nearly zero for values of λ where the kernel product is most considerable.

Consider, for example, $f(\lambda)$ constant ($= 2$), $n = 100$, $k = 10$ with linear tapering. The integrals are easily evaluated numerically. $\mathcal{E}(z_r) = 1.9192$ for $1 \leqslant r \leqslant 44$, and 0.9596 for $r = 0$ or 45. Here are some values for C:

r	C	r	C	r	C
$r = 1$,	$C = ^{-}0.0381$	$r = 8$,	$C = 0.0036$	$r = 16$,	$C = 0.0012$
2	$^{-}0.0319$	10	0.0002	18	0.0003
4	$^{-}0.0136$	12	$^{-}0.0018$	20	$^{-}0.0010$
6	0.0006	14	$^{-}0.0003$	22	$^{-}0.0004$

As r runs from 23 to 44, the same values for C are met in the opposite order with reverse sign. Thus z_1 and z_{44} are distributed as the sum of squares of independent normal variables having zero means and variances 0.9215 and 0.9977, a distribution that differs insensibly from an exponential distribution.

8. The numerical results at the end of the last exercise, concerning the empirical spectrum of a sample of white noise, may be obtained by a more rudimentary and direct method, without referring to random sinusoids. If the given series $\{x_t\}$ $(1 \leqslant t \leqslant n)$ is realized from uncorrelated variables all having zero mean and unit variance, let the coefficients $\{a_r\}$ and $\{b_r\}$ be defined as explained above, after tapering, with reference to a time origin in the middle of the series. Then it is easy to show that for any integer r, where $1 \leqslant r < \frac{1}{2}(n-k)$, a_r is uncorrelated with b_r,

$$\text{var}(a_r) = \left[2/(n-k)\right] \Sigma_t W_t^2 \cos^2\left[2\pi r(t - \tfrac{1}{2}(n+1))/(n-k)\right],$$
$$\mathcal{E}(z_r) = \left[2/(n-k)\right] \Sigma_t W_t^2.$$

9. Since the distribution of elements of an empirical (line) spectrum $\{z_r\}$ is nearly exponential, with skewness (γ_1) coefficient equal to 2, a plot of $\{z_r\}$ against $\{r\}$ will show isolated high points even if the true spectral density is constant. Because ratios of elements are relevant to judgments of significant inequality, sometimes a plot of $\{\ln z_r\}$ against $\{r\}$ is preferred, but then the distribution is skew in the opposite direction ($\gamma_1 = {}^{-}1.14$). For an approximately symmetrical distribution, $\{z_r^{0.3}\}$ may be plotted against $\{r\}$ instead.

Let $Y = X^p$, where the random variable X has an exponential distribution with density e^{-x} $(x > 0)$, and p is constant, $0 < p < 1$. Consider choosing p so that Y has a nearly symmetric density. Show that

(a) the median = the mode if $p = 1 - \ln 2 = 0.307$;
(b) the mean = the mode if $(1-p)^p = p!$, or $p = 0.302$;
(c) the mean = the median if $(\ln 2)^p = p!$, or $p = 0.290$.

When $p = 0.3$, the mean, median and mode of the distribution for Y are nearly equal.

10. (a) The function *GAIN* (in Appendix 3) calculates the gain function $G(\lambda)$, or $|G(\lambda)|$, at frequencies $F = \{\lambda\}$, of an m-point linear filter with weights $W = \{w_j\}$ $(j = 0, 1, \ldots, m-1)$. If the filter is symmetric, $G(\lambda)$ is given by (19) and (20). Otherwise

$$|G(\lambda)| = \sqrt{(\Sigma_j w_j \cos 2\pi\lambda j)^2 + (\Sigma_j w_j \sin 2\pi\lambda j)^2}\,,$$

and a warning message (relevant to *GAINLAG*) is given if $G(\lambda)$ almost vanishes at any λ in F, within the APL comparison tolerance.

(b) The function *GAINLAG* calculates $G(\lambda)$ and $l(\lambda)$ for any linear filter W at frequencies $F = \{\lambda\}$ for which $G(\lambda) \neq 0$. From (18), we can choose

$$l(\lambda) = \frac{1}{2\pi\lambda} \tan^{-1} \frac{\Sigma_j w_j \sin 2\pi\lambda j}{\Sigma_j w_j \cos 2\pi\lambda j} + \frac{M(\lambda)}{2\lambda} \qquad (0 < \lambda \leqslant \tfrac{1}{2}),$$

where $\tan^{-1}$ stands for the principal value in $(-\frac{1}{2}\pi, \frac{1}{2}\pi)$ of the inverse tangent, and $M(\lambda)$ is an integer-valued function chosen to make $l(\lambda)$ continuous. If $\Sigma_j w_j \neq 0$, we can choose

$$l(0) = \frac{\Sigma_j w_j\, j}{\Sigma_j w_j}\,.$$

Finally from (19),

$$G(\lambda) = (\Sigma_j w_j \cos 2\pi\lambda j) \cos\left[2\pi\lambda\, l(\lambda)\right] + (\Sigma_j w_j \sin 2\pi\lambda j) \sin\left[2\pi\lambda\, l(\lambda)\right].$$

Note that the phase-shift function $2\pi\lambda l(\lambda)$ may be taken to be continuous for $0 \leqslant \lambda \leqslant \frac{1}{2}$, and $l(\lambda)$ itself will then be continuous for $0 < \lambda \leqslant \frac{1}{2}$. Examples to think about:

$$\text{(i)} \qquad m = 2,\ w_j = (-1)^j; \quad \text{(ii)} \qquad m \geqslant 2,\ w_j = \frac{2(j+1)}{m(m+1)}.$$

Whether $2\pi\lambda l(\lambda)$ can be chosen to be a continuous function tending to 0, or one tending to $\pm\frac{1}{2}\pi$, as $\lambda \to 0$, depends on which is the first nonzero member of the sequence: $\Sigma_j w_j$, $\Sigma_j w_j j$, $\Sigma_j w_j j^2$, ...

11. Look at chapter 46 (Time series: trend and seasonality) in Kendall and Stuart (1966), and particularly see the moving averages in (46.7) and (46.8). Examine the gains of some of these moving averages.

12. A geometric moving average $\{y_t\}$ is sometimes calculated for a given time series $\{x_t\}$:

$$y_t = (1-p)(x_t + px_{t-1} + p^2x_{t-2} + \ldots) \quad (t \text{ integer})$$

for some fixed p, $0 < p < 1$. Since the length is infinite, in practice the moving average must be truncated, but ignore that. An attractive feature is easy updating:

$$y_{t+1} = (1-p)x_{t+1} + py_t.$$

Show that the gain of the filter is

$$G_y(\lambda) = \frac{1-p}{\sqrt{1 - 2p\cos(2\pi\lambda) + p^2}}.$$

Residuals may be defined: $z_t = x_t - y_t$; the residual filter has weights p, $-(1-p)p$, $-(1-p)p^2$, ... Show that the gain of this filter is

$$G_z(\lambda) = \frac{2p\sin(\pi\lambda)}{\sqrt{1 - 2p\cos(2\pi\lambda) + p^2}}.$$

(Because the lag of the moving-average filter is not 0 except when $\lambda = \frac{1}{2}$, the gains of the two filters do not add to 1 in general.) Specimen values:

	$p = \frac{1}{2}$		$p = \frac{3}{4}$		$p = \frac{7}{8}$	
λ	$G_y(\lambda)$	$G_z(\lambda)$	$G_y(\lambda)$	$G_z(\lambda)$	$G_y(\lambda)$	$G_z(\lambda)$
0	1	0	1	0	1	0
0.1	0.75	0.47	0.42	0.78	0.21	0.91
0.2	0.52	0.61	0.24	0.84	0.11	0.93
0.3	0.40	0.65	0.18	0.85	0.08	0.93
0.4	0.35	0.66	0.15	0.86	0.07	0.93
0.5	0.33	0.67	0.14	0.86	0.07	0.93

13. Economic and other time series may refer to equally spaced times less than a year apart. Many published series are quarterly or monthly, and may be expected to show a more or less regular seasonal effect with frequency 1/4 or 1/12, respectively, relative to a time unit equal to the spacing of the times. A regular seasonal effect need not be simply sinusoidal, but if it is expressed as a sum of sinusoids the frequencies will be positive integer multiples of the basic frequency just mentioned. Some kinds of daily series may be expected to show a weekly effect with frequency 1/7; an hourly series may be expected to show a diurnal effect with frequency 1/24; and so on. Published economic series are often "seasonally adjusted"; consider, for present purposes, an unadjusted series, such as some of those in *Business Statistics*.

Suppose that a seasonal effect with frequency $1/m$ is expected, where m is an integer not less than 2. How may one isolate the effect, and see whether it is progressively changing? To isolate an additive seasonal effect, subtract from the given series a symmetric moving average of odd length having zero gain at frequencies (not exceeding $\frac{1}{2}$) that are positive integer multiples of $1/m$. A constant additive seasonal effect is not absorbed by such a moving average and so remains intact in the residuals. The cosine-weighted moving average of length $2m - 1$ will serve, but unless it is modified in some way at the ends of the series, the moving-average series and the residual series are shorter than the given series by $2(m - 1)$ elements. The shortest possible suitable moving average is the uniform moving average of length m, if m is odd; or when m is even, a nearly uniform moving average of length $m + 1$, having $m - 1$ middle weights equal to $1/m$ and first and last weights equal to $1/2m$. Then the moving-average series and the residual series are shorter than the given series by only $m - 1$ or m elements, according to whether m is odd or even. (Cosine weights and the nearly uniform weights coincide when $m = 2$, which is not a very interesting case. Usually in practice $m \geqslant 4$.)

Print out the residual series so that it appears in successive rows of a rectangular table having m columns. The column averages are estimates of the (average) seasonal effect. If from all the entries in each column the corresponding column average is subtracted, final residuals are obtained. The given series (with some elements discarded from each end) has now been expressed as the sum of moving average, constant seasonal effect and final residuals. The final residuals should be examined for evidence of trends in the columns, which would indicate that the seasonal effect was progressively changing.

Usually economic time series do not have a nearly constant additive seasonal effect until after logarithms have been taken. The U.S. monthly money-supply series, M_1 to M_5, after logarithms have been taken, show seasonal effects that are gradually changing—the winter values tending to become lower, the summer values higher. The U.S. monthly series of electric power production or consumption (in logarithms) have shown, since the 1950's, a striking change in July and August, presumably due to increase in air-conditioning.

A time series, together with its moving average, may be plotted against time, using two different symbols, by means of the function *TDP*.

14. In order that a moving average with weights $\{w_j\}$ should not absorb a constant seasonal effect with integer period m, the sum of w_j over all values of j satisfying $j \equiv k \pmod{m}$ must be the same for all k $(0 \leqslant k \leqslant m - 1)$. This property is found not only in cosine weights of length $2m - 1$ and in the uniform or nearly uniform odd-length weights mentioned in the previous exercise, but also in appropriate mixtures of uniform and cosine weights; the symmetric filter (iv) of Chapter 10 was an example of such a mixture. The sine weights (ii) of Chapter 10, however, do not have the property. If it is desired to use cosine weights as far as possible, but to extend the moving average so that as few readings as possible are lost at either end of the series, mixed uniform/cosine weights are convenient for the extension.

For example, the upper half of the table below shows total electric power production by utilities in the U.S., in millions of kilowatt hours, bimonthly from Jan.–Feb. 1969 to Nov.–Dec. 1976. The figures are taken from *1977 Business Statistics*, where a monthly series is given going back to 1947. The grouping into pairs of months makes the data easier to look at without loss of the main features. The lower half of the table shows 100 times the natural logarithm of the power production divided by the number of days in the two months. For four of the month pairs the number of days is 61, but 62 for July–Aug. and 59 or (in a leap year) 60 for Jan–Feb.

Electric power production by utilities

	J–F	*M–A*	*M–J*	*J–A*	*S–O*	*N–D*
1969	231723	224941	234426	267844	239600	243648
1970	247742	240494	249229	283327	254642	254098
1971	260782	254958	266969	291783	268862	270581
1972	281876	272194	283268	320668	291100	298217
1973	302429	287027	307983	350483	310091	301107
1974	299277	291847	309466	351740	304054	309233
1975	311190	301747	315638	356509	309827	321955
1976	334984	317317	330702	372777	328605	352074

100 × logarithm of daily power production

	J–F	*M–A*	*M–J*	*J–A*	*S–O*	*N–D*
1969	827.6	821.3	825.4	837.1	827.6	829.3
1970	834.3	828.0	831.5	842.7	833.7	833.5
1971	839.4	833.8	838.4	845.7	839.1	839.7
1972	845.5	840.3	844.3	855.1	847.1	849.5
1973	854.2	845.6	852.7	864.0	853.4	850.4
1974	853.2	847.3	853.2	864.4	851.4	853.1
1975	857.1	850.6	855.1	865.7	853.3	857.1
1976	862.7	855.7	859.8	870.2	859.2	866.1

The left half of the following table shows a moving average of the logarithmic version of the data, and the right half shows residuals from the moving average. The moving average has been calculated with 11-point cosine weights from Nov.–Dec. 1969 to Jan.–Feb. 1976. The moving average could easily have been carried back, without change of weights, before Nov.–Dec. 1969, using data from before 1969, but as an example of technique we pretend earlier figures are not available. The 11-point cosine weights are approximately as follows:

0.07 0.25 0.5 0.75 0.93 1 0.93 0.75 0.5 0.25 0.07 ÷6.

The moving-average values for Sept.–Oct. 1969 and for Mar.–Apr. 1976 have been calculated with 9-point mixed cosine/uniform weights, approximately these:

$$0.15\ \ 0.5\ \ 0.85\ \ 1\ \ 1\ \ 1\ \ 0.85\ \ 0.5\ \ 0.15\ \ \div 6;$$

and the moving-average values for July–Aug. 1969 and for May–June 1976 have been calculated with 7-point nearly uniform weights:

$$0.5\ \ 1\ \ 1\ \ 1\ \ 1\ \ 1\ \ 0.5\ \ \div 6.$$

	Moving average of 100 × log daily power						Residuals from the moving average					
	J–F	*M–A*	*M–J*	*J–A*	*S–O*	*N–D*	*J–F*	*M–A*	*M–J*	*J–A*	*S–O*	*N–D*
1969				828.6	829.7	830.7				8.5	¯2.1	¯1.5
1970	831.7	832.7	833.5	834.4	835.3	836.2	2.5	¯4.7	¯2.0	8.3	¯1.7	¯2.8
1971	837.1	838.0	838.9	839.9	840.9	842.1	2.3	¯4.2	¯.5	5.8	¯1.8	¯2.3
1972	843.4	844.7	846.1	847.5	848.8	850.1	2.1	¯4.4	¯1.8	7.6	¯1.8	¯.7
1973	851.3	852.3	852.9	853.3	853.4	853.5	2.9	¯6.6	¯.2	10.7	¯.0	¯3.0
1974	853.5	853.6	853.8	854.1	854.6	855.0	¯.4	¯6.3	¯.6	10.2	¯3.2	¯1.9
1975	855.4	855.8	856.4	857.0	857.8	858.6	1.7	¯5.2	¯1.2	8.7	¯4.5	¯1.5
1976	859.5	860.4	861.5				3.2	¯4.7	¯1.7			

The residuals from the moving average do not show any obvious trend in any column. (As was remarked at the end of the last exercise, a striking trend appears in the July–Aug. column if the whole published series is examined.) The seasonal effect may be estimated simply as the average of the seven entries in each column, namely:

$$2.1\ \ {}^{-}5.2\ \ {}^{-}1.2\ \ 8.5\ \ {}^{-}2.2\ \ {}^{-}2.0.$$

After these column means have been subtracted, final residuals are obtained (not reproduced).

Any production series relating to calendar months may be expected to show some association with the varying numbers of weekdays, weekend days and public holidays that fall in the months. Suppose, for example, that it is desired to perceive association of this bimonthly series of electric power production with the number of weekend days (Saturdays and Sundays) in each month pair. The number of weekend days in each month pair is counted (reference may be made to the blue pages of the telephone directory), from which a convenient index of weekendedness can be calculated:

$$\frac{\text{number of weekend days}}{\text{number of days}} - \frac{2}{7}.$$

This series is treated in the same way as the logarithmic power production series—the same kind of moving average is subtracted, and then the column means from the table of residuals to obtain a table of final residuals. The regression coefficient of log power production on weekendedness is simply calculated as the scalar product of the two series of final residuals, divided by the sum of squares of the final residuals for weekendedness. In principle, the relation between power production and weekendedness could vary with the season of the year, and separate regression coefficients could be calculated from the final residuals in each column. In the present instance, with data for so few years, little association is to be seen between power production and weekendedness; in any case, for this purpose, the grouping of the data into month pairs was unfortunate, because weekendedness varies more between single months than between pairs of consecutive months.

Consider how to forecast the course of electric power production after 1976, using information relating to 1976 and previous years. There are at least three kinds of methods, differing in the kinds of information invoked. One kind of method is to extrapolate the power-production series in light of its past behavior, using no information except the series itself. Another kind of method is to search for other series such that the power-production series has a strong regression relation with the other series after they have been substantially lagged. Then knowledge of the other series up to 1976 may permit good forecasting of power production for an interval after 1976 that is as long as the lags. (Quite possibly no such other series can be found.) A third kind of method is to ponder all available information—technological and political, mostly not quantitative—that seems to bear on the future of electric power production, and then guess.

Proceeding by the first method, it is natural to consider trend and seasonal effect separately. Because the seasonal effect (on the logarithmic scale) was roughly constant in 1969–1976, it seems plausible to forecast that the seasonal effect will continue as estimated above. The behavior of the moving average in 1969–1976 was more complicated. There was a nearly linear trend up to early 1973, implying a growth in production of about $6\frac{1}{2}\%$ per year, or doubling in about 11 years (a trend maintained since the 1950's, in fact); then there was a slowing, presumably caused by the oil crisis of late 1973, a brief embargo by Arab states on exportation of oil to the U.S.A.; and in 1975–1976 the growth rate returned close to its former value, much as though nothing had happened. Without taking political or other information into account, the obvious forecast of trend is to extrapolate its recent behavior. Thus one may subtract from the logarithmic daily production figures the estimated constant seasonal effect, fit a straight line to (say) the last 11 values, extrapolate the line, add on the seasonal effect, exponentiate and multiply by the number of days to obtain a predicted future course of the series. That is shown in the table below, together with what actually happened, as reported in the monthly *Survey of Current Business* up to the issue for December 1980, no further biennial supplement having appeared by then.

Electric power production by utilities

	J–F	*M–A*	*M–J*	*J–A*	*S–O*	*N–D*
1977	359148	325526	348399	393791	342394	350207
1978	370947	332906	362592	408232	361218	367998
1979	395849	352485	364819	407324	360397	366323
1980	388735	356104	365163	431486		

Forecast of power production after 1976

	J–F	*M–A*	*M–J*	*J–A*	*S–O*	*N–D*
1977	346000	336100	353200	399500	356600	360900
1978	366900	356400	374600	423600	378100	382600
1979	389000	377900	397200	449100	401000	405700
1980	419500	400700	421100	476200		

All the data given here (1969–1980), together with a moving average for the whole of this series calculated in the manner explained above, are shown plotted on the next page. The points representing the moving average have been connected

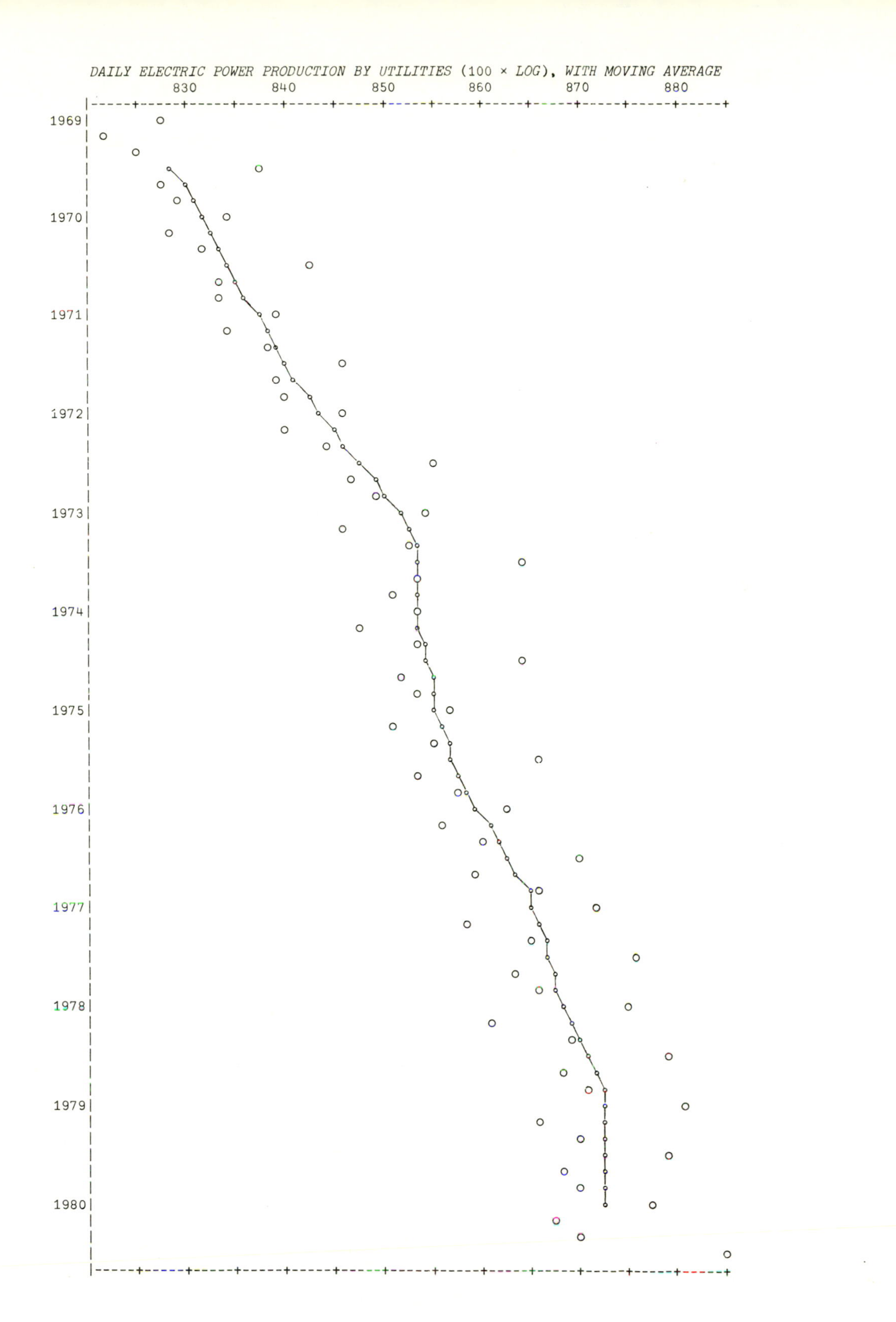
DAILY ELECTRIC POWER PRODUCTION BY UTILITIES (100 × LOG), WITH MOVING AVERAGE
830
840
850
860
870
880
1969
1970
1971
1972
1973
1974
1975
1976
1977
1978
1979
1980

by hand to make them easier to see—because of the coarse grain of the plot the moving average does not seem as smooth as it would if it were accurately plotted.

Evidently the linear projection of the trend was wide of the mark. After 1976 no unusual clairvoyance was needed (except in Detroit) to predict that a fuel shortage would return; the only question was when. Even the seasonal effect seems to have changed a little since 1976, the peak in Jan.-Feb. having grown similar in size to that for July-Aug. The moral is not to make forecasts.

15. Take samples from a pair of stationary random sequences having some simple relation, and examine the association of the sample series with the phase-difference plot yielded by *PREHAR*, and possibly also by the regression calculation of *HAR*1. Examine the same data with different amounts of prewhitening and tapering.

(a) If the two random sequences have constant spectral density (they are white noise), there is no need to prewhiten the series. When the interrelation of the series involves lags, tapering sometimes makes the trend in the phase-difference plot clearer. For example, to generate 50 observations from two random sequences, one lagged 4 time units after the other with constant gain at all frequencies:

```
X1←RNORMAL 54
X0←(¯4↓X1)+0.25×RNORMAL 50
X1←4↓X1
FT←2 FFT (10 5 ρX0),[2.5] 10 5 ρX1
PREHAR

Y0←6 TAPER X0
Y1←6 TAPER X1
FT←2 FFT (11 4 ρY0),[2.5] 11 4 ρY1
PREHAR
```

Here the series are named *X*0 and *X*1. First the phase-difference plot is obtained for the series directly, and then the series are tapered (circularized) to length 44, under the names *Y*0 and *Y*1, and another phase-difference plot is obtained. How the two plots compare will depend on the accident of the random numbers. Often the second plot shows the relation between the series a little more clearly.

Note that the functions described in *HOWHAR* presuppose that the length of the series, after any tapering, is even. A few small changes would be needed if the series had odd length, as explained in *HOWHAR*.

(b) A more interesting pair of random sequences, also with constant spectral density, is sampled in the commands below. *X*1 is white noise. *X*0 is equal to a two-point moving average of *X*1 (this constitutes a red noise), to which is added independent blue noise consisting of the first difference of a white noise. The lag is 4.5 time units, the same at all frequencies, but the correlation between the series decreases with increasing frequency. The trend in the phase-difference plot is linear, clear at low frequencies, vanishing at high frequencies.

```
X0←(1↓X0)-¯1↓X0←RNORMAL 61
X0←0.5×X0+¯4↓(1↓X1)+¯1↓X1←RNORMAL 65
X1←5↓X1
FT←2 FFT (10 6 ρX0),[2.5] 10 6 ρX1
PREHAR

Y0←6 TAPER X0
Y1←6 TAPER X1
FT←2 FFT (9 6 ρY0),[2.5] 9 6 ρY1
PREHAR
```

(c) A pair of random sequences with frequency-dependent lag may be sampled (using the function *MA* in Appendix 3):

```
X0← 1 2 1 0 ¯1 2 ¯1 MA X1←RNORMAL 66
X1←¯1↓5↓X1
X0←0.25×X0+RNORMAL 60
```

Use the function *GAINLAG* to verify that the squared gain of the filter ¯1 2 ¯1 0 1 2 1 is not very far from constant, and the lag is close to 5 in the frequency interval $(0,\frac{1}{4})$, but the phase shift $2\pi\lambda l(\lambda)$ is roughly constant for frequencies in $(\frac{1}{4},\frac{1}{2})$. Thus $X0$ has a nearly constant spectral density, and the phase-difference plot indicates a lag close to 4 in the frequency interval $(0,\frac{1}{4})$, but no lag at higher frequencies.

16. The function *HAR*1 can be tried out on any of the above sampled data sets. An interesting test is provided by the following data:

```
X0←¯5↓ 1 1 1 MA X1←RNORMAL 107
X1←¯1↓6↓X1
X0←X0+0.5×RNORMAL 100
```

The lag of $X0$ behind $X1$ is constant at 5 time units, but the gain function of the filter 1 1 1 changes sign as the frequency passes through $\frac{1}{3}$. The trend line in the phase-difference plot should therefore side-step by π at frequency $\frac{1}{3}$, and the calculation in *HAR*1 may become unstable near that frequency.

Whereas $X1$ is a white noise, the process generating $X0$ has a decidedly nonconstant spectral density. One might prefer to prewhiten it before doing the harmonic analysis. The type of unsymmetric filter used on the data in Chapter 10 is not satisfactory. Because the lag of an unsymmetric filter varies with frequency, phase relations between the series may be disturbed unless the same filter is applied to both series; but here $X0$ needs prewhitening while $X1$ does not. A symmetric filter with weights $(-a, 1, -a)$ could be applied to $X0$, where a is something like one-half of the first serial correlation coefficient of $X0$; and then $X1$ could be brought to the same length by dropping its first member and last member. This prewhitening, followed by tapering with (say) $k = 8$, could be compared with no prewhitening and tapering with $k = 10$. (The phase-difference plots are almost the same, except for some predictable redistribution of the symbols representing the weights.)

17. (a) Examine the function *HAR*1. In [8] A is determined for given B. In [10] B is improved by one step of Newton-Raphson iteration. Since the value of G is not needed in these steps, its calculation can be postponed to [11].

When *HAR*1 is understood, *HAR*2 will be seen to be similar. In [8]–[14] the two A's and two G's are found, given the two B's; this is a linear least-squares calculation followed by a polar transformation. In [16]–[22] the two B's are improved by one step of Newton-Raphson iteration: this involves determining the matrix of second derivatives of the objective function S with respect to the two B's.

(b) Write a program *HARP*, analogous to *HAR*1 and *HAR*2, for harmonic regression of a given series on any number p of predictor series. After exhaustive debugging, send a copy to the author of this book (or to his heirs and assigns in perpetuity).

18. (a) From the displayed expression in Exercise 10 (a) above, show that the square of the gain function of an m-point linear filter (of which the first and last elements of the vector of weights are both nonzero) can be expressed as a polynomial of degree $m - 1$ in $\cos 2\pi\lambda$ (the coefficient of the highest-degree term being nonzero).

Consequently, the spectral density of a stationary "moving-average" random sequence, equal to an m-point linear filtering of white noise, is a polynomial of degree $m - 1$ in $\cos 2\pi\lambda$; and the spectral density of a stationary "autoregressive" random sequence, such that an m-point linear filtering of the sequence is equal to white noise, is the reciprocal of a polynomial of degree $m - 1$ in $\cos 2\pi\lambda$.

(b) Consider how to estimate m, given a time series supposed to be realized from a stationary moving-average sequence equal to m-point filtering of white noise. An eye judgment can be made by plotting the estimated spectral density against the cosine of the angular frequency and assessing the degree of polynomial needed to fit it. It is obviously desirable to perform some averaging or smoothing of the raw line spectrum, since the latter is so variable and skew in distribution. The sort of broad sine-weighted moving average of the raw spectrum illustrated in Chapter 10 may be used and is sometimes satisfactory. The very high correlation between neighboring values, however, can make judgment difficult. Sometimes the plot may be easier to interpret if the averages do not overlap. For example, the raw spectrum can be grouped in fours: $z_1 + z_2 + z_3 + z_4$, $z_5 + z_6 + z_7 + z_8, \ldots,$ are plotted against $\cos(5\pi/n)$, $\cos(13\pi/n), \ldots,$ where n is the length of the series after tapering. The sum of four such adjacent ordinates, if the spectral density is nearly constant, is distributed nearly like a χ^2 variable with 8 d.f., having coefficient of variation 0.5 and skewness coefficient (γ_1) 1, neither intolerably large.

(c) Consider how to estimate m, given a time series supposed to be realized from a stationary autoregressive sequence such that an m-point filtering could reduce it to white noise—this is said to be an autoregressive sequence of order $m - 1$. Now we want to plot the reciprocal of the estimated spectral density against the cosine of the angular frequency. Because the reciprocal of a χ^2 variable has a much more skew distribution than the χ^2 variable itself, the estimated spectral density should have a good many equivalent degrees of freedom. To obtain a plot with nearly independent points, the raw spectrum can be grouped in twelves: $1/(z_1 + \ldots + z_{12})$, $1/(z_{13} + \ldots + z_{24})$, $\ldots$, are plotted against $\cos(13\pi/n)$, $\cos(37\pi/n)$, $\ldots$. The reciprocal of the sum of twelve such adjacent ordinates, if the spectral density is nearly constant, has coefficient of variation 0.32 and skewness 1.41. Here, as in (b), if the spectral density is far from flat, some prewhitening should be done and then its effect on the estimated spectral density should be allowed for.

Chapter 11

Regression: Public School Expenditures

A typical example of material to which simple regression methods can be applied is information in the *Statistical Abstract of the United States 1970* concerning public school expenditures and other quantities possibly related thereto. Table 181 (p. 122) lists estimated per-capita expenditures during 1970, in dollars, on public school education in each of the fifty states together with the District of Columbia (which for present purposes will be referred to as a state). The schools are elementary and secondary schools, and some other programs under the jurisdiction of local boards of education are also included; state universities are not included. The total estimated expenditures have been divided by the estimated resident population size of each state to give per-capita expenditures (not expenditures per student).

Conditions and expenditure rates for public school education vary greatly between neighboring school districts. Many inequalities have been merged together to form aggregate information for states. The per-capita expenditures shown vary considerably, from $112 for Alabama to $372 for Alaska. We can hardly expect these figures to be closely related to other aggregate information about the states, but it is interesting to see whether some relation can be detected. One would certainly expect public education expenditure to be related to the wealth of the state. It should also have some relation to need, that is, to the the numbers of persons of school age living in the state. It might further be related to the degree of urbanization of the state—urban areas permit economies resulting from operation on a larger scale than in rural areas, but also tend to have higher salary scales and other costs.

As a measure of the first of these variables, wealth, we have taken the per-capita personal income in 1968, listed in the same table as the school expenditures. Personal incomes reflect regional differences not only in living standards and conditions but also in living cost—food prices, land values, etc. For need we have taken the proportion, per thousand of the population in each state, of persons below the age of 18 in 1969, obtained by division from the total population and the population under 18 given for 1969 in Table 25 (p. 25) of the same *Statistical Abstract*. The under-eighteens include, of course, children below school age, and a small proportion of students in public education programs are over 18, so this variable is not a perfect measure of the student body, but it is one that comes readily to hand. For urbanization we have taken the proportion per thousand of the population in each state classified as urban in the 1970 census (*Statistical Abstract of the United States 1971*, Table 17, p. 18). We shall refer to the four variables as *SE*70 (school expenditures), *PI*68 (personal incomes), *Y*69 (young persons) and *URBAN*70 (proportion urban). For computing, the variables were entered as 51-element vectors under those names.

Figure B:29 shows the data, as used, except for *Y*69, which is shown rounded to the nearest tenth but has been used without rounding. The states are listed in a standard order, in nine groups and four regions named as shown. The names of states are given in the two-letter abbreviations approved by the Post Office. The last column of the table shows a one-letter label for each state; all the letters of the alphabet and then all but one of the same underscored are used. The labels are useful in plotting.

It is natural to explore the relation between these variables by performing linear least-squares regression of *SE*70 as "dependent" variable on *PI*68, *Y*69 and *URBAN*70 as "independent" variables. (Everyone agrees that the names, independent and dependent, are unfortunate here, but no others are so widely used. "Predictor" and "predicend", or "explanatory" and "explained", often seem preferable. Usage in this book is not consistent.)

A. Examination of the Data

Preliminary plots

Before any regression is calculated, there can be some interest in making scatterplots of the variables. If there were only one "independent" as well as one "dependent" variable, a scatterplot of them would most certainly be valuable, as with Fisher's data in Figure B:1. Here we have four variables in all. Four-dimensional plots are a problem to execute and, if executed, a problem to comprehend. The six possible scatterplots of the variables in pairs can, however, be executed and looked at easily enough. What can be learned from them?

In looking at such plots, we become aware of the marginal distribution of each variable. We notice, in particular, that *SE*70 has one strikingly

			SE70	PI68	Y69	URBAN70	
NORTHEAST REGION:	NEW ENGLAND STATES:	ME	189	2824	350.7	508	A
		NH	169	3259	345.9	564	B
		VT	230	3072	348.5	322	C
		MA	168	3835	335.3	846	D
		RI	180	3549	327.1	871	E
		CT	193	4256	341.0	774	F
	MIDDLE ATLANTIC STATES:	NY	261	4151	326.2	856	G
		NJ	214	3954	333.5	889	H
		PA	201	3419	326.2	715	I
NORTH CENTRAL REGION:	EAST NORTH CENTRAL STATES:	OH	172	3509	354.5	753	J
		IN	194	3412	359.3	649	K
		IL	189	3981	348.9	830	L
		MI	233	3675	369.2	738	M
		WI	209	3363	360.7	659	N
	WEST NORTH CENTRAL STATES:	MN	262	3341	365.4	664	O
		IA	234	3265	343.8	572	P
		MO	177	3257	336.1	701	Q
		ND	177	2730	369.1	443	R
		SD	187	2876	368.7	446	S
		NB	148	3239	349.9	615	T
		KS	196	3303	339.9	661	U
SOUTH REGION:	SOUTH ATLANTIC STATES:	DE	248	3795	375.9	722	V
		MD	247	3742	364.1	766	W
		DC	246	4425	352.1	1000	X
		VA	180	3068	353.0	631	Y
		WV	149	2470	328.8	390	Z
		NC	155	2664	354.1	450	A
		SC	149	2380	376.7	476	B
		GA	156	2781	370.6	603	C
		FL	191	3191	336.0	805	D
	EAST SOUTH CENTRAL STATES:	KY	140	2645	349.3	523	E
		TN	137	2579	342.8	588	F
		AL	112	2337	362.2	584	G
		MS	130	2081	385.2	445	H
	WEST SOUTH CENTRAL STATES:	AR	134	2322	351.9	500	I
		LA	162	2634	389.6	661	J
		OK	135	2880	329.8	680	K
		TX	155	3029	369.4	797	L
WEST REGION:	MOUNTAIN STATES:	MT	238	2942	368.9	534	M
		ID	170	2668	367.7	541	N
		WY	238	3190	365.6	605	O
		CO	192	3340	358.1	785	P
		NM	227	2651	421.5	698	Q
		AZ	207	3027	387.5	796	R
		UT	201	2790	412.4	804	S
		NV	225	3957	385.1	809	T
	PACIFIC STATES:	WA	215	3688	341.3	726	U
		OR	233	3317	332.7	671	V
		CA	273	3968	348.4	909	W
		AK	372	4146	439.7	484	X
		HI	212	3513	382.9	831	Y

Figure B:29

outlying value, 372 for Alaska, much higher than the next highest figure, 273 for California. The other variables do not have any such distributional peculiarity. Of the three plots of *SE*70 against each other variable in turn, that against *PI*68 shows a fairly strong relation, whether or not the Alaska observation is included. This plot demonstrates that there is "something there", that there is a regression effect to be observed. Otherwise, because the three "independent" variables may be correlated, these direct plots of the dependent against each independent variable separately are not very informative. More interesting are the remaining three plots of the "independent" variables against each other in pairs. We see that *Y*69 has little correlation with either *PI*68 or *URBAN*70, but the two latter are substantially correlated. In undesigned observations we cannot expect the "independent" variables to be perfectly orthogonal; but the less their correlation the more surely can their linear effects be disentangled. These two-variable plots are not reproduced here.

Three variables may be simultaneously plotted, after a fashion, by letting two of them define the abscissas and ordinates of points at which symbols are printed whose size indicates the value of the third variable. We call such a plot a *triple scatterplot*. Figure B:30 shows a triple scatterplot in which *PI*68 is abscissa, *Y*69 is ordinate and *URBAN*70 is roughly indicated by size and blackness of symbol. The correlation of the latter with the abscissa *PI*68 can be easily seen, as also the comparative lack of correlation of *Y*69 with either of the other variables.

In general, plots of "independent" variables against each other may be valuable in showing peculiarities in the joint distribution affecting what can be determined by regression analysis. If the observations are scattered in something like a uniform way throughout a simple region of the space of independent variables, we shall be able to determine, with whatever precision is available, the behavior of the regression function in that region. Any great unevenness in the distribution, as for example a large gap in the middle of the points, will limit our results in a way that we should preferably understand. In the present instance, nothing very peculiar is evident. One might perhaps judge that the upper right-hand point in the plot in Figure B:30 (representing Alaska) is somewhat outlying from the other points, which seem by themselves to form a satisfactory roughly uniform cloud.

As in the two preceding chapters, when computing is mentioned the emphasis is on what is computed and on the result, not on how the computing is done. However, any reader intending to use the programs listed in Appendix 3 will probably like to see examples of their execution, and so the top of Figure B:30 shows the call of the function *TSCP* and the three questions put to the user. In the plot itself, the stars indicate that two or more points fell on top of each other.

```
      PI68 TSCP Y69
SIZE?  TYPE TWO NUMBERS, SUCH AS:  25 25
⎕:
      35 34
SAME SCALES?  Y OR N.
N
SYMBOLS TO BE USED?
⎕:
      '.∘○⊖⎕⍠'[¯2+⌈URBAN70÷125]

EXTREME ABSCISSAS ARE:  2081  4425
ABSCISSA UNIT STEP IS 68.941
EXTREME ORDINATES ARE:  326.18  439.72
ORDINATE UNIT STEP IS 3.4404
```

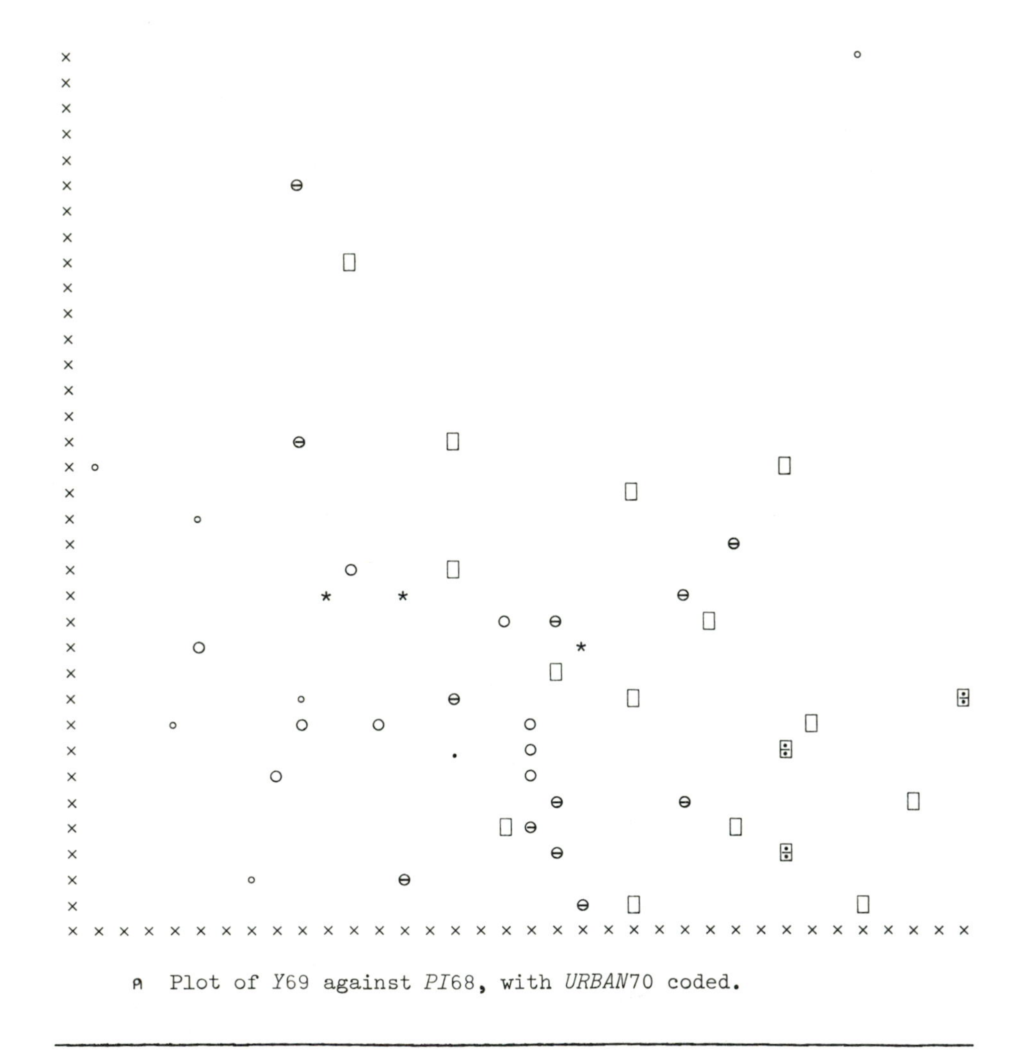

```
      ⍝ Plot of Y69 against PI68, with URBAN70 coded.
```

Figure B:30

Least-squares regression

Suppose we entertain the possibility that observations of a "dependent" variable $\{y_i\}$ are related to some "independent" variables $\{x_i^{(1)}\}, \{x_i^{(2)}\}, \ldots$ by a linear structure,

$$y_i = \beta_0 + \beta_1 x_i^{(1)} + \beta_2 x_i^{(2)} + \cdots + \epsilon_i \quad (i = 1, 2, \ldots, n), \tag{1}$$

where the β's (regression coefficients) are constants and the ϵ's are independently drawn from a normal population $N(0, \sigma^2)$; and we accordingly decide to make a least-squares regression calculation. Perhaps we are mainly interested in the values of the estimated regression coefficients. These can be succinctly expressed in APL with the divide-quad function: if Y stands for the given vector of y's, and if X is a matrix whose columns are the given vectors of x's, together with a column of ones, then the estimated regression coefficients are given by the command

$$B \leftarrow Y ⌹ X$$

The divide-quad algorithm is of high quality from a numerical analyst's point of view—Householder transformations are applied directly to the right argument, without formation of the product of X by its transpose—and execution of the primitive function is fast. Hence immediately we have the residuals RY, and the residual sum of squares, RSS:

$$RSS \leftarrow RY +.\times RY \leftarrow Y - X +.\times B$$

With designed experimental observations we know (or should know) what regression coefficients can be estimated, and how estimates of their standard errors can be obtained from the residual sum of squares. Sometimes there is a good deal of confidence that the entertained linear structure is an adequately close reflection of the truth. Then the procedure just indicated is natural and satisfactory.

With undesigned observations, such as those on school expenditures, the situation is often quite different. There are no grounds, previous experience or theoretical considerations, for expecting any particular linear structure to be adequate. Even if the structure is adequate, there may appear, within limits of observational error, to be linear dependences or "multicollinearities" between the independent variables (especially if there are many of them), and therefore not all the regression coefficients can be usefully estimated by simple least squares. We need to feel our way, to see whether the regression coefficients in an entertained structure can be estimated and whether the structure fits. Determining the coefficients in a given structure is no longer the main objective, but rather perceiving a satisfactory theoretical description for the data and estimating whatever parameters may appear in it.

Sometimes, particularly in the physical sciences, there are regression coefficients or other parameters of a hypothetical structure that can be thought of, naively perhaps, as existing and having meaning apart from the rest of the structure, and in particular apart from a postulated normally-distributed homoscedastic system of unexplained "errors". We may think that a mountain top has a height, a latitude and a longitude, definable independently of the manner in which measurement may be attempted and therefore without reference to a distribution of errors. Be that as it may, the regression coefficients of *SE*70 on the explanatory variables here assembled seem to have little or no meaning of interest except as partial descriptions of a pattern present in a particular set of data. For an entertained structure, we shall be interested to see not only the estimated regression coefficients and the residual sum of squares but also various other kinds of output, mostly graphical, that can show whether the structure fits well, can suggest how it might be changed to fit better, and can make us qualitatively aware of the behavior of the data as a whole.

In order not to be caught unawares by multicollinearity, an attractive procedure is to perform the least-squares regression calculation in stages, using what is sometimes called the modified Gram-Schmidt algorithm. The independent variables are introduced into the fitted linear structure one by one or in small groups. At each stage the vector of residuals, that is, the orthogonal complement with respect to all the independent variables so far used, is calculated not only for the dependent variable but also for all the independent variables not so far used; and then at the next stage regression calculations are made on these residuals, not on the original variables. If at any stage the residual vector of an independent variable not so far used has all its elements small, of the order of the observational round-off error for that variable, the variable should be discarded as being effectively a linear combination of the variables already used.

A set of APL programs for the procedure is described and listed in Appendix 3 and illustrated in the following figures. When "stepwise" regression is carried out in batch mode, the programmer usually formulates a complete strategy for including or discarding independent variables in the fitted structure. With interactive computing, it is satisfactory, and perhaps safer, for the user to make the choices himself.

Specimen calculations

At the top of Figure B:31 we see the matrix *X* of independent variables formed, the first column consisting of 1's, the remaining three columns of *PI*68, *Y*69 and *URBAN*70, respectively. The first two rows and the last row are printed out as a check, and then the variables that went into *X* are

```
      ρX←1,PI68,Y69,[1.5] URBAN70
51  4
      X[1 2 51 ;]
     1          2824            350.7157        508
     1          3259            345.8856        564
     1          3513            382.8715        831
      )ERASE PI68 Y69 URBAN70
      )COPY ASP REGRESSIONGROUP
SAVED  15.42.59 08/15/73

      X REGRINIT SE70

INITIAL SUM OF SQUARES (ISS), NUMBER OF READINGS AND MEAN SQUARE ARE
2073394  51  40654.78

      REGR 1

RESIDUAL SUM OF SQUARES (RSS), D.F. (NU) AND MEAN SQUARE ARE
107901  50  2158.02
POSSIBLE REDUCTIONS IN  RSS  ARE
0  48087  10476  7482

      B
196.3137  0  0  0

      REGR 2

RESIDUAL SUM OF SQUARES (RSS), D.F. (NU) AND MEAN SQUARE ARE
59814  49  1220.691
POSSIBLE REDUCTIONS IN  RSS  ARE
0  0  19538  7682

      B
17.71003  0.05537594  0  0

      REGR 3

RESIDUAL SUM OF SQUARES (RSS), D.F. (NU) AND MEAN SQUARE ARE
40276  48  839.083
POSSIBLE REDUCTIONS IN  RSS  ARE
0  0  0  6786

      B
¯300.998  0.0611792  0.8359011  0

      REGR 4

RESIDUAL SUM OF SQUARES (RSS), D.F. (NU) AND MEAN SQUARE ARE
33490  47  712.551
POSSIBLE REDUCTIONS IN  RSS  ARE
0  0  0  0

      B
¯286.741  0.08064685  0.817113  ¯0.105797
```

Figure B:31

erased. The function *REGRINIT* initializes some global output variables and displays the initial sum of squares, etc. (not particularly interesting here). The function *REGR* is used to bring one or more independent variables (columns of *X*) into the fitted regression relation. *REGR* is first executed with argument 1; regression is performed on the first column of *X*, causing the mean to be subtracted from *SE*70 and from the last three columns of *X*. The sum of squares displayed (107901) is the total sum of squares of *SE*70 about the mean. The "possible reductions in *RSS*" show how the residual sum of squares will be reduced if any one further independent variable is introduced into the regression. The first reduction is zero; nothing further will be achieved by doing regression on column 1 of *X*, because that has been done already. The other three possible reductions (48087, 10476, 7482) are substantially larger than the mean square just above (2158.02), especially the reduction corresponding to column 2 of *X*, that is, *PI*68—we have already remarked that a simple scatterplot of *SE*70 against *PI*68 showed a strong relation. After the first execution of *REGR*, the global vector *B* listing the regression coefficients is asked for. Its first element (196.31) is the mean of *SE*70, and its other elements are zero because the second, third and fourth columns of *X* have not yet been used.

We may now in subsequent calls of *REGR* introduce any or all of the remaining columns of *X*. What has been done in Figure B:31 is to introduce the columns singly, each time choosing the one that will most reduce the residual sum of squares. As it happens, that policy leads to introduction of the columns in their order of numbering, column 2, then column 3, then finally column 4. After each execution of *REGR*, the regression coefficients *B* are asked for.

At this point we can construct the following analysis-of-variance table for *SE*70:

	Sum of squares	D.f.	Mean square
For *PI*68	48087	1	48087
For *Y*69	19538	1	19538
For *URBAN*70	6786	1	6786
Residual	33490	47	713
Total about mean	107901	50	2158

The ratio of the residual mean square to the gross mean square about the mean is 0.330—about two-thirds of the variation in public school expenditure from state to state has been accounted for by the linear regression on the three explanatory variables.

The terms in the above analysis of variance relate to introduction of the explanatory variables seriatim in the order shown. These variables are not orthogonal, and the individual reductions in residual sum of squares due to introducing them in a different order would be different, as can be determined by repeating the calculation with different orders. However, whatever the order of introduction, the reduction due to *PI*68 is always greater than that due to *Y*69, which in turn is always greater than that due to *URBAN*70. That the variables are not orthogonal is seen from the print-outs of *B*. Consider for example the second element of *B*, the coefficient of *PI*68 in the fitted regression. At the first print-out this is zero, because the second column of *X* has not yet been used. Thereafter it is 0.055, 0.061, 0.081, in the successive print-outs. If the second column of *X* had been orthogonal to the other columns, these three values would have been equal. Had they decreased towards zero, instead of increasing, there would have been a suggestion that the second column of *X* was a less important variable, in presence of the others, than it appeared at its introduction; and therefore perhaps the calculation should be redone with the other variables introduced first.

The function *REGR* has several other global output variables besides *B*, and some are shown in Figure B:32, which directly continues the terminal session of the previous figure. *RX* is a matrix of the same size as *X*, the "residual *X*-matrix". Because the columns of *X* were introduced in order of numbering, 1, 2, 3, 4, the columns of *RX* are related to the columns of *X* as follows. The first column of *RX* is the same as the first column of *X*; the second column of *RX* is equal to the second column of *X* minus its projection on the first; the third column of *RX* is equal to the third column of *X* minus its projection on the first two columns of *RX* (or of *X*); and the fourth column of *RX* is equal to the fourth column of *X* minus its projection on the first three columns of *RX* (or of *X*). The columns of *RX* always span the same linear space as those of *X*, and when all columns of *X* have been introduced into the regression the columns of *RX* are orthogonal.

It would be tedious to have the whole of *RX* printed out, but we should know something of its behavior. At the top of Figure B:32 we see the first two rows and the last row of *RX*. Individual columns can be examined quickly with the function *SHOW*. As an example, *SHOW* is applied to the third column of *RX*. We see the three lowest and three highest values in the column. The highest value is 87.23, with index number 50, so that it relates to Alaska; the lowest value, ¯35.37, with index number 26, relates to West Virginia. The root-mean-square value of entries in the column is 23.42; and a frequency distribution of the values is shown having six cells of equal width, of which the extreme cells are centered at the extreme values.

The elements in any column of *RX* ought to be large relative to the observational error for that variable. Not only should the root-mean-square value of elements in the column be larger than the standard deviation of observational error, but preferably the same should be true even if a few of the largest elements were deleted. In the present case, if the source material is accurate to anywhere near the number of digits in its tabulation,

```
      RX[1 2 51 ;]
     1          ¯401.2941        ¯10.95346      ¯84.11833
     1            33.70588       ¯12.76357     ¯109.0202
     1           287.7059         25.98573      117.8096

      SHOW RX[;3]
3 LOWEST, 3 HIGHEST (WITH RANKS) AND R.M.S. VALUES
¯35.37 (26), ¯31.45 (37), ¯31.35 (9), 50.53 (45), 58.66 (43),
      87.23 (50);  23.42
FREQUENCY DISTRIBUTION:  6  22  16  4  2  1

      STATEDISPLAY ⌊0.5+RY

N.E.:  15  ¯30  18  ¯39  5  ¯60
M.A.:  37  3  21
E.N.C.:  ¯34  ¯19  ¯43  0  ¯1
W.N.C.:  51  37  1  ¯11  ¯12  ¯47  9
S.A.:  ¯2  15  ¯6  ¯2  9  ¯15  ¯14  ¯21  31
E.S.C.:  ¯17  ¯2  ¯24  ¯19
W.S.C.:  ¯1  ¯12  ¯8  ¯20
MTN.:  43  ¯2  33  0  29  17  11  ¯36
PAC.:  2  51  51  16  ¯10


      VARIANCE
ESTIMATED VARIANCE MATRIX OF B IS:  XPXI×RSS÷NU

      +S←(1 1 ⍉V←XPXI×RSS÷NU)*0.5
64.90336  0.009298499  0.1597494  0.03428252

      (1 1 ↓V)÷(1↓S)∘.×1↓S
   1              0.09324518    ¯0.6784222
   0.09324518     1              0.03811034
  ¯0.6784222      0.03811034     1


      STRES

      SHOW DIAGQ
3 LOWEST, 3 HIGHEST (WITH RANKS) AND R.M.S. VALUES
0.47 (50), 0.79 (45), 0.8 (43), 0.98 (15), 0.98 (25), 0.98 (14);  0.92
FREQUENCY DISTRIBUTION:  1  0  0  3  13  34

      STATEDISPLAY ⌊0.5+SRY

N.E.:  15  ¯30  20  ¯39  5  ¯61
M.A.:  37  3  21
E.N.C.:  ¯33  ¯19  ¯42  0  ¯1
W.N.C.:  50  36  1  ¯11  ¯12  ¯46  8
S.A.:  ¯2  15  ¯6  ¯2  9  ¯15  ¯14  ¯20  31
E.S.C.:  ¯16  ¯2  ¯24  ¯19
W.S.C.:  ¯1  ¯12  ¯8  ¯20
MTN.:  42  ¯2  32  0  32  17  12  ¯37
PAC.:  2  50  51  23  ¯9
```

Figure B:32

all columns of *RX* are clearly significantly different from zero. (*REGR* has a trap built into it, preventing use of an independent variable if the corresponding column of *RX* is too near zero—see *HOWREGRESSION* in Appendix 3.)

Another item of global output is *RY*, the residuals of the dependent variable (here *SE*70). *RY* is displayed in the figure, its elements being rounded to the nearest integer and arranged according to the nine groups of states (New England, Middle Atlantic, etc.) by a function named *STATE-DISPLAY*.

Sometimes it is convenient to supplement the analysis-of-variance table by obtaining estimated standard errors, or an estimated variance matrix, for the regression coefficients *B*. This may be done by executing the function *VARIANCE*, which causes a global output variable *XPXI* to be formed. When (as here) all columns of *X* have been brought into the regression, *XPXI* is equal to the inverse of the matrix product of *X*-tranpose and *X*; it is calculated from *RX* rather than by direct matrix inversion. *XPXI* multiplied by the residual mean square is the usual estimated variance matrix for *B*, on the assumption that the structure (1) is exactly correct and that all needed independent variables are linearly independent and have been brought into the regression. In the figure, the square roots of the diagonal elements of this estimated variance matrix are named *S* and displayed; they are estimated standard errors of *B* (on the assumption just mentioned). In the following line the correlation matrix of the second, third and fourth elements of *B* is asked for. It will be seen that the correlation of greatest magnitude is ¯0.68 between the second and fourth elements of *B*, due to the correlation between *PI*68 and *URBAN*70 already noted.

The next thing shown in Figure B:32 is execution of a function *STRES* causing a global output variable *SRY* of "standardized residuals" to be formed and also another global output variable named *DIAGQ*. The residuals *RY* can be expressed as the inner product of the dependent variable (*SE*70) and a symmetric idempotent matrix, say *Q*, expressible as a function of *X* and *XPXI*. Then, on the assumption just mentioned that the structure (1) is exactly correct, etc., $Q\sigma^2$ is the variance matrix of *RY*. The output variable *DIAGQ* is the vector of diagonal elements of this *Q*, and so proportional to the variances of the elements of *RY*. It can be shown that the sum of the elements in *DIAGQ* must be equal to the number of residual degrees of freedom, ν say (here 47). If all members of *DIAGQ* were equal, they would be equal to ν/n (here 0.92). The output variable *SRY* is found from *RY* by dividing each element by the corresponding member of the square root of (n/ν) times *DIAGQ*. If all members of *DIAGQ* were equal, *SRY* would be just the same as *RY*.

DIAGQ is of some interest. Its elements necessarily lie between 0 and 1, both limits included; and their average value is ν/n, which is usually (as here) close to 1. If any observation has values of the independent variables far removed from the other observations, this observation will have a relatively large effect on the estimated regression coefficients, and the corresponding element of *DIAGQ* will be usually low. In the figure, *SHOW* is applied to *DIAGQ*, and one very low value, 0.47, is seen for the fiftieth state, Alaska. Figure B:30 showed that the combination of independent-variable values for Alaska was unusual. Thus Alaska has had a large weight in determining the regression relation. Are school expenditures for Alaska consistent with those of other states? The modest value of the standardized residual for Alaska, 23, suggests that they are. We return to this question below.

The principal use to be made of the residuals, *RY* or *SRY*, of the dependent variable is in scatterplots showing how well the assumed regression structure fits. Figure B:33 shows the standardized residuals of *SE*70 plotted against the fitted values *on the same scale.* Because the plot is a good deal broader than it is high, we see at a glance that the fitted regression relation accounts for the greater part of the total variation in *SE*70. The symbols used in plotting indicate multiplicity as follows: the three characters, circle, circle overstruck with minus, circle overstruck with star, denote a single point, two coincident points, three or more coincident points, respectively. (As it happens, the third character is not needed here.) From the plot we see that the marginal distribution of the residuals is reasonably symmetric and normal-looking, and the residuals do not show any obvious relation with the fitted values—no definite suggestion of, say, a quadratic regression on the fitted values, or a progressive change in variance as we go from low to high fitted values. The outlying high fitted value relates to Alaska.

Another way of displaying the relation of residuals to fitted values is illustrated by the triple scatterplot in Figure B:34. We divide the independent variables into two groups; *PI*68, being the most important variable, has been taken as one group, and *Y*69 and *URBAN*70 together as the other group. The abscissas in the plot are the contributions of *PI*68 to the fitted values, and the ordinates are the combined contributions of *Y*69 and *URBAN*70 to the fitted values, on the same scale. The plotting symbols represent the standardized residuals *SRY* coded on a seven-point scale from large-negative to large-positive:

$$\underline{M} \quad M \quad - \quad \circ \quad + \quad P \quad \underline{P}$$

(The letter *M* stands for minus and *P* for plus.) We see that *PI*68 has a greater effect on fitted values than the other two independent variables together, and also that one state (Alaska again) is represented by a point far

```
      (SE70-RY) SCATTERPLOT SRY
SIZE?  TYPE TWO NUMBERS, SUCH AS:  25 25
⎕:
      35 30
SAME SCALES?  Y OR N.
Y

EXTREME ABSCISSAS ARE:  135.15   355.71
ABSCISSA UNIT STEP IS 6.4872
EXTREME ORDINATES ARE:  ¯64.872  51.898
ORDINATE UNIT STEP IS 6.4872


x               o         o o
x
x                   o o       o
x         o           o o
x                                                                     o
x               o o       o
x             o   o           o
x   o         o   o       o
0 o o - - - - ⊖ o - o - o o - - o - - o - - - - - - - - - - - - - - - -
x   o                       o         o
x         o o o   o   o
x     o o     ⊖           o
x o
x                     o o
x                       o       o       o
x                   o
x
x                                     o
x
x x x x x x x x x x x x x x x x x x x x x x x x x x x x x x x x x x x x

      A  Plot of standardized residuals of SE70 against fitted values.
```

Figure B:33

from the others. The residuals do not seem to have a quadratic regression relation with abscissas and ordinates; we do not see, as sometimes in such plots, a saddle effect, one pair of opposite corners having mostly positive residuals, and the other pair mostly negative. There is some suggestion that the lower right part of the plot has more extreme residuals than the upper left part, i.e. that there is a progressive change in variance of the residuals from upper left to lower right.

The residuals may also be plotted against each of the dependent variables in turn. Sometimes it is interesting to make what is called a "partial residual" or "component plus residual" plot. Considering one of the independent variables, say *PI*68, we subtract from the dependent variable its fitted response to the *other* independent variables, obtaining values for *SE*70 that are "corrected" for departures of the other independent variables

```
      CODE←'M̲M-∘+PP̲'
      Z←CODE[⌊4.5+SRY×3÷⌈/|SRY]
      +/CODE∘.=Z
1  5  12  17  7  8  1

      (X[;2]×B[2]) TSCP X[; 3 4]+.×B[3 4]
SIZE?  TYPE TWO NUMBERS, SUCH AS:  25 25
⎕:
      35 35
SAME SCALES?  Y OR N.
Y
SYMBOLS TO BE USED?
⎕:
      Z

EXTREME ABSCISSAS ARE:  167.83  356.86
ABSCISSA UNIT STEP IS 5.5599
EXTREME ORDINATES ARE:  174.65  308.09
ORDINATE UNIT STEP IS 5.5599


×                                                             +
×
×
×
×
×
×
× -               P
×
×         -
×                   - +   -   +
×                 -       P
×                 - o -
×       o -                       P
×                 -     +     +       P             o   M
×             o                     -   *   o   o
×               o             *     *             +
×
×                                     *     M
×                                   o P         o
×                         o                               M       M̲
×                                 P     +
×                                                   M   P̲
×                                                       o             o
×                                           o                 P
× × × × × × × × × × × × × × × × × × × × × × × × × × × × × × × × × × × ×

      ⍝ Contribution of Y69 and URBAN70 to fitted value for SE70 against
      ⍝ that of PI68 to same, with standardized residuals of SE70 coded.
```

Figure B:34

from (say) their means. These corrected values are then plotted against the selected independent variable. We have noted that public education expenditure should be related to the wealth of the state. A plausible conjecture, worth testing, is that for two states having the same values for *Y*69 and *URBAN*70 the school expenditures would be proportional to the values for *PI*68. If that were not so, perhaps the school expenditures would be proportional to a power of *PI*68 not very different from 1. It is therefore interesting to see the corrected values for *SE*70 plotted against *PI*68 with the origin included in the plot. Such a plot appears in Figure B:35, where the origin is marked with a small circle and the states are identified by the one-letter code given at the right side of Figure B:29. (The states are identified in this plot, but were not in Figure B:33—clearly either plot could have been made in the style of the other, at the whim of the user.) We notice two things in the plot: first, school expenditures seem to be roughly proportional to *PI*68, perhaps more closely proportional to a power of *PI*68 just above 1; second, the dispersion (in linear measure) of *SE*70 about the regression curve seems to increase proportionately with *PI*68. Similar "partial residual" plots can be made for *Y*69 and *URBAN*70—not reproduced here. That for *Y*69 also indicates rough proportionality of the two variables, but not the proportional dispersion. That for *URBAN*70 shows a reasonably linear relation between the variables, but the regression line does not go anywhere near the origin.

Evaluation

We have seen quite a lot of computer output, too much perhaps, concerning the linear regression of *SE*70 on *PI*68, *Y*69 and *URBAN*70. What are we to make if it?

The regression function that we have fitted, linear in the three independent variables, seems reasonably satisfactory. However, the last plot shown suggests forcefully that our assumption of constant residual variance is false, and that it would be fairer to try to relate the logarithms of *SE*70 and *PI*68. Presumably the logarithm of *Y*69 should also be taken, but there is no clear need to transform *URBAN*70. Accordingly, it is reasonable to repeat all the foregoing calculations after taking logarithms of the dependent variable and of the first two independent variables. The output is mostly not reproduced here, but the findings are briefly reported.

It seems (as we shall consider in some detail below) that now the residual variance is indeed nearly constant. When our former kind of analysis-of-variance table is constructed anew, a slightly surprising feature appears. Having expressed the data on what is apparently a fairer scale, we might have expected that the ratio of residual mean square (47 d.f.) to gross mean

```
      XX←X[; 3 4]- 51 2 ρ(+⌿X[; 3 4])÷51

      (0,X[;2]) TSCP 0,SE70-XX+.×B[3 4]
SIZE?  TYPE TWO NUMBERS, SUCH AS:  25 25
⎕:
      35 32
SAME SCALES?  Y OR N.
N
SYMBOLS TO BE USED?
⎕:
      '∘',51↑A̲

EXTREME ABSCISSAS ARE:  0  4425
ABSCISSA UNIT STEP IS 130.15
EXTREME ORDINATES ARE:  0  307.98
ORDINATE UNIT STEP IS 9.9348
```

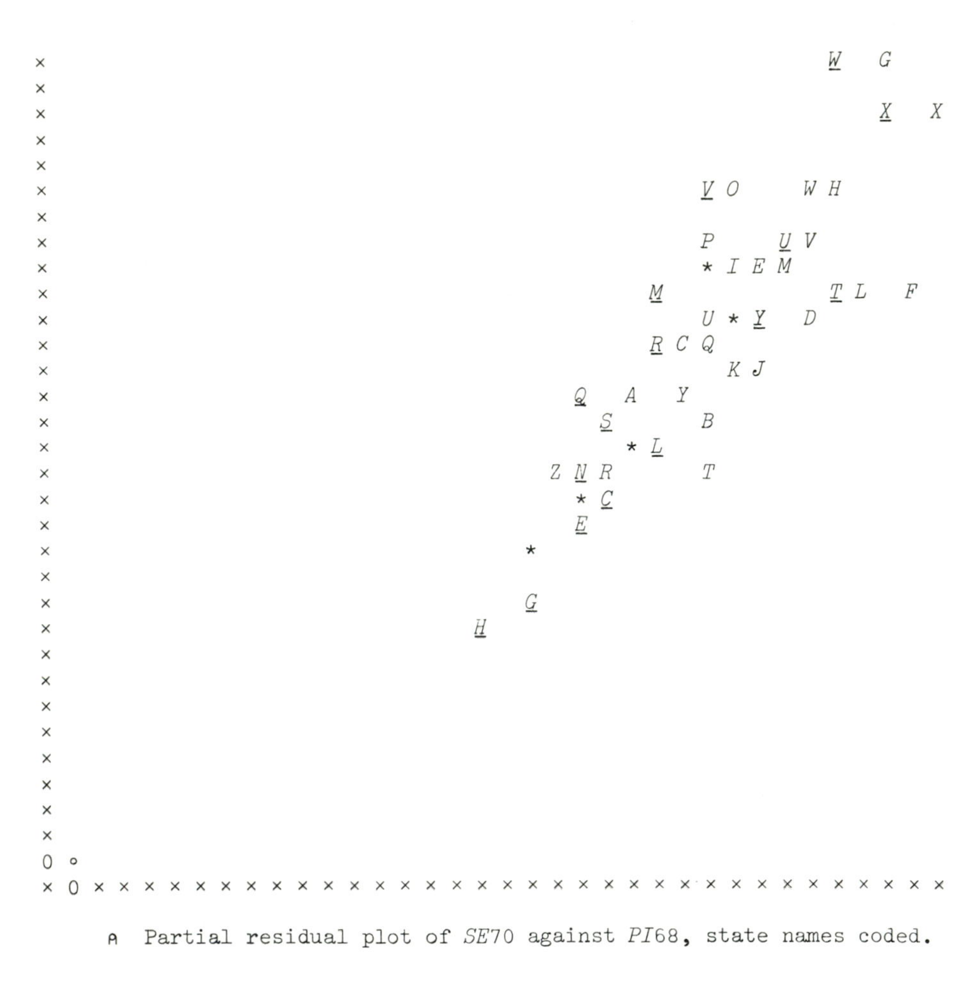

⍝ Partial residual plot of SE70 against PI68, state names coded.

Figure B:35

square about the mean (50 d.f.) would be a little smaller than before, but it is almost exactly the same, 0.331. Plots of the residuals against log*PI*68 and log*Y*69 suggest a slightly curved relation. Perhaps we should consider a regression function that is quadratic in the independent variables rather than linear. There would then be six more regression coefficients, corresponding to the three squares and the three products of the independent variables in pairs. So we start again with an *X*-matrix of size 51-by-10. (For numerical stability the means may be subtracted from each independent variable before the squares and products are formed.)

If the second-degree terms are introduced into the regression one at a time (after the constant and first degree terms), we find that only three are required to achieve nearly all the possible reduction in residual sum of squares: the terms are log*PI*68 squared, log*Y*69 squared, and the product of log*PI*68 and *URBAN*70. These seem an odd assortment. For the purpose of judging whether the second-degree terms are needed, it is probably fairer to treat all six together rather than pick out any subset. We get then the following analysis of variance for log*SE*70:

	Sum of squares	D.f.	Mean square
For log*PI*68 (linear)	13166	1	13166
For log*Y*69 (linear)	3799	1	3799
For *URBAN*70 (linear)	1227	1	1227
Quadratic terms	1811	6	302
Residual	6392	41	156
Quadratic + residual	8203	47	175
Total about mean	26394	50	528

(To be precise, the dependent variable here is 100 times the natural logarithm of *SE*70; and the first two independent variables have been defined similarly.)

The ratio of the mean square for quadratic terms (6 d.f.) to the residual mean square (41d.f.) comes at about the upper 10% point of Fisher's variance-ratio (*F*) distribution. Neither fish nor fowl! One prefers a linear relation to a quadratic one, in general, because it is simpler to think about and would probably extrapolate better (if that were an issue). On the other hand, there seems here to be no theoretical reason to suppose that the regression surface should be flat, and if it is slightly curved a quadratic expression could well be adequate. In fitting the full quadratic function of the three independent variables we encounter other instances of the phenomenon noticed when just a linear function was fitted, that one state (Alaska) had considerably greater weight in the fitting than the other states. Now besides Alaska there are two other states that have large individual weights in the fitting of the quadratic regression, as indicated by low values of *DIAGQ*—0.028 for Alaska, 0.22 for Massachusetts, 0.36 for West Virginia. No doubt the quadratic regression has been unsatisfactorily estimated.

In the absence of a compelling reason either to retain the quadratic terms or to reject them, let us try to proceed doing both.

In the logarithmic regression analysis, then, the residual variation is 30% (with quadratic terms) or 33% (without them) of the gross variation about the mean. The regression coefficients of log*SE*70 on log*PI*68 and log *Y*69 are 1.28 and 1.35, with conventional estimated standard errors 0.15 and 0.29, respectively, and correlation coefficient 0.12—these results from the analysis without quadratic terms are completely analogous to those in Figure B:32. The fitted relation can be expressed:

$$SE70 \propto (PI68)^{1.28} \times (Y69)^{1.35} \times \exp(^{-}0.00045 \times URBAN70). \qquad (2)$$

The conventional estimated standard error of the multiplier inside the exponential is 0.00017. One may similarly express the fitted relation when the quadratic terms are included; the more complicated formula has little direct interest.

The exponents of *PI*68 and *Y*69 in (2) are larger than 1, but not very significantly so. It could be argued that exponents of 1 would represent a just and equitable allocation of resources. With that in mind, one may perform a linear regression of log*SE*70, with log*PI*68 and log*Y*69 both subtracted, on *URBAN*70. The residual mean square is 183 (49 d.f.), the plot of residuals against the independent variable shows nothing to cause concern, and corresponding to (2) we have

$$SE70 \propto (PI68) \times (Y69) \times \exp(^{-}0.00025 \times URBAN70). \qquad (3)$$

The residuals from any of these fitted regression relations are interesting to study. Those of largest magnitude indicate states whose rate of expenditure for public school education is most different from a pattern abstracted from all the states. The author was amused to notice in Figure B:32 that his state of residence, Connecticut, had the largest residual, a negative one, and that Minnesota, Oregon and California had the next largest residuals, all positive. Is it fair to stigmatize Connecticut for outstanding lack of support of public school education, and compliment those other states for their fine performance? In a sense, no, for no account has been taken of the source of funding. But we may properly ask, in which states are school expenditures most out of line?

Any such evaluation would be better based on the logarithmic type of analysis. The residual variation being more nearly constant, the fitted values can be accepted as a norm to which all kinds of states, richer or poorer, having more or fewer children, and being more or less urbanized, seem to conform impartially. Figure B:36 shows standardized residuals for the three variant analyses already mentioned, (i) with quadratic terms fitted,

Standardized residuals of log *SE*70

Analysis (i), quadratic regression on three variables
N.E.: 7 ¯15 8 ¯22 ¯8 ¯15
M.A.: 14 ¯2 3
E.N.C.: ¯15 ¯6 ¯14 5 2
W.N.C.: 22 16 ¯3 ¯9 ¯10 ¯25 3
S.A.: 6 11 3 0 ¯1 ¯10 ¯3 ¯9 13
E.S.C.: ¯9 3 ¯2 6
W.S.C.: 11 ¯4 ¯10 ¯9
MTN.: 18 1 15 1 1 8 ¯2 ¯10
PAC.: 3 19 21 ¯1 ¯6

Analysis (ii), linear regression on three variables
N.E.: 9 ¯15 12 ¯20 1 ¯25
M.A.: 18 3 10
E.N.C.: ¯17 ¯9 ¯19 0 0
W.N.C.: 22 18 ¯1 ¯2 ¯3 ¯26 4
S.A.: ¯1 7 0 ¯1 7 ¯7 ¯3 ¯11 14
E.S.C.: ¯11 ¯4 ¯20 ¯5
W.S.C.: ¯1 ¯4 ¯10 ¯13
MTN.: 21 1 15 ¯1 21 9 9 ¯15
PAC.: 2 24 22 ¯4 ¯4

Analysis (iii), linear regression on one variable
N.E.: 9 ¯14 17 ¯21 ¯3 ¯21
M.A.: 19 2 8
E.N.C.: ¯17 ¯6 ¯17 4 2
W.N.C.: 24 19 ¯3 ¯1 ¯1 ¯26 4
S.A.: 5 10 2 ¯2 2 ¯8 ¯6 ¯12 9
E.S.C.: ¯14 ¯10 ¯26 ¯9
W.S.C.: ¯7 ¯6 ¯16 ¯16
MTN.: 23 0 18 ¯2 20 8 7 ¯9
PAC.: 3 23 23 15 ¯2

```
×     o
×               o           o o                             o
× o         o
×       o                   o
× o               ⊖                     o             o
×   o                       o
×       o   ⊖
0 o - - o - - - - - - ⊛ - ⊖ - - ⊖ - o - - - - - - - - - - - - - - - - - -
×             o               ⊖   o   o o o                             o
×                   o   o
×   o             o             o
×               o             o         o
×       o         o o     o
×
×           o     o
× × × × × × × × × × × × × × × × × × × × × × × × × × × × × × × × × × × ×
```

Standardized residuals of log *SE*70 in Analysis (ii) against log *Y*69.

Figure B:36

(ii) without quadratic terms fitted, i.e. logarithmic deviations from relation (2) above, (iii) linear regression on *URBAN*70, assuming regression coefficients on log*PI*68 and log*Y*69 both equal to 1, i.e. logarithmic deviations from relation (3) above. Standardized rather than unadjusted residuals are shown, for no strong reason—the two sorts are almost the same. The three analyses yield patterns of residuals that are mostly similar but have a few noticeable differences also. The largest residual in all three is negative, for Nebraska. Thereafter the analyses give somewhat discrepant rankings. If we think there may be some truth in all three analyses and therefore average their results, we obtain Massachusetts and Connecticut as the next lowest states after Nebraska. Analyses (ii) and (iii) would place Alabama with these, but not analysis (i), which gives Alabama a very small residual. As for the largest positive residuals, the three analyses agree in giving highest residuals to Minnesota, California, Oregon, Montana, in various orders. Again there is a state, New Mexico, that comes close to those in analyses (ii) and (iii), but not in analysis (i).

The actual magnitudes of these residuals can be interpreted directly as approximate percentages by which *SE*70 is above or below its fitted value.

The tables of residuals can be examined from a slightly different point of view, for evidence of regional effects. There is a preponderance of positive residuals among the Mountain and Pacific states, and of negative residuals among the southern states. Any attempt to measure a regional effect will be somewhat unsatisfactory. The *Statistical Abstract of the United States* arranges the states in four regions, as shown in Figure B:29, corresponding roughly with everyday terminology (except that the second region is named North Central instead of Mid West), but hardly denoting sharp cultural or economic separation. For want of anything better, one can modify the previous analyses by estimating separate means for each region rather than a common mean, thereby reducing the number of residual degrees of freedom by 3. The regression coefficients are changed, and the principal regional effect that emerges is a difference between the West Region and the other three together, school expenditures being higher in the west. Conceivably such an apparent regional effect could be caused by regional biases in the data. Conceivably it could be real and due to cultural differences or even to some inherent geographical factor. No attempt has been made to pursue the matter.

A summary may be helpful. We began by noticing that the per-capita school expenditure figures varied considerably from state to state, the highest being nearly three-and-a-half times the lowest. School budgets result from careful deliberation by many governing bodies, local, state, federal. Many decisions are made, by persons who are not compelled to think alike. Differences of public sentiment and political power, perhaps even idiosyncratic opinions of single influential persons, could result in wide differences in school expenditure rates. Nevertheless, all those responsible for drawing up the budgets will be conscious of similar problems, how to make best use of seemingly inadequate resources. One might hope that some of the observed variation in expenditures could be rationalized by relation to resources, needs and physical conditions, without reference to politics or opinion. We accordingly chose three readily accessible variables that in some degree indicated resources, needs and conditions. Regression of the school expenditure rates on these variables accounted for about two-thirds of the total variance. The fitted regression relation given above at (2) seems plausible. There are still sizable percentage deviations by some states from the common pattern, as shown by the residuals in Figure B:36. Of our three explanatory variables, probably the least satisfactory is the proportion urban (*URBAN*70), since areas classified as urban in the census include both city centers and suburbs, in which school conditions are very different. Perhaps the tightness of the regression relation would be increased if proportions of inner-city and suburban populations could be brought in separately.

B. Some Technical Questions

For the remainder of this chapter, let us turn to some technical statistical questions prompted by the foregoing study: (a) should Alaska be omitted? (b) how may tests of goodness of fit of the structure (1) be made? (c) what is typical behavior of *DIAGQ*? (d) would the method of least squares be better replaced by a "robust" method of fitting?

Should Alaska be omitted?

Alaska has had a greater effect than any other state in determining the fitted regression on the three explanatory variables—either as at first in Figure B:32, or in the logarithmic versions, both analysis (i) with quadratic terms and analysis (ii) without. Alaska is an untypical state, influenced by special conditions and forces; and the thought occurs that perhaps it should be excluded from the analysis. On reflection, however, one realizes that Alaska is not unique in its uniqueness. There are other states that must surely also be considered extraordinary, in other ways and for other reasons: Hawaii, the District of Columbia, California, Florida, New York, North Dakota, ... Perhaps not every state is equally extraordinary, but it is hard to argue that any particular state is too different from all others to be included, without opening the door to erosion of many of the rest for similar reasons.

So let us not reject Alaska because it is Alaska, but ask merely whether it conforms with the pattern visible in the rest of the data. The easiest thing to do is to repeat analyses (i) and (ii) with Alaska omitted, to see what difference there is. The principal difference is that the standard errors of the regression coefficients are a bit increased. For example, in analysis (ii), the numerical coefficients (two exponents and a multiplier) on the right side of (2) are replaced by 1.31, 1.42, ¯0.00048, with conventional estimated standard errors 0.18, 0.36, 0.00020. The analysis of variance now goes:

	Sum of squares	D.f.	Mean square
For log*PI*68 (linear)	10543	1	10543
For log *Y*69 (linear)	2096	1	2096
For *URBAN*70 (linear)	1059	1	1059
Quadratic terms	1792	6	299
Residual	6390	40	160
Quadratic + residual	8182	46	178
Total about mean	21879	49	447

The ratio of residual mean square to gross mean square about the mean is now 36% for analysis (i) and 40% for analysis (ii). The fitted values for the states other than Alaska are very little changed from their previous values. The residual for Alaska, i.e. the difference between the observed value of log*SE*70 and the extrapolated fitted value from the regression with Alaska omitted, is ¯10 for analysis (i) and ¯6 for analysis (ii), both modest values.

In sum then, when Alaska is omitted the rest of the data tell nearly the same story, but with reduced precision.

With most sorts of data (perhaps with all), we cannot be absolutely certain that every observation is sound and acceptable at face value. Possibly something has occurred which, if we knew of it, would convince us that an affected observation ought not to be treated like the others but should be excluded. Because we cannot be completely confident about every observation, a fitted regression relation should preferably not depend critically on just one or two readings but rather should seem to pervade all the readings. If on the contrary one or two readings largely determined some coefficient in the fitted regression, we should know about that. Often in designed experiments, each observation is planned to give the same amount of information about the treatment effects. The only way such an observation can have a great influence on an estimated coefficient is by having a large residual (of the dependent variable). But in undesigned studies, one observation can have a large effect because of an outlying combination of independent variables, as here with Alaska. If we were sure that the observation was fully trustworthy, we should congratulate ourselves on having such an informative observation. But because a slight doubt lurks, it seems wise to check whether the specially informative observation is consistent with the others before resting content with its information. Alaska has passed this test.

Significance tests

If suitable scatterplots are made of the residuals *RY* (or *SRY*), as illustrated in Figures B:33, 34, 35, there is often little need to make formal significance tests. Figure B:33 seemed to show that the residuals of *SE*70 were not related in any interesting way to the fitted values, and Figure B:35 showed a striking progressive change in residual variance with *PI*68. In neither case did a calculated test seem called for. On the other hand, in the logarithmic analysis (ii), the plot of the residuals (*SRY*) against log *Y*69 in Figure B:36 gives a slight suggestion that the residual variance may be decreasing as *Y*69 increases. There is no a-priori impossibility that the residual variance should depend on *Y*69 or any other explanatory variable. In a borderline case a calculated test is interesting.

It is common practice among statisticians, as R. A. Fisher said, to be disinclined to introduce a complication into a simple theoretical structure unless a significance test has indicated a "significant" discrepancy between the data and that structure. Such use of significance tests is philosophically unclear and has been disparaged by some authorities. However, the whole process by which theoretical descriptions or "models" of data are chosen is philosophically unclear, and those who disapprove of significance tests have cast little light on it. Most persons actively engaged in statistical analysis of data think in terms of significance tests when considering goodness of fit of a theoretical structure. Moreover, the test statistics are generally suggested by appearances in scatterplots, and not chosen in advance. Rightly or wrongly, we follow that practice here. How one reacts to a particular test may depend on how important the effect tested is judged to be. Although in the above logarithmic analysis the ratio of the mean square for quadratic terms to the residual mean square did not achieve the traditional distinction of a 5% significance level, there seemed to be some merit in including the quadratic terms in the regression when calculating the residuals; and for just a few states inclusion of the terms did make a noticeable difference. But in regard to constancy of residual variance, some moderate nonconstancy probably does not matter much for our purposes. To act as though we believed the residual variance to be constant seems reasonable unless we find rather strong evidence to the contrary.

For theoretical discussion of residuals, it is convenient to have some ordinary algebraic notation. Consider fitting the structure (1) by least squares, where there are p regression coefficients $\beta_0, \beta_1, \ldots, \beta_{p-1}$, corresponding to linearly independent vectors $\{1\}, \{x_i^{(1)}\}, \ldots, \{x_i^{(p-1)}\}$. Let $\{z_i\}$ be the residuals of $\{y_i\}$, which we have previously denoted by *RY*.

The matrix Q or $\{q_{ij}\}$ yields $\{z_i\}$ from $\{y_i\}$:

$$z_i = \Sigma_j q_{ij} y_j = \Sigma_j q_{ij} \epsilon_j.$$

(Suffixes i, j, k will always range from 1 to n, the number of observations.) Q has rank $\nu = n - p$. We have already remarked that the average value of the diagonal elements $\{q_{ii}\}$ is ν/n. If ν is close to n, the ith row of Q will (for most if not all values of i) have its diagonal element q_{ii} close to 1 and all the others close to 0; and so z_i mostly reflects ϵ_i. But if ν is considerably less than n, z_i does not so closely reflect ϵ_i, being an average of all the ϵ's with not so very unequal weights. Then peculiarities in the behavior of the ϵ's, departure from the assumed independent $N(0, \sigma^2)$ distribution, can no longer be seen easily by looking at the z's.

Sometimes we are unsure how many x-variables to bring into the regression, how many regression coefficients should be estimated. When our purpose is to examine residuals in order to observe goodness of fit of the assumed structure, it is better to err on the side of estimating too few regression coefficients rather than too many. It is better not to estimate small regression effects, but rather leave them as small biases in the residuals, so that the unexplained variations in the data can be more directly observed. Tukey has suggested a rule of thumb, to be used when regression coefficients have some natural sequence or order of importance: continue estimating coefficients and subtracting the effects from the residuals only so long as the ratio of residual mean square to the number of residual degrees of freedom diminishes. In the logarithmic analysis above, the linear effects of the independent variables were naturally estimated first, and then the quadratic effects. After only the linear terms had been fitted, the ratio of residual mean square to residual degrees of freedom was 3.71, and after the quadratic terms it was 3.80, slightly larger. Therefore, for testing goodness of fit, the residuals without quadratic terms fitted, analysis (ii), are the better ones.

When the structure (1) is exactly correct, the residuals $\{z_i\}$, given the β's, the x-values and σ^2, have a singular joint normal distribution with zero means and variance matrix

$$\{\text{cov}(z_i, z_j)\} = \sigma^2 \{q_{ij}\}.$$

Thus the moment-generating function of the joint distribution for z_i and z_j, for any i and j (equal or unequal), is

$$\mathcal{E}(\exp(z_i t_1 + z_j t_2)) = \exp\left[\tfrac{1}{2}\sigma^2(q_{ii} t_1^2 + 2q_{ij} t_1 t_2 + q_{jj} t_2^2)\right].$$

By expanding the right side as a Taylor series one may determine any moments in the joint distribution of z_i and z_j. For example, the variance matrix of the squared residuals is

$$\left\{\text{cov}(z_i^2, z_j^2)\right\} = 2\sigma^4 \{q_{ij}^2\}.$$

Let $\{x_i\}$ be any given vector, perhaps one of the independent variables used in the regression, perhaps something else. Consider the possibility that the structure (1) is correct except that the ϵ's are not homoscedastic; that is, we assume that the ϵ's are independent and ϵ_i has the distribution $N(0, \sigma^2 \exp[\alpha(x_i - \bar{x})])$, where α is a constant close to 0, σ is a positive constant, and $\bar{x}$ is some average or central value of the x's, to be chosen conveniently. We could try to estimate α by doing some kind of regression of $\{z_i^2\}$ on $\{x_i\}$. To begin with, suppose we decide arbitrarily to use the statistic $\Sigma_i z_i^2(x_i - \bar{x})$. By virtue of the idempotency of Q we have

$$q_{ii} = \Sigma_j q_{ij}^2,$$

$$\mathcal{E}(z_i^2) = \sigma^2\left[q_{ii} + \alpha \Sigma_j q_{ij}^2(x_j - \bar{x}) + O(\alpha^2)\right].$$

Let us define $\bar{x}$ as follows:

$$\nu\bar{x} = \Sigma_i q_{ii} x_i.$$

Then

$$\mathcal{E}\{\Sigma_i z_i^2(x_i - \bar{x})\} = \alpha\sigma^2 \Sigma_i \Sigma_j q_{ij}^2(x_i - \bar{x})(x_j - \bar{x}) + O(\alpha^2).$$

This suggests that we take as an estimate of α

$$a = \frac{\Sigma_i z_i^2(x_i - \bar{x})}{s^2 \Sigma_i \Sigma_j q_{ij}^2(x_i - \bar{x})(x_j - \bar{x})}, \tag{4}$$

where s^2 is the residual mean square,

$$\nu s^2 = \Sigma_i z_i^2.$$

Naturally we suppose that the denominator on the right side of (4) does not vanish, that is, that $s^2 > 0$ and the members of $\{x_i\}$ are not all equal and the double sum does not vanish.

On the null hypothesis that $\alpha = 0$, this statistic a, being a homogeneous function of $\{z_i\}$ of degree zero, is independent of s^2. (When independent $N(0, \sigma^2)$ variables are expressed in polar coordinates, the radius vector is independent of the angles; that fact permitted Fisher, 1930, to determine exact moments of the distribution under the null hypothesis of his measures, g_1 and g_2, of skewness and kurtosis.) The expectation of the square of the numerator on the right side of (4) can be written

$$\mathcal{E}\big(\Sigma_i \Sigma_j z_i^2 z_j^2(x_i - \bar{x})(x_j - \bar{x})\big) = 2\sigma^4 \Sigma_i \Sigma_j q_{ij}^2(x_i - \bar{x})(x_j - \bar{x}).$$

Since $\mathcal{E}(s^4) = (\nu + 2)\sigma^4/\nu$, we have

$$\mathcal{E}(a) = 0,$$

$$\mathcal{E}(a^2) = \frac{2\nu}{(\nu + 2)\Sigma_i \Sigma_j q_{ij}^2(x_i - \bar{x})(x_j - \bar{x})}.$$

One can similarly determine higher moments of the distribution of a, on the null hypothesis—for example,

$$\mathcal{E}(a^3) = \frac{8\nu^2 \Sigma_i \Sigma_j \Sigma_k q_{ij} q_{ik} q_{jk}(x_i - \bar{x})(x_j - \bar{x})(x_k - \bar{x})}{(\nu + 2)(\nu + 4)\left[\Sigma_i \Sigma_j q_{ij}^2 (x_i - \bar{x})(x_j - \bar{x})\right]^3}.$$

The right side here, being an odd function of $\{x_i - \bar{x}\}$, is a measure of asymmetry of the x's.

The double sum appearing in the denominator of the definition of a, and in these expressions for its variance and third moment, is computationally a bit expensive to evaluate. When ν is close to n, the double sum is well approximated by its diagonal terms,

$$\Sigma_i q_{ii}^2 (x_i - \bar{x})^2,$$

since the off-diagonal terms are not only small but tend to cancel each other. Similarly for the triple sum in the numerator of the expression for $\mathcal{E}(a^3)$.

Applying this test to our logarithmic analysis (ii), to see whether the residual variance changes with *Y*69, we find, using the approximating single sum in the denominator,

$$a = ^{-}0.0296, \quad \text{s.e.}\,(a) = 0.0366,$$

or with the exact double sum,

$$a = ^{-}0.0285, \quad \text{s.e.}\,(a) = 0.0359.$$

Thus the visual impression is confirmed that evidence of association is only weak. The standard error has been calculated on the assumption that the fitted regression structure is correct, without quadratic terms.

By contrast, Figure B:35 suggested a strong relation between residual variance in our first analysis and *PI*68. Testing this, we find with the approximating single sum,

$$a = 0.000987, \quad \text{s.e.}\,(a) = 0.000392,$$

or with the double sum,

$$a = 0.000947, \quad \text{s.e.}\,(a) = 0.000384.$$

Either way, a differs from zero by about two and a half standard errors.

We proposed the test statistic appearing as numerator in the definition of a, equation (4), arbitrarily. Would some other statistic have served better? Would the unadjusted residuals $\{z_i\}$ have been better replaced by standardized residuals, for example? The statistic chosen in (4) is probably not optimal in general. The variance matrix of the squared residuals being proportional to $\{q_{ij}^2\}$, one could by inverting this matrix determine the linear combination of the squared residuals constituting a minimum-variance unbiased estimate of $\alpha\sigma^2$, for α close to 0. That hardly seems to be worth the trouble. In one special case the expression (4) can be shown to be best of its kind, namely when the dependent variable consists of several samples, the rth being of size n_r ($n_r \geqslant 1$ for all r, $\Sigma_r n_r = n$), and the independent variables are effectively indicators for these samples, so that when all regression coefficients have been estimated the mean of each sample has been found, and the fitted values are just the sample means. Let the x-variable chosen to compare the residual variance with be itself an indicator of the samples, in the sense that it takes the same value for every member of the rth sample, for each r. Then among functions that are linear in the squared residuals, our statistic as^2 defined at (4) can be shown to be the minimum-variance unbiased estimate of $\alpha\sigma^2$ when α is close to 0. The result holds however unequal the n_r's may be. This seems to dispose of any suggestion that the unadjusted residuals would generally be better replaced by standardized residuals, and affords some ground for hope that our statistic is always reasonably good.

We have considered a particular test for heteroscedasticity. In similar fashion one may develop tests related to other features seen in the scatterplots. One may make a quadratic regression of residuals on fitted values, this being Tukey's "one degree of freedom for nonadditivity". One may test for dependence of residual variance on the fitted values, adapting the heteroscedasticity test just considered by using the fitted values instead of an x-vector. One may test for nonnormality in the distribution of the ϵ's by summing third or fourth powers of the residuals. The fact that, under the null hypothesis, the residuals have (in general) unequal variances and are unequally correlated need not inhibit us from formulating and using the same kinds of test statistics as if no regression parameters had been estimated and the residuals were identical with the "unexplained variation", the ϵ's. Summary information about a number of such tests is given in Appendix 2.

Behavior of *DIAGQ*

Usually in designed experiments the regression structure intended to be fitted has the property that the diagonal elements $\{q_{ii}\}$ of Q are equal. In unplanned studies, as we have seen, that is not so—we know merely that

$$0 \leqslant q_{ii} \leqslant 1 \quad (1 \leqslant i \leqslant n), \quad \Sigma_i q_{ii} = \nu.$$

Typically how unequal are the diagonal elements of Q? Can one formulate plausible conditions under which they are likely to be nearly equal?

The simplest case of interest is as follows. There is just one independent variable $\{x_i\}$ and we consider the simple linear regression structure,

$$y_i = \beta_0 + \beta_1 x_i + \epsilon_i \quad (1 \leqslant i \leqslant n),$$

where β_0 and β_1 are constants, the ϵ's are independent $N(0, \sigma^2)$, and let us suppose the x's are drawn independently from some distribution. Then

$$q_{ii} = 1 - \left\{ \frac{1}{n} + \frac{(x_i - \bar{x})^2}{\Sigma_j (x_j - \bar{x})^2} \right\} \quad (1 \leqslant i \leqslant n). \tag{5}$$

To a statistician the most natural distribution to postulate for the x's, at least as a point of reference if not as the most likely eventuality, is a normal one. Let us try to say something about the least diagonal element, $\min_i q_{ii}$, in that case, and in particular find its median value.

For any fixed i, q_{ii} is essentially a beta variable. We may write

$$q_{ii} = 1 - \frac{1}{n}\left\{1 + \frac{(n-1)t_{n-2}^2}{(n-2) + t_{n-2}^2}\right\},$$

where t_{n-2} is a random variable having Student's distribution with $(n-2)$ degrees of freedom. Since the right side is a monotone function of t_{n-2}^2, we can determine the distribution of q_{ii} from Student's distribution. For any constant u between 0 and $(n-1)/n$,

$$P(u) \equiv \text{prob}(q_{ii} < u) = \text{prob}\left(t_{n-2}^2 > (n-2)\left[\frac{n-1}{nu} - 1\right]\right).$$

Now $nP(u)$ is the expected number of diagonal elements of Q that are less than u. Let u_0 be such that

$$nP(u_0) = \tfrac{1}{2}.$$

Since the number of diagonal elements less than u_0 is a (random) nonnegative integer, and there is positive probability that the number is not less than 2, the probability that no diagonal element of Q is less than u_0 exceeds $\frac{1}{2}$. Therefore the median of $\min_i q_{ii}$ exceeds u_0; for at the median the probability that no diagonal element of Q is to the left is just $\frac{1}{2}$.

The actual distribution of the number of diagonal elements of Q that are less than a given u cannot be determined easily, because the diagonal elements are not independent. It is plausible to conjecture, however, that if $nP(u) < 1$ and n is not very small the distribution will be close to a Poisson distribution. The Poisson distribution for which the probability of 0 is $\frac{1}{2}$ has its mean equal to $\ln 2 = 0.693$. Let us therefore define u_1 thus:

$$nP(u_1) = \ln 2.$$

Then u_1 is an approximate value for the median of $\min_i q_{ii}$.

A cruder approximation when n is large is obtained by writing

$$q_{ii} \approx 1 - (1 + \chi_1^2)/n, \tag{6}$$

where χ_1^2 is a random variable having the tabulated chi-squared distribution with one degree of freedom, the square of a $N(0,1)$ variable. This is obtained from (5) by setting $\bar{x} = 0$ and $\Sigma_j (x_j - \bar{x})^2 = n\,\text{var}(x_i)$. Hence

$$P(u) \approx \text{prob}(\chi_1^2 > n(1-u) - 1).$$

An approximation for the median of $\min_i q_{ii}$ is u_2 defined by

$$n\,\text{prob}(\chi_1^2 > n(1 - u_2) - 1) = \ln 2.$$

For example, when $n = 51$, our lower bound to the median of the least member of *DIAGQ*, under the assumption that the one independent variable is normally distributed, is

$$u_0 = 0.854,$$

and our approximate values for the median are

$$u_1 = 0.865, \quad u_2 = 0.861.$$

In fact in our analysis (iii) above, the two smallest members of *DIAGQ* were both approximately 0.88. The independent variable *URBAN*70 does not show any striking nonnormality.

Suppose now that there are $(p - 1)$ independent variables and we fit a linear regression structure in those variables. If we postulate that the independent variables have a joint nonsingular normal distribution, we can in principle write similar expressions to the above, but the only easy one corresponds to (6), namely

$$q_{ii} \approx 1 - (1 + \chi_{p-1}^2)/n, \tag{7}$$

and hence

$$n\,\text{prob}(\chi_{p-1}^2 > n(1 - u_2) - 1) = \ln 2.$$

For example, when $n = 51$ and there are three jointly-normal independent variables, our approximation to the median of $\min_i q_{ii}$ is

$$u_2 = 0.771.$$

In Figure B:32 the least member of *DIAGQ* was 0.47, and in the logarithmic analysis (ii) it was 0.51, all other members being greater than 0.77. As we have seen, the independent variables do not appear to be a sample from a normal distribution, having one strongly outlying point.

For our logarithmic analysis (i) the three independent variables were joined by six functions of them to make a quadratic regression. Our preceding results can be extended to this case. For quadratic regression on $(p - 1)$ independent variables having a joint nonsingular normal distribution, we have, corresponding to (7),

$$q_{ii} \approx 1 - \{(p + 1) + \chi_{p-1}^4\}/2n,$$

and hence

$$n\,\text{prob}\left(\chi_{p-1}^2 > \sqrt{2n(1 - u_2) - (p + 1)}\right) = \ln 2.$$

For quadratic regression on three jointly-normal independent variables, with $n = 51$, u_2 so defined is negative ($^-0.167$). Evidently the approximations made are too crude here. The corresponding result for quadratic regression on only one normal independent variable, with $n = 51$, is $u_2 = 0.607$, and a small simulation suggests that this is close to the true median of $\min_i q_{ii}$.

Robust regression

The method of least squares is sensitive to nonnormality in the distribution of errors. A single extreme outlier among the dependent-variable values can throw the whole thing out, utterly changing the estimated regression coefficients and the analysis of variance from what they would have been had the outlier been closer in or omitted from the calculation. Therefore it is inadequate, in regression calculations, to see only the minimal sufficient statistics for the implied linear structure (with its assumption of independent homoscedastic normal errors). All too often the minimal statistics are all that are yielded by computer regression programs. The other kinds of output illustrated above, including scatterplots of the residuals, offer powerful protection against misunderstanding at little cost.

In recent years, as was remarked briefly in the last chapter, there has been considerable interest in modifying or replacing the method of least squares so as to achieve insensitivity to outliers or to a leptokurtic distribution of errors, yet retaining tolerable efficiency when the errors are normal. In some fashion, reduced weight, possibly zero weight, is given to the more outlying dependent-variable values, as compared with the inlying ones. Possibly also reduced weight may be assigned to observations for which the independent variables are outlying. The idea is to fit the regression relation so that most of the observations come close to it, little or no attention being paid to a small proportion of outlying observations.

Perhaps the most satisfying way of doing robust regression is to fit a fully defined structure, just like (1) except that the ϵ's are postulated to have a distribution belonging to some family of which the members have various shapes, including normality as a special case. For example, one can postulate that the ϵ's are independently drawn from a common distribution of Karl Pearson's Types II or VII, having probability differential element proportional to

$$\left\{1 + \frac{4\alpha\epsilon^2}{(2-\alpha)^3\sigma^2}\right\}^{-1/\alpha} \frac{d\epsilon}{\sigma},$$

where σ is a scale parameter and α is a shape parameter, either positive but less than 2 (for a Type VII distribution) or negative (for a Type II distribution). The normal distribution is the limit as $\alpha \to 0$. If $\alpha > 0$, ϵ may take any real value, but if α is negative ϵ is restricted to the finite interval for which the expression in braces is positive. This is Jeffreys's form of Pearson's distributions, in which the parameters σ, α and any one of the regression coefficients are in his terminology mutually orthogonal. When $\alpha = 1$ we have the Cauchy distribution (or, apart from scale, the Student distribution with 1 degree of freedom). Jeffreys suggested the distribution with $\alpha = 1/4$ (Student distribution with 7 degrees of freedom) as representing errors of measurement in astronomy. The uniform distribution is the limit as $\alpha \to -\infty$. The interval $-\frac{1}{2} < \alpha < 1$ corresponds to a great range of distribution shapes from short-tailed to long-tailed.

One may proceed to fit this structure by examining the likelihood function. In order to obtain results that are intelligible in the same way as those of a least-squares analysis when normality is assumed, one may integrate out the two nuisance parameters α and σ with respect to some suitable prior distribution to obtain a marginal likelihood function for the regression coefficients. (For example, one could let σ and α have independent prior distributions, that for $\ln\sigma$ uniform over the real line, that for α uniform over $(-\frac{1}{2}, 1)$.) If the marginal likelihood can be well approximated by a multivariate Student density, the information about the regression coefficients can be expressed finally in the form of estimated coefficients, together with estimated standard errors (or a complete variance matrix) to be understood as though based on a normal-theory residual mean square having a stated number of degrees of freedom.

A by-product of this approach is direct information about the shape of the error distribution—we see the range of values for α that are reasonably compatible with the data. When positive values for α are more compatible than negative values, observations having a large residual in the dependent variable do not have much effect on the determination of the regression coefficients—really extreme outliers have almost no effect. In this approach (as in others mentioned below), outliers among the independent-variable values are *not* discounted but allowed to contribute fully.

This treatment is computationally expensive, enough to give one pause. It is a rather desperate way to determine whether the least-squares analysis was much colored by the implied assumption of normality. To postulate the Type II-VII error distribution, with the extra parameter α, is much weaker than postulating normality ($\alpha = 0$); and the Pearson distributions seem intrinsically quite plausible, unlike some other nonnormal distributions that have been considered in this context. However, even this weaker assumption, like the normality assumption, may be false, for it still implies that the errors are homoscedastic and symmetrically distributed (not skew). The analysis may be more trustworthy than a least-squares analysis, but cannot be claimed to be wholly unassailable. If the residuals of the dependent variable in the least-squares analysis strongly indicated a long-tailed error distribution, the analysis based on the Type II-VII assumption would be likely to give parameter estimates, fitted values and residuals rather different from those of least squares, more like what would be yielded by least squares if some oberservations were omitted. What is most interesting is to compare the two analyses when the least-squares residuals do *not* strongly indicate nonnormality. Does *assuming* normality when there is an *appearance* of normality really have much effect on our judgment?

The data at hand are a suitable test case for this question. Let us consider the logarithmic analysis (ii), linear regression of $\log SE70$ on log*PI*68, log*Y*69 and *URBAN*70. The scatterplots of residuals (of which only one has been reproduced here in Figure B:36) did not show any departure from normality. In fact one can calculate a kurtosis statistic based on the sum of fourth powers of the residuals. This is done in Appendix 2, Figure B:47. An equivalent normal deviate of ¯0.74 is obtained. Thus platykurtosis, rather than leptokurtosis, is suggested, but not strongly.

The likelihood function for the Type II-VII assumption about the error distribution has been examined, after numerical integration with respect to the nuisance parameters σ and α. Independent uniform prior distributions for $\ln\sigma$ and for α have been supposed, respectively over the real line and over the interval $(-\frac{1}{2}, 1)$. The prior mean for α is $\frac{1}{4}$, the value suggested by Jeffreys; to this extent the prior distribution for α has a slant towards long-tailedness. There is a technical reason for preferring that the distribution for α should not extend below $-\frac{1}{2}$. Our procedure involves integration with respect to σ of the likelihood function and also of its first and second partial derivatives with respect to the regression coefficients. More computational effort is required for this integration when $\alpha < -\frac{1}{2}$ because the second derivatives are then unbounded. Other computational difficulties are encountered if α is allowed to range much above 1 towards its theoretical upper limit of 2. The interval $(-\frac{1}{2}, 1)$ for α is for practical purposes very broad, far more so than its measure in this parametrization suggests, but is indeed less broad than the theoretical extreme of $(-\infty, 2)$, and it does not necessarily include all possibilities of practical interest.

The present material favors values for α at the lower end of the interval. In fact, if α were not restricted, the likelihood function would be maximized for α in the neighborhood of ¯0.7. The interval $(-\frac{1}{2}, 1)$ has nevertheless been adhered to. Contributions from values of α in the top third of this interval have been very small.

The results of the analysis can be summed up by saying that they are almost indistinguishable from those of least squares. With an extra digit to show the differences more clearly, our least-squares results were expressed by equation (2) with the numerical coefficients

1.275 1.354 ¯0.000448

having estimated standard errors (derived from a residual mean square with 47 degrees of freedom)

0.146 0.294 0.000169

and correlation matrix

$$\begin{bmatrix} 1 & 0.12 & ^{-}0.68 \\ 0.12 & 1 & 0.02 \\ ^{-}0.68 & 0.02 & 1 \end{bmatrix}$$

(In addition to the above three regression coefficients, there was also a fourth, the intercept term or proportionality factor, omitted from (2) because not very interesting, and we continue to omit it.)

The marginal likelihood function for the Type II-VII analysis can be approximated by a multivariate Student density, and therefore signifies roughly the same as if different observations had been obtained for which we could assert normality of errors, yielding these numerical coefficients in place of those in (2):

$$1.260 \quad 1.371 \quad {}^{-}0.000436$$

having estimated standard errors (derived from a residual mean square with 54 degrees of freedom)

$$0.146 \quad 0.294 \quad 0.000170$$

and correlation matrix

$$\begin{bmatrix} 1 & 0.18 & {}^{-}0.68 \\ 0.18 & 1 & {}^{-}0.01 \\ {}^{-}0.68 & {}^{-}0.01 & 1 \end{bmatrix}$$

One of these numerical coefficients has changed by about a tenth of its estimated standard error, and the other two by less than that. The standard errors themselves are almost unchanged. The residuals of the dependent variable after subtracting the fitted values, when rounded to the nearest integer, are mostly the same as the least-squares residuals, a few differing by one unit. We may surely conclude that the implied assumption of normality in the least-squares method colored our results very little.

Quicker methods

The robust regression methods usually considered are much faster than the foregoing Type II-VII method. They can sometimes be expressed as minimizing a function of the residuals other than the sum of squares. Usually in principle the methods are iterative, but with ingenious choice of starting values one iterative step may suffice. In effect, the assumption of normality is replaced by an assumption that the error distribution is longer-tailed. The results will be particularly good if indeed the error distribution is longer-tailed in the right amount.

Theoretical considerations are usually in terms of sampling variances and efficiencies. Let us review a few relevant facts. Consider for simplicity estimation of a single location parameter; and suppose that the errors, if they are not normally distributed, have a common distribution of Pearson's Type II or VII, as above. Then the sample mean (the least-squares estimate of the population median) is fully efficient only when $\alpha = 0$ (normality). If $\alpha \geqslant \frac{2}{3}$ (Student distribution with 2 or less degrees of freedom) the error distribution does not have a finite variance, and so neither does the sample mean, but except when n is very small the maximum-likelihood estimate has a finite variance; thus the large-sample efficiency of the mean is 0. If $\alpha \leqslant -1$, the maximum-likelihood estimate is "superefficient", Fisher's measure of information being infinite, and so again the large-sample efficiency of the mean is 0. The sample median does much better than the mean when α is above, say, $\frac{1}{2}$, but decidedly less well when α is negative, as is indicated in the following brief table of large-sample efficiencies relative to the maximum-likelihood estimate. The ratio of the large-sample variance of the median to that of the mean increases as α decreases, reaching 3 for the uniform distribution ($\alpha = -\infty$).

Efficiencies of the mean and the median of large samples from a Pearson Type II or VII distribution

$\alpha =$	-1	$-\frac{3}{4}$	$-\frac{1}{2}$	$-\frac{1}{4}$	0	$\frac{1}{4}$	$\frac{1}{2}$	$\frac{3}{4}$	1
Mean:	0	0.39	0.70	0.92	1	0.89	0.50	0	0
Median:	0	0.18	0.35	0.50	0.64	0.74	0.81	0.84	0.81

These large-sample efficiencies are somewhat misleading for two reasons. First, they do not reflect small-sample behavior—when the number of observations is 2, for example, the mean and the median are equal to each other (and possibly to the maximum-likelihood estimate) and are therefore equally efficient (possibly fully efficient). Second, the zero efficiency of the mean when $\alpha \leqslant -1$ signifies differently from the zero efficiency when $\alpha \geqslant \frac{2}{3}$. For when $\alpha \leqslant -1$, the sample mean and its usual estimated standard error are good and usable in the ordinary way, though indeed much less good than what could be achieved if the value of α were correctly known. But when the error distribution is long-tailed and $\alpha \geqslant \frac{2}{3}$, the mean and its estimated standard error will sometimes be grossly misleading—the sample mean is a bad estimate, catastrophically bad. This is the zero efficiency that must be guarded against. There is a general well-founded belief that error distributions with longer-than-normal tails do occur in practice. Less is known (because the matter is not commonly inquired into) about the occurrence of shorter-than-normal tails. It is long tails that have the limelight.

To fit a family of error distributions, as we did above with the Pearson Type II-VII family, is to attempt to deal uniformly fairly and correctly with all members of the family considered. The quicker robust methods are designed as an acceptable compromise for some set of longer-tailed distributions—a rather broad set. Their performance for shorter-tailed distributions is not generally considered.

Huber's method, to minimize the sum of a function ρ of the residuals defined in terms of a positive number k by

$$\rho(z) = \tfrac{1}{2}z^2 \quad \text{if } |z| \leqslant k,$$
$$= k|z| - \tfrac{1}{2}k^2 \quad \text{if } |z| \geqslant k,$$

was the first to be established and studied, and though now it is perhaps not the most highly regarded it can still be thought of as the type of such methods. If the observations are very numerous, k can be chosen in accordance with the apparent distribution shape of the residuals. But for small data sets one may prefer to choose k iteratively so that a prearranged proportion (such as 20%) of the residuals exceed k in magnitude.

Applied to the present data, with k chosen by the 20% rule, Huber's method yields these numerical coefficients in place of those in (2):

$$1.296 \quad 1.325 \quad {}^{-}0.000464$$

As could be expected, the changes from least squares are in the opposite direction to those for the Type II-VII method. Estimated standard errors are somewhat larger than before,

$$0.164 \quad 0.328 \quad 0.000182$$

and the correlation matrix is

$$\begin{bmatrix} 1 & {}^{-}0.01 & {}^{-}0.67 \\ {}^{-}0.01 & 1 & 0.08 \\ {}^{-}0.67 & 0.08 & 1 \end{bmatrix}$$

We have remarked that in most regression studies we are not sure beforehand what sort of regression relation should be fitted nor which of the available variables should appear in it, and therefore we try several versions. When the method of least squares is abandoned, we lose the convenient arithmetic possibility of doing regression by stages, the residuals of one stage serving as input to the next. Moreover, we (possibly, presumably) lose our familiar way of gauging by an F test whether introduction of further independent variables has perceptibly improved the fit. Consider applying Huber's method (with the 20% rule) successively to the present data, first fitting only a constant term, then linear regression on log*PI*68, then linear regression on that and on log*Y*69, finally linear regression on all three variables. We may construct something looking like an analysis of variance, relating to the minimized sum of rhos of residuals of log*SE*70:

	Sum of rhos	
Reduction due to log*PI*68	5657	(k reduced to 21.95)
Further reduction due to log*Y*69	1591	(k reduced to 20.20)
Further reduction due to *URBAN*70	644	(k reduced to 18.31)
Final	3996	(k = 18.31)
Initial	11887	(k = 27.35)

The ratios of these terms to each other are not greatly different from those of the corresponding sums of squares in the least-squares analysis (ii). If we wished to test the significance of the regression on one or more independent variables, would it be near enough correct to divide the terms by the usual numbers of degrees of freedom and refer the ratio to the F distribution?

Another question concerning the Huber analysis is, does the distribution shape of the residuals change progressively as more independent variables are introduced into the regression relation? With the method of least squares, if the error distribution is not normal, the marginal distribution of each residual becomes more nearly normal by the central limit effect, as more independent variables are introduced. Is there a case, with Huber's method, for progressively changing the proportion of residuals to exceed k in magnitude, as the number of independent variables increases, and if so how?

One may ask corresponding questions concerning any other method of "robust" fitting, including the Type II-VII method. We recommend such questions, as Graunt puts it, to the attention of the curious.

To bring these ruminations on robustness to a conclusion, we offer the following suggestions:

(a) The method of least squares should never be used blind (though it often is). If after some reasonable examination of goodness of fit the implied theoretical structure seems satisfactory, and is plausible in light of any available considerations apart from the material at hand, the least-squares analysis should be accepted—it is often the simplest and cheapest analysis available, and the one best supported by theory.

(b) If the examination of goodness of fit indicates that the implied theoretical structure is satisfactory except that the error distribution is longer-tailed than normal, a quick robust method, such as Huber's, may be illuminating.

(c) Indiscriminate replacement of least squares by any robust method, without appropriate examination of goodness of fit, perhaps with the notion that goodness of fit does not matter, would be a misfortune for statistical science.

(d) When data must be processed blind, without human consideration, for the sake of controlling an immediate action, and not to improve understanding, robust methods are indeed safer than simple least squares. Such automatic data processing is not what is understood here by statistical analysis.

Notes for Chapter 11

This chapter, together with Appendix 2, illustrates and continues several previous discussions of regression (Anscombe, 1961, 1967, 1973). Other treatments of these topics have been given by Draper and Smith (1966), Daniel and Wood (1971),

Ehrenberg (1975), Mosteller and Tukey (1977), Hoaglin and Welsch (1978), Belsley, Kuh and Welsch (1980).

On "partial residual" plots see Larsen and McCleary (1972) and Wood (1973).

The procedure for robust regression by fitting a Pearson Type II or VII distribution of errors has been previously described in detail (Anscombe, 1967, sections 2.3 and 2.4).* The reciprocal of the shape parameter here denoted by α was there denoted by m. Integration of the likelihood function with respect to $\ln(\sigma_0^2/\sigma^2)$ has been by two-point Gaussian quadrature over equal intervals. Integration with respect to α has been by Simpson's rule over values from $^-0.5$ to 1 by steps of 0.125. Otherwise, the procedure indicated there has been followed exactly. See Jeffreys (1948) for his parametrization of the Type II and VII distributions and discussion of robust estimation.

On other robust methods see Huber (1964, 1972) and Andrews, Bickel, Hampel, Huber, Rogers and Tukey (1972). The estimated standard errors quoted for Huber's method have been calculated as suggested by Huber (1972, foot of p. 1061).

*An error that has come to light in that paper occurs in line 8 on page 20, where "not constant" should read "positive and nearly constant", that is, not changing rapidly.

Exercises for Chapter 11

1. This exercise and the next two refer to assessing distribution shape in homogeneous samples, in principle a simpler matter than assessing distributional properties of regression residuals.

(a) A sample of observations from a one-dimensional distribution or population can be represented by the empirical distribution function $\hat{F}(x)$, defined for any x as the proportion of the sample values not greater than x. $\hat{F}(x)$ is a step function, rising by $1/n$ wherever x is equal to one of the observations, provided that the value does not occur more than once in the data. The step function is not conveniently implemented on a typewriter terminal. If $x_{(j)}$ is the jth order statistic, that is, the jth member when the sample is arranged in nondecreasing order, and if $x_{(j)}$ is not equal to any other order statistic, the midpoint of the jump in $\hat{F}(x)$ at $x = x_{(j)}$ is $(j - \frac{1}{2})/n$. A convenient graphical representation of $\hat{F}(x)$ is had by plotting the points $\{(x_{(j)}, (j - \frac{1}{2})/n)\}$. If X is the vector of data, the plot can be obtained, using the function *SCATTERPLOT* in Appendix 3, with the instruction

$$X[\Delta\!\!|\,X]\ SCATTERPLOT\ (^{-}0.5 + \iota\rho X) \div \rho X$$

(b) For the purpose of assessing normality of the distribution, the ordinates may be transformed to "equivalent normal deviates". The functions *NIF* and *INIF* in Appendix 3 give the normal integral function and its inverse. The plot can be obtained with the instruction

$$X[\Delta\!\!|\,X]\ SCATTERPLOT\ INIF\ (^{-}0.5 + \iota\rho X) \div \rho X$$

(c) If the sample is large in size and the values are mostly distinct, it may be pleasanter to see selected equivalent normal deviates plotted against corresponding percentage points of the sample. The function *QUANTILES* in Appendix 3 gives percentage points of a sample. The equivalent normal deviates may conveniently be chosen to be equally spaced. For example,

$$P \leftarrow NIF\ Y \leftarrow 0.5 \times {}^{-}5 + \iota 9$$
$$(P\ QUANTILES\ X)\ SCATTERPLOT\ Y$$

(d) Suppose we have independent samples from two distributions or populations. Call the samples $X1$ and $X2$. They need not be equal in size. To compare the distribution shapes, having chosen a set of proportions P, execute

$$(P\ \ QUANTILES\ \ X1)\ \ SCATTERPLOT\ \ P\ \ QUANTILES\ \ X2$$

If, as at (c) above, P corresponds to equally spaced equivalent normal deviates, this plot permits us to see simultaneously the similarity of the distributions and their normality. If the distributions are the same apart from location and scale, the points should lie on a straight line, and if the distributions are normal the points should be equally spaced, approximately. (Quantile-quantile plots were described by Wilk and Gnanadesikan, 1968.)

Such empirical distributions or quantile plots are difficult to make judgments of significance about, because the points are high correlated. One may, for example, make a plot as at (c) for a large sample from some nonnormal distribution, such as a logistic distribution, and be pleased to see that the points lie on an elegant curve, clearly not a straight line. Then one may repeat the exercise with a large sample from a normal distribution, or perhaps make a plot as at (d) for two samples from the same distribution, and be vexed to see another elegant curve—not far from a straight line, perhaps, but deviating smoothly from straightness. Mere shape, without regard to magnitude, is of dubious import in any of these empirical distributions.

2. In 1(a) the ordinate at $x_{(j)}$ was taken to be $(j - \frac{1}{2})/n$, which would seem a natural choice. Other choices have been suggested, and in particular $(j - \frac{1}{3})/(n + \frac{1}{3})$ can be claimed to be approximately median-unbiased. Except when n is quite small there is little difference between these choices. The instruction in 1(b) becomes

$$X[\text{⍋}X]\ SCATTERPLOT\ INIF\ (^{-}1 + 3 \times \iota\rho X) \div 1 + 3 \times \rho X$$

and the definition of *QUANTILES* may be similarly modified.

The property of median-unbiasedness can be stated thus. Let $X_1, X_2, \ldots, X_n$ be independent identically distributed random variables with continuous distribution function $F(x)$, and let $X_{(j)}$ be the jth order statistic ($1 \leqslant j \leqslant n$). Then for any j the median of the distribution of $F(X_{(j)})$ is approximately $(j - \frac{1}{3})/(n + \frac{1}{3})$.

To prove this, note that $\{F(X_j)\}$ are independently uniformly distributed over the interval $(0, 1)$; and without loss of generality we may take $F(x) = x$ and write $X_{(j)}$ for $F(X_{(j)})$. Then $X_{(j)}$ has probability differential element proportional to

$$x^{j-1}(1 - x)^{n-j}\,dx = x^{j-1/3}(1 - x)^{n-j+2/3}\left[x^{-2/3}(1 - x)^{-2/3}\,dx\right].$$

For any positive a and b, let $G(x) = x^a(1 - x)^b$, and consider a monotone 1-to-1 transformation, $x = q(y)$. $G(x)$ is maximized at $x = a/(a + b)$, and so $G(q(y))$ is maximized at $y = q^{-1}(a/(a + b))$. Let q satisfy

$$dy = x^{-2/3}(1 - x)^{-2/3}\,dx.$$

Then it can be shown (Anscombe, 1964) that $d^3G/dy^3 = 0$ at the maximum where $dG/dy = 0$. Thus $G(q(y))$ is nearly symmetric about its maximum, and when it is normalized to be a probability density function the median is approximately the mode, $q^{-1}(a/(a + b))$, and in the x-scale the median is approximately $a/(a + b)$.

When $j = 1$, the extreme case, the above approximate median for $X_{(1)}$ is at $0.667/(n + 0.33)$. The true median is m where

$$\int_m^1 n(1 - x)^{n-1}\,dx = \tfrac{1}{2},$$

or $m = 1 - \exp[-(\ln 2)/n]$, which is approximately $0.693/(n + 0.35)$.

3. Let $\{X_i\}(i = 1, 2, \ldots, n)$ be independent identically distributed random variables having mean 0 and variance 1. Let $\bar{X}$ be the average of $\{X_i\}$, defined by $n\bar{X} = \Sigma_i X_i$. We wish to estimate the skewness measure $\mathcal{E}(X_i - \mathcal{E}X_i)^3/[\text{var}(X_i)]^{3/2}$.

(a) An obvious statistic, relying on the known mean and variance, is

$$T_1 = \Sigma_i X_i^3/n.$$

Show that T_1 is unbiased and that if the distribution of X_i is normal

$$\text{var}(T_1) = 15/n.$$

(b) Had the mean of the distribution not been known to be zero, we might have suggested (for $n \geqslant 3$)

$$T_2 = n\Sigma_i\left(X_i - \bar{X}\right)^3/\left[(n-1)(n-2)\right].$$

Show that T_2 is unbiased and that if the distribution of X_i is normal

$$\text{var}(T_2) = 6n/\left[(n-1)(n-2)\right].$$

Thus when n is large, under normality, $\text{var}(T_1)/\text{var}(T_2) = 2.5$—an unusual and striking instance of a large effect produced by a small change in definition of a statistic.

4. In Figures B:32 and 36, vectors of 51 residuals are shown displayed on several lines according to groups of states. The function that was used, called *STATEDISPLAY*, is not reproduced. Write a suitable definition for such a function.

5. Given $\{(x_i, y_i)\}$ $(i = 1, 2, \ldots, n)$, assume that for each i

$$y_i = \alpha + \beta x_i + \epsilon_i,$$

where α and β are constant, $\{x_i\}$ are fixed and $\{\epsilon_i\}$ are independent identically distributed $N(0, \sigma^2)$. Then the maximum-likelihood or least-squares estimate of β is

$$b = (\Sigma_i(y_i - \bar{y})(x_i - \bar{x}))/\Sigma_i(x_i - \bar{x})^2,$$

where $\bar{x}$ and $\bar{y}$ are the means, provided the $\{x_i\}$ are not all equal.

(a) Show that b is distributed $N(\beta, \sigma^2/\Sigma_i(x_i - \bar{x})^2)$.

(b) Suppose someone intended to calculate b from the above formula, but by accident omitted the step of subtracting $\bar{x}$ from $\{x_i\}$. When he wanted to use $\{x_i - \bar{x}\}$, he actually used $\{x_i\}$. The estimate of β calculated was therefore

$$b^* = (\Sigma_i(y_i - \bar{y})x_i)/\Sigma_i x_i^2.$$

What is the distribution of b^*? If, still with the same misunderstanding, an attempt were made to test the significance of the deviation of the estimate of β from 0, would the significance be underestimated or overestimated? (Treat σ as known —unless you prefer to treat σ as unknown, in which case consider explicitly how σ would be estimated. Should not direct calculation of terms in the analysis of variance, and checking whether the sums of squares added up correctly, have led to detection of the error?)

6. A test is carried out twice a day for four consecutive days, each day at noon and at 8:00 p.m., so that the intervals between consecutive tests are alternately 8 and 16 hours. Let x_i and y_i denote respectively the time (elapsed from some zero epoch) and a reading made in the ith test ($i = 1, 2, \ldots, 8$). For convenience let time be measured in 8-hour units centered at 4:00 a.m. on the third day, so that $\{x_i\} = {}^-5, {}^-4, {}^-2, {}^-1, 1, 2, 4, 5$. Cubic regression of y on x is to be calculated. Show that the first three orthogonal polynomials in $\{x_i\}$, found by Gram-Schmidt orthogonalization, are

$$\{x_i\}, \ \{x_i^2 - 23/2\}, \ \{x_i^3 - (449/23)x_i\}.$$

Multiply the second by 2/3 and the third by 23/6 to obtain polynomials with smallest-possible integer values. (Orthogonal polynomials for equally spaced values of the argument were tabulated by Fisher and Yates, 1938.)

7. Consider fitting by least squares a linear regression relation,

$$y_i = \Sigma_r x_{ir} \beta_r + \epsilon_i.$$

Here the suffix i takes values $1, 2, \ldots, n$, the suffix r (and later s) takes values $1, 2, \ldots, p$, where $p < n$; $\{y_i\}$ is the vector of given values of the "dependent" variable, $\{x_{ir}\}$ is the matrix of given values of the explanatory variables, usually including a column of 1's, $\{\beta_r\}$ is the vector of regression coefficients to be estimated, and $\{\epsilon_i\}$ is a vector of independent identically distributed "errors", perhaps distributed $N(0, \sigma^2)$.

(a) We have seen that a possible method of estimating $\{\beta_r\}$ by least squares is through Gram-Schmidt orthogonalization. The columns of $\{x_{ir}\}$ are replaced by mutually orthogonal vectors spanning the same linear subspace of Euclidean n-dimensional space (say n-space) as that spanned by the columns of $\{x_{ir}\}$. This implies a change of axes in the r-space of vectors $\{\beta_r\}$, after which each regression coefficient is estimated directly as if the others were not there. An implementation of the procedure is given in *REGR*, described in the documentary variable *HOWREGRESSION* (in Appendix 3), and its execution has been illustrated in Chapter 11. The reader may like to examine *REGR* and other functions in the group, verifying that the procedures yield the results claimed. One check is to apply the functions to a very small data set and print out all global variables at each step. What happens if *REGR* is executed twice in succession with the same argument? Note that the functions can be used to perform regression of several dependent variables simultaneously on the same set of explanatory variables. The procedure of simple regression of a dependent variable on one independent variable, with intercept, referred to in Exercise 5 above, and implemented in the function *FIT*, is a special case of Gram-Schmidt orthogonalization.

(b) Another possible method, also numerically very stable, uses Householder reflexions. There is an orthogonal change of axes in n-space, so that the regression concerns only p coordinates of the dependent variable, the remaining coordinates consisting of what are termed uncorrelated residuals. An implementation is given in *HHT*, described in *HOWHOUSEHOLDER*; a similar procedure is used in the IBM implementation of the primitive domino (divide-quad) function.

(c) It is easy to show that the least-squares estimate $\{b_r\}$ of $\{\beta_r\}$ satisfies the so-called normal equations,

$$\Sigma_s S_{rs} b_s = T_r,$$

where $S_{rs} = \Sigma_i x_{ir} x_{is}$, $T_r = \Sigma_i x_{ir} y_i$. Gauss's original method was to solve these equations by what is now known as Gaussian elimination. When n is very large, and when the matrix $\{x_{ir}\}$ is well conditioned, so that (roughly) no nonzero vector $\{\lambda_r\}$ exists for which $\Sigma_r x_{ir} \lambda_r$ nearly vanishes for all i, proceeding by way of the normal equations is attractive, because $\{S_{rs}\}$ and $\{T_r\}$ can be accumulated in one pass through the data without need to hold all the data simultaneously in the computer's fast memory. However, if $\{x_{ir}\}$ is not so well conditioned, computational round-off error usually has a more serious effect on the results than it does for either of the preceding methods. One should not be too dogmatic about this; Ling (1974) has pointed out that there are conditions under which the normal equations perform better than Gram-Schmidt. Working in APL, the obvious way to solve the normal equations, if they are formed, is through the divide-quad function.

(Subscript notation is used here, for the sake of conformity with notation elsewhere in the book and particularly in Appendix 2. In these exercises, the matrix operations go well in APL, and the reader is invited to so express them.)

8. When the matrix $\{x_{ir}\}$ is ill conditioned, the solution procedures referred to in the previous exercise break down in the sense that at some stage a divisor is almost or exactly zero. Suppose, for example, that there is a nonzero vector $\{\lambda_r\}$ such that $\Sigma_r x_{ir} \lambda_r = 0$ for all i. Then if $\{b_r\}$ satisfies the normal equations so does $\{b_r + k\lambda_r\}$, for any constant k.

(a) One way of handling this situation, in finding a solution $\{b_r\}$, is to delete one or more columns of $\{x_{ir}\}$ so that the remaining columns are nearly enough linearly independent. This is equivalent to constraining each member of $\{b_r\}$ corresponding to an omitted column to be zero. If the nature of the linear dependence is known in advance, this can be done in advance. Otherwise, it can be done during the computation with a catch in the program to prevent use of a too small divisor—as in *REGR*.

(b) When results of a designed experiment are analyzed, $\{x_{ir}\}$ may consist entirely of 1's, 0's or other small integers and may have much symmetry. Suppose there is essentially just one nonzero vector $\{\lambda_r\}$ specifying linear dependence, as above. (The vectors specifying linear dependence of the columns of $\{x_{ir}\}$ constitute a linear space, and we now assume it is a 1-space.) Then to constrain any of the regression coefficients to be zero may spoil the symmetry. Let $\{l_r\}$ be any arbitrary vector such that $\Sigma_r l_r \lambda_r \neq 0$. Then there is just one solution $\{b_r\}$ of the normal equations that satisfies the constraint $\Sigma_r l_r b_r = 0$. The solution can be found by choosing a vector $\{k_r\}$ and adding $k_r \Sigma_s l_s b_s$ to the left side of the rth normal equation, so that the equations become

$$\Sigma_s (S_{rs} + k_r l_s) b_s = T_r.$$

Possibly $\{k_r\}$ and $\{l_s\}$ can be chosen so that the matrix $\{S_{rs} + k_r l_s\}$ is not only nonsingular but triangular, that is, it has only zeroes on one side of the main diagonal. The equations are then not only uniquely soluble but easy to solve, perhaps much easier to solve than if one regression coefficient had been set zero, as at (a).

(c) Illustrate the foregoing by determining the least-squares fit of an additive structure to a two-way table. Let the table, with k rows and l columns, be denoted by $\{y_{ij}\}$ $(i = 1, 2, \ldots, k; j = 1, 2, \ldots, l)$, and let the additive structure be

$$y_{ij} = \alpha_i + \beta_j + \epsilon_{ij},$$

where $\{\alpha_i\}$ is the vector of row constants, $\{\beta_j\}$ the vector of column constants, and $\{\epsilon_{ij}\}$ the matrix of independent identically distributed "errors". Such a table could represent observations in an experiment in randomized blocks, the rows indicating blocks and the columns treatments. (For this and the next exercise, see Fisher, 1925, sections 48 and 49, and 1935, chapters 4 and 5.)

Show that this structure can be expressed in the same style as that at the beginning of Exercise 7, by destructuring $\{y_{ij}\}$ and $\{\epsilon_{ij}\}$ into vectors of length kl (as with the APL monadic comma) and catenating $\{\alpha_i\}$ and $\{\beta_j\}$ to form a vector of regression coefficients of length $k + l$. The kl-by-$(k + l)$ matrix of explanatory variables then consists entirely of 1's and 0's. One regression parameter is redundant; clearly any constant could be added to each member of $\{\alpha_i\}$ and subtracted from each member of $\{\beta_j\}$ without altering the additive "outer product" $\{\alpha_i + \beta_j\}$. Obtain the least-squares estimates of $\{\alpha_i\}$ and $\{\beta_j\}$ by two methods. First set one parameter, say β_1, equal to 0, and estimate all the others. (Estimation by stages, as mentioned in Exercise 10 below, may be done, with $\{\alpha_i\}$ estimated at the first stage and $\{\beta_j\}$, for $j \neq 1$, at the second stage.) Hence find the fitted values, estimating $\{\alpha_i + \beta_j\}$. Then repeat the exercise, this time imposing the symmetrical constraint $\Sigma_j \beta_j = 0$ (or alternatively $\Sigma_i \alpha_i = 0$), and observe that with this the normal equations can easily be adapted into triangular form and solved. Verify that the same fitted values are obtained as before, but more easily.

Usually still more redundancy is introduced into the fitted structure, so that it becomes

$$y_{ij} = \mu + \alpha_i + \beta_j + \epsilon_{ij},$$

where μ is a general mean; and then it is convenient to have two constraints, $\Sigma_i \alpha_i = \Sigma_j \beta_j = 0$. The estimates of $\{\alpha_i\}$ and of $\{\beta_j\}$ are called the row effects and the column effects, respectively. Row effects and column effects are "orthogonal" in the sense that either are unaffected by inclusion or exclusion of the other from the fitted structure—they correspond to projection of the vector of observations onto orthogonal linear subspaces of kl-space.

9. Observations $\{y_{ij}\}$ $(i = 1, 2, \ldots, a; j = 1, 2, \ldots, a)$ are obtained in a Latin-square experiment. The treatment for the (i, j)th observation is numbered $t(i, j)$; t takes values $1, 2, \ldots, a$, and (for example) $t = 1 +$ the residue of $(i - 1) + (j - 1)$ mod a. For any i, as j goes from 1 to a, t takes all its values once each; and similarly for any j, as i goes from 1 to a. Show how to fit by least squares the structure

$$y_{ij} = \mu + \alpha_i + \beta_j + \gamma_{t(i,j)} + \epsilon_{ij},$$

and construct an analysis of variance. For the case $a = 4$, exhibit the projection matrix Q, of size 16-by-16, and verify that all its off-diagonal elements are equal in magnitude. What is the correlation of any pair of residuals? Consider also $a = 3$, and verify that some pairs of residuals have correlation 1.

10. Let the linear regression structure at the beginning of Exercise 7 be partitioned into two pieces:

$$y_i = \Sigma_r x_{ir}^{(1)} \beta_r^{(1)} + \Sigma_s x_{is}^{(2)} \beta_s^{(2)} + \epsilon_i,$$

where r takes values $1, 2, \ldots, p_1$, and s takes values $1, 2, \ldots, p_2$ $(p_1 + p_2 = p)$. Show that the fitting of the regression relation can be done in stages, as follows. First set $\{\beta_s^{(2)}\}$ equal to zero, or act as though the second term on the right side above were absent; and determine the least-squares estimate of $\{\beta_r^{(1)}\}$ and the residual sum of squares explicitly as functions of $\{y_i\}$. Denote the estimate of $\{\beta_r^{(1)}\}$ by $\{b_r^{(1)*}\}$ and the residual sum of squares by S^*. Now in S^* replace $\{y_i\}$ by $\{y_i - \Sigma_s x_s^{(2)} \beta_s^{(2)}\}$, and minimize with respect to $\{\beta_s^{(2)}\}$. Denote the minimized value of S^* by S, and the estimate of $\{\beta_s^{(2)}\}$ by $\{b_s^{(2)}\}$. Then S is the same residual sum of squares and $\{b_s^{(2)}\}$ is the same estimate of $\{\beta_s^{(2)}\}$ as if the fitting had been done all at once. In general, $\{b_r^{(1)*}\}$ is not the same estimate of $\{\beta_r^{(1)}\}$ as if the fitting had been done all at once, but it can be corrected by replacing $\{y_i\}$ by $\{y_i - \Sigma_s x_{is}^{(2)} b_s^{(2)}\}$ in the expression found at stage 1.

Fisher's analysis of covariance (1932, 1934, section 49.1) is a striking instance of least-squares fitting by stages. Gram-Schmidt orthogonalization of a regression relation may be regarded as repeated application of the same principle to the individual regression coefficients.

11. Consider fitting an additive structure to a two-way table, as at Exercise 8(c) above, except that one entry, say y_{11}, is missing and there are altogether $kl - 1$ observations. Do this by two alternative methods, and reconcile the answers.

(a) Insert an arbitrary value for y_{11}, say $y_{11} = 0$. Let $\{x_{ij}\}$ be defined: $x_{11} = 1$, $x_{ij} = 0$ if $(i, j) \neq (1, 1)$. Fit the linear structure

$$y_{ij} = \mu + \alpha_i + \beta_j - \gamma x_{ij} + \epsilon_{ij},$$

with constraints $\Sigma_i \alpha_i = \Sigma_j \beta_j = 0$. Here γ is one extra parameter, and its estimate is interpreted as an estimate of the missing observation. Fit this structure by stages, first all parameters except γ, then γ. This is known as the covariance method of handling a missing value.

(b) Fit the usual additive structure to the $kl - 1$ actual observations. Use the constraints to make the matrix of the normal equations nearly triangular and solve by Gaussian elimination.

12. Let $\{y_{ij}\}$ be a given two-way table (matrix) and $\{x_i\}$ a given vector associated with rows of the table $(i = 1, 2, \ldots, k;\ j = 1, 2, \ldots, l)$. Suppose the elements of $\{x_i\}$ are not all equal and $\Sigma_i x_i = 0$. Consider fitting by least squares the relation

$$y_{ij} = \alpha_i + \beta_j + x_i \gamma_j + \epsilon_{ij},$$

where $\{\alpha_i\}$, $\{\beta_j\}$, $\{\gamma_j\}$ are vectors of parameters to be estimated and $\{\epsilon_{ij}\}$ is a matrix of independent identically distributed "errors". Clearly there is redundancy in the parameters: a constant K could be added to each β and subtracted from each α, or K added to each γ and Kx_i subtracted from α_i for each i, all with no effect.

Show that the fitting can be accomplished as follows. Perform ordinary linear regression of each column of $\{y_{ij}\}$ on $\{x_i\}$, and let $\{z_{ij}\}$ be the residuals; that is, $z_{ij} = y_{ij} - b_j - x_i c_j$, where $b_j = (\Sigma_i y_{ij})/k, c_j = (\Sigma_i x_i y_{ij})/\Sigma_i x_i^2$. Then estimate $\{\alpha_i\}$ with two constraints, $\Sigma_i \alpha_i = \Sigma_i x_i \alpha_i = 0$, by $\{a_i\}$ where $a_i = (\Sigma_j z_{ij})/l$. (These seem to be the most elegant constraints to use, so that $\{\alpha_i\}$ has zero mean and zero regression on $\{x_i\}$.) An analysis of variance can be obtained:

	Sum of squares	Degrees of freedom
Fitting $\{\beta_j\}$ and $\{\gamma_j\}$	$(\Sigma_{ij} y_{ij}^2) - \Sigma_{ij} z_{ij}^2$	$2l$
Subsequent fitting of $\{\alpha_i\}$	$l\Sigma_i a_i^2$	$k - 2$
Residual	$(\Sigma_{ij} z_{ij}^2) - l\Sigma_i a_i^2$	$(k-2)(l-1)$
Gross about zero	$\Sigma_{ij} y_{ij}^2$	kl

Each sum of squares should be computed positively, not by subtraction as shown above; and then verifying that the terms sum to the indicated total is a check on correctness.

A data set to which this kind of analysis seems appropriate is the table, given in recent issues of the *Statistical Abstract of the United States*, of crime rates (per hundred thousand of the population) in each state in each of seven broad categories of crime (from Murder to Motor Vehicle Theft). After the data are transformed to logarithms or to a low positive power (such as the 0.2 power) a simple additive structure fits fairly well, but there are substantial residuals. On the whole, crimes involving theft of property seem relatively higher in the more highly urban states. That suggests bringing in a measure of urbanness, such as *URBAN*70 in Chapter 11, and regressing separately each type of crime on that variable. After that is done, the state effects, denoted by $\{\alpha_i\}$ above, show a geographic pattern, suggesting regional differences in hot-bloodedness and contempt for law, over and above what may be associated with variation in urbanness.

13. Let $\{y_{ij}\}$ be a given two-way table and $\{w_{ij}\}$ an associated matrix of nonnegative weights, such that the row sums $\{\Sigma_j w_{ij}\}$ and the column sums $\{\Sigma_i w_{ij}\}$ are all positive ($i = 1, 2, \ldots, k$; $j = 1, 2, \ldots, l$). Typically the weights are small nonnegative integers. Consider estimating row parameters $\{\alpha_i\}$ and column parameters $\{\beta_j\}$ so as to minimize the weighted sum of squares,

$$S = \Sigma_{ij} w_{ij}(y_{ij} - \alpha_i - \beta_j)^2.$$

Suppose $k \geqslant l$, and let us entertain the possibility that k is large, say 1000. There is at least the usual one degree of redundancy in this parametrization. If there are enough zeroes among the weights, there may be further redundancy.

The structure can be fitted by a general procedure for equal-weighted least-squares regression. The matrix $\{y_{ij}\sqrt{w_{ij}}\}$ is destructured into a vector of length kl, and a kl-by-$(k + l)$ matrix of explanatory variables is set up whose entries are either zeroes or square roots of weights; and then the Gram-Schmidt or some other method can be applied. Clearly if k is large space of order k^2 is needed.

A more economical procedure takes advantage of the mutual orthogonality of the $\{\alpha_i\}$ and the mutual orthogonality of the $\{\beta_j\}$, and fits by stages. The matrix of the normal equations can be partitioned into four submatrices

$$\begin{pmatrix} A & W \\ W' & B \end{pmatrix}$$

where A is a k-by-k diagonal matrix with $\{\Sigma_j w_{ij}\}$ in the main diagonal, W is the k-by-l matrix of weights $\{w_{ij}\}$, W' is its transpose, and B is a l-by-l diagonal matrix with $\{\Sigma_i w_{ij}\}$ in the main diagonal. At least one constraint can be imposed; this could be to set one parameter to zero, or possibly the more symmetric condition,

$$\Sigma_{ij} w_{ij} \beta_j = 0.$$

The latter leads to a simple solution of the equations if $\{w_{ij}\}$ is an outer product of the form $\{u_i v_j\}$, and may be expected to help numerical stability if $\{w_{ij}\}$ is not very different from such an outer product. Divide each of the first k normal equations by the diagonal element of A, and then subtract a suitable multiple of the constraint coefficients from the top right submatrix, so that its (i, j)th element becomes

$$(w_{ij}/\Sigma_j w_{ij}) - (\Sigma_i w_{ij})/\Sigma_{ij} w_{ij}.$$

Call this submatrix U. U vanishes if W is the outer product just mentioned. Now use the first k equations to eliminate $\{\alpha_i\}$ from the last l equations. The l-by-l matrix of the resulting equations to estimate $\{\beta_j\}$ is then

$$B - W'U.$$

Thus the solution is accomplished by forming these l equations, solving them to estimate $\{\beta_j\}$ (possibly not uniquely), and back-substituting into the first k equations to estimate $\{\alpha_i\}$.

Write a program to implement this procedure.

14.* Another method of dealing with the problem of the previous exercise is iterative. First the weighted mean, $(\Sigma_{ij} w_{ij} y_{ij})/\Sigma_{ij} w_{ij}$, is subtracted from every member of $\{y_{ij}\}$; call the result $\{z_{ij}\}$. Then, for each i, the weighted mean, $(\Sigma_j w_{ij} z_{ij})/\Sigma_j w_{ij}$, is subtracted from every entry in the ith row of $\{z_{ij}\}$. Name the result $\{z_{ij}\}$ again. Then, for each j, the weighted mean, $(\Sigma_i w_{ij} z_{ij})/\Sigma_i w_{ij}$, is subtracted from every entry in the jth column of $\{z_{ij}\}$; and the result is named $\{z_{ij}\}$ again. And so on, back to rows, back to columns, in alternation, always subtracting row means and column means from the entries in the residual matrix. Whenever the process is stopped, $\{y_{ij} - z_{ij}\}$ is taken to be the estimate of $\{\alpha_i + \beta_j\}$. Consider convergence as the number of iterations tends to infinity, under the condition of infinitely precise computation.

(a) Show that the residual sum of squares, $\Sigma_{ij} w_{ij} z_{ij}^2$, converges.

(b) Show that the reduction in the residual sum of squares caused by one step in the iterative procedure (subtracting row means, or subtracting column means, as the case may be) can be expressed as a weighted sum of squares of the means that are

*This exercise results from discussion with John A. Hartigan and David B. Pollard.

subtracted. Deduce that

$$\Sigma_j w_{ij} z_{ij} \to 0 \quad \text{for each } i,$$

$$\Sigma_i w_{ij} z_{ij} \to 0 \quad \text{for each } j.$$

(c) Show that in the limit S is minimized.

15. A set of data has the following structure. There are cells, which either form a row, or are arranged in a rectangular two-dimensional array, or are arranged in a rectangular array in three or more dimensions. In each cell there are a nonnegative number of observations. For definiteness, suppose that the cells form a two-dimensional array indexed by (i, j) $(i = 1, 2, \ldots, k; j = 1, 2, \ldots, l)$. Then in the (i, j)th cell there are n_{ij} observations denoted by $\{y_{ijr}\}$ $(r = 1, 2, \ldots, n_{ij})$.

One might consider fitting an additive structure,

$$y_{ijr} = \alpha_i + \beta_j + \epsilon_{ijr},$$

where $\{\alpha_i\}$ and $\{\beta_j\}$ are row and column parameters and $\{\epsilon_{ijr}\}$ are errors that may be supposed to have some pattern of correlation within cells. From the observations in each cell it is convenient to calculate two quantities,

$$\bar{y}_{ij} = (\Sigma_r y_{ijr})/n_{ij}, \quad v_{ij} = \Sigma_r (y_{ijr} - \bar{y}_{ij})^2,$$

provided $n_{ij} \geqslant 1$, and say $\bar{y}_{ij} = v_{ij} = 0$ if $n_{ij} = 0$. Then an analysis can be performed on $\{\bar{y}_{ij}\}$ with weights $\{n_{ij}\}$, as in the two previous exercises; and $\{v_{ij}\}$ provides information about within-cell variance. Let us turn attention now to the representation of the data. If the $\{n_{ij}\}$ are not all equal, the data do not constitute a perfect rectangular array and therefore cannot be represented directly, without modification, as a single APL variable. There are, in fact, many ways of representing such data in terms of APL variables, as was remarked at the end of Chapter 4.

If few cells are empty, a concise representation is in terms of two APL variables. One lists all the observations as a vector, all the observations in the first cell, then all those in the second cell, and so on, taking the cells row by row. The other variable is the matrix $\{n_{ij}\}$; its dimension vector shows how the cells are arranged, and its elements tell how many observations are in each cell. The first variable is the content of the data set, the second may be called its dimension array. The two variables should have related names. For example, the dimension array could be assigned always the same name as the content vector but prefixed with an underscored D, provided the name of the content vector was not underscored—we are thinking of the representation of several similar data sets. Then if the name of the content vector was used as argument of a function, both the content vector and the dimension array could be accessed through the execute function.

The data can be represented as a single APL variable constructed out of the observations $\{y_{ijr}\}$ together with some additional numbers used as markers or dummies. Suppose the observations are necessarily nonnegative. Then negative numbers can be used for this purpose. The data can be represented as a vector, like the content vector just considered, but with a marker inserted after the observations in each cell have been listed. The marker could be ¯1 at the end of a cell listing if the next cell to be listed was in the same row, ¯2 if the next cell was in the next row, ¯3 if all cells had been listed (or the next cell would be in the next plane). Another possibility is to introduce into each cell extra dummy observations (¯1) so that every cell has as many observations in it as the most populous cell had at the outset. Now the data form a perfect three-dimensional rectangular array.

Each of these representations requires that all observations have been assigned to their proper cells before the listing begins. Another kind of representation, permitting the observations to be entered in any order, that can be economical if many cells are empty, is as a matrix with three columns, one column listing the observations, the other two columns showing the row number and the column number of the cell to which each observation belongs.

Write a program that accepts a data set in some one of these forms, checks that any consistency requirements are satisfied, and then (a) translates the representation into some one of the other forms and (b) obtains the matrix of cell means $\{\bar{y}_{ij}\}$ and the matrix of within-cell sums of squares $\{v_{ij}\}$. Consider the relative convenience of the forms of data representation (a) for entering the data, (b) for storing the data, (c) for statistical analysis.

16. A linear constraint introduced to counteract redundancy in the parametrization is a different matter from a constraint, perhaps reflecting supplementary information, that does not serve to counteract redundancy. The first kind of constraint is consistent with the normal equations, which correctly characterize any set of parameter values minimizing the sum of squares; thus the constraint may be combined with the normal equations. The second kind of constraint is, in general, not consistent with the normal equations, which do not characterize parameter values that minimize the sum of squares conditionally on the constraint.

Consider, for example, fitting the usual additive structure to a two-way table with the constraint that the first two column effects are equal ($\beta_1 = \beta_2$). (Assume that there are at least three columns.) If columns represent treatments, this constraint might be imposed because the first two treatments were in fact the same. Determine the least-squares estimates of the row effects and the column effects, and see how they and the residual sum of squares compare with what would have been obtained if this constraint had not been imposed.

17. Exercise 12 was concerned with fitting an additive structure to a two-way table together with regressions in each column on a variable associated with the rows. If presence of such a variable were suspected but its values were not available a priori, the values could be treated as additional parameters to be estimated. Then the structure to be fitted could be written:

$$y_{ij} = \alpha_i + \beta_j + \gamma_i \delta_j + \epsilon_{ij},$$

where $\{\alpha_i\}$, $\{\beta_j\}$, $\{\gamma_i\}$, $\{\delta_j\}$ were vectors of parameters and $\{\epsilon_{ij}\}$ the errors. In other contexts a simple multiplicative structure, without an additive component, might be entertained for a two-way table:

$$y_{ij} = \alpha_i \beta_j + \epsilon_{ij}.$$

In Fisher's first published analysis of a two-way table, the data were obviously badly fitted by a simple additive structure, and a simple multiplicative structure was fitted by least squares instead (Fisher and Mackenzie, 1923).

Structures nonlinear in the parameters sometimes behave much like linear structures, and were indeed what Gauss had in mind in developing the theory of linear least squares. It is possible, however, for a nonlinear structure to behave qualitatively differently from a linear one. Consider a two-way table with (say) 5 rows and 5 columns, consisting entirely of zeroes except for a single element equal to one. What is the least-squares additive representation? Note that the residual corresponding to the one outlier is considerably larger than any other residual. By contrast, there is a multiplicative representation of the table that fits exactly—all residuals are zero.

McNeil and Tukey (1975) have made interesting comparisons of fitting nonlinear structures to two-way tables both by least squares and by robust methods.

18. Given a k-by-l table $\{y_{ij}\}$ or Y, consider fitting a simple multiplicative structure by least squares, minimizing

$$S = \Sigma_{ij}(y_{ij} - \alpha_i \beta_j)^2.$$

The parametrization is redundant, because all elements of $\{\alpha_i\}$ can be multiplied by a nonzero constant and all elements of $\{\beta_j\}$ divided by the same constant, without effect. A convenient constraint is that one of these two parameter vectors is of unit length, say $\Sigma_j \beta_j^2 = 1$. The matrix $\{\alpha_i \beta_j\}$ that minimizes S may be described as the closest (in the sense of least squares) rank-1 approximation to Y.

(a) Suppose $k \geqslant l$. Show that the fitting may be done as follows. The table is multiplied on the left by its transpose to form a l-by-l matrix U. The largest characteristic root of U is found, say λ_1, and the corresponding vector, say $\{b_j\}$, normalized to have unit length, $\Sigma_j b_j^2 = 1$. The vector is the desired estimate of $\{\beta_j\}$. The corresponding estimate $\{a_i\}$ of $\{\alpha_i\}$ satisfies: $a_i = \Sigma_j y_{ij} b_j$, $\Sigma_i a_i^2 = \lambda_1$. The residual sum of squares, the minimized value of S, is equal to $(\Sigma_{ij} y_{ij}^2) - \lambda_1$. If the largest root of U is simple, $\{b_j\}$ is unique apart from sign and the fitted values $\{a_i b_j\}$ are unique.

The function *JACOBI*, described in *HOWJACOBI*, finds roots and vectors of a symmetric matrix by Jacobi's method. (In recent years faster algorithms have been developed for this purpose, but their order of magnitude is the same.) Having chosen a suitable tolerance C, and if Y has at least as many rows as columns, one may proceed:

```
T←C JACOBI (⍉Y)+.×Y
+\□←ROOTS
(,Y)+.*2
```

The sum of all the roots should be equal to the sum of squares of the elements of Y. If the roots appear in decreasing order,

```
A←Y+.×B←T[2;;1]
+/,Z×Z←Y−A∘.×B
+/1↓ROOTS
```

The residual sum of squares should be equal to the sum of all roots except the first.

(b) Once an additive structure has been fitted by least squares to a two-way table, the residuals have zero row effects and column effects, and nothing is gained by repeating the fitting process on the residuals. With multiplicative structures the situation is different. When the fitting described above has been done, the residual matrix $\{y_{ij} - a_i b_j\}$ can be treated the same way, so that further vectors $\{\gamma_i\}$ and $\{\delta_j\}$ are chosen to minimize

$$\Sigma_{ij}(y_{ij} - a_i b_j - \gamma_i \delta_j)^2.$$

Show that the desired values of $\{\gamma_i\}$ and $\{\delta_j\}$ can be obtained from the second largest root λ_2 of U above, and its corresponding vector $\{d_j\}$. The estimate $\{c_i\}$ of $\{\gamma_i\}$ satisfies: $c_i = \Sigma_j y_{ij} d_j$, $\Sigma_i c_i^2 = \lambda_2$. Show that $\{c_i\}$ is orthogonal to $\{a_i\}$ and $\{d_j\}$ to $\{b_j\}$. The residual sum of squares is equal to $(\Sigma_{ij} y_{ij}^2) - (\lambda_1 + \lambda_2)$.

Suppose the given table Y has not been subjected to any linear constraints; for example, it is not a table of residuals from the fitting of an additive structure to a two-way table. Then at the first stage the number of independent parameters fitted was $k + l - 1$. At this second stage the number of further independent parameters fitted can be reckoned to be $k + l - 3$, because of the orthogonality conditions.

The same operation can be done for each nonzero root of U, after which Y is represented exactly as the sum of rank-1 matrices, of which the sum of the first m constitutes the best rank-m approximation (in the sense of least squares). This complete representation is known as the singular-value decomposition of Y (see Good, 1969). If all roots of U are positive, the number of independent parameters fitted is

$$(k + l - 1) + (k + l - 3) + (k + l - 5) + \cdots + (k + 1 - l) = kl.$$

Because the structures being fitted are nonlinear, the components into which $\Sigma_{ij} y_{ij}^2$ is analyzed do not have the same distributional properties, on any null hypothesis, as the terms in an ordinary analysis of variance; but by way of a crude approximation the components may be thought of as having degrees of freedom equal to these numbers of successive independent fitted parameters.

(c) Sometimes it is of interest to test the null hypothesis that the given table consists only of independent "errors" identically distributed in a normal distribution $N(0, \sigma^2)$. Then U behaves like the variance matrix of a sample of $(k + 1)$ independent observations from a l-dimensional isotropic normal distribution with nonzero means. A suitable approximate test of isotropy or sphericity is given by Anderson (1958, section 10.7.4), good when k is much larger than l. It has been implemented in the function *ISOTROPY*; the first argument is taken equal to k, the second is the vector of roots of U.

19. Let $\{a_i b_j\}$ be the least-squares rank-1 approximation of a matrix $\{y_{ij}\}$ or Y, of which not every element is zero. Let U be as defined in the previous exercise.

(a) Show that U has at least one positive root.

(b) If there is a nonzero vector $\{c_i\}$ such that $\Sigma_i c_i y_{ij} = 0$ for all j, then $\Sigma_i c_i a_i = 0$. If there is a nonzero vector $\{d_j\}$ such that $\Sigma_j d_j y_{ij} = 0$ for all i, then $\Sigma_j d_j b_j = 0$, and U has at least one zero root. (If $\{c_i\} = \{1\}$, Y has had its column means subtracted, and if $\{d_j\} = \{1\}$, its row means.)

(c) Show by counterexample that $\Sigma_{ij} y_{ij} = 0$ does not imply $\Sigma_{ij} a_i b_j = 0$.

20. Let $\{y_{ij}\}$ or Y be a k-by-l table ($k \geqslant l$) having the additive structure:

$$y_{ij} = \alpha_i + \beta_j + \epsilon_{ij},$$

where $\{\alpha_i\}$ and $\{\beta_j\}$ are constant and $\{\epsilon_{ij}\}$ are independent $N(0,\sigma^2)$ errors. Let Z be the residuals after fitting the additive structure by least squares. Let H_r denote an orthonormal r-by-r matrix of which all entries in the last row are equal to $1/\sqrt{r}$. Let T be the $(k-1)$-by-$(l-1)$ matrix obtained from $H_k Y H_l'$ (where the prime denotes transpose) by deleting the last row and the last column. Then T is a set of "uncorrelated residuals", independent $N(0,\sigma^2)$ variables, and $H_k Z H_l'$ is equal to T bordered by a last row and last column of zeroes. Show that the joint distribution of the $l-1$ nonzero roots of $Z'Z$ is the same as that of all roots of $T'T$.

Show that the mixed additive and multiplicative structure displayed at the beginning of Exercise 17 can be correctly fitted by least squares by the procedure just indicated—first the additive structure is fitted, then the multiplicative structure is fitted to the residuals Z from the additive structure. (Proceeding in the reverse order, first fitting the multiplicative structure and then the additive structure to the residuals, does not in general yield an equivalent result.) The isotropy test can be applied to the nonzero roots of $Z'Z$ by the function *ISOTROPY*, the first argument being taken equal to $k-1$, if k is sufficiently greater than l.

As an example, having fitted a simple additive structure to the (transformed) crime rates mentioned in Exercise 12, do the singular-value decomposition of the table of residuals. It will be found, with recent versions of the crime-rate data, that there are two relatively large and not very unequal roots. The corresponding vectors represent linear combinations of two intelligible contrasts: one between the primarily violent crimes, Murder, Rape and Assault, and the primarily stealing crimes, Burglary, Larceny and Motor Vehicle Theft; the other between Robbery and the rest. When the whole additive and two-term multiplicative structure has been fitted, each state is characterized by three parameters, the additive row effect and the "loadings" of the violence factor and of the robbery factor. The violence-factor loading is strongly regional, being high in the south, and the additive row effect and the robbery-factor loading both seem related to urbanness.

21. Fisher (1935, end of chapter 4) quoted some of the observations in an experiment on barley varieties carried out at six sites in Minnesota in 1931 and 1932. This was perhaps the first randomized agricultural field experiment conducted elsewhere than at Rothamsted and under Fisher's direction. The paper by Immer, Hayes and Powers (1934) gives the yields of all the 360 plots, together with detailed instructions for calculation of the analysis of variance. Doing that amount of arithmetic with a desk calculator was quite a feat; both calculation and proof-reading seem to have been done with great care. The paper gives no information about several matters vital to the interpretation of the data—the character of the sites, cultivation and manurial procedures, weather. (There is one puzzle in the paper, namely that the crops are first described as having been grown in 1930 and 1931, but thereafter in all tables and discussion the years are referred to as 1931 and 1932. The same discrepancy occurs in Fisher's book. Presumably the later dates are correct. Barley in Minnesota is apparently always sown in the spring, so that the whole growing process occurs within one calendar year.)

There were 6 sites (experiment stations) distributed across Minnesota and 2 years. At each site in each year an experiment was laid out in 3 randomized blocks (replications), the same 10 varieties being tested in each block. Thus the data set may be thought of as a four-dimensional array with dimension vector 6 2 3 10 . The factor Blocks is "nested" within Stations and Years; the other three factors are "crossed". Once the data set is entered in the computer, the analysis of variance can be calculated effortlessly with the functions *ANALYZE* and *EFFECT* encountered in Chapter 7.

Immer and coauthors seem to think the analysis of variance to be evidently a good summary of the data. Since every interaction calculated appears to be substantial, the results are in fact not easy to assimilate. What is presumably of particular interest is to understand the main effect of Varieties and the interaction of Varieties with Stations and Years. The latter is represented in the analysis of variance by three distinct terms (Varieties × Stations, Varieties × Years, Varieties × Stations × Years, having respectively 45, 9 and 45 degrees of freedom). For this purpose it is convenient to sum the data over Blocks, and arrange the result as a 12-by-10 matrix whose rows correspond to combinations of Stations and Years (let us say, Experiments) and whose columns refer to Varieties. Tukey's test shows that a simple additive structure is not a good way of representing row effects and column effects. The additive structure is more effective after the table has been transformed to logarithms, but then the residual variability is clearly not constant. A simple multiplicative structure (fitted by least squares to the untransformed table) is more satisfactory and just as intelligible as the simple additive structure in describing the main effects of Varieties and Experiments. The residuals from the multiplicative structure seem to be not far from isotropic; at any rate, their singular-value decomposition is not dominated by the first one or two terms. The relative performance of the varieties has differed somewhat with place and year, and only with information about local conditions and weather could one hope to obtain a clear and plausible description of the interaction. Immer *et al.* try a different approach, thinking of variance components associated with sampling. "Insofar as these six stations constitute a random sample of conditions to be found over the entire state and that these 2 years are a random sample of weather conditions to be encountered in future years, general recommendations may be drawn up for the entire state with reasonable assurance that the variety or varieties recommended will prove to be consistently superior in most places of the state and in most years." To this one may reasonably retort: nonsense! Some contemporary background information on barley in Minnesota is given by Immer, Christensen, Bridgford and Crim (1935).*

22. The adaptive regression procedure described in *HOWT7*, in which the errors are assumed to have a distribution belonging to a specified class of symmetric distributions of Pearson's Type VII or II, does not necessarily lead to results close to those yielded by traditional normal-law methods.

*I am indebted to Dr. Donald Rasmusson (University of Minnesota, Department of Agronomy and Plant Genetics) for helpful discussion of these matters.

A modest example of historic interest is the set of ten readings of differences in length of sleep (in hours) induced by two soporific drugs, quoted by Student (1908) to illustrate what we now call his t-test. The readings are as follows, one reading being somewhat outlying from the others:

1.2 2.4 1.3 1.3 0.0 1.0 1.8 0.8 4.6 1.4

Ten readings give miserably little information about distribution shape. The evidence against normality is mildly "significant". Not surprisingly, what we make of the readings in light of an unshakable faith in normality differs from what is conveyed by the adaptive analysis when the shape parameter α is supposed uniformly distributed over $(-\frac{1}{2}, 1)$—and could well differ still further from what would flow from other assumptions.

To apply the functions of *HOWT7*, set X equal to a 10-by-1 matrix of ones and let therefore XX be empty. Then the vector *BETA* of regression parameters has one member, the median of the distribution of the readings. The quantity denoted by q in Anscombe (1967), or by Q in *HOWT7*, should be chosen larger than $3 \div N$ (that is, 0.3), the value suggested when the readings appear to be normally distributed—the value 0.8 is satisfactory. After integration with respect to the scale parameter σ and the shape parameter α, the marginal likelihood L is a function of the one location parameter β and can be tabulated easily.

L turns out to be noticeably unsymmetrical, but we may attempt to approximate it by a symmetric Student density as if we had had different data known to be normally distributed. L is very roughly the same as the marginal likelihood function for the mean of normally distributed data for which the sample mean was 1.28 and the estimated standard error was 0.21 with 1 degree of freedom. The corresponding statement when, following Student, we assume the actual data to be normally distributed is that the sample mean is 1.58 and the estimated standard error is 0.39 with 9 degrees of freedom.

In one respect the two analyses are in good agreement. Student was interested in assessing the probability that the center of the frequency distribution of the readings (our β) was positive, given a uniform prior distribution for the parameter. The marginal likelihood L for the adaptive analysis (the exact L, not the symmetric approximation to it) is sufficiently unequal in its tails, higher in the upper tail than in the lower, that this probability is very close indeed to 1, as it is also for the conventional normal-law analysis.

Chapter 12

Contingency Tables and Pearson-Plackett Distributions

The preceding two chapters have addressed various forms of regression analysis, topics that are among those most commonly encountered in statistical analysis of data. We turn now to the more primitive subject of categorical (qualitative, attribute) variables and their association in contingency tables.

Qualitative variables have always been important in statistical practice, in some kinds of statistical practice at least, ever since 1662. Much of the statistical material collected in the nineteenth century was expressed as simple counts and frequencies of attributes or categories. During the great fifty-year period in the development of statistical methodology, from 1890 to 1940, when so much of modern statistical science took shape, attention was mainly directed towards quantitative variables, as very successful techniques related to patterns of normal variation were worked out—regression, correlation, analysis of variance. In Udny Yule's *An Introduction to the Theory of Statistics* (1911), which was apparently the first book in English professing to deal with statistical theory, pride of place was given to qualitative variables; the first of the three parts is headed The Theory of Attributes, and the third part, Theory of Sampling, begins with a chapter on sampling of attributes. Today Yule's account of association of categorical variables does not seem very helpful. Curiously, the one technique for analysis of contingency tables that is now widely known and used, namely the application of Karl Pearson's χ^2 test (1900) to testing association between categorical variables, was not clearly understood and readily available until it was presented in Fisher's *Statistical Methods for Research Workers* (1925). Elucidation of the number of degrees of freedom in the test called for algebraic insight similar to that needed for the analysis of variance being developed at the same time. Appreciation of a reasonable significance test for association had to wait until enthusiasm for arbitrary measures of association had waned.

During the last twenty-five years or so, greatly increased theoretical attention has been paid to categorical variables and contingency tables, stimulated by the series of papers of Goodman and Kruskal (1954 onwards). Although many ideas have been explored, there seems to have been no broad consensus in regard to any of them, and the subject may be said to be confused. Perhaps unevenness in statistical experience of categorical variables has hindered progress. Whereas many statisticians (and persons engaged in statistical work) almost never encounter categorical variables and see little reason to be interested in contingency tables, there are others who deal all the time with categorical variables and almost never have occasion to perform a regression or analysis of variance.

Below we make some general comments on categorical variables.Then the χ^2 test for association is considered, together with two alternative tests for association, one based on a likelihood ratio, the other on the probability of the data. Finally, we present an approach to studying association between variables having ordered categories, using methods due to John Tukey (and hinted at by Yule). The key idea is to fit Pearson-Plackett distributions having constant crossproduct probability ratios and examine the residuals.

A. Independent Categorical Variables

Contingency tables

A contingency table may be defined as a rectangular table in which the entries are counts (nonnegative integers) and the rows and columns refer to two modes of classification. The cross classification, or double polytomy, can arise in different ways that should sometimes be distinguished. One possibility is that each classification refers to a categorical variable, and the table has been formed by observing a pair of categorical variables for each of a set of individuals. For example, each of a set of 707 patients with cancer of the stomach was classified according to the portion of the stomach involved and according to the ABO blood group:

	Blood group			
Site	O	A	B or AB	Total
(a) Pylorus and antrum	104	140	52	296
(b) Body and fundus	116	117	52	285
(c) Cardia	28	39	11	78
(d) Extensive	28	12	8	48
Total	276	308	123	707

(Data supplied by Dr. Colin White; see White and Eisenberg, 1959. The sites (a), (b), (c) are regions of the stomach; the category (d) refers to cases where the involvement was so extensive that the site could not be specified. The blood groups B and AB have been combined because there were few cases in the AB category.)

Another possibility is that one classification is by an observed categorical variable, but the other classification relates to provenance of the individuals, indicating a set of samples or levels of a treatment factor. For example, Graunt (1662, chapters 8 and 12) noted that the ratio of males to females in his aggregated reports of christenings in London, 1628–1662, was slightly different from the ratio in ninety years of christenings in a country parish. His figures can be set out in a table as follows:

	Christenings male	Christenings female	Total
London	139782	130866	270648
Country parish	3256	3083	6339
Total	143038	133949	276987

Another well-known example comes from the 1954 poliomyelitis vaccine trials (Francis *et al.*, 1955, Table 2b). A part of the test of the Salk vaccine was a randomized double-blind trial offered to about three-quarters of a million children, of whom about 420 thousand elected to participate. Reported cases of polio were eventually classified either as confirmed, paralytic and nonparalytic being distinguished, or as doubtful or not polio. (For a brief description of the trials see Meier, 1972.)

	Polio reported			Polio	
	Paralytic polio	Nonparalytic polio	Doubtful or not polio	not reported	Total
Vaccinated	33	24	25	200663	200745
Placebo	115	27	20	201067	201229
Not inoculated or incomplete placebo	121	36	25	338596	338778
Incomplete vaccinations	1	1	0	8482	8484
Total	270	88	70	748808	749236

In the second and third of these tables, the rows refer to a treatment or source, the columns to a response variable; whereas in the first table both rows and columns refer to what may be called response variables. (Plackett, 1974, speaks of "factors" and "responses" in making this distinction.) If we call the two categorical variables in the first table responses, we do not mean to imply that we view them symmetrically. Conceivably blood group influences the site of a stomach cancer, but not the other way round.

We have been thinking of a contingency table in two dimensions, but in principle a crossclassification may be a rectangular array in any number of dimensions, two or more. A three-dimensional contingency table may possibly represent (i) the joint frequency of occurrence of three categorical "response" variables for a set of individuals, or (ii) the occurrence of two response variables for a set of samples or levels of a factor, or (iii) the occurrence of one response variable for a double set of samples or levels of two factors.

Variables with arbitrary categories

Categorical variables differ in the degree of apparent arbitrariness of the categories. At one extreme, consider the variable, sex. Almost any human being (and many sorts of animal) can upon proper examination be classified unambiguously as either male or female. Of course, more than just sex could be observed; for example, a variable could have three categories, "child", "adult male", "adult female", but in that case the variable is not simply sex but sex-and-maturity. Or again, if adequate examination is not possible, there could be three categories, "male", "female", "sex not determined", and now the variable could be described as ascertained sex. However, if sex is precisely what is required and if it can be observed, then it is a variable with just two well-defined categories, and that is that.

Consider now the answer given to a question by a respondent who is invited to check one of (say) five boxes. If the question concerns the respondent's attitude towards a given assertion, the five boxes might be labeled:

1. Strongly agree.
2. Mildly agree.
3. No opinion, indifferent or confused.
4. Mildly disagree.
5. Strongly disagree.

If the respondent does indeed answer the question by checking just one of the boxes, then we have a variable with five defined categories. The categories cannot be subdivided, but they could possibly be amalgamated. For example, categories 1 and 2 could be pooled into an "agree" category, and 4 and 5 into a "disagree" category, leaving three categories in all; and again for some purpose it might be interesting to reduce the original five categories to only two, namely (1, 2, 4, 5), "decided", versus 3, "undecided". In general, when a variable has more than two categories, there may be a case for reducing the number of categories, either by pooling, or possibly by omitting the individuals that fall in some category.

Sometimes it appears that the categories of a variable are arbitrary from the start. However they are defined, they could have been defined differently—categories could have been subdivided and reformed in many ways, and there is no clearly best or right way. In a study of road accidents, information may be sought on the variable, type and make of car. A full description of an individual car would contain many items: the maker's name, the name of the model, its year, type of body, number of doors, overall weight and dimensions, color, and a list of possibly numerous optional features (engine size, type of brakes, type of transmission, power-assisted this and that, air conditioning, radio, etc.). Very likely, no two cars in the study will be identical in all these respects. If kind of car is to be brought into consideration, grouping into a few categories will probably be deemed necessary, but that can be done in many very different ways. Categorization might reasonably, for various purposes, be based primarily on maker's name (particularly distinguishing domestic cars from foreign), or on overall weight, or on size, or on original cost, or on age, . . . , or on any combination of these qualities.

This may be judged the most characteristic and challenging situation with categorical variables. The categories are arbitrary, they might have been subdivided, and they may now be pooled.

Sometimes categorial variables may be said to have *ordered* categories. We take this to mean simply that the categories are listed in an order, and that if any pooling of categories is to be done attention shall be restricted to the pooling of adjacent categories. The restriction usually seems appropriate if the categories are defined as intervals of some quantitative variable, such as income or weight or date; and also if the categories represent the degree or intensity of some effect that conceivably might have been measured on a linear scale but was not. The five response-boxes for a question, mentioned above, seem to be linearly ordered from strong agreement to strong disagreement, and might therefore be declared to be ordered categories. On closer examination one may not be so sure—perhaps some respondents answer most questions by either strong agreement or strong disagreement, whereas others of less aggressive temperament answer most questions by either mild agreement or mild disagreement, and others again, disapproving of the nature of the questions, answer most of them by professing no opinion. Thus an initial declaration that categories are ordered, like any other decision in statistical analysis of data, could be thought better of later.

Ordering of categories is not necessarily linear; it could be circular. The variable, birthday, could be arbitrarily categorized by, say, the twelve months, January, February, . . . , December; or more finely than that or less finely; but for most purposes each category would be chosen to represent an interval of dates, and the ordering is circular, December 31 and January 1 being adjacent. We shall, however, consider only linear ordering below.

The name "contingency table" comes from Karl Pearson's peculiar use of "contingency" (1904), when he wished to refer to categorical variables whose categories were not ordered. This restriction of the context of the word "contingency" does not now seem useful and has been forgotten. More recently, the term "contingency table" is sometimes restricted to a crossclassification where there is some notion of independence of the individuals, residual variability being compared with the variability of a multinomial distribution. We turn to this matter now.

Stochastic independence and definitions of no association

Traditional (and recent) theory about the statistical analysis of contingency tables rests on assumptions, or tests hypotheses, of stochastic independence of individuals. For two observed categorical variables (single set of individuals and bivariate response), the possibility is entertained that each individual independently has a probability p_{ij} of being classified in the (i, j)th cell, that is in the ith category of the first variable and the jth category of the second; we have

$$p_{ij} \geqslant 0 \quad (\text{all } i \text{ and } j), \quad \Sigma_{ij} p_{ij} = 1.$$

For a set of samples or treatments and one observed categorical variable (one factor and one response, shown as rows and columns respectively), the corresponding entertained possibility is that each individual in the ith sample independently has a probability p_{ij} of being classifed in the jth category; we have

$$p_{ij} \geqslant 0 \quad (\text{all } i \text{ and } j), \quad \Sigma_j p_{ij} = 1 \quad (\text{all } i).$$

We shall always suppose that the number of rows and the number of columns of a contingency table, and the number of categories in any classification, are not less than 2.

Of the various assumptions or hypotheses that can be made about the matrix of probabilities $\{p_{ij}\}$, the best known is the hypothesis of no association between the classifications. For the bivariate response case, the probability matrix is the outer product of two vectors, or equivalently (if all probabilities are positive)

$$\ln p_{ij} = \alpha_i + \beta_j \quad (\text{all } i \text{ and } j), \tag{1}$$

where $\{\alpha_i\}$ and $\{\beta_j\}$ are vectors. In the special case that each variable has only two categories (the table is a double dichotomy), this definition is equivalent to unit value for the fourfold ratio,

$$\frac{p_{11} p_{22}}{p_{12} p_{21}} = 1.$$

For the one factor and one response case, we say there is no association if all rows of the probability matrix are the same.

Suppose the observed count in the (i, j)th cell is n_{ij}, and let $n_{i.}$, $n_{.j}$ and N be the ith row total, the jth column total and the grand total:

$$n_{i.} = \Sigma_j n_{ij} \quad (\text{all } i), \quad n_{.j} = \Sigma_i n_{ij} \quad (\text{all } j), \quad N = \Sigma_{ij} n_{ij}.$$

If the hypothesis of no association is to be tested, the probability distribution of whatever test statistic is chosen may be conditioned on sufficient statistics for the parameters, in order to eliminate those parameters from the test. In the bivariate response case, both sets of marginal totals, $\{n_{i.}\}$ and $\{n_{.j}\}$, constitute such a set of sufficient statistics for the parameters $\{\alpha_i\}$ and $\{\beta_j\}$. In the one factor and one response case, the column totals $\{n_{.j}\}$ are sufficient statistics, and the row totals $\{n_{i.}\}$ are probably regarded as fixed. In either case, if both sets of marginal totals are conditioned on, the probability distribution for the interior of the table, on the hypothesis of no association, is hypergeometric:

$$\text{prob}(\{n_{ij}\} | \{n_{i.}\}, \{n_{.j}\}) = \frac{(\Pi_i n_{i.}!)(\Pi_j n_{.j}!)}{N!(\Pi_{ij} n_{ij}!)} . \tag{2}$$

A three-dimensional contingency table may be considered similarly. For three observed categorical variables (single set of individuals and trivariate response), the possibility is entertained that each individual independently has a probability p_{ijk} of being classified in the (i, j, k)th cell;

$$p_{ijk} \geqslant 0 \quad (\text{all } i, j, k), \quad \Sigma_{ijk} p_{ijk} = 1.$$

And similarly for one factor and bivariate response, and for two factors and univariate response.

Various definitions have been considered for lack of association between the classifications. Particular attention has been paid to linear expressions for $\ln p_{ijk}$, the "log linear model". A definition of no three-variable association in the trivariate response case is that

$$\ln p_{ijk} = \alpha_{ij} + \beta_{ik} + \gamma_{jk} \quad (\text{all } i, j, k), \tag{3}$$

where $\{\alpha_{ij}\}$, $\{\beta_{ik}\}$, $\{\gamma_{jk}\}$ are matrices. In the special case that each variable has only two categories, this definition is equivalent to equality of two fourfold ratios,

$$\frac{p_{111}\,p_{122}}{p_{112}\,p_{121}} = \frac{p_{211}\,p_{222}}{p_{212}\,p_{221}},$$

or unit value for the eightfold ratio,

$$\frac{p_{111}\,p_{122}\,p_{212}\,p_{221}}{p_{112}\,p_{121}\,p_{211}\,p_{222}} = 1, \tag{4}$$

agreeing with Bartlett's definition (1935) for the 2-by-2-by-2 table.

Let us pause a moment to consider the plausibility of the assumption that individuals are assigned to categories independently. For the three two-dimensional tables quoted above, if any probabilistic machinery is introduced to describe the phenomena, independence of individuals seems a natural hypothesis or working assumption. The site of one person's stomach cancer and his blood group would usually seem causally independent of the same for another person; only rather far-fetched reasons (related persons, observer bias) can be suggested for stochastic dependence. However, that is not always so. For example, the *Statistical Abstract of the United States 1970* (p. 526) shows a tabulation of the number of scientists, categorized by discipline in nine categories, employed in each of the fifty states (and other areas). The state in which any one scientist is employed is not wholly independent of where others are employed. For if there exists in a state an educational institution or an industrial or governmental or whatever organization that requires a particular type of scientist, very probably several of them are needed—scientists are found in related clusters. If an attempt is made to analyze such a table by typical methods for contingency tables, implying a basic multinomial distribution, the residual variability appears too large. The table can indeed be instructively analyzed by taking logarithms of the entries and applying methods commonly used for two-way tables—for example, an additive analysis followed by a multiplicative analysis of the residuals. Since a contingency table (as defined above) is a particular kind of two-way table, general methods for two-way tables can always be resorted to.

Granted that we think in terms of independent individuals and multinomial sampling, we need still to consider the plausibility of any special relations that are postulated to hold between the probabilities. Two such possible relations have been mentioned already, the definition (1) of no association between two variables, and the definition (3) of no three-variable association; and we shall consider below other relations that may possibly subsist between the probabilities (namely that they constitute a Pearson-Plackett distribution). How reasonable are (1) and (3)? There seems to be complete agreement that relation (1) should be taken as the definition of no association between two variables. In view of what has been said about arbitrariness of categories, we note that if a matrix of probabilities $\{p_{ij}\}$ satisfies (1) it will still do so after any pooling of categories that leaves a crossclassification with at least two categories each way. Relation (3) is a different matter, however. If the array $\{p_{ijk}\}$ of probabilities for a three-dimensional table (trivariate response) satisfies this relation, at least one of the variables having more than two categories, the array of probabilities will in general not continue to satisfy (3) after pooling of categories that leaves a crossclassification with at least two categories each way. Thus relation (3) is not really interesting for variables having arbitrary categories, because it depends on the particular choice of categories. Whereas Bartlett's definition (4) of no three-variable association in a 2-by-2-by-2 table seems to be reasonable and widely accepted, the relation (3) is not an entirely satisfactory generalization of it. For *ordered* categories, another (incompatible) generalization is possible, as we shall see.

The χ^2 test for association

Karl Pearson's famous 1900 paper on χ^2 was written before the era of mathematical precision. What he thought he was doing is unclear. One interpretation is that he intended to use the probability of the sample as test statistic, but developed his χ^2 statistic as an approximation to (a monotone function of) the sample probability, and then the tabulated χ^2 distribution as an approximation to the sampling distribution of the χ^2 statistic. The two successive approximations somewhat obscure the argument. The χ^2 test was the first significance test having broad applicability that was not readily interpretable in terms of inverse probability, and undoubtedly it contributed much to Fisher's exploration of the concept of significance test.

The probability density of the observations under some hypothesis is a possible statistic for testing conformity with that hypothesis. Its use is vaguely seen in the earliest probabilistic reasoning. A probability density is the derivative of the probability measure with respect to some base measure, and the latter is arbitrary. The significance level yielded by the test depends much on this base measure. When the distribution of the observations (under the hypothesis) is continuous, there is generally no base measure that commands attention as the right one, no best scale in which the observations should be expressed—unless perhaps a scale can be chosen in which all possible observations have the same precision. However, when the observations are counts there is just one base measure that is virtually always used and commands respect as the most natural, namely counting measure on the integers; and then at least the probability of the observations is an attractive test statistic.

To construct a test for association in a contingency table, we may proceed, as suggested above, by conditioning on the marginal totals to eliminate nuisance parameters, therefore accepting (2) as the probability distribution for the table, and we may use the right side of (2) itself as the test statistic. In principle the test can be carried out as follows. All possible tables consistent with the given margins are enumerated. For each the right side of (2) is calculated. These probabilities are permuted into nondecreasing order, and a list is made of all the *different* values of the probabilities and the multiplicity with which each has been obtained. Let the different values (in strictly ascending order) be $P_1, P_2, \ldots,$ and the corresponding multiplicities $r_1, r_2, \ldots$. Then the probability, under the hypothesis of no association and conditionally on the given margins, that a table has test statistic (probability) P_s is $r_s P_s$, and the probability that it has test statistic (probability) less than P_s is $r_1 P_1 + r_2 P_2 + \cdots + r_{s-1} P_{s-1}$.

Suppose that P_s is the probability of the given table. Then the significance level may be defined to be

$$\alpha = r_1 P_1 + r_2 P_2 + \cdots + r_{s-1} P_{s-1} + \tfrac{1}{2} r_s P_s. \tag{5}$$

The significance level shows how the given table is ranked relative to all tables that might have been obtained under the hypothesis of no association (given the margins), the ranking being with respect to the test statistic, which is here the probability (2). The test statistic has a discrete distribution under the hypothesis being tested, and the ranking of the observed table relative to all possible tables can be completely specified by quoting two probabilities, say the probability that a table has the observed value of the statistic, and also the probability that a table has a value of the statistic less than that observed—from which the third relevant probability, that a table has a value of the statistic greater than that observed, can be immediately deduced. Traditionally the first two probabilities have been summed to give as significance level the probability that a table has a value of the test statistic less than or equal to that observed—or of course, at whim, the first and third probabilities may be summed. Such a definition treats the two tails of the distribution of the statistic unsymmetrically, for no good reason. If just one number is to be quoted as significance level, the definition (5) above is more satisfactory. Then under the hypothesis being tested

$$\mathcal{E}(\alpha) = \tfrac{1}{2}.$$

Our test has ordered the tables by their rarity under the hypothesis being tested. If the significance level α obtained for a given table is near 0, we conclude that the table is of great rarity, under the hypothesis, and we have grounds for suggesting that the hypothesis is false. If α is not near 0, the pattern of the table is judged to be not very unusual, under the hypothesis. A value of α very close to 1 might cause suspicion that the entries had been somehow manipulated into too-perfect proportionality, since the tables whose rows (or columns) are nearly proportional have the highest probabilities.

For a 2-by-2 table, the test just described is nearly Fisher's well-known exact test (1934, section 21.02). In fact, Fisher considered tables departing from expectation in one specified direction rather than in both directions, and so the above test can be called a two-sided version of Fisher's one-sided test. For larger tables such a test is rarely used, because the enumeration of all possible tables is so vast. Even for Graunt's 2-by-2 table quoted above, there are as many as 6340 possible tables with the same margins; for the stomach cancer table the number exceeds 10^{10}.

Pearson's χ^2 statistic is obtained when all the factorials in the right side of (2) are approximated by Stirling's formula. Some notation is needed, for the marginal proportions (all of which are supposed to be positive),

$$p_i = n_{i.}/N \quad (\text{all } i), \quad q_j = n_{.j}/N \quad (\text{all } j),$$

the expected values and standardized residuals,

$$m_{ij} = N p_i q_j, \quad z_{ij} = (n_{ij} - m_{ij})/\sqrt{m_{ij}} \quad \text{(all } i \text{ and } j\text{)},$$

and Pearson's statistic

$$\chi^2 = \Sigma_{ij} z_{ij}^2.$$

On applying Stirling's formula and judiciously expanding a logarithm, we obtain an approximation to the probability of the table, expression (2),

$$\left[(2\pi N)^{\nu}(\Pi_i p_i)^{l-1}(\Pi_j q_j)^{k-1}\right]^{-1/2}\exp(-\tfrac{1}{2}\chi^2), \tag{6}$$

where k and l are the numbers of rows and columns and $\nu = (k - 1)(l - 1)$, the number of degrees of freedom in the table when the margins are fixed. (This expression is asymptotic for a sequence of tables in which $N \to \infty$ with $\{p_i\}$ and $\{q_j\}$ constant and $\{z_{ij}\}$ bounded.)

Expression (6) is monotone in χ^2, which therefore serves as an approximation to the test statistic (2). (A monotone one-to-one transformation of a test statistic does not alter the ordering induced on the sample space and therefore makes no difference to the significance level, that is, no essential difference to the test.) To Karl Pearson χ^2 no doubt seemed easier to calculate than the exact probability, though today we scarcely notice any difference. (Pearson did not quote an expression for the exact probability in his 1900 paper, which in any case related to testing agreement of data with a given multinomial distribution rather than no association in a contingency table; and whether he consciously preferred his χ^2 statistic to the exact probability is a matter for conjecture.) The sampling distribution of the χ^2 statistic, under the hypothesis being tested, is discrete and no easier to determine precisely than that of the probability (2). The immense saving of effort comes in approximating the distribution of the χ^2 statistic by what we refer to here as the tabulated χ^2 distribution with ν degrees of freedom.

The χ^2 statistic and approximating tabulated χ^2 distribution seem to constitute a very satisfactory general-purpose test for association in a contingency table, in a wide range of circumstances. A good feature of the χ^2 statistic is that it is the sum over the cells of the table of an expression,

$$(n_{ij} - m_{ij})^2/m_{ij}, \tag{7}$$

that can be directly interpreted as a measure of discrepancy between the observed count n_{ij} and its expected value m_{ij} on the hypothesis being tested. A computer program should do what was inevitable in Pearson's day and display the individual contributions to χ^2, or better still their square roots $\{z_{ij}\}$.

Two limitations on the use of χ^2 for testing goodness of fit are recognized. One is that the number of degrees of freedom, ν, should not be too large. Goodness-of-fit tests with ν between 1 and, say, 20 often seem helpful. Such a test with ν equal to, say, 2000 may perhaps not seem so helpful; of the 2000 modes of departure from the hypothesis being tested, some are almost certainly much more interesting and likely than others, and a more focused test may be preferable.

The other limitation is that the cell expectations $\{m_{ij}\}$ should not be too low; 5, or sometimes 2, is quoted as a desirable lower bound. The reason usually given is to ensure that the distribution of the statistic should be well approximated by the tabulated χ^2 distribution, but a more fundamental matter is that usually when some cell expectations are low the χ^2 statistic has unreasonable behavior which we should prefer not to approximate even if we could. Specifically, if the (i, j)th cell has small expectation m_{ij} and if the marginal totals $n_{i.}$ and $n_{.j}$ are both large, the count n_{ij}, on the hypothesis being tested, is distributed nearly in a Poisson distribution, nearly independently of the counts in other such cells. Then the contribution (7) to χ^2 has (nearly) expectation 1, variance $2 + (1/m_{ij})$, third central moment $8 + (22/m_{ij}) + (1/m_{ij}^2)$; the smaller m_{ij} is the larger the dispersion. Now however small m_{ij} is, the count in the cell can be informative concerning goodness of fit. If $m_{ij} = 0.01$ and $n_{ij} = 1$, the cell provides quite strong evidence against the hypothesis being tested. Thus a good test of goodness of fit will not ignore cells with small expectation. But there is a respect in which cells with low expectation, say below 2, are less informative than cells with expectation greater than, say, 3. That is, if $m_{ij} > 3$, the count n_{ij} can differ "significantly" from m_{ij} either by being too low or by being too high. But if $m_{ij} < 2$, the lowest possible count of 0 is not improbably low, and the only way for the count to differ "significantly" from m_{ij} is by being too high. We may expect that a good test of goodness of fit will give less weight to cells with expectation below 2 than to cells with expectation above 3, and that is not at all how the χ^2 statistic behaves.

The above remarks are based on the assumption that the row totals $\{n_{i.}\}$ and the column totals $\{n_{.j}\}$ are not small, say not less than 25, and considerably larger than any cell expectation m_{ij} being referred to. By contrast, it is actually possible for a cell count to be totally uninformative, in which case it should contribute nothing to a good test of goodness of fit. Suppose that a contingency table has just two columns and the column totals are equal, and suppose that one row total is equal to 1. Then there are only two possibilities for the entries in that row, namely 0 1 and 1 0, and these have equal probabilities. The contribution from the row to any reasonable statistic measuring association (and in this respect the χ^2 statistic behaves reasonably) is fixed and conveys no information.

Alternatives to the χ^2 statistic

Fisher (1950) recommended that when some cell expectations were small the χ^2 statistic should be replaced by a likelihood ratio statistic, the ratio of the likelihood of the hypothesis being tested to the maximum likelihood obtainable by free adjustment of the probabilities $\{p_{ij}\}$ of the cells. Minus twice the logarithm can be expressed as the sum over all cells of

$$\left.\begin{array}{ll} 2\left\{n_{ij}\ln\dfrac{n_{ij}}{m_{ij}} - (n_{ij} - m_{ij})\right\} & (\text{for } n_{ij} \geqslant 1), \\ 2m_{ij} & (\text{for } n_{ij} = 0). \end{array}\right\} \quad (8)$$

Since the sum over all cells of $(n_{ij} - m_{ij})$ necessarily vanishes, this term may be dropped from the above expression without altering the statistic. But it should not be dropped if we wish the expression to serve, like (7), as a measure of discrepancy between n_{ij} and m_{ij} in the individual cell. For each nonnegative integer value for n_{ij}, the expression (8), like Pearson's (7), is nonnegative for all positive values for m_{ij} and approaches 0 when m_{ij} approaches n_{ij}. The behavior of this statistic is much better than that of the χ^2 statistic when some cells have low expectation; if it errs it does so in the opposite direction of giving little weight to cells whose expectation is below 1. When no cell expectation is less than, say, 1.5 and all marginal totals are large enough (perhaps not less than 25), the distribution of the statistic seems to be fairly well approximated by the tabulated χ^2 distribution with the usual number of degrees of freedom. The statistic appears in some recent computer packages, but without display of the individual contributions (8).

Possibly more effective, but little considered, as an alternative to Pearson's χ^2, is direct use of the probability of the sample, which we have suggested was perhaps what Pearson himself really wished to work with. If the contingency table is of size 2-by-2 or 3-by-2, so that ν is either 1 or 2, and if the marginal totals are not very great, enumeration of all possible tables consistent with the margins is feasible, and the expression (5) for the significance level can be evaluated directly. Otherwise, we may need a convenient approximation to the distribution of the statistic. In any case, also, it will be helpful to have the statistic expressed as the sum over all cells of a measure of discrepancy between n_{ij} and m_{ij}, similar to (7) or (8). Minus twice the logarithm of the probability of the sample can be expressed as the sum over all cells of

$$2\{\ln(n_{ij}!) - n_{ij}\ln m_{ij} + f(m_{ij})\},$$

where the term $n_{ij}\ln m_{ij}$ could be dropped because its sum over all cells depends only on the fixed marginal totals, but for present purposes it should not be dropped, and the function $f(\cdot)$ can be chosen arbitrarily without affecting the test. We might choose $f(\cdot)$ to facilitate some approximation to the distribution of the statistic, or instead we might choose $f(\cdot)$ so that the above expression would have similar properties to (7) and (8) as a measure of discrepancy between n_{ij} and m_{ij}. For the latter purpose a good if not uniquely best choice for $f(\cdot)$ yields the statistic as the sum over all cells of

$$2\left\{\ln\frac{n_{ij}!}{(m_{ij}-\frac{1}{2})!}-(n_{ij}+\tfrac{1}{2}-m_{ij})\ln m_{ij}\right\}. \qquad (9)$$

For each nonnegative integer value for n_{ij}, the expression (9) vanishes when $m_{ij} = n_{ij} + \frac{1}{2}$, is slightly less than 0 when m_{ij} is a little greater than $n_{ij} + \frac{1}{2}$, but elsewhere is positive. We shall be nearly but not quite correct if we assert that (9) is always nonnegative, and close to 0 when m_{ij} is close to $n_{ij} + \frac{1}{2}$. Expression (9) seems to be a reasonable measure of discrepancy in the individual cell, except perhaps when the expectation m_{ij} is close to 0, say less than 0.1. The notation $(m_{ij} - \frac{1}{2})!$ is intended to mean the same as $\Gamma(m_{ij} + \frac{1}{2})$. We repeat that the sum over all cells of the expression (9) is a test statistic exactly equivalent to the probability of the table (2), if the distribution is handled exactly, without approximation.

Three goodness-of-fit statistics have been mentioned, Pearson's χ^2, the likelihood-ratio statistic defined as the sum over cells of (8), and the probability-of-the-sample statistic defined as the sum over cells of (9). To establish a simple approximation to the distribution of one of these statistics, or to establish conditions under which the tabulated χ^2 distribution with ν degrees of freedom is a good approximation—to do these things convincingly is hard, and much of the published literature is disappointing.* It is easy to do some computing, easy to do asymptotics, not easy to associate computing with enough theoretical understanding that general conclusions can be drawn plausibly.

The one theoretical insight that everyone has is that the asymptotic distribution of each test statistic is the tabulated χ^2 distribution with ν degrees of freedom, as all cell expectations tend to infinity. For the χ^2 statistic, but not for the other two, concise expressions for some moments can be found. As for cells with low expectation, the principal theoretical insight comes from recognizing that the count in a cell for which the row total and the column total are both much larger than the cell expectation has approximately a Poisson distribution, and is approximately independent of the count in another such cell. The behavior of (7) or (8) or (9) for a single Poisson-distributed cell can be studied numerically, and any moment of its distribution can be determined empirically as a function of the cell expectation. This consideration goes some way to bridge the gap between the real world and asymptotics. The following brief table gives the mean, variance and third central moment of each of the three statistics for a single Poisson-distributed cell. The limiting values of the three moments as the cell expectation increases are 1, 2, 8, respectively.

*For example, Larntz (1978) has computed with too little theoretical support, and Williams (1976) has theorized without computing.

Behavior of goodness-of-fit statistics for a single Poisson-distributed cell

	Cell expectation									
	0.02	0.05	0.1	0.2	0.5	1	2	5	10	20
Pearson's χ^2 *statistic* (7)										
Mean	1.00	1.00	1.00	1.00	1.00	1.00	1.00	1.00	1.00	1.00
Variance	52.00	22.00	12.00	7.00	4.00	3.00	2.50	2.20	2.10	2.05
μ_3	3608.	848.0	328.0	143.0	56.00	31.00	19.25	12.44	10.21	9.10
Likelihood-ratio statistic (8)										
Mean	0.16	0.30	0.47	0.70	1.01	1.15	1.14	1.05	1.02	1.01
Variance	0.69	0.86	0.86	0.73	0.73	1.36	2.23	2.27	2.09	2.04
μ_3	4.23	4.10	3.66	3.36	3.00	2.66	5.81	9.88	8.64	8.24
Probability-of-sample statistic (9)										
Mean	2.84	2.04	1.51	1.11	0.86	0.85	0.92	0.98	0.99	1.00
Variance	1.23	1.84	2.26	2.46	2.19	1.78	1.67	1.89	1.96	1.98
μ_3	9.80	11.59	11.81	11.01	9.40	8.08	6.68	7.15	7.74	7.89

From this Poisson-cell table we may infer (approximately) the first three moments of the distribution of one of the statistics for a contingency table in which a few cells have small expectations but all other cells have very large expectations and all marginal totals are very large. The number of cells with low expectations should be not greater than the number of degrees of freedom ν. Then the moments of the statistic are equal to those of the tabulated χ^2 distribution with ν degrees of freedom, corrected by adding, for each cell with low expectation, the difference between the moments as given in the Poisson-cell table and the limiting values of 1, 2, 8. Suppose, for example, that in a 3-by-2 contingency table two cells in the same column have expectations 1 and 2, and all four remaining cells have very large expectations. Then the mean of the probability-of-the-sample statistic is $2 + (0.85 - 1) + (0.92 - 1) = 1.77$, its variance is $4 + (1.78 - 2) + (1.67 - 2) = 3.45$, and its third central moment is similarly 14.76. In order that these values shall be nearly correct, the expectations in the four remaining cells must indeed be very large. If they are not so very large, perceptibly different values for the moments are found. For a contingency table having these cell expectations and marginal totals

1	24	25
2	48	50
22	528	550
25	600	625

the probability-of-the-sample statistic has in fact mean 1.87, variance 3.13, third central moment 12.79. (These values have been found by enumerating the 351 possible tables consistent with the margins and for each calculating the probability of the table and the value of the test statistic.)

The above rule gives, on the whole, slightly better estimates if the corrections are made for every cell in the contingency table and not just for the two with lowest expectations—we get 1.76 for the mean, 3.41 for the variance, 14.53 for the third moment. The rule often gives, as here, for the probability-of-the-sample statistic, too low a value for the mean and too high values for the second and third moments.

We shall see what use can be made of information about moments of one of the goodness-of-fit statistics. But let us first consider briefly how to assess the effectiveness of a continuous approximation (such as the tabulated χ^2 distribution) to the discrete distribution of a goodness-of-fit statistic.

Sometimes it is important, in comparing a continuous distribution with a discrete distribution, to make a "continuity correction". The one chosen here is the symmetric definition of the significance level, namely that the significance of an observed value k of a statistic T (for which large values are significant) is reckoned to be

$$\alpha = \text{prob}(T > k) + \tfrac{1}{2}\text{prob}(T = k),$$

which is compared with the integral from k to ∞ of the density function of the continuous approximation. If instead the usual unsymmetric definition of significance level is adopted, namely $\text{prob}(T \geqslant k)$, a possible continuity correction consists of determining the next lower value k' that could be assumed by T, and comparing the significance level with the integral from $\frac{1}{2}(k + k')$ to ∞ of the approximating density function—an inconvenient procedure if k' is hard to find.

How close should a good approximation to the significance level be? An error of 0.01 is less serious in a significance level of 0.05 than in a significance level of 0.01. We are interested in practice in a broad range of significance probabilities, surely 0.1 to 0.001 or so. The simple probability scale is inconvenient. The equivalent normal deviate, such that the integral from the deviate to ∞ of the standard normal density function is equal to α, is a much more convenient measure of significance. We shall concentrate on equivalent normal deviates between 1 and 3, corresponding to α between 0.16 and 0.0013; the value 1.64 corresponds to $\alpha = 0.05$.

The first thing we notice when we examine the distribution of one of the goodness-of-fit statistics is that the distribution is not only discrete but rough. That is, the attainable values of the statistic are distributed unevenly, with irregular gaps between them. If an approximating density function were to reflect all attainable significance levels correctly, it could not be as smooth as one of the (tabulated) χ^2 densities but would have numerous small wobbles. Any feasible approximating distribution will not have such wobbles, and so there is a minimum feasible precision. Somewhat different approximating distributions may do about equally well; no one is clearly the best. We might expect that the roughness of the exact distribution would be greater the smaller the value of ν and the smaller the lowest cell expectation, and to some extent computations bear that out. A candidate for being called the best smooth approximating density function is the density of a random variable $a + bX$, where X has the tabulated χ^2 distribution with some number (usually noninteger) of degrees of freedom, the parameters a, b and the degrees of freedom being chosen so that the random variable has the same first three moments as the statistic. At least, this approximating density always seems to do about as well as possible for equivalent normal deviates between 1 and 3 (or indeed between 0 and 4), though it may do poorly for some of the smallest attainable values of the statistic, corresponding to negative equivalent normal deviates.

Now for the Pearson χ^2 statistic the first three moments can (with much trouble) be determined in readily computable form—though programming the calculation would be a pity because it would encourage people to use the statistic in conditions where it behaved unreasonably. For the other two statistics no such expressions for the moments can be found, but the procedure indicated above for estimating the moments from the moments of single Poisson-distributed cells with the same expectations seems often to work fairly well.

For example, for the 3-by-2 table with marginal totals and cell expectations shown above, there are 26 possible values of the probability-of-the-sample statistic having equivalent normal deviates (for the significance level) lying between 1 and 3. The approximate distribution fitted to the correct first three moments gives equivalent normal deviates with errors ranging from ¯0.09 to 0.10. The approximate distribution fitted to the estimated first three moments gives equivalent normal deviates with errors ranging from ¯0.11 to 0.10. The usual approximation (tabulated χ^2 with 2 degrees of freedom) gives equivalent normal deviates with errors ranging between ¯0.22 and ¯0.04.

As another example, consider a contingency table having these marginal totals and cell expectations:

5	20	25
5	20	25
15	60	75
25	100	125

The probability-of-the-sample statistic has first three moments equal to 2.07, 3.71, 13.97; and the estimated moments are 1.94, 3.71, 13.88. There are 42 possible values of the statistic having equivalent normal deviates between 1 and 3. Errors in approximate equivalent normal deviates range as follows:

For the distribution fitted to the correct first three moments, between ¯0.11 and 0.15.

For the distribution fitted to the estimated first three moments, between ⁻0.08 and 0.17.

For the usual tabulated χ^2 with 2 degrees of freedom, between ⁻0.17 and 0.10.

In both of these examples, the error in the equivalent normal deviate when the approximating distribution is fitted either to the correct or to the estimated moments is usually less than 0.1 in magnitude, but occasionally is larger. Since in the first example the moments are rather different from those of the tabulated χ^2 with 2 degrees of freedom, the latter gives a perceptibly poorer approximation. But in the second example the moments are closer, and there is nothing to choose between the approximations. A third example illustrates the ill effect of a cell with very low expectation:

0.1	24.9	25
5	1245	1250
19.9	4955.1	4975
25	6225	6250

The probability-of-the-sample statistic has first three moments 2.54, 4.05, 18.21; and the estimated moments are 2.48, 4.11, 18.75. There are 23 possible values of the statistic having equivalent normal deviates between 1 and 3. Errors in approximate equivalent normal deviates range as follows:

For the distribution fitted to the correct first three moments, between ⁻0.11 and 0.21.

For the distribution fitted to the estimated first three moments, between ⁻0.12 and 0.22.

For the usual tabulated χ^2 with 2 degrees of freedom, between 0.06 and 0.35.

We have just been considering the accuracy of continuous approximations to the distribution of the probability-of-the-sample statistic used to test association in 3-by-2 contingency tables. Similar results can be obtained for the likelihood-ratio statistic, except that the distribution of the statistic seems to be rougher and therefore less accurately approximated by any smooth continuous distribution; typical errors in equivalent normal deviates are twice as large as what we have seen. A possible reason for the difference can be suggested. The roughness in the distribution of a goodness-of-fit statistic comes partly from unequal spacing of the attainable values of the statistic and partly from unequal probabilities associated with those values. For the probability-of-the-sample statistic, the probabilities are a monotone function of the values, and probabilities for neighboring values are nearly equal; the roughness may be said to come mainly from the unequal spacing of the values. But for the likelihood-ratio statistic (or for Pearson's χ^2), not only are the attainable values unevenly spaced, but adjacent values may have sharply unequal probabilities, and the two kinds of unevenness seem to act independently. It is reasonable to conjecture that the probability-of-the-sample statistic has the smoothest distribution of any goodness-of-fit statistic.

Consideration of examples such as these, together with the theoretical results already noted, suggest the following conclusions regarding tests for association in a contingency table (or agreement of data with a given multinomial distribution, or other such problems of goodness of fit). See also the exercises at the end of the chapter.

Tentative conclusions. The Pearson χ^2 statistic behaves unsatisfactorily when some cell expectations are small, say below 3, and its use should then be discouraged. Either the likelihood-ratio statistic or the probability-of-the-sample statistic may reasonably be used without any restriction on the smallness of cell expectations. The distribution can be determined either by enumerating (some of) the tables consistent with the given margins, or by random sampling. If an easy continuous approximation to the distribution is desired, restrictions are advisable on cell expectations and marginal totals, such as perhaps these:

(i) no marginal total should be less than 25,
(ii) no cell expectation should be less than 0.4,
(iii) only a few cell expectations should be less than 5.

Then the distribution of the probability-of-the-sample statistic is fairly well approximated by the usual tabulated χ^2 distribution with the usual number of degrees of freedom ν; and the same for the likelihood-ratio statistic except that the minimum cell expectation should be raised to 1.5. Sometimes the approximation can be noticeably improved (and never noticeably worsened in the upper tail) by fitting a linear function of a χ^2 variable to the estimated first three moments of the statistic, as explained above. Provided this is done, restriction (ii) on the cell expectation becomes less important, and can be relaxed, the larger ν is, for both the probability-of-the-sample statistic and the likelihood-ratio statistic. The quality of approximation is less good for the likelihood-ratio statistic than for the probability-of-the-sample statistic.

Examination of the given tables

The ordinary approximate χ^2 test is adequate for the three tables quoted at the beginning of the chapter. The function *CONTINGENCY* displays individual standardized residuals, which were called $\{z_{ij}\}$ above, before summing their squares to form the χ^2 statistic. Figure B:37 shows the function applied to the three given tables.

```
      ⍝ WHITE-EISENBERG TABLE, STOMACH CANCER SITE AND BLOOD GROUP

      M1
 104  140   52
 116  117   52
  28   39   11
  28   12    8

      CONTINGENCY M1
THE MATRIX OF EXPECTED FREQUENCIES IS NAMED  EF .
LEAST MEMBER OF  EF  =  8.3508
STANDARDIZED RESIDUALS (SR):
 ¯1.07        0.97        0.07
  0.45       ¯0.64        0.34
 ¯0.44        0.86       ¯0.7
  2.14       ¯1.95       ¯0.12

CHI-SQUARED  =  12.654
DEGREES OF FREEDOM  =  6

      ⍝ GRAUNT TABLE, SEX OF CHRISTENINGS IN LONDON AND COUNTRY

      M2
 139782  130866
   3256    3083

      CONTINGENCY M2
THE MATRIX OF EXPECTED FREQUENCIES IS NAMED  EF .
LEAST MEMBER OF  EF  =  3065.5
STANDARDIZED RESIDUALS (SR):
  0.05       ¯0.05
 ¯0.31        0.32

CHI-SQUARED  =  0.19806
DEGREES OF FREEDOM  =  1

      ⍝ FRANCIS TABLE, SUMMARY OF 1954 POLIOMYELITIS VACCINE TRIAL

      M3
      33       24       25  200663
     115       27       20  201067
     121       36       25  338596
       1        1        0    8482

      CONTINGENCY ¯1 0 ↓M3
THE MATRIX OF EXPECTED FREQUENCIES IS NAMED  EF .
LEAST MEMBER OF  EF  =  18.97
STANDARDIZED RESIDUALS (SR):
 ¯4.67        0.09        1.38        0.07
  4.9         0.69        0.23       ¯0.1
 ¯0.18       ¯0.6        ¯1.24        0.02

CHI-SQUARED  =  50.293
DEGREES OF FREEDOM  =  6
```

Figure B:37

For the stomach cancer table, the χ^2 statistic is 12.65 with 6 degrees of freedom, falling just above the 5% point of the tabulated distribution. Rather than exclaim "significant!", let us more tamely remark that χ^2 is large enough to suggest some association between the variables. Two of the individual standardized residuals are substantial, one over 2 in magnitude, one just below 2. Under the hypothesis being tested, these residuals are distributed as though they were independent random variables with zero mean and unit variance observed subject to linear constraints on rows and columns (we refer to Fisher's decomposition of the distribution of the table into constrained Poisson variables). Marginally each standardized residual has variance less than 1, and an easy rule of thumb is that any of them exceeding 2 in magnitude can be regarded as "significant". Here it appears that the ratio of blood groups O to A among the site (d) patients is unlike that for the other patients. If the last row of the table is omitted, the rest yields a χ^2 value of 3.64 with 4 degrees of freedom, clearly a middling value.

In general, blood groups, determined genetically, are not expected to be noticeably associated with diseases whose causes are mainly environmental. Whenever an association between a disease and a genetic variable can be definitely established, there is great interest. Many such associations are looked for, most often without success. If the present data had resulted from a haphazard search for associations, not guided by theory or past experience, the suggestion of association that we have found could be readily dismissed as insufficiently striking to warrant attention. In fact, however, previous studies had suggested that the incidence of stomach cancer differed between persons in blood group O and persons in blood group A; and this investigation of site and blood group was designed to throw further light on the matter. Other variables besides those already mentioned were examined (including sex and national origin of patients, and method of diagnosis of the cancer), and the authors concluded that there was indeed some relation between blood group and site.

As for Graunt's records for christenings, he says that in the country parish "there were born 15 Females for 16 Males, whereas in London there were 13 for 14, which shews, that London is somewhat more apt to produce Males, then the country. And it is possible, that in some other places there are more Females born, then Males." His arithmetic here is not impeccable; the ratios are more like 18 females for 19 males in the country parish, and in London 15 for 16. Clearly he thinks the difference is, as we should say, significant. We approach these records now with a very different expectation from Graunt's. Modern experience is that a little over 51% of births are male, the ratio being apparently nearly the same for all human societies. We therefore expect to find no real difference between Graunt's ratios for London and the country, and that is what the χ^2 test indicates.

When Graunt had shown the way, others examined the constancy of sex ratios. Arbuthnot (1710) listed the numbers of male and female christenings in London each year from 1629 to 1710, the earlier part of this material being presumably copied from Graunt. (The table is not reproduced here.) He was concerned to show that the ratio varied less than it would if sex were determined by blind chance and births constituted a sequence of what we now call Bernoulli trials, with constant probabilities for each sex. He had, however, no idea of how a binomial variable behaves when the number of trials is large. In fact his table shows more variability, not less, than the Bernoulli-trials hypothesis would imply: $\chi^2 = 170$ with 81 degrees of freedom. A possible explanation is counting or copying errors in the data.

The Salk vaccine table, with the last row omitted, is also shown treated by *CONTINGENCY* in Figure B:37. To see the direct evidence of efficacy of the vaccine, just the first two rows should be compared. The difference in incidence of paralytic polio is highly significant, and large enough to be medically important. The third row of the table is also interesting, however, because it demonstrates a striking difference in incidence of paralytic polio between the placebo group of volunteers and the nonvolunteers.

Such an experiment aims to measure the magnitude of treatment effects. In the present case an even more important purpose is to dispel decisively the skeptic's doubt whether the treatment has any effect at all.

B. Associated Variables with Ordered Categories

We have just seen some contingency tables for which the χ^2 test for association is interesting and relevant. Other cases may be different: the variables are obviously associated, no one expected them not to be associated, and there is no interest whatever in the traditional test for association. If a contingency table is then formed and examined, the purpose is to understand the association, its kind and degree. This was the starting point for Goodman and Kruskal's 1954 paper on measurement of association.

In general, measuring association between qualitative variables seems a rather unsatisfactory task, unless some information, or some indicator, is available about how the association will go. There are ways of defining association between a pair of qualitative variables, whose categories are not ordered or related, when nothing more is given. But consideration is likely to be more profitable if, for example, a further variable can be found that explains the association, or if some relation between the categories of the given variables can be drawn on.

The rest of the chapter is devoted to a special topic, examining association between variables having (more or less arbitrary) *ordered* categories. Four contingency tables are used as examples to test the general ideas. First, we take the example of this kind quoted by Goodman and Kruskal

(1954), material published by Kiser and Whelpton (1950, p. 402). The study of social and psychological factors affecting fertility was carried out on a carefully selected (nonrandom) sample of couples in Indianapolis in 1941–42. In this table, rows refer to the educational attainment of the wife (years completed in school) and columns to four categories of fertility planning (from the most effective, labeled "Number and spacing planned", to the least effective, "Excess fertility"). Goodman and Kruskal pooled some rows, but all are shown here.

Education of the wife	Fertility-planning status: Number and spacing planned	Number planned	Quasi-planned	Excess fertility	Total
College 1 +	102	35	68	34	239
High School 4	156	53	177	103	489
High School 3	35	27	38	19	119
High School 2	35	29	73	94	231
High School 1	26	24	43	49	142
Grade School 6	49	37	52	80	218
Total	403	205	451	379	1438

Next, we take a table by Gilby (1911, p. 106) that has been quoted by Kendall (1943). Schoolboys were classified according to their clothing in five categories and according to the teacher's estimate of dullness (lack of intelligence) in seven categories. The categories were defined in some detail and briefly labeled as follows:

I. Very well clad
II. Well clad
III. Clothing poor but passable
IV. Clothing insufficient
V. Clothing the worst

A. Mentally defective
B. Slow dull
C. Slow
D. Slow intelligent
E. Fairly intelligent
F. Distinctly capable
G. Very able

There were so few cases in clothing category V and in dullness category A that these were amalgamated with the next categories. There appears to be a (modest) negative association between dullness and quality of clothing, that is, a positive association between intelligence and clothing.

Clothing	Dullness: A, B	C	D	E	F	G	Total
I	33	48	113	209	194	39	636
II	41	100	202	255	138	15	751
III	39	58	70	61	33	4	265
IV, V	17	13	22	10	10	1	73
Total	130	219	407	535	375	59	1725

Stuart's tables (1953) referring to left-eye and right-eye vision of men and women employed in ordnance factories in 1943–46 have been much discussed. In the table for men quoted here, unaided distance vision for each eye is categorized in four grades.

Right eye	Left eye Highest grade	Second grade	Third grade	Lowest grade	Total
Highest grade	821	112	85	35	1053
Second grade	116	494	145	27	782
Third grade	72	151	583	87	893
Lowest grade	43	34	106	331	514
Total	1052	791	919	480	3242

Last we present a table concerning social mobility (Glass, 1954) that was used by Goodman (1972) to illustrate a great variety of analytic methods, but not the one tried here. From a stratified random sample of adult civilians in Britain in 1949, father-son pairs were examined in regard to occupation. Occupations were classified in seven status categories, from 1, "professional and high administrative", to 7, "unskilled manual".

Father's status category	Son's status category 1	2	3	4	5	6	7	Total
1	50	19	26	8	18	6	2	129
2	16	40	34	18	31	8	3	150
3	12	35	65	66	123	23	21	345
4	11	20	58	110	223	64	32	518
5	14	36	114	185	714	258	189	1510
6	0	6	19	40	179	143	71	458
7	0	3	14	32	141	91	106	387
Total	103	159	330	459	1429	593	424	3497

Tukey (1971) has pointed to a valuable way of studying frequencies for a qualitative variable whose categories are ordered. The number of categories is reduced to two by pooling of adjacent categories, in all possible ways. If there are k categories initially, we consider the first versus the union of the second to the kth, and then the first and second versus the third to the kth, and so on, up to the first $k - 1$ categories versus the last. For each of these $k - 1$ dichotomies we calculate the "log odds", the difference of logarithms of the frequencies in the two cells. The $k - 1$ values of log odds may well prove to have an orderly intelligible relation with other variables present—for example, a linear relation with a quantitative variable, leading naturally to a theoretical description in terms of a logistic distribution.

If one has two qualitative variables with ordered categories, the same approach leads to the formation of all possible double dichotomies by pooling of adjacent categories. If the numbers of categories are k and l respectively, there are $(k - 1)(l - 1)$ possible double dichotomies. For each the difference of log odds for the first variable is found for the two categories of the second, that is, the logarithm of the fourfold ratio of frequencies for the table,

$$\ln \frac{n_{11} n_{22}}{n_{12} n_{21}} .$$

The $(k - 1)(l - 1)$ values of the log fourfold ratio are examined for any interesting pattern. The simplest thing that could happen would be that they were all equal, to within sampling variation. It is often interesting to consider a specific hypothesis that the fourfold ratios of cell probabilities are constant. If the hypothesis fits, an economical description of association has been obtained; if it does not, then how it does not will probably be interesting to examine. Distributions having constant fourfold ratios of probabilities have been studied in generality by Plackett (1965). Much earlier they were considered briefly by Karl Pearson (1913). We refer to them as Pearson-Plackett distributions, or more succinctly as Plackett distributions.

Pearson-Plackett distributions

Let X and Y be any two random variables with distribution functions $G(x)$ and $H(y)$:

$$G(x) = \text{prob}(X \leqslant x), \quad H(y) = \text{prob}(Y \leqslant y).$$

Then a joint distribution for X and Y can be defined by the condition that the fourfold probability ratios are constant. Let the joint distribution function be denoted by $F(x, y)$:

$$F(x, y) = \text{prob}((X \leqslant x) \wedge (Y \leqslant y)).$$

For any arbitrary point (x, y), let A, B, C, D denote the probability measures of the four quadrants with vertex at (x, y):

$$\left.\begin{aligned} A &= F(x, y), \\ B &= G(x) - F(x, y), \\ C &= H(y) - F(x, y), \\ D &= 1 + F(x, y) - (G(x) + H(y)). \end{aligned}\right\}$$

Then the condition is that

$$\frac{AD}{BC} = \psi, \tag{10}$$

where ψ is a positive constant, for all (x, y) for which neither $G(x)$ nor $H(y)$ assumes the value 0 or 1.

If X and Y have purely discrete distributions, so that $G(x)$ and $H(y)$ are step functions, the joint distribution is also purely discrete; if X and Y are absolutely continuous, so is the joint distribution. In the special case that $\psi = 1$, X and Y are independent. Complete dependence is approached as $\psi \to 0$ or ∞, whenever X and Y have continuous distributions and sometimes when they do not.

The above results can be proved by tedious elementary arguments. Provided $\psi \neq 1$, equation (10) is quadratic to determine $F(x, y)$ when $G(x)$, $H(y)$ and ψ are given. There is only one acceptable root, namely

$$F(x, y) = \frac{1 + (\psi - 1)(G + H) - \sqrt{[1 + (\psi - 1)(G + H)]^2 - 4\psi(\psi - 1)GH}}{2(\psi - 1)}. \tag{11}$$

If $\psi = 1$, the equation is linear and its root is

$$F(x, y) = G(x)H(y).$$

To show that the function $F(x, y)$ so defined is indeed a bivariate distribution function, one must verify correct limiting behavior as $x \to \pm\infty$ and $y \to \pm\infty$ (easily done) and also show that the second difference

$$F(x + \xi, y + \eta) - F(x + \xi, y) - F(x, y + \eta) + F(x, y)$$

is nonnegative for all (x, y) and all positive ξ, η. If $G(x)$ and $H(y)$ are absolutely continuous, this nonnegativity condition is simply that the density function $\partial^2 F/\partial x \partial y$ exists and is nonnegative for almost all (x, y). By differentiating the relation $AD - \psi BC = 0$ and simplifying, one may obtain the following explicit expression for the density function:

$$\frac{\partial^2 F}{\partial x \partial y} = \frac{G'(x)H'(y)}{K^2}\left\{\psi + \frac{2AD}{K}(\psi - 1)^2\right\}, \tag{12}$$

where $K = A + D + (B + C)\psi$; and this is evidently nonnegative. Having established the result for absolutely continuous marginal distributions, we may deduce it for other marginal distributions by suitably mapping an absolutely continuous distribution into them.

The only type of bivariate distribution with which most of us feel much familiarity (other than the joint distribution of a pair of independent random variables) is the bivariate normal distribution. Naturally we ask whether Plackett distributions resemble bivariate normal distributions. In a sense, all Plackett distributions having the same value for ψ are similar to each other. Continuous Plackett distributions with the same ψ can be transformed into each other by monotone transformations of each variable separately, and discrete distributions can be obtained from continuous ones by grouping. To express this another way, if for some fixed ψ equation (11) has been solved for various pairs of values for G and H, the results can be immediately used in constructing the Plackett distribution for any other pair of marginal distributions for which the distribution functions assume the same values.

If therefore one understands the character of a Plackett distribution defined for a particular pair of continuous marginal distributions, the understanding carries over to other marginal distributions, for the same ψ. Let us look at Plackett distributions for some kind of standard marginal distribution. Two obvious candidates for a standard marginal distribution are the uniform distribution over the unit interval and the standard normal distribution having zero mean and unit variance. The latter turns out to be the more illuminating.

Consider then the Plackett distribution having standard normal margins and some value for ψ. What is it like, and in particular how does it compare with a bivariate normal distribution having the same margins? Except when $\psi = 1$, the case of independence, such a Plackett distribution is not bivariate normal, but neither is it very unlike a bivariate normal distribution. To match a distribution of each sort for comparison, let us make the central ordinate of each density function the same. The central ordinate of the Plackett distribution is easily seen from (12) to be

$$\frac{1}{2\pi}\,\frac{1+\psi}{2\sqrt{\psi}},$$

and that of a bivariate normal distribution with $N(0, 1)$ margins and correlation coefficient ρ is

$$\frac{1}{2\pi}\,\frac{1}{\sqrt{1-\rho^2}}.$$

The central ordinates are the same if

$$\rho = \frac{\psi - 1}{\psi + 1}. \tag{13}$$

Figure B:38 shows ordinates of the bivariate normal distribution with $\rho = 0.6$. The display is intended to suggest, as well as can be done with a typewriter terminal, what contours of constant probability density look like. They are, as is well known, similar ellipses (concentric and coaxial). In the figure the center is marked, and the three zones show where the base-2 logarithm of the ratio of the probability density to the density at the center is between ¯1 and ¯2 (innermost zone), between ¯4 and ¯6 (middle zone), and between ¯8 and ¯10 (outermost zone).

Figure B:39 shows, in the same style, ordinates of the Plackett distribution with $\psi = 4$, having the same central ordinate as the bivariate normal distribution.* It will be seen that near the center the distributions are closely alike. Further out they differ in a readily understandable way.

* Pearson (1913) presents the same distribution.

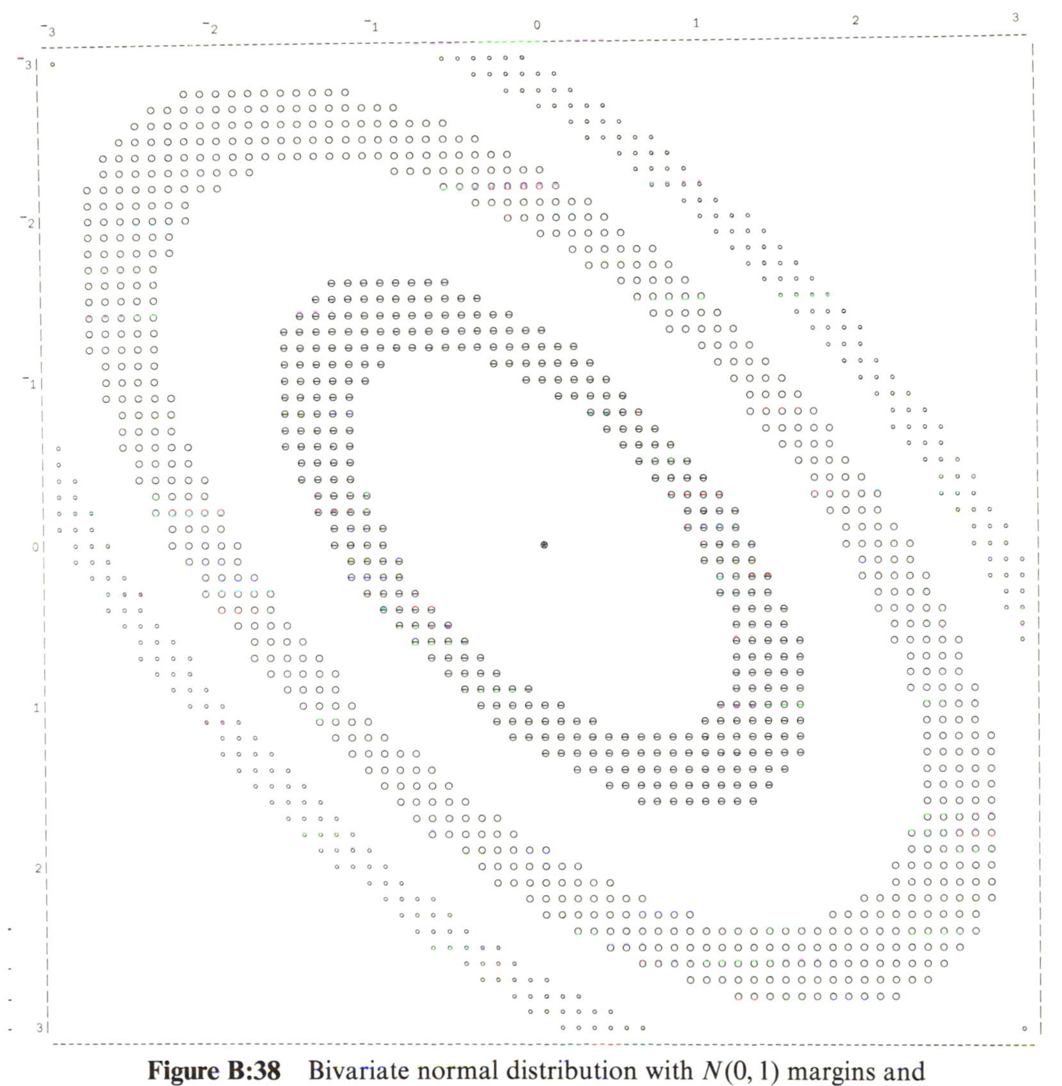

Figure B:38 Bivariate normal distribution with $N(0, 1)$ margins and correlation coefficient 0.6

The contours of the Plackett distribution are approximately elliptical, they are concentric and coaxial, but the ellipticity diminishes with distance from the center. In fact, it can be proved that the contours approach circularity as distance from the center tends to infinity and the probability density tends to zero. (To prove it, consider a large circle centered at the origin. As the radius tends to infinity, the ratio of the maximum to the minimum values of the density function at points on the circle remains bounded; and the logarithmic derivative of the density function at any point on the circle, in the direction of the radius, tends to infinity. The circle is therefore close to a constant-density contour.)

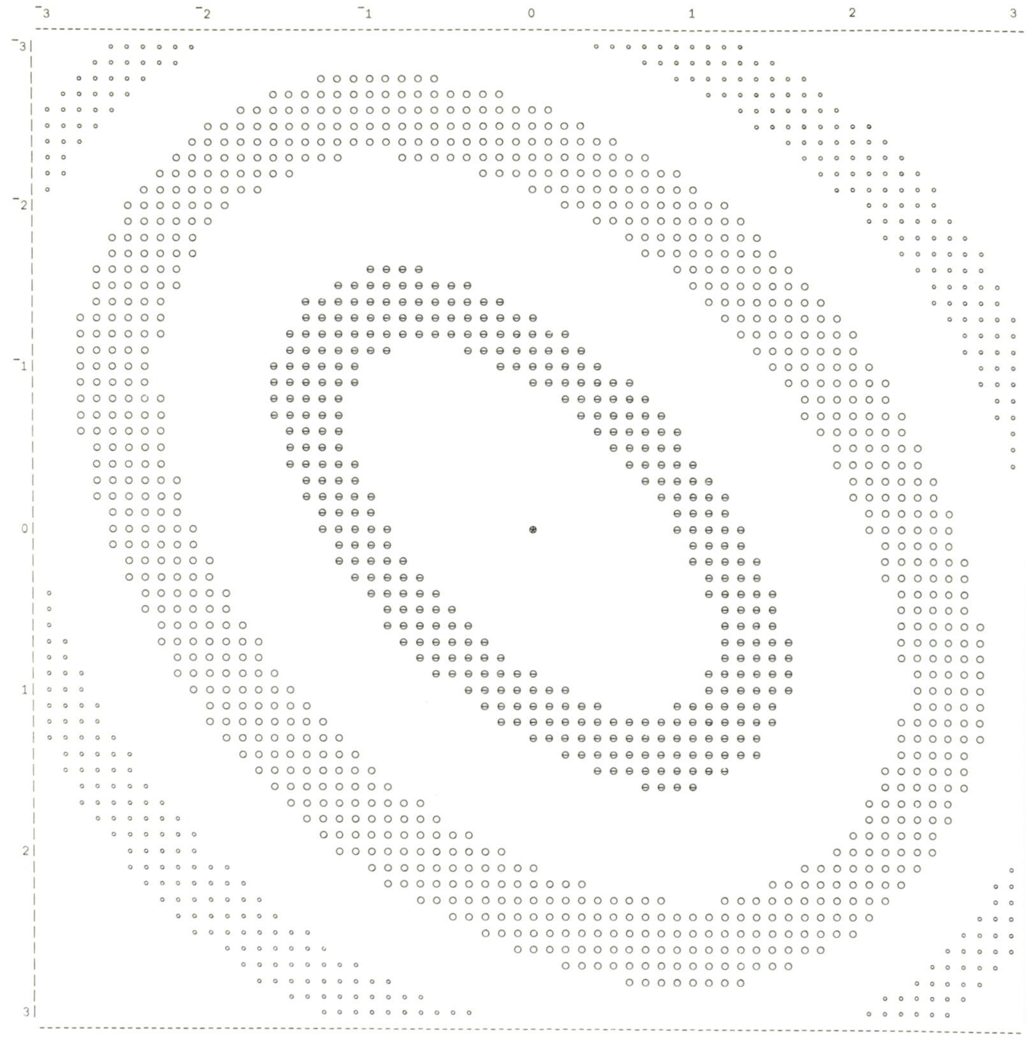

Figure B:39 Bivariate Plackett distribution with $N(0, 1)$ margins and crossproduct ratio 4

Thus the central part of a Plackett distribution having normal margins is like a bivariate normal distribution, and outer parts are like less highly correlated bivariate normal distributions. The two figures perhaps exaggerate the magnitude of the difference in shape of the distributions, because the eye is caught by the outer contours where the density is low and observations are improbable. If a random sample from, say, the Plackett distribution with $\psi = 4$ were supposed to come from a bivariate normal distribution and were fitted accordingly (by moments), a value for ρ in the neighborhood of 0.4 would be chosen, and the sample size would have to be very large for poor fit to be likely to be detectable.

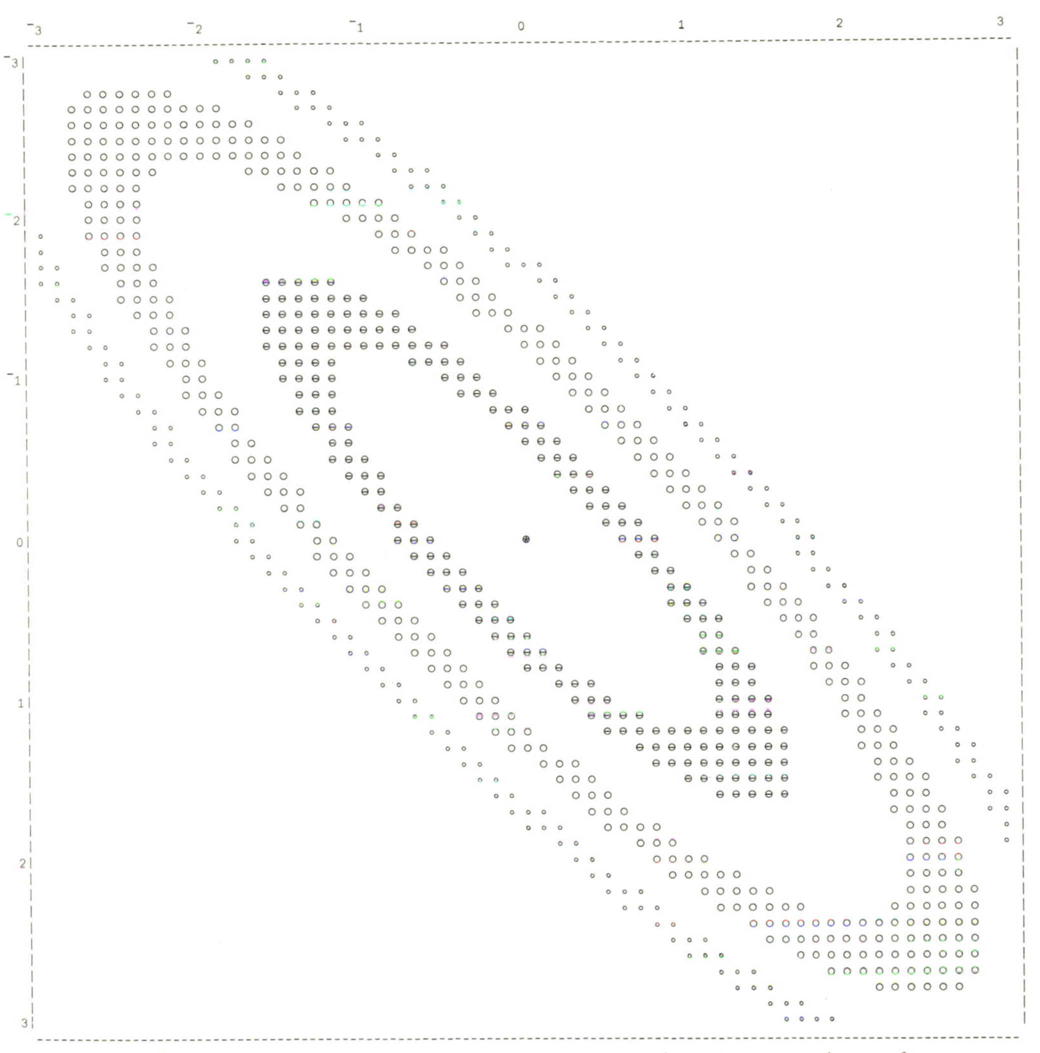

Figure B:40 Bivariate normal distribution with $N(0, 1)$ margins and correlation coefficient 0.875

Figures B:40 and B:41 similarly compare a bivariate normal distribution having $\rho = 0.875$ and a Plackett distribution with $\psi = 15$—again matched, not by moments or other plausible averaging, but so that the centers agree. The difference in appearance is now much greater; but still a substantial sample size would be needed if a random sample from one of these distributions were to be seen not to be well fitted by a distribution of the other type. A normal distribution having $\rho = 0.7$, say, would not seem so glaringly different overall from the Plackett distribution with $\psi = 15$ as the one shown with $\rho = 0.875$.

Plackett's 1965 paper was concerned with the construction of bivariate

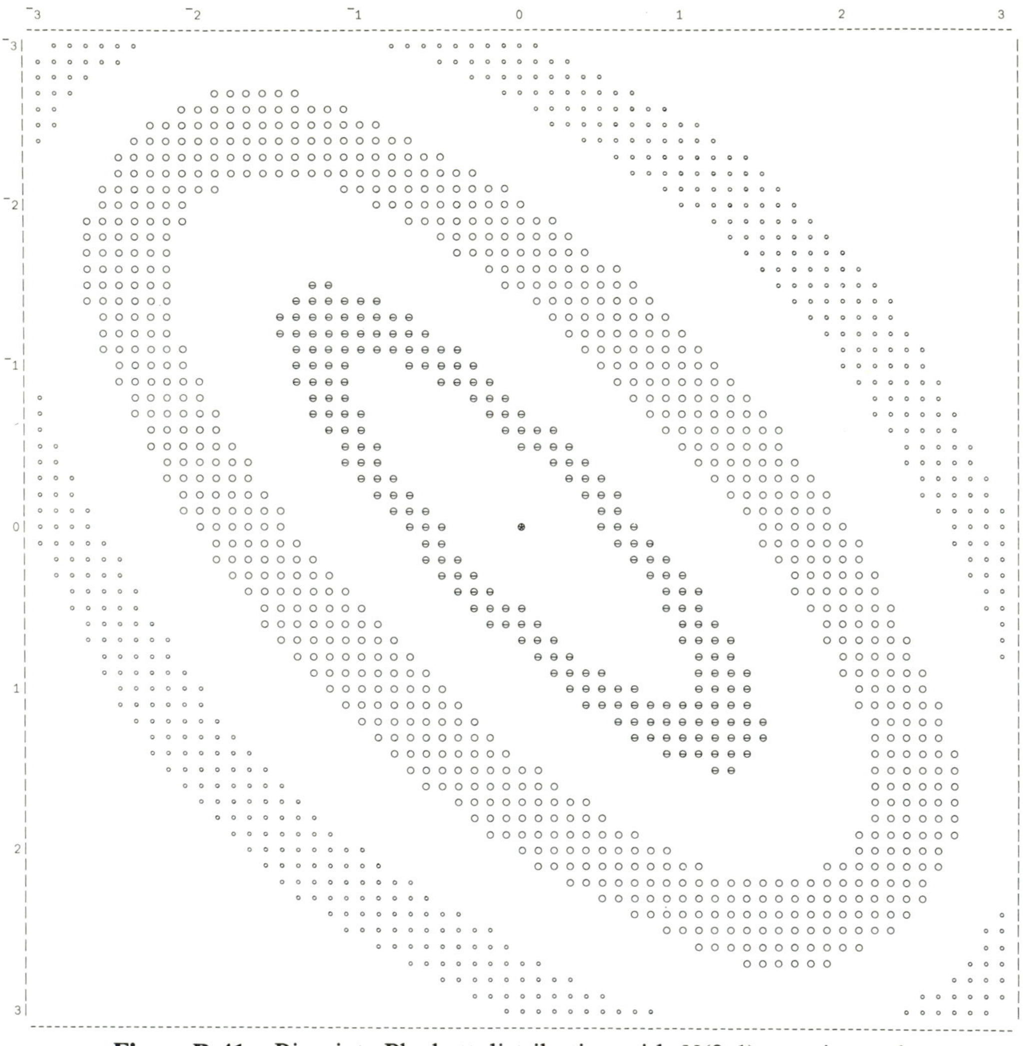

Figure B:41 Bivariate Plackett distribution with $N(0, 1)$ margins and crossproduct ratio 15

distributions from given marginal distributions. Distributions in more than two dimensions can be constructed similarly. Suppose for example that three random variables X, Y, Z have given univariate distribution functions. Bivariate distributions for these variables in pairs can be defined as Plackett distributions having constant fourfold probability ratios, ψ_{XY}, ψ_{XZ}, ψ_{YZ}, say. Then a joint distribution for all three variables can be determined by the condition that the eightfold probability ratio defined at (4) above is constant, ψ_{XYZ}, say. This condition is a quartic equation for the trivariate distribution function when all the univariate and bivariate marginal distribution functions are known, and is readily solved by a few steps of Newton-Raphson approximation from a suitable starting value.*

If three variables are jointly normally distributed, the three correlation coefficients of the variables in pairs, say ρ_{XY}, ρ_{XZ}, ρ_{YZ}, must satisfy a consistency condition, namely that the correlation matrix is nonnegative, or

$$1 + 2\rho_{XY}\rho_{XZ}\rho_{YZ} \geqslant \rho_{XY}^2 + \rho_{XZ}^2 + \rho_{YZ}^2.$$

Similarly the four crossproduct ratios for a three-dimensional Plackett distribution must in general satisfy consistency conditions, in order that the probability measure determined by the joint distribution function should be properly nonnegative. For example, if the variables are independent in pairs, $\psi_{XY} = \psi_{XZ} = \psi_{YZ} = 1$, and if the marginal distributions are continuous, there is an upper bound for $|\ln \psi_{XYZ}|$ of roughly 3.33, so that the density shall not be negative at the center. However, when the marginal distributions are discrete, the variables being independent in pairs still, this bound may be exceeded; and in particular when the univariate marginal distributions are over just two values apiece there is no bound for $|\ln \psi_{XYZ}|$.

Particular interest attaches to trivariate distributions with no three-variable association, $\psi_{XYZ} = 1$. Then if the univariate margins are symmetrical (for example, uniform or normal), the distribution is symmetric about its center, in the sense that densities are unaltered by reflection in the center point. When the univariate margins are normal, the whole distribution is nearly jointly normal, just as was the case in two dimensions. But if ψ_{XYZ} differs from 1, the distribution has a twist (even when the univariate margins are normal) that makes it qualitatively unlike a trivariate normal distribution.

A Plackett distribution can be applied as a theoretical description to a crossclassification of response-variables having ordered categories, by postulating that the cell probabilities constitute a Plackett distribution with discrete margins. (To conform exactly with the above account, the categories should have numerical values, but only the order of the values is important, because of the invariance of Plackett distributions to monotone transformation of the variables, and so we may as well think of the value of a category as equal to its numbering.) Thus in two dimensions we suppose there are vectors of marginal probabilities $\{p_i\}$ and $\{q_j\}$, say, and a fourfold ratio ψ, such that $\{p_{ij}\}$ are the probabilities in the Plackett distribution so defined.

*There is an interval within which the trivariate distribution function at any point (x, y, z) must lie, when all the univariate and bivariate marginal distributions are given, so that the probability of each octant having (x, y, z) as vertex should be positive and therefore the eightfold ratio of probabilities should be positive and finite. The midpoint of this interval is a suitable starting value for iterations to achieve a given value for the eightfold ratio.

It is possible, alternatively, to view the matter differently, as Karl Pearson did when he treated 2-by-2 tables by "tetrachoric *r*". Namely, we may suppose that each category of an observed variable corresponds to some interval of an unobserved underlying variable having a continuous probability distribution. We are free to suppose any form for this distribution, and Pearson assumed it was normal, and that the joint distribution of two such underlying variables was jointly normal. As it turns out, joint normality is computationally inconvenient, because probabilities of rectangular regions in a bivariate normal distribution are troublesome to calculate. If instead we postulate that the joint distribution is of the Plackett type, the computing becomes considerably easier. There is now, however, nothing gained by thinking of unknown intervals of unobserved variables with continuous distributions. What can be done is precisely the same as if we supposed the observed variables to have a discrete Plackett distribution. Nevertheless it is proper to see the very close affinity between the treatment of contingency tables illustrated below and Pearson's method of tetrachoric *r*.

Application to contingency tables

The function *FOURFOLD* fits a Plackett distribution to the contingency table formed by crossclassifying a set of individuals with respect to two response-variables having ordered categories. As Tukey has suggested, the log fourfold ratio of frequencies is first calculated for every 2-by-2 table that can be formed by pooling adjacent rows and columns. The print-out may be examined for trend or pattern. Then a Plackett distribution is fitted by a conditional-maximum-likelihood method, the individual fitted values (expected frequencies) for the cells of the original table are computed, and the standardized residuals are displayed (observed count minus fitted value, divided by the square root of the latter). Finally, just as with the function *CONTINGENCY*, a χ^2 statistic for goodness of fit is obtained by summing squares of standardized residuals. Several details in this procedure call for comment.

In forming the log fourfold ratios of frequencies, something must be done about the possibility that a frequency is zero—the computer does not tolerate infinite values. The simplest device is to add a small positive constant to each of the four frequencies before the fourfold ratio is calculated.* Just what the constant is, matters little. There is something to be

*Another simple device would be to work with some function of the fourfold ratios, whether of frequencies or of probabilities, which did not, like the logarithm, become infinite as the argument approached 0 or ∞. For example, the right side of equation (13) above goes from $^-1$ to 1 as ψ goes from 0 to ∞, and the corresponding function of frequencies is familiar as Yule's Q. However, for the following estimation procedure that transformation is less convenient than the logarithmic.

said for approximately unbiasing the estimation of $\ln\psi$ when cell frequencies are not small. For that purpose the right constant is $\frac{1}{2}$, which has therefore been adopted in *FOURFOLD*. We digress for a moment to consider this matter.

Let n be a random variable having a Poisson distribution with mean λ, where λ is large. Then it is easy to prove that

$$\mathcal{E}\left(\ln\left(n+\tfrac{1}{2}\right)\right) = \ln\lambda + O(\lambda^{-2}),$$

$$\operatorname{var}\left(\ln\left(n+\tfrac{1}{2}\right)\right) = \lambda^{-1} + \tfrac{1}{2}\lambda^{-2} + O(\lambda^{-3}),$$

$$\mathcal{E}\left(\left(n+\tfrac{1}{2}\right)^{-1}\right) = \lambda^{-1} + \tfrac{1}{2}\lambda^{-2} + O(\lambda^{-3}).$$

Thus $(n+\frac{1}{2})^{-1}$ is an approximately unbiased estimate of the variance of $\ln(n+\frac{1}{2})$, which is an approximately unbiased estimate of $\ln\lambda$.

We are, however, concerned not with Poisson variables but with multinomial variables, for which we may hope to find similar results. Let n_{11}, n_{12}, n_{21}, n_{22} be random variables having a multinomial distribution with total number N and positive probabilities p_{11}, p_{12}, p_{21}, p_{22} (summing to 1). Let

$$L = \ln\frac{\left(n_{11}+\frac{1}{2}\right)\left(n_{22}+\frac{1}{2}\right)}{\left(n_{12}+\frac{1}{2}\right)\left(n_{21}+\frac{1}{2}\right)}, \quad \psi = \frac{p_{11}p_{22}}{p_{12}p_{21}},$$

$$V = \Sigma_{ij}\left(n_{ij}+\tfrac{1}{2}\right)^{-1} \quad (i, j = 1,2).$$

Then it is not hard to prove that, as $N \to \infty$ with $\{p_{ij}\}$ fixed,

$$\mathcal{E}(L) = \ln\psi + O(N^{-2}),$$

$$\mathcal{E}(V) = \left(\Sigma_{ij}p_{ij}^{-1}\right)N^{-1} + \left\{\left(\tfrac{1}{2}\Sigma_{ij}p_{ij}^{-2}\right) - \Sigma_{ij}p_{ij}^{-1}\right\}N^{-2} + O(N^{-3}).$$

With more effort it is possible to prove* that

$$\operatorname{var}(L) = \mathcal{E}(V) + O(N^{-5/2}).$$

Thus V is an approximately unbiased estimate of the variance of L, which is an approximately unbiased estimate of $\ln\psi$.

In *FOURFOLD* the expressions L and V are computed for each 2-by-2 table. A weighted average of the L-values, with weights inversely proportional to V, is formed, as though the L-values were independent (which they are not); and this, together with values on either side at a suitable distance, are taken as the three initial values for $\ln\psi$ in the following estimation procedure.

If the given table has k rows and l columns, a Plackett distribution fitted to it has $k + l - 1$ parameters, namely $k - 1$ marginal row probabilities (defining k row probabilities summing to 1), $l - 1$ marginal column probabilities, and the constant fourfold probability ratio ψ. To estimate all these

*I am indebted to Diccon R. E. Bancroft for this result.

parameters by maximum likelihood would be a formidable task. What is much easier is to estimate the marginal row probabilities directly from the observed row totals, and the marginal column probabilities from the observed column totals, in the usual way (that would be maximum-likelihood estimation under the hypothesis of no association, $\psi = 1$), and then, with these parameter values inserted in the likelihood function, choose the one remaining parameter ψ to maximize what may be called the conditional likelihood function. The maximization is done iteratively: at each stage, the conditional likelihood function is evaluated for three equal-spaced values for $\ln\psi$, and the vertex of the parabola passing through the three points is determined; in successive stages the trial values for $\ln\psi$ are usually closer spaced, the middle one being the estimate. The process ordinarily converges rapidly—the successive triples of trial values for $\ln\psi$ are printed out, so that the user may watch the convergence. There is just one kind of situation, of little practical interest, in which the process does not converge, namely when because of zeroes in the given table the true maximum-likelihood value for $\ln\psi$ is $\pm\infty$. In the limited number of iterations that are executed, divergent behavior of $|\ln\psi|$ can be recognized easily.

At the final stage, the estimate of $\ln\psi$ is printed, together with a conventional standard error. The latter is just the reciprocal square root of minus the second derivative of the log conditional likelihood function at its maximum. The estimated parameter values are used in computing the expected frequencies and standardized residuals.

A little experimentation suggests that this piecemeal fitting procedure yields estimates of all the parameters that are very close (but not in general identical) to those given by proper maximum likelihood. Presumably the fitting is near enough efficient for the resulting χ^2 statistic to behave like the tabulated χ^2 distribution with $(k - 1)(l - 1) - 1$ degrees of freedom, on the hypothesis that individuals were distributed multinomially with probabilities constituting a Plackett distribution—that is, provided all expected frequencies are not too small. If some expected frequencies are considered to be too small, the standardized residuals and the χ^2 statistic can be easily recomputed after suitable pooling of rows or columns both in the original table and in the table of expected frequencies (as illustrated below). Alternatively, the χ^2 statistic can be replaced by one of the other statistics discussed above; see Exercise 8 at the end of this chapter.

In Figures B:42–45 we see *FOURFOLD* applied to the four tables with ordered categories, shown above. For comparison, *CONTINGENCY* is executed first in each case, although the hypothesis of no association is clearly implausible, and in itself the large χ^2 value obtained is of no interest.

For the Kiser-Whelpton table on education and fertility, the log fourfold ratios of frequencies do not show any very obvious pattern. The weighted average of these log ratios is 0.62, from which as starting value the

```
⍝ KISER-WHELPTON TABLE, EDUCATION OF WIFE AND FERTILITY PLANNING

      M4
 102   35   68   34
 156   53  177  103
  35   27   38   19
  35   29   73   94
  26   24   43   49
  49   37   52   80

      CONTINGENCY M4
THE MATRIX OF EXPECTED FREQUENCIES IS NAMED  EF .
LEAST MEMBER OF  EF  =  16.965
STANDARDIZED RESIDUALS (SR):
   4.28         0.16        ¯0.8        ¯3.65
   1.62        ¯2            1.91       ¯2.28
   0.29         2.44         0.11       ¯2.21
  ¯3.7         ¯0.69         0.06        4.24
  ¯2.19         0.83        ¯0.23        1.89
  ¯1.55         1.06        ¯1.98        2.97

CHI-SQUARED  =  116.25
DEGREES OF FREEDOM  =  15


      FOURFOLD M4
LOG FOURFOLD RATIOS (LFR):
    0.8          0.73         0.88
    0.76         0.44         0.8
    0.84         0.6          0.99
    0.5          0.25         0.62
    0.34         0.14         0.58
STAGE 0:  LOG PSI  =  0.55009  0.61907  0.68805
STAGE 1:  LOG PSI  =  0.64129  0.6635  0.68572
STAGE 2:  LOG PSI  =  0.65791  0.66346  0.66902
STAGE 3:  LOG PSI  =  0.66347
   (S.E.  =  0.083984)

THE MATRIX OF EXPECTED FREQUENCIES IS NAMED  EF .
LEAST MEMBER OF  EF  =  17.233
STANDARDIZED RESIDUALS (SR):
   0.63        ¯0.48         0.38       ¯0.99
   0.04        ¯2.65         2.08       ¯0.28
   0.77         2.35        ¯0.22       ¯2.25
  ¯2.48        ¯0.41        ¯0.49        2.96
  ¯0.51         1.5         ¯0.62        0.1
   1.57         2.46        ¯2.22       ¯0.38

CHI-SQUARED  =  56.214
DEGREES OF FREEDOM  =  14

      EF[1 6 ;]
  95.787       37.951       64.968       40.294
  39.149       24.747       70.621       83.484
```

Figure B:42

```
      ⍝ GILBY TABLE, CLOTHING AND INTELLIGENCE RATING OF SCHOOLBOYS

      M5
 33   48  113  209  194   39
 41  100  202  255  138   15
 39   58   70   61   33    4
 17   13   22   10   10    1

      CONTINGENCY M5
THE MATRIX OF EXPECTED FREQUENCIES IS NAMED  EF .
LEAST MEMBER OF  EF  =  2.4968
STANDARDIZED RESIDUALS (SR):
  ¯2.16         ¯3.64         ¯3.03          0.84          4.74          3.7
  ¯2.07          0.48          1.86          1.45         ¯1.98         ¯2.11
   4.26          4.2           0.95         ¯2.34         ¯3.24         ¯1.68
   4.9           1.23          1.15         ¯2.66         ¯1.47         ¯0.95

CHI-SQUARED  =  174.82
DEGREES OF FREEDOM  =  15


      FOURFOLD M5
LOG FOURFOLD RATIOS (LFR):
  ¯0.57         ¯0.8          ¯0.89         ¯0.94         ¯1.24
  ¯1.26         ¯1.15         ¯1.07         ¯0.84         ¯0.91
  ¯1.44         ¯1.07         ¯1.19         ¯0.63         ¯0.57
STAGE 0:  LOG PSI  =  ¯1.0616  ¯0.97207  ¯0.88255
STAGE 1:  LOG PSI  =  ¯1.0243  ¯1.0019  ¯0.97953
STAGE 2:  LOG PSI  =  ¯1.002
   (S.E.  =  0.080726)

THE MATRIX OF EXPECTED FREQUENCIES IS NAMED  EF .
LEAST MEMBER OF  EF  =  0.96294
STANDARDIZED RESIDUALS (SR):
  1.79          0.16          0.02         ¯0.36         ¯0.58          0.55
 ¯2.11         ¯0.22          0.26          0.96          0.25         ¯0.78
  0.45          0.6          ¯0.68         ¯0.47          0.48          0
  1.24         ¯0.82          0.43         ¯1.45          1             0.04

CHI-SQUARED  =  17.144
DEGREES OF FREEDOM  =  14

      ⌊/,EF← 2 5 6 POOL EF
8.2529
      SR←(( 2 5 6 POOL M5)-EF)÷EF*0.5
      0.01×⌊0.5+100×SR
  1.79          0.16          0.02         ¯0.36         ¯0.32
 ¯2.11         ¯0.22          0.26          0.96         ¯0.03
  0.45          0.6          ¯0.68         ¯0.47          0.45
  1.24         ¯0.82          0.43         ¯1.45          0.96

      ⍝ CHI-SQUARED WITH 11 D.F.
      +/,SR×SR
15.824
```

Figure B:43

conditional-maximum-likelihood procedure reaches 0.66 as the estimate for $\ln\psi$, or nearly 2 for ψ. The χ^2 value of 56.2 with 14 degrees of freedom shows that a Plackett distribution does not fit the data well. Eight of the twenty-four standardized residuals exceed 2 in magnitude. There does not seem to be a systematic pattern to them; four pairs of residuals exceeding 2 in magnitude and of opposite sign can be found next to each other in the same row or in the same column. The constant-fourfold-ratio hypothesis seems satisfactory, but residual variability exceeds that of a multinomial distribution. Possibly the large residual variance can be explained away as a consequence of tight constraints in the representative sampling, so that the 1438 families studied should not be thought of as *independently* governed by the same set of cell probabilities. At any rate, the association in the table can be described reasonably well as arising from a constant ψ-value of about 2, with excess apparently-random residual variability, the variance about 4 times as large as for stochastic independence of individual families.

It can be claimed that even this rather weak description tells us more about the association in the table than the measure γ given by Goodman and Kruskal (1954). Their measure has an attractive direct probabilistic interpretation, but in calculating it they are fitting no hypothetical structure to the table, and from the value obtained we can infer nothing about whether the measured association pervades the table uniformly, or whether different parts of the table are differently associated. If nothing is fitted, there is no goodness of fit.

The constant-ψ hypothesis also fits the Gilby table for clothing and intelligence, and this time the residual variability is about right. The estimate for $\ln\psi$ is $^-1.00$, or 0.37 for ψ. The calculated χ^2 is 17.1 with 14 degrees of freedom. One of the expected frequencies is as low as 0.96, and the χ^2 value is open to question. After execution of *FOURFOLD*, the last two columns, labeled F and G, are pooled. Now χ^2 is 15.8 with 11 degrees of freedom, the smallest expected frequency being 8.25; the significance level is about 15%, unremarkable.

The other two tables tell a different story. The constant-ψ hypothesis does not fit. In trying the hypothesis and failing, we learn something interesting about the behavior of the tables.

Stuart's table for men's vision yields standardized residuals, when a Plackett distribution is fitted, that not only are mostly large in magnitude but also have a striking pattern of signs—positive in the main diagonal, negative immediately above and below the main diagonal, and positive again elsewhere. Clearly the constant-ψ hypothesis is wrong in a systematic way. The given frequencies are too small next to the main diagonal, too large far from it. Such an effect could be caused by observer bias. Presumably both eyes were tested before anything was recorded. It would not be surprising if the oculist tended to report that the eyes were similar

```
      ⍝ STUART TABLE, MEN'S DISTANCE VISION IN RIGHT AND LEFT EYES

      M6
 821  112   85   35
 116  494  145   27
  72  151  583   87
  43   34  106  331

      CONTINGENCY M6
THE MATRIX OF EXPECTED FREQUENCIES IS NAMED  EF .
LEAST MEMBER OF  EF  =  76.101
STANDARDIZED RESIDUALS (SR):
    25.93         ¯9.04         ¯12.36         ¯9.68
    ¯8.65         21.95          ¯5.15         ¯8.25
   ¯12.79         ¯4.53          20.73         ¯3.93
    ¯9.59         ¯8.16          ¯3.29         29.22

CHI-SQUARED  =  3304.4
DEGREES OF FREEDOM  =  9

      FOURFOLD M6
LOG FOURFOLD RATIOS (LFR):
     3.4           2.39           1.99
     2.46          2.97           2.49
     1.85          2.34           3.44
STAGE 0:  LOG PSI  =  2.64  2.7193  2.7987
STAGE 1:  LOG PSI  =  3.182  3.2614  3.3407
STAGE 2:  LOG PSI  =  3.2491  3.2689  3.2888
STAGE 3:  LOG PSI  =  3.2688
   (S.E.  =  0.062741)

THE MATRIX OF EXPECTED FREQUENCIES IS NAMED  EF .
LEAST MEMBER OF  EF  =  10.819
STANDARDIZED RESIDUALS (SR):
    0.46         ¯5.39          5.19          7.35
   ¯5             4.7          ¯2.4           1.25
    3.38         ¯2.27          2.1          ¯3.66
    9.1           2.05         ¯4.19          0.64

CHI-SQUARED  =  303.85
DEGREES OF FREEDOM  =  8
```

Figure B:44

unless they were clearly different. Conceivably, also, there may be some physiological effect tending to bring or keep the eyes close to equality, so that small inequalities in vision are avoided, but large inequalities are not prevented.

Stuart's companion table for women's vision, not reproduced here, shows the same kind of behavior. For men, the estimate of $\ln\psi$ is 3.27, for women 3.40, or the estimates of ψ are 26 and 30 respectively. One could try correcting for the possible observer bias or physiological effect just mentioned, by transferring a small proportion, about 7%, of the count in

```
      ⍝ GLASS TABLE, SOCIAL MOBILITY, FATHER'S STATUS AND SON'S STATUS

      M7
 50   19   26    8   18    6    2
 16   40   34   18   31    8    3
 12   35   65   66  123   23   21
 11   20   58  110  223   64   32
 14   36  114  185  714  258  189
  0    6   19   40  179  143   71
  0    3   14   32  141   91  106

      CONTINGENCY M7
THE MATRIX OF EXPECTED FREQUENCIES IS NAMED  EF .
LEAST MEMBER OF  EF  =  3.7995
STANDARDIZED RESIDUALS (SR):
 23.7          5.42        3.96        ¯2.17       ¯4.78       ¯3.39       ¯3.45
  5.51        12.71        5.27        ¯0.38       ¯3.87       ¯3.46       ¯3.56
  0.58         4.88        5.69         3.08       ¯1.51       ¯4.64       ¯3.22
 ¯1.09        ¯0.73        1.3          5.09        0.78       ¯2.54       ¯3.89
 ¯4.57        ¯3.94       ¯2.39        ¯0.94        3.9         0.12        0.44
 ¯3.67        ¯3.25       ¯3.68        ¯2.59       ¯0.6         7.41        2.08
 ¯3.38        ¯3.48       ¯3.73        ¯2.64       ¯1.36        3.13        8.62

CHI-SQUARED  =  1361.7
DEGREES OF FREEDOM  =  36


      FOURFOLD M7
LOG FOURFOLD RATIOS (LFR):
    3.67         2.94         2.77         2.3          1.81         1.99
    3.27         2.9          2.61         2.17         1.79         2.01
    2.78         2.46         2.07         1.71         1.48         1.29
    2.62         2.12         1.67         1.47         1.26         1.23
    4.23         2.23         1.6          1.25         1.16         0.95
    3.28         2.3          1.57         1.18         1.06         1.2
STAGE 0:  LOG PSI  =  1.6423  1.6908  1.7393
STAGE 1:  LOG PSI  =  1.4468  1.4953  1.5438
STAGE 2:  LOG PSI  =  1.4789  1.491  1.5032
STAGE 3:  LOG PSI  =  1.4911
   (S.E.  =  0.053348)

THE MATRIX OF EXPECTED FREQUENCIES IS NAMED  EF .
LEAST MEMBER OF  EF  =  2.8765
STANDARDIZED RESIDUALS (SR):
  9.69         0.28       ¯0.19        ¯3.28       ¯2.86       ¯0.47       ¯1
  0.78         5.08        0.63        ¯2.12       ¯2.01       ¯0.38       ¯0.89
 ¯2.12         0.31        0.2         ¯0.58        0.44       ¯0.27        2.2
 ¯2.06        ¯2.29       ¯1.78         0.85        0.4         2.35        1.41
 ¯2.45        ¯1.32        0.58         0.03       ¯0.33       ¯0.76        2.94
 ¯2.05        ¯0.37        0.58         1.96        0.33        1.5        ¯2.9
 ¯1.7         ¯0.8         0.86         2.91        2.23       ¯1.62       ¯1.65

CHI-SQUARED  =  241.3
DEGREES OF FREEDOM  =  35
```

Figure B:45

any diagonal cell into each of the adjacent cells in the same row or column, making two such transfers from the first and last cells in the main diagonal, and four from each of the other main-diagonal cells. We then find that the Plackett distribution fits much less badly, if not exactly well, with $\ln\psi$ roughly 2.5 (or $\psi = 12$) for both men and women. Perhaps that is a juster measure of association between eyes than the preceding values.

As for the social mobility table, the log fourfold ratios of frequencies show a strong smooth trend from left to right and some trend up and down as well. Association seems to be less intense among the higher-numbered status categories than among the lower-numbered. The large χ^2 obtained when the Plackett distribution is fitted confirms what could be guessed from the empirical fourfold ratios, that the simple constant-ψ hypothesis is inadequate. Whether further understanding would result from fitting a probabilistic structure having more parameters, may be doubted.

Prospect

Under the influence of flexible computing the traditional areas of statistical methodology assume fresh interest. Even so, the reader may feel that this chapter has been unduly old-fashioned. To devote such attention to testing independence or measuring association of just two qualitative variables, or even of three, is perhaps unreasonable. Statistical material does sometimes involve only two or three qualitative variables, prompting the sorts of questions we have been considering. But more typically today, with our vastly increasing capacity for amassing and organizing facts of all sorts, when qualitative variables appear they appear in great numbers. We may have, for example, categorized responses made by several thousand persons to a battery of a couple of hundred questions; or a mass of categorical information about each of some large group of hospital patients. Then questions concerning independence or association of variables by twos and threes may seem rather idle, not of prime importance. A more relevant objective of statistical analysis could be to perceive clusters among the individuals, if clusters occur. That is, we could seek to make a selection of the variables (or to devise a set of weights for the variables), so that when pairs of individuals were compared with respect to the selected set of variables (or with respect to all the variables, with the assigned weights) the individuals would be seen mostly to fall into relatively tight clusters—as tight as possible. An objective of this kind was utterly unthinkable before the computer age. Now it is certainly thinkable, if not readily achievable. We have been concerned here with reconsideration of old topics. The computer also permits the formulation and exploration of altogether new topics.

Notes for Chapter 12

Plackett (1974) and Bishop, Fienberg and Holland (1975) have surveyed the literature of contingency tables, and Cochran's review article on χ^2 (1952) is still useful. Consequently no attempt is made here to cite relevant literature systematically.

For a discussion of significance tests assessing goodness of fit see Anscombe (1963). (In that paper a distinction was made between *chances* and *probabilities*, the latter relating to personal attitudes, the former to frequency phenomena. Though the distinction makes for clarity, it flies so in the face of general usage that here the word "probability" does double duty.) A very early and striking example of the use of a probability to assess significance is the passage from Cardano quoted by David (1962, chapter 6).

Behavior of the Pearson χ^2 statistic when some cells have small expectations has been considered by many writers. Comparison with the probability-of-the-sample statistic has been made by, among others, Neyman and Pearson (1931), Shanawany (1936), Freeman and Halton (1951), Tate and Hyer (1973).

In the investigation of alternatives to Pearson's χ^2, I was assisted by Gary W. Oehlert. We tabulated a selection of distributions of the probability-of-the-sample statistic and of the likelihood-ratio statistic, together with continuous approximations; one of us considered goodness of fit of counts to a given multinomial distribution, the other independently considered association in a contingency table. The tentative conclusions reported above are based on the two sets of calculations. Possibly a more detailed presentation of this material will be made elsewhere.

The treatment of contingency tables with ordered categories by reduction, through pooling of adjacent categories, to all possible 2-by-2 tables, of which the crossproduct ratios are calculated, is based on chapter 29x of Tukey (1971). Only a little of that material reappears in the regular published version of the book (Tukey, 1977, chapter 15, sections E and F). The approach was adumbrated by Yule (1900, sections 27 and 48; 1912, sections 41 and 57), but was not recognized as a viable technique, perhaps because it entailed too much calculation. Only vague mention was made of it in Yule's *Introduction to the Theory of Statistics*. It was in rebuttal of Yule's paper (1912) that Pearson considered distributions with constant fourfold ratio of probabilities (Pearson and Heron, 1913; Pearson, 1913). I am indebted to Dr. Peter McCullagh for this reference.

In regard to the possibility of representing curves with a typewriter terminal, as in Figures B:38–41, smoother results could be obtained with the aid of a special typeball for plotting. Very smooth results can be obtained with graphical display devices, other than a typewriter terminal.

Benzécri's book on data analysis (1973) and Hartigan's on clustering (1975) are striking examples of developments in statistical methodology, unthinkable before the computer age, alluded to at the end of the chapter.

Exercises for Chapter 12

1. Haldane (1937) found the first four moments of the χ^2 statistic relevant to several problems of testing goodness of fit, including the following: the counts $\{n_i\}$ $(i = 1, 2, \ldots, k)$ are hypothesized to have a given multinomial distribution with positive total number N and positive probabilities $\{p_i\}$. If $m_i = Np_i$, we have $N = \Sigma_i n_i = \Sigma_i m_i$, and

$$\chi^2 = \Sigma_i (n_i - m_i)^2 / m_i.$$

Then

$$\mathcal{E}(\chi^2) = k - 1,$$

$$\text{var}(\chi^2) = 2(k-1) + (\Sigma_i 1/m_i) - (k^2 + 2k - 2)/N.$$

Of the three terms on the right side of this expression for var(χ^2), the first two sum to the estimated variance found by our suggested procedure from variances for independent Poisson-distributed cells. The third term, $(k^2 + 2k - 2)/N$, is therefore the error in the estimated variance; it depends on k and N but not otherwise on the expectations $\{m_i\}$.

Haldane (1940) found the first two moments of χ^2 used to test the hypothesis of no association in a contingency table with given margins. In the notation of this chapter,

$$\mathcal{E}(\chi^2) = \frac{(k-1)(l-1)N}{N-1},$$

$$\text{var}(\chi^2) = \frac{2N}{N-3}(K - \sigma)(L - \tau) + \frac{N^2}{N-1}\sigma\tau,$$

where K, L, σ, τ are defined by

$$(N-1)K = (k-1)(N-k), \quad (N-1)L = (l-1)(N-l),$$

$$(N-2)\sigma = (N\Sigma_i 1/n_{i.}) - k^2, \quad (N-2)\tau = (N\Sigma_j 1/n_{.j}) - l^2.$$

This version of var(χ^2) was given by Dawson (1954). A similar expression can in principle be found for the third moment. (The writer does not know whether such an expression has been published.)

2. Examine the behavior of the expression (9) as a function of its two arguments. Verify that if $n_{ij} = 0$ the minimum value of $^-0.006$ is attained when $m_{ij} = 0.555$; if $n_{ij} = 1$ the minimum value of $^-0.0004$ is attained when $m_{ij} = 1.525$; if $n_{ij} = 2$ the minimum value of $^-0.0001$ is attained when $m_{ij} = 2.515$, approximately.

3. Asymptotic expressions for moments of the probability-of-the-sample statistic (9) can be found. We denote (9) by T, write $n_{ij} = m_{ij} + \delta$, and drop the subscripts from m_{ij}. Then as $m \to \infty$ with $\delta = O(\sqrt{m})$,

$$T = 2\left\{\ln \frac{(m+\delta)!}{(m-\frac{1}{2})!} - (\delta + \tfrac{1}{2})\ln m\right\}$$

$$= \frac{1}{4m} + \frac{\delta}{m}\left(1 - \frac{1}{6m}\right) + \frac{\delta^2}{m}\left(1 - \frac{1}{2m} + \frac{1}{6m^2}\right) - \frac{\delta^3}{3m^2}\left(1 - \frac{1}{m}\right)$$

$$+ \frac{\delta^4}{6m^3}\left(1 - \frac{3}{2m}\right) - \frac{\delta^5}{10m^4} + \frac{\delta^6}{15m^5} + o(m^{-2}).$$

For a single Poisson-distributed cell, $\mathcal{E}(\delta) = 0$, $\mathcal{E}(\delta^2) = \mathcal{E}(\delta^3) = m$, $\mathcal{E}(\delta^4) = 3m^2 + m$, etc., and we find

$$\mathcal{E}(T) = 1 - \frac{1}{12m} - \frac{1}{12m^2} + O(m^{-3}) \sim 1 - \frac{1}{12(m-1)}.$$

The expression on the right is quite a good approximation even when m is as small as 2. Similarly, with much effort,

$$\text{var}(T) \sim 2 - \frac{1}{3(m - 3/2)},$$

$$\mu_3(T) \sim 8 - \frac{2}{m-2}.$$

In the computer program *CTG*2 mentioned below, slightly modified expressions are used, to be more accurate when m is between 5 and 10:

$$\mathcal{E}(T) \approx 1 - \frac{1}{12(m-1.16)}, \quad \text{var}(T) \approx 2 - \frac{1}{3(m-1.97)},$$

$$\mu_3(T) \approx 8 - \frac{2}{m - 2.71}.$$

Similar expressions can in principle be found for moments of the sum of (9) over the cells of a given multinomial distribution, or over the cells of a contingency table with given margins, under the hypothesis of no association. Computer assistance with the algebra is called for.

4. Williams (1976) has obtained asymptotic expressions for the expectation of the likelihood-ratio statistic for several problems of testing multi-dimensional contingency tables, and in particular the expectation of the sum of (8) over the cells of a two-dimensional table to test the hypothesis of no association. He carries the expansion to the term of order m^{-1} (in the notation of the previous question); the error is $O(m^{-2})$. For the single Poisson-distributed cell, Williams's type of expansion gives for the expectation of (8)

$$1 + \frac{1}{6m} + O(m^{-2}).$$

Williams's purpose is to approximate the distribution of the likelihood-ratio statistic by the random variable aX, where X has the tabulated χ^2 distribution with the usual number ν of degrees of freedom and the multiplier a is chosen by setting νa equal to the approximate expectation of the statistic. This approximate distribution is no doubt better than the distribution of X itself if each cell expectation is high enough, perhaps greater than 5; but consideration of the single Poisson-distributed cell suggests that the reverse may be true if any cell expectation is less than 2. He does not warn the reader of this.

5. Consider the smoothness of the distribution of a goodness-of-fit statistic when one cell has small expectation and all other cells have very large expectations.

(a) For the multinomial problem specified in the first exercise above, consider the sum over all k cells of either (7) or (8) or (9), where n_{ij} is replaced by n_i and m_{ij} by m_i. Consider a sequence of such problems in which m_1 is constant and $\{m_i\}$ for $2 \leqslant i \leqslant k$ tend to infinity in constant proportion. Then the statistic can be expressed as $T + U$, where T is the contribution from the first cell, having a discrete distribution which is asymptotically that for a Poisson-distributed cell, and U is the sum of contributions from all other cells, having asymptotically the tabulated χ^2 distribution with $\nu - 1$ degrees of freedom, where $\nu = k - 1$; asymptotically T and U are independent.

The smoothness of the asymptotic distribution of $T + U$ depends on ν. If the discrete random variable T is added to an absolutely continuous random variable having small dispersion and a single mode, the sum may have a multimodal density; but if the absolutely continuous component has a large dispersion and smooth density, so will the sum. The asymptotic distribution of U has a large dispersion and smooth density when ν is large. An easy but not very satisfactory way to describe smoothness is in terms of the number of continuous derivatives. The asymptotic distribution of $T + U$ has the same number of continuous derivatives as that of U, namely the integer part of $\frac{1}{2}(\nu - 4)$. Thus if $\nu = 4$ or 5, the density is continuous but not its first derivative; if $\nu = 6$ or 7, the density has a continuous first derivative but not second derivative; and so on.

(b) For the problem of testing association in a contingency table, a similar argument applies, but defining a suitable sequence of problems in which one cell expectation is constant and all others tend to infinity is a little more complicated. For a k-by-l table, if we start with a set of positive row totals $\{n_{i.}\}$ and column totals $\{n_{.j}\}$, where $N = \Sigma_i n_{i.} = \Sigma_j n_{.j}$, let a table in the sequence have the following row totals, column totals and grand total. For a large number r, the grand total is Nr, the first row total is $n_{1.}\sqrt{r}$, and the other row totals are $\{n_{i.}c\}$ $(2 \leqslant i \leqslant k)$ where

$$(N - n_{1.})c = Nr - n_{1.}\sqrt{r}.$$

Similarly the first column total is $n_{.1}\sqrt{r}$ and the other column totals are $\{n_{.j}c\}$ $(2 \leqslant j \leqslant l)$ where now

$$(N - n_{.1})c = Nr - n_{.1}\sqrt{r}.$$

Then as $r \to \infty$ the same statements as at (a) hold, except that now $\nu = (k - 1)(l - 1)$.

6. An oversimplified representation of the "roughness" of a portion of the distribution function of a goodness-of-fit statistic, when ν is not too small, may be had as follows.

A random walk in a plane consists of horizontal steps to the right alternating with vertical steps upward. The lengths of all steps are mutually independent and (say) all have expectation 1. The horizontal step-lengths are exponentially distributed (like intervals between events in a Poisson process), and therefore have variance 1. The vertical step-lengths are identically distributed with nonnegative variance v. Let the random walk start from the origin, and let (x_n, y_n) be the point reached after n horizontal and n vertical steps, x_n being the sum of the first n horizontal steps and y_n that of the first n vertical steps. Then results such as these can be proved for given n:

(a) the variance of the distance of (x_n, y_n) from the expected path having equation $y = x$ is equal to $\frac{1}{2}n(1 + v)$;
(b) the variance of the distance of (x_n, y_n) from the line joining the origin to (x_{2n}, y_{2n}) is approximately one half of the answer to (a), provided that n is large enough for $x_n + y_n$ to be approximately normally distributed.

Such variances, being proportional to $1 + v$, are least when $v = 0$.

An exponential distribution for intervals between attainable values of a goodness-of-fit statistic seems (empirically) to be approximately correct, when values in a not-too-broad interval are considered. For the probability-of-the-sample statistic, the probabilities associated with these values are nearly equal, approximating the condition $v = 0$ in the above representation.

7. Execution of the function *CONTINGENCY* has been illustrated in the figures. The definition is given in Appendix 3. The argument X is the contingency table, denoted above by $\{n_{ij}\}$. The global output variable *EF* is the matrix of cell expectations, denoted above by $\{m_{ij}\}$, and *SR* is the standardized residuals, denoted above by $\{z_{ij}\}$.

Also shown in Appendix 3 is a function *CTG*2, similar to *CONTINGENCY* except that the probability-of-the-sample statistic and also the likelihood-ratio statistic are calculated instead of Pearson's χ^2. This function calls on three other functions, *LFACT* (giving logarithms of factorials), *CSIF* (the tabulated χ^2 integral function) and *INIF* (inverse of the normal integral function). In line [1] of *CTG*2 the "name classification" of each of these three names is determined; if it is 3, a function of that name is in the active workspace. If the classification is not 3 for one or more of the three names, line [2] tells the user to copy the missing functions, and execution is terminated; otherwise line [2] is passed over. Lines [3]–[9] are almost identical to lines [1]–[6] and [10] of *CONTINGENCY*. Lines [10]–[26] are concerned with the probability-of-the-sample statistic, and [27]–[31] with the likelihood-ratio statistic.

Lines [11]–[14] and [28]–[31] correspond, more or less, to lines [7]–[9] of *CONTINGENCY*. The contributions from every cell to the test statistic are calculated and summed; and so-called standardized residuals are obtained from the individual contributions by first replacing by zero any contribution that is negative, then taking square roots, and affixing a sign according to whether the observed cell count is below (negative sign) or above (positive sign) the value that would minimize the contribution, given the cell expectation. These standardized residuals are to be interpreted the same way as those given by *CONTINGENCY*, and the sum of their squares is the test statistic. (For the likelihood-ratio statistic, the precaution of replacing negative contributions by zero is almost certainly vacuous, because negative values can conceivably arise only through computational round-off error.)

Lines [15]–[26] are concerned with approximately assessing the significance level of the probability-of-the-sample statistic, by the procedure of estimating the first three moments of the distribution of the statistic (under the hypothesis of no association) and fitting a linear function of a variable having the tabulated χ^2 distribution. The calculation is omitted if any marginal total is less than 25 (line [15]). The local variable W defined in line [17] is a three-element vector consisting of a first approximation to the estimated first three cumulants of the distribution of the statistic. Values for the tabulated χ^2 distribution with ν degrees of freedom are corrected as shown above in Exercise 3 for all cells whose expectations are not less than 5. If no cell has expectation less than 5, there is no more correcting to be done, and line [18] is a branch to line [24]. Otherwise corrections are made for cells having expectation below 5. It would be possible to develop rational approximations to the required cumulant corrections, but this has not been done, and lines [19]–[23] evaluate the corrections directly from their definition by brute force. Line [24] redefines W, so that the approximating continuous distribution is that of a random variable equal to $W[1] + W[3] \times$ a χ^2 variable with $W[2]$ degrees of freedom.

Apply *CTG*2 to the three contingency tables given at the beginning of Chapter 12. It will be seen that the results are practically indistinguishable from those in Table B:37. For the Francis table relating to the polio vaccine trial, there is no longer any obvious need to omit the last row of the table, since a minimum cell expectation of 0.8 is no cause for alarm. Letting the last row remain increases ν from 6 to 9, the probability-of-the-sample statistic from 52.1 to 53.5, and the likelihood-ratio statistic from 52.3 to 55.8. Judgments are unchanged.

8. *FOURFOLD* can be modified so that at the end goodness of fit is tested as in *CTG*2 rather than by Pearson's χ^2. Using the "canonical representation" or other editing facility, omit lines [29]–[31] of *FOURFOLD*, line [32] being therefore renumbered [29], and then add on lines [10]–[26] (or [10]–[31]) of *CTG*2, which will become lines [30]–[46] (or [30]–[51]) of the new version of *FOURFOLD*. The naming of local variables in the two functions has been chosen to permit this blending.

Apply the modified *FOURFOLD* to the four contingency tables discussed in the later part of Chapter 12. For the Gilby table, there is no need to pool the last two columns, as was done in Figure B:43. Again, judgments are unchanged.

9. In Appendix 3 the documentary variable *HOWPD* describes functions *PD*2 and *PD*3 that yield discrete Plackett distributions in two and three dimensions having given univariate marginal distributions and given crossproduct probability ratios. The functions may be used for projects such as these:

(a) Obtain the probabilities for a bivariate Plackett distribution having a given value for ψ, such as 4, and marginal distributions formed from a standard normal distribution by grouping in intervals (all equal except the extreme intervals) such as these: below ¯3.1, ¯3.1 to ¯2.9, ¯2.9 to ¯2.7, . . . , 2.9 to 3.1, above 3.1. (The function *NIF* in Appendix 3 can be used to obtain the marginal probabilities.) Take a random sample, say of size 500, from this distribution: indirect use can be made of the function *SAMPLE* mentioned in Chapter 7. Make a scatterplot, representing the categories by the numerical values ¯3.2, ¯3.0, ¯2.8, . . . , 3.0, 3.2. Does the plot suggest a bivariate normal distribution? Normality could be appropriately tested with the kurtosis statistic of Mardia (1970); see *MARDIA* in Appendix 3.

(b) Verify the statements made in Chapter 12 concerning consistency conditions for the four crossproduct probability ratios in a trivariate Plackett distribution. Explore further.

10. Goodman (1979) has proposed a method of describing association in a two-dimensional contingency table having ordered categories, by first computing the matrix of crossproduct ratios for all 2-by-2 subtables that can be obtained from the given table as the intersection of two adjacent rows and two adjacent columns. If the crossproduct ratios seem to differ only by sampling variation, they are summarized by a single average value; otherwise, they are summarized in terms of row effects and column effects, in various possible ways.

(a) If X is a table (matrix) of positive counts or of positive probabilities, write a one-line APL program to find the matrix of crossproduct ratios for all 2-by-2 subtables of X formed from pairs of adjacent rows and adjacent columns—we refer to these crossproduct ratios as *Goodman ratios*. (Do a divide-reduction of a three-dimensional array consisting of four matrices stacked against each other, namely X with the last row and last column dropped, X with the last row and first column dropped, X with the first row and first column dropped, and X with the first row and last column dropped, respectively.)

(b) For what discrete bivariate distributions are Goodman crossproduct ratios of adjacent probabilities all equal (and different from 1)? For a bivariate normal distribution the mixed second derivative of the logarithm of the density function is constant. It follows that if a bivariate normal distribution is converted into a discrete distribution by grouping both variables in small *equal* intervals the Goodman ratios will be (approximately) equal. If the grouping is in very unequal intervals and the correlation coefficient is large, the Goodman ratios will be very unequal. What is the general form of density function of an absolutely continuous distribution for which the mixed second derivative of the logarithm of the density is constant?

(c) We have seen that a bivariate Plackett distribution with normally distributed margins (and constant crossproduct ratio different from 1) resembles a bivariate normal distribution, except that near the center the correlation seems to be higher (in magnitude) than farther from the center. We should expect that if such a distribution were converted into a discrete distribution by grouping both variables in small equal intervals the Goodman ratios would be more different from 1 near the center than farther out. Investigate the matter empirically using the functions *PD*2 and *NIF* mentioned in the preceding exercise and the function obtained in (a) above.

(d) Goodman illustrates his procedure with three sets of data: (i) a table associating mental health and socioeconomic status, (ii) the Glass social-mobility table reproduced in Chapter 12 above, (iii) a modified version of the same in which the largest (fifth) category in both variables is split.* Apply *FOURFOLD* to the first of these tables and verify that the Plackett distribution fits well. Goodman finds that, after omission of the main diagonal, his second table has nearly equal values for his (local) crossproduct ratios, whereas we found that the Plackett (aggregated) crossproduct ratios were obviously not constant—association was stronger between the lower-numbered categories than between the higher-numbered. Goodman's finding seems to be an accident of the marginal distributions in the table (and of his omission of the main diagonal). If the table had been obtained by grouping continuous variables in intervals of equal width, the marginal distributions must have been skew. (If it had been obtained by grouping a bivariate normal distribution the intervals must have been unequal, but then the Goodman ratios could not have been constant.) When in the third table the category with largest frequency is split, Goodman finds that his ratios are no longer nearly equal. Apply *FOURFOLD* to the whole of Goodman's third table and verify that the association in the table varies in the same way as for the second table, as one would expect. The splitting has made no perceptible difference to the import of the data. When categories are as arbitrary as the categories in these tables, a method of analysis that is bound to be disturbed by splitting of a category or by pooling of two adjacent categories is unsatisfactory. (See also McCullagh, 1980.)

11. In *FOURFOLD* the association of two variables having ordered categories is assessed by fitting a Plackett distribution. In principle, any other family of bivariate distributions that included independence might be considered instead. Karl Pearson thought in terms of a bivariate normal distribution made discrete by grouping.

A grouped bivariate normal distribution can be fitted to a given contingency table by a conditional-maximum-likelihood method similar to that in *FOURFOLD*, line [13] onwards, provided the probabilities in the cells can be computed for given marginal probabilities and a given correlation coefficient ρ. If ψ_0 is the crossproduct ratio of probabilities in a bivariate normal distribution for the quadrants whose vertex is at the mean, it can be shown that

$$\rho = \sin \frac{\pi(\sqrt{\psi_0} - 1)}{2(\sqrt{\psi_0} + 1)} .$$

The crossproduct ratio of probabilities, when the vertex of the quadrants is not at the mean, is greater than this value ψ_0 if ρ is positive (or lower if ρ is negative, or in either case farther from 1 if $\rho \neq 0$). A first estimate for ρ, with which to start the iterations, can therefore be had by substituting for ψ_0 in the above equation the empirical fourfold ratio obtained by pooling adjacent rows and columns of the given table to make a 2-by-2 table whose marginal totals are most nearly equal. Note that if the given table is better fitted by a grouped bivariate normal distribution than by a Plackett distribution, the log fourfold ratios of frequencies seen at the

*There is a slight discrepancy between tables (ii) and (iii) as published—the total counts differ by 1.

beginning of the output of *FOURFOLD* should appear to have values of lower magnitude in the middle of the display than round the edges—the opposite effect to that shown in Figure B:44 for the Stuart data, where the middle entry 2.97 is higher than six of the eight entries surrounding it.

The basic problem consists, then, of evaluating the probability content of given rectangular regions for a given bivariate normal distribution. We may as well suppose the distribution to have zero means and unit variances. Then we wish to be able to evaluate the following integral, for any given limits a, b, c, d, finite or infinite, and any given ρ less than 1 in magnitude:

$$\int_a^b dx \int_c^d dy \left(2\pi\sqrt{1-\rho^2}\right)^{-1} \exp\left[-\tfrac{1}{2}(x^2 - 2\rho xy + y^2)/(1-\rho^2)\right].$$

When the limits of integration are finite the evaluation can be done straightforwardly by quadrature. For the sake of also managing infinite limits, however, it is better to change variables from x and y to u and v defined thus:

$$x = \sqrt{1+|\rho|}\,\Phi^{-1}(u), \quad y = \sqrt{1+|\rho|}\,\Phi^{-1}(v),$$

where $\Phi^{-1}(\ )$ is the inverse of the standard normal integral function,

$$\Phi(t) = \int_{-\infty}^{t} \left(\sqrt{2\pi}\right)^{-1} \exp\left[-\tfrac{1}{2}z^2\right] dz.$$

The limits of integration are transformed to α, β, γ, δ:

$$\alpha = \Phi\left(a/\sqrt{1+|\rho|}\right), \quad \beta = \Phi\left(b/\sqrt{1+|\rho|}\right), \quad \text{etc.}$$

The integral becomes

$$\int_\alpha^\beta du \int_\gamma^\delta dv \sqrt{(1+|\rho|)/(1-|\rho|)}\, \exp\left[-\tfrac{1}{2}|\rho|\left(\Phi^{-1}(u) \pm \Phi^{-1}(v)\right)^2/(1-|\rho|)\right],$$

where the $\pm$ sign means $+$ if $\rho < 0$, $-$ if $\rho > 0$. Each interval, (α, β) and (γ, δ), may be divided into equal subintervals of length not exceeding (say) 0.025, and two-point Gaussian quadrature performed in each. The error will be less than 1 in the fourth decimal place unless ρ is close to ± 1. The least favorable situation, when $\rho > 0$, occurs when either $a = c = -\infty$ ($\alpha = \gamma = 0$) or $b = d = \infty$ ($\beta = \delta = 1$); in this case, finer subintervals could be used. (See also Amos, 1969, and Sowden and Ashford, 1969.)

Appendix 1

Two Occasional Papers

I. A Century of Statistical Science, 1890–1990

Extract from a keynote address delivered on August 21, 1969, to the International Symposium on Statistical Ecology, New Haven, Connecticut (Anscombe, 1971).

First I make the apology always needed when one tries to present a neat view of history. That can only be done by molding such facts as halfway agree with one's view and by turning a blind eye to all the rest. Please don't be incensed by the injustice of it all. What I shall say has, I believe, a grain of truth in it, and you shouldn't expect more.

1. Let me begin with the period of roughly fifty years, from about 1890 to about 1940. This half-century saw the formation of the science of statistics as most of us think of it today—a remarkably coherent development from infancy all the way to maturity. The center of the development was in England. The stimulus for the development came mainly from the biological sciences. (In fact, the words "statistics" and "biometry" were nearly synonymous, and an aspiring statistician would let it be known he was at least a little bit of a biologist, a biologist at heart.)

At the start of that half-century a principal stimulus to the development to come was the statistical work of Francis Galton, a gentleman amateur of science. By the end of the half-century, a standard of mathematical professionalism had been set by Jerzy Neyman. What a long journey from Galton to Neyman! And, as we look at it now, how fast and direct the journey seems to have been!

One might perhaps break the half-century into two quarters, and characterize the first quarter, dominated by Karl Pearson, as a period of rapid and far-reaching exploration, and the second quarter, dominated by R. A. Fisher, as a period of continued rapid exploration coupled with explicit development of basic concepts and awareness of the identity and scope of the science of statistics. But the two quarters are of a piece. The second quarter appears as the natural projection of the first.

By about 1940 a point of balance or synthesis had been reached. There was a certain generally received body of doctrine, which has been expounded in the majority of the vast number of textbooks of statistics that have appeared since—even in many of the more recent ones.

2. Let us turn now to the next quarter-century, from about 1940 to about 1965. We might call this a period of broadening—broadening in many ways. The special association of statistics with the biological sciences ceased, because problems arising in many other fields of endeavor also claimed the attention of statisticians—in the physical sciences, in the social sciences, in the humanities, in engineering, in industry, in business. The renewed active association with the social sciences was pleasing and natural, for statistics was originally a social science. In addition to these ever broadening external stimuli, there was the internal stimulus of development of mathematical theory for its own sake, with little or no concern for possible use. I prefer not to dwell on that.

Another kind of broadening was ideological. Under the impact of decision theory and the neo-Bayesian movement, there was a gradually increasing awareness that statistical problems are of many kinds, of many *more* kinds than the 1940 received doctrine recognized. I believe that perception of diversity and toleration of antithetic techniques were the main achievement of the period at the conceptual level.

3. Well now, I have summarily reviewed three quarters of a century. What about the fourth quarter that we have only recently entered, the period from about 1965 to about 1990? I am sure this will prove to be the period of the computer. Although computers (I mean, of course, the high-speed stored-program kind) have been harnessed for statistical purposes ever since their appearance round about 1950, and have become steadily more important, only recently have they begun to have considerable impact on basic ideas and methods of statistics, and they will surely have in future much greater impact still. We can distinguish three radically different kinds of ways in which computers may impinge on the science of statistics.

First, since statistical theory is mathematical, there are the ways in which computers impinge on mathematics. A typical example is the development of approximations. A little quick computation may suffice to show that a contemplated approximation is terrible, and that no-one should waste his time trying to derive good theoretical bounds to its accuracy. Such computations may suggest a modified approximation, in turn suggesting a different mathematical approach. Computers can be used to tabulate or graph particular functions and so give the investigator a vivid understanding of the character of a class of functions. Again, a computer may be used to explore a class of combinatorial possibilities. Perhaps some day computer techniques of theorem proving will be turned to account in statistics.

A second kind of impingement of computers on statistics is in simulating or sampling stochastic processes. A good many years ago Maurice Bartlett simulated a measles epidemic, showing that a relatively simple theory could account for qualitative features of real epidemics. Such simulations will surely become a common tool in the elaboration of mathematical theories of random phenomena.

A third kind of impingement of computers on statistics is in the analysis of data. The techniques of the Fisherian era were conditioned by the computing resource of that era, the desk calculator. At first when computers were used for statistical analysis, the aim was to permit those same techniques to be used on a larger scale, faster, more easily and more cheaply. We are only beginning to realize that the computer invites us to do things that were not done in the dask-calculator era, and therefore invites us to reconsider what statistical analysis is all about. After relative stagnation in the third quarter-century, statistical analysis seems likely to be transformed almost out of recognition during the fourth quarter. The subject has come to life again.

4. I should like to consider in a little detail how it is that the computer is encouraging a revolution in statistical analysis. The computer can do two things for the statistician very easily:

(a) make extensive numerical calculations,
(b) form character arrays constituting tables, graphs, and charts of various kinds.

Numerical calculation. The great power in numerical calculation affects statistical analysis in at least two ways:

(i) Most methods of statistical analysis are based on some kind of model or structure, in the sense that if that model were true the method of analysis would be entirely satisfactory. For example, the customary analysis of variance of a set of observations arranged as a row-column crossclassification (in which we calculate row means and column means and find sums of squares between rows and between columns and the residual sum of squares) would be very satisfactory if indeed each observation was equal to a row constant plus a column constant plus an independent random "error" normally distributed with zero mean and constant variance. Now in fact we never *know* that the implied model is true, so it behooves us to try to see whether the model fits. There isn't any foolproof procedure, but what is sometimes done today is to calculate a complete set of residuals after fitting the model and then examine various sorts of display of the residuals or make significance tests based on them. Such examination of residuals should presumably be regarded as a routine part of the analysis. (With a desk calculator, obtaining the residuals at least doubles the effort of performing a standard analysis of variance, which is a good reason why Fisher did not advocate it.)

With this idea of seeing whether the analysis fits, goes the idea that we cannot know *in advance* what the right method of analysis will prove to be. There is no surely best procedure of analysis. The computation can hardly be a one-shot affair, but will generally involve several attempts. Each body of data is likely to present unique features, either peculiarities in the numerical material or peculiarities in background information, to which the statistical treatment should be sensitive.

(ii) We can now contemplate methods of statistical analysis involving far more computation than the methods advocated before 1940. Laplace and Gauss defended least squares by saying that anything other than linear methods of combination of observations would be impracticable; and that remained nearly true until computers arrived. But now far larger computations are feasible. If anything useful can be learned from a large calculation, we should consider making it. Alternatives to the method of least squares, computationally much more expensive, yet feasible and possibly more appropriate sometimes, are currently being experimented with.

Character displays. It is difficult for even the most acute mind to grasp a complicated set of relationships by studying the sort of table in which the results of statistical analysis are often presented. Such material can be far more readily assimilated if presented in graphical or semi-graphical form. A semi-graph is a character display in which some quantities are represented by symbols (e.g. by numerals) and others are represented by position. Charts and diagrams, including simple scatterplots, look as if they would be easy to make by hand, and indeed they *can* be so made, but the work is often dreadfully tedious. The tedium may have contributed to the common notion that purely numerical analyses are *better* than graphical ones, that taking a look is somehow cheating. Until recently, good ideas for graphical analyses came mainly from persons who were *not* professional statisticians. Now, above all because of the imaginative and challenging recent work of John Tukey, statisticians are beginning to recognize the immense potential of good visual methods. Methods of the future will no doubt include computer-made stereoscopic color movies with multichannel sound, presenting to our senses relations in inconceivably many dimensions. But the modest chart, stationary, monochrome and mute on a piece of paper, has a power of communication that has probably not yet been fully realized and is certainly not widely enough appreciated.

5. Since computers seem destined to have such profound effects on the science of statistics, one may ask why the effects have not become more evident sooner. The slowness, I think, is not just due to the conservatism or timidity of statisticians. Many formidable problems with computers and computing have been encountered and have had to be tackled. I should like to mention two of these problems—not necessarily the two most important, but certainly problems that have taken time to be dealt with. The problems are not peculiar to statistical work; they arise in many other connections.

One is the problem of size, the capacity of the computer to handle large amounts of material at once. There is a strange phenomenon here. If we take some kind of task that used to be performed without computer aid, with paper and pencil and a lot of ingenuity, and think about performing similar but larger tasks with a computer, we soon find ourselves contemplating tasks that are utterly beyond the capacity of the vastest computer we now have unless perhaps a great deal of ingenity is again brought to bear. Although our large computers seem so very large and powerful, and can be used so effectively for many purposes, yet it is surprisingly easy to conceive of projects that are far too large for straightforward execution. This will always be so. Present-day large cores do, however, offer us a wonderful, if not infinite, opportunity for extensive computation.

Another problem is that of communication between man and computer. Can a computer be controlled as a pianist controls his instrument? Informing the computer of the user's desires has in the past been anything but easy. Fortran was a great improvement over its predecessors, but something an order of magnitude better than simple Fortran programming and library program packages is needed.

The need has now apparently been met, in a variety of ways. Which way will emerge as the most generally successful cannot yet be seen, though some of us have our opinions. Better ways of communication must become available and be adopted.

II. Directions of Statistical Research

An opinion solicited by the U.S. Army Research Office, submitted June 2, 1977.

1. There is an enormous current literature of statistical science (theory, methodology). Journals issued in the U.S. include

Journal of the American Statistical Association
Annals of Statistics
Biometrics
Technometrics
Communications in Statistics
Journal of Multivariate Analysis

Journals issued abroad include

Journal of the Royal Statistical Society, series A, B, C (U.K.)
Biometrika (U.K.)
Journal of Applied Probability (U.K.)
Review of the International Statistical Institute (Netherlands)
Bulletin of the International Statistical Institute (various countries)
Sankhya (India)
Annals of the Institute of Statistical Mathematics (Japan)

There are many others—the list could easily be tripled. New books are constantly being published, from elementary textbooks (mostly bad) to research monographs (often good). And then there are all the theses and research reports.

[The journals were listed above partly to define the domain of discourse. These remarks do not refer to the compilation of statistical information, of the kind so admirably provided in *Historical Statistics of the U.S.* and the annual *Statistical Abstract of the U.S.*]

In such a flood of activity, much too large for any one person to read, how can one identify the most promising lines of new development? Can one guess what ideas and themes now being proposed and explored will be seen fifty years hence to have been important and fruitful? How will this period in the history of statistical science be characterized? Only the foolhardy would try to answer!

2. Though much research in statistical science is pursued for its own sake, I take the primary purpose of the subject to be discovery of how to deal with the statistical problems that arise, rarely or frequently, in almost every kind of human activity. Statistical science is primarily a practising subject, like medicine. Just as medical research is generally expected to have some bearing on the medical treatment of

patients, so statistical research should be expected to have some bearing on statistical practice.

3. Some statisticians concern themselves with statistical problems arising in a particular substantive field. For example, there are peculiar statistical and ethical problems in the planning of experiments to test medical treatments (the topic of an excellent symposium* at Rockefeller University, May 27, 1977). There is in principle no limit to the number of such special studies, directed towards a particular field of inquiry, that may prove to be of the greatest importance for that field. I can make no attempt to point to promising studies of this sort.

4. What of the main body of statistical research, not clearly oriented towards a particular substantive field? Most of this seems to be at several removes from raw practical problems. Much of it can be said to look backwards rather than forwards, in the sense that theoretical problems already perceived are given a more precise, or more elegant, or more general, treatment. It can indeed transpire that a broad new idea or viewpoint emerges slowly from studies that seemed at first to be quite narrow.

Below are some central topics of statistical theory that seem to me to have great current interest, and promise substantial further development. I think they will in retrospect appear important.

(a) *Robustness studies* (G. E. P. Box, J. W. Tukey, P. J. Huber, P. J. Bickel and others). The current enthusiasm for abolishing traditional least-squares methods seems to me excessive and therefore impermanent, but a "robustness" attitude is surely here to stay. One should not just assume, without examination, as has often been done in the past, that the most convenient mathematical model will fit. Procedures should be used that are not sensitive to likely inaccuracy in the model.

(b) *Clustering methods* (R. R. Sokal, R. Sibson and N. Jardine, J. A. Hartigan). It seems that multivariate data, whether quantitative or categorical, are often better approached in a clustering frame of mind than with the multivariate normal expectations of traditional multivariate analysis. Clustering methods are useful for much else besides what is called "numerical taxonomy". Often a classification problem must be solved before a statistical inquiry can be profitably undertaken.

(c) *Regression analysis for time series* (E. J. Hannan, E. Parzen, J. W. Tukey, G. E. P. Box, G. M. Jenkins and others). Time series have long been of interest to economists. Now great amounts of time-series material are gathered by physical scientists, for example in environmental monitoring. One sort of approach to studying the interrelation of time series ("time domain" analysis) is presented in the well-known book of Box and Jenkins. Sometimes "frequency domain" analysis is more insightful, but seems to have been little used in practice.

(d) *Graphical techniques* (E. Anderson, J. W. Tukey, R. Gnanadesikan and others). Not many years ago scorned as unrigorous and childish, or even unfair, graphical methods of presenting statistical material are now seen to be of central importance for correct understanding. There is still plenty of room for development of new and more effective methods.

*See *Science*, **198** (1977), 675–705.

(e) *Analysis of large data sets*. We have moved into an era when many immense data sets are available and invite study. Such data sets have often only a low density of interesting information in them, for the purpose of the study. Beyond the purely computing difficulty of handling large data sets, there are difficulties in statistical procedure, for example in devising adequate graphical displays. A scatterplot with a million points is a different object to consider (less easy) than one with only a hundred points. This topic is still in its infancy. Much will depend on its growing up fast.

5. The five topics listed above have at least one element in common, namely computing. Very broadly, one may say that during the past decade or so the science of statistics has been revolutionized (and it will continue to be revolutionized) by the impact of the computer. This revolution is just as profound as the one that occurred between about 1920 and 1950 when much more powerful mathematics was brought to bear on statistics. I have refrained from listing "Statistical computing" as a topic, because I think it is many topics.

If one were looking back rather than forward and trying to assess the most important things that have happened in statistical science since, say, 1950, other topics than the above would be mentioned—for example, the great gain in our understanding of general principles of statistical inference. While of course there may some day be further major developments in inference, I don't see much to be excited about just now, and I therefore did not list inference as a current topic.

Appendix 2

Tests of Residuals

When a linear regression relation between some variables has been fitted by least squares, steps should be taken to assess the adequacy of the fitted regression structure, and to ascertain whether conditions are such as to make the method of least squares effective and appropriate. On the whole the most satisfactory procedures are graphical. Scatterplots of the residuals against values of explanatory variables or against the fitted values can reveal whether the linear relation fitted is satisfactory and how nearly the errors conform to the theoretical ideal of independent identically distributed normal random variables. Sometimes graphical methods can be usefully supplemented by calculated significance tests. In Chapter 11, a heteroscedasticity test was considered in some detail, as an example.

Tests of significance of regression coefficients, and Tukey's nonadditivity test, are well known. They are "exact" tests associated with terms in an analysis of variance. The residuals may also be used for other tests, notably tests for nonnormality, for heteroscedasticity, and (in the event that the data have a temporal order) for serial correlation. These tests are not associated with terms in an analysis of variance; it is difficult or impractical to make the tests "exact", but not hard to carry them out approximately. Such tests can be made on residuals from any linear-least-squares fitting. The needed formulas assume a particularly simple form, not calling for matrix computation, when the data consist of a simple two-way (rectangular) table and the linear structure to be fitted is the sum of row and column constants, leading to Fisher's analysis of variance for such a table.

Some tests of residuals were specified in a previous publication (Anscombe, 1961). Most results given in that paper seem to be accurate; the most serious error that has come to light is in the first part of equation (42) on p. 8, which should read

$$E(g_2^3) \sim \frac{1728}{n^2 c^6},$$

with n^2 in the denominator instead of n. (An innocuous error is that equation (7) on p. 3 of the paper presupposes Design Condition 2 as well as 1.)

The purpose of this appendix is to give summary information about a variety of tests, and show possible implementations of them as computer programs. In the first part, residuals from any fitted linear regression are considered in general. In the second part, equivalent results are given for the special case of the two-way table. The third part presents something analogous for stationary time series. What may be said to correspond to regression residuals is the innovations in an autoregressive representation of the series—they are autoregression residuals. Three versions of one sort of test, for kurtosis, are presented, bearing some resemblance to Mardia's multivariate kurtosis test. A fourth part presents Tukey's rule of thumb for deciding how many regression coefficients to estimate in order to have informative residuals.

Material has been drawn from two unpublished reports (Anscombe, Bancroft and Glynn, 1974; Anscombe and Chang, 1980).

A. Tests Based on Regression Residuals

The following notation was introduced in Chapter 11. The vector of values of the dependent variable is $\{y_i\}$ $(i = 1, 2, \ldots, n)$. In the regression, $\{y_i\}$ is projected onto a fixed p-dimensional linear subspace. Ordinarily a vector of 1's, denoted by $\{1\}$, is included in the subspace (this property was referred to as Design Condition 1 in Anscombe, 1961), in which case the subspace can be defined by p linearly independent vectors $\{1\}$, $\{x_i^{(1)}\}, \{x_i^{(2)}\}, \ldots, \{x_i^{(p-1)}\}$. The possibility is entertained that for all i

$$y_i = \beta_0 + \beta_1 x_i^{(1)} + \cdots + \beta_{p-1} x_i^{(p-1)} + \epsilon_i, \tag{1}$$

where the β's are constants and $\{\epsilon_i\}$ are independent identically distributed random variables having the normal distribution $N(0, \sigma^2)$. This is taken as the null hypothesis for the tests. The residuals, after estimation of the β's by least squares, are denoted by $\{z_i\}$, and Q or $\{q_{ij}\}$ is the projection matrix, which we regard as fixed: $z_i = \Sigma_j q_{ij} y_j$. On the null hypothesis

$$z_i = \Sigma_j q_{ij} \epsilon_j, \tag{2}$$

and $\{z_i\}$ are jointly normally distributed with variance matrix $Q\sigma^2$. The residual degrees of freedom are $\nu = n - p$, and $\Sigma_i q_{ii} = \nu$. Under Design Condition 1 every row of Q sums to zero, $\Sigma_j q_{ij} = 0$. We denote an n-by-n identity matrix by $\{\delta_{ij}\}$, so that $\delta_{ij} = 1$ if $i = j$, $\delta_{ij} = 0$ otherwise. Then ordinarily all elements of $\{|\delta_{ij} - q_{ij}|\}$ are small. Some formulas simplify if (Design Condition 2) $q_{ii} = \nu / n$ for all i, and simplify still further if in addition all off-diagonal elements of Q are equal in magnitude (Design Condition 2 with equal-magnitude correlations). The residual mean square is s^2, where

$$\nu s^2 = \Sigma_i z_i^2. \tag{3}$$

Definitions of test statistics. Measures of nonnormality can be constructed from third and fourth moments of the residuals. One may either form analogs of the Pearson statistics for a homogeneous sample, usually denoted by $\sqrt{b_1}$ and b_2, or form analogs of Fisher's statistics, g_1 and g_2, as follows.

$$\sqrt{b_1} = \frac{\Sigma_i z_i^3}{n}\left(\frac{n}{\Sigma_i z_i^2}\right)^{3/2}, \quad b_2 = \frac{\Sigma_i z_i^4}{n}\left(\frac{n}{\Sigma_i z_i^2}\right)^2. \tag{4}$$

$$g_1 = \frac{\Sigma_i z_i^3}{s^3 \Sigma_{ij} q_{ij}^3} = \frac{\nu\sqrt{\nu/n}}{\Sigma_{ij} q_{ij}^3}\sqrt{b_1}\,. \tag{5}$$

$$g_2 = \left(\frac{\Sigma_i z_i^4}{s^4} - \frac{3\nu\Sigma_i q_{ii}^2}{\nu+2}\right)\Big/ D = \frac{\nu^2}{nD}\left(b_2 - \frac{3n\Sigma_i q_{ii}^2}{\nu(\nu+2)}\right), \tag{6}$$

where

$$D = \left(\Sigma_{ij} q_{ij}^4\right) - \frac{3\left(\Sigma_i q_{ii}^2\right)^2}{\nu(\nu+2)}\,.$$

The definitions of g_1 and g_2 are chosen as follows. Suppose that (1) holds as stated except that $\{\epsilon_i\}$ are independent identically distributed variables having mean 0, variance σ^2, and third and fourth moments μ_3, μ_4. Then s^2 is the multiple of $\Sigma_i z_i^2$ that is an unbiased estimate of σ^2; g_1 is the multiple of $\Sigma_i z_i^3$ that is an unbiased estimate of μ_3, divided by s^3; g_2 is the linear combination of $\Sigma_i z_i^4$ and s^4 that is an unbiased estimate of $\mu_4 - 3\sigma^4$, divided by s^4.* Thus g_1 and g_2 are estimates, calculated from the residuals, of skewness and kurtosis in the distribution of $\{\epsilon_i\}$, whereas $\sqrt{b_1}$ and b_2 can be said to measure skewness and kurtosis in the distribution of the residuals. If ν is substantially less than n, the two distributions may be rather different.

When Q is given, g_1 and g_2 are known linear functions of $\sqrt{b_1}$ and b_2, and significance tests based on either style of definition are equivalent. The Pearsonian definitions do not refer to Q, and they lead to slightly simpler formulas for moments under the null hypothesis. But the Fisherian definitions yield values that may be regarded as more interesting measures of nonnormality, and they are preferred here (though we shall revert to b_2-like measures when we come to time-series innovations, partly because there is then no fixed projection matrix Q).

A heteroscedasticity statistic, denoted by a, measuring progressive change in residual variance in relation to an "independent" variable $\{x_i\}$, was given in Chapter 11, equation (4). It was remarked that $\{x_i\}$ could be replaced by the vector of fitted values, $\{Y_i\}$ say, to obtain a measure of progressive change in residual variance in relation to the expectation $\{\mathcal{E}(y_i)\}$. That statistic will be denoted by h.

Tukey's test (1949), "one degree of freedom for nonadditivity", can be expressed in terms of regression of $\{z_i\}$ on the squared fitted values, $\{Y_i^2\}$. Under Design Condition 1, any constant can be subtracted from the fitted values before squaring, without affecting the test (except through round-off error). The regression

*The statistics are not in general optimal. In a tour de force Pukelsheim (1980) has determined third-degree and fourth-degree polynomials in $\{z_i\}$ that are unbiased estimates of μ_3 and μ_4 and have minimum variance on the null hypothesis.

coefficient is

$$f = \frac{\Sigma_i z_i Y_i^2}{\Sigma_{ij} q_{ij} Y_i^2 Y_j^2}. \tag{7}$$

The test cannot be made if the denominator vanishes.

When the observations $\{y_i\}$ are equally spaced in time, there may be interest in examining $\{z_i\}$ for serial correlation or other serial structure. A first serial correlation coefficient may be defined:

$$r_1 = \left(\frac{\Sigma_i z_i z_{i+1}}{s^2} + E \right) / G, \tag{8}$$

where

$$E = -\Sigma_i q_{i(i+1)}, \quad G = \Sigma_{ij}(q_{ij} q_{(i+1)(j+1)} + q_{i(j+1)} q_{(i+1)j}),$$

and in each of these summations i or j runs from 1 to $n - 1$. The definition is chosen as follows. Suppose that (1) holds as stated except that $\{\epsilon_i\}$ are jointly normally distributed and form a stationary Markov sequence with marginal variance σ^2 and first serial correlation coefficient ρ, where ρ is near 0. Then r_1 is an unbiased estimate of $\rho\sigma^2$, divided by s^2.

Distributions on the null hypothesis. To make tests we require information about the sampling distributions for these statistics on the null hypothesis. The nonadditivity statistic leads to Tukey's exact test. His one-degree-of-freedom term is equal to the regression coefficient f multiplied by the numerator on the right side of (7). This term is subtracted from the residual sum of squares, $\Sigma_i z_i^2$, and compared with the remaining sum of squares having $\nu - 1$ degrees of freedom by the variance-ratio (F) test. For the other statistics, exact expressions for a few moments can be obtained and used in approximate tests.

For g_1 and g_2 we find

$$\mathcal{E}(g_1) = \mathcal{E}(g_2) = \mathcal{E}(g_1 g_2) = \mathcal{E}(g_1^3) = \mathcal{E}(g_1 g_2^2) = 0,$$

$$\operatorname{var}(g_1) = \frac{3\nu^2}{(\nu+2)(\nu+4)} \frac{\{2\Sigma_{ij} q_{ij}^3 + 3\Sigma_{ij} q_{ii} q_{ij} q_{jj}\}}{(\Sigma_{ij} q_{ij}^3)^2}, \tag{9}$$

$$\operatorname{var}(g_2) = \frac{24\nu^3}{(\nu+2)(\nu+4)(\nu+6)} \frac{D + 3F}{D^2}, \tag{10}$$

where

$$F = \Sigma_{ij} q_{ii} q_{ij}^2 q_{jj} - (\Sigma_i q_{ii}^2)^2/\nu,$$

$$\mathcal{E}(g_1^2 g_2) = \frac{36\nu^4}{(\nu+2)(\nu+4)(\nu+6)(\nu+8)(\Sigma_{ij} q_{ij}^3)^2 D}$$
$$\times \Big\{ 6\Sigma_{ijk} q_{ij} q_{ik}^2 q_{jk}^2 + 6\Sigma_{ijk} q_{ij}^2 q_{ik} q_{jk} q_{kk} + 3\Sigma_{ijk} q_{ii} q_{ik} q_{jj} q_{jk} q_{kk}$$
$$+ 4\Sigma_{ijk} q_{ii} q_{ik} q_{jk}^3 + 6\Sigma_{ijk} q_{ii} q_{ij} q_{jk}^2 q_{kk}$$
$$- \frac{3(\nu+4)}{\nu(\nu+2)} (\Sigma_k q_{kk}^2)(2\Sigma_{ij} q_{ij}^3 + 3\Sigma_{ij} q_{ii} q_{ij} q_{jj}) \Big\}, \tag{11}$$

$$\mathcal{E}(g_2^3) = \frac{864\nu^5}{(\nu+2)(\nu+4)(\nu+6)(\nu+8)(\nu+10)D^3}$$

$$\times \Big\{ 2\Sigma_{ijk}\, q_{ij}^2 q_{ik}^2 q_{jk}^2 + 2\Sigma_{ijk}\, q_{ii} q_{ij} q_{jj} q_{ik} q_{jk} q_{kk} + 4\Sigma_{ijk}\, q_{ii} q_{ij} q_{ik} q_{jk}^3$$

$$+ 3\Sigma_{ijk}\, q_{ii} q_{ij}^2 q_{jk}^2 q_{kk} - \frac{5\nu^2 + 34\nu + 60}{\nu^2(\nu+2)^2} (\Sigma_i q_{ii}^2)^3$$

$$- \frac{4(\nu+5)}{\nu(\nu+2)} (\Sigma_i q_{ii}^2)(D + 3F) \Big\}. \qquad (12)$$

These expressions can be evaluated by ordinary matrix operations (matrix products, direct products, sum reductions) on n-by-n matrices, if there is room in the active workspace for several such matrices to be held at once—the total number of operations being independent of n. Expressions for two fourth-degree moments, $\mathcal{E}(g_1^4)$ and $\mathcal{E}(b_2^4)$, have been obtained (Anscombe, 1961; Anscombe and Glynn, 1975), but evaluation seems to call in general for the setting up of n-by-n-by-n arrays, scarcely practicable if n is large enough to be interesting (50 or more, say).

Under Design Conditions 1 and 2 some terms in the above expressions vanish. For various special designs simple non-matrix expressions for the moments can be found. That is illustrated below for the two-way table. Another example occurs under Design Conditions 1 and 2 with equal-magnitude correlations; the moments depend just on n and ν. We conjecture that these reduced expressions for $\mathcal{E}(g_1^2 g_2)$ and $\mathcal{E}(g_2^3)$ are a fairly good approximation to the true values even when Design Condition 2 and equal-magnitude correlations do not obtain.

A different sort of approximation to the moments is asymptotic for large n. That possibility is not explored systematically here, but one type of asymptotic result is referred to below in connection with time-series innovations, and is now briefly mentioned. Consider a sequence of regression problems, in which $n \to \infty$, p is constant, and the differences between Q and an identity matrix are $O(n^{-1})$ in some sense. An unnecessarily strong condition is that there is a constant C such that for all n, i and j,

$$|\delta_{ij} - q_{ij}| < C/n.$$

An adequate weaker condition is that the independent variables $\{x_i^{(1)}, \ldots, x_i^{(p-1)}\}$ are a random sample from some $(p-1)$-dimensional distribution having positive-definite variance matrix and finite moments of all orders. Then one may establish results such as these, with possibly a probabilistic interpretation of the symbols $O(n^{-1})$:

$$\Sigma_{ij}\, q_{ij}^4 = n - 4p + O(n^{-1}),$$

$$\Sigma_{ij}\, q_{ii} q_{ij}^2 q_{jj} = n - 3p + O(n^{-1}),$$

$$\Sigma_{ijk}\, q_{ii} q_{ij} q_{ik} q_{jk}^3 = n - 5p + O(n^{-1}),$$

$$\Sigma_{ijk}\, q_{ii} q_{ij} q_{jj} q_{ik} q_{jk} q_{kk} = n - 4p + O(n^{-1}).$$

(The last result is the most delicate to prove.) With these it is easy to deduce moments of the asymptotic distribution for b_2:

$$\mathcal{E}(b_2) \sim \frac{3n}{n+2}, \quad \text{var}(b_2) \sim \frac{24}{n+15}, \quad \gamma_1(b_2) \sim \frac{6\sqrt{6}}{\sqrt{n+29}}, \tag{13}$$

where $\gamma_1(b_2)$ means the standardized third moment or skewness measure of the distribution for b_2, $\mathcal{E}(b_2 - \mathcal{E}b_2)^3/[\text{var}(b_2)]^{3/2}$. The results are correct within a factor $1 + O(n^{-2})$. It is remarkable that no properties of Q appear in these results, not even the value of p.

Consider now how g_1 and g_2 are distributed. Whereas the marginal distribution for g_1 is symmetric, that for g_2 is very skew unless n is very large. It appears that, when $\nu \geqslant 30$, the marginal distribution for g_2 is well enough determined for practical purposes by first approximating g_2 by a linear function of the reciprocal of a χ^2 variable with A degrees of freedom (say), where A is chosen to make the standardized third moment correct and the coefficients of the linear function to make the first two moments correct, and then approximating the χ^2 variable by the Wilson-Hilferty cube-root transformation (Anscombe and Glynn, 1975). We find

$$A = 6 + \frac{8}{m_{03}}\left[\frac{2}{m_{03}} + \sqrt{1 + \frac{4}{m_{03}^2}}\right], \tag{14}$$

where $m_{03} = \mathcal{E}(g_2^3)/[\text{var}(g_2)]^{3/2} = \gamma_1(b_2)$ as defined above; and the equivalent normal deviate is

$$\text{e.n.d.}(g_2) = \frac{(1 - 2/(9A)) - \left([1 - (2/A)]/\left[1 + x\sqrt{2/(A-4)}\,\right]\right)^{1/3}}{\sqrt{2/(9A)}}, \tag{15}$$

where $x = g_2/\sqrt{\text{var}(g_2)}$. If, for example, e.n.d.$(g_2) = 1.96$, the approximation tells us that g_2 is at the upper $2\frac{1}{2}\%$ point of its distribution; if e.n.d.$(g_2) = -1.96$, g_2 is at the lower $2\frac{1}{2}\%$ point.

To complete the joint distribution for g_1 and g_2, after the marginal distribution for g_2, the conditional distribution for g_1, given g_2, is needed. We make the crude assumption that $\text{var}(g_1 \mid g_2)$ is linear in g_2. Thus we suppose

$$\text{var}(g_1 \mid g_2) = \text{var}(g_1)\left[1 + \frac{m_{21}\, g_2}{\sqrt{\text{var}(g_2)}}\right], \tag{16}$$

where $m_{21} = \mathcal{E}(g_1^2\, g_2)/[\text{var}(g_1)\sqrt{\text{var}(g_2)}\,]$; and the distribution for $g_1 \mid g_2$ is treated as normal. The sum of squares of the e.n.d.'s for g_2 and $g_1 \mid g_2$, having approximately the tabulated χ^2 distribution with 2 degrees of freedom, makes a combined test for nonnormality (like the one suggested by D'Agostino and Pearson, 1973, 1974). Shenton and Bowman (1977) have indicated that the distribution for $g_1 \mid g_2$ tends to be platykurtic, actually bimodal when n is small and g_2 is positive and large. Treating the distribution as normal is then "conservative", underestimating the significance of large values. When g_2 is negative, the right side of (16) is capable of being much less than the unconditional var(g_1), even sometimes negative —clearly a grain of salt is in order.

The first three moments of the distribution for the heteroscedasticity statistic a were given in Chapter 11. It seems that usually the distribution is nearly symmetric and only mildly leptokurtic, and no great error is committed in regarding the distribution as normal. Similarly for the statistic h, when the fixed explanatory variable $\{x_i\}$ is replaced by the fitted values $\{Y_i\}$. The moments relate to the conditional distribution for h, given $\{Y_i\}$.

All these tests refer to the same null hypothesis expressed at (1) above. One sort of departure from the null hypothesis could cause more than one of the statistics to be large; the tests are by no means independent. We have just considered how g_1 behaves conditionally on g_2. One could similarly consider how a and h behaved conditionally on g_2, to obtain a heteroscedasticity test that did not presuppose normality. That is not done here. Bickel (1978) has developed an elegant asymptotic theory of such tests.

Our last test statistic is the serial correlation coefficient r_1 defined at (8). We find

$$\mathcal{E}(r_1) = 0, \quad \operatorname{var}(r_1) = \frac{\nu G - 2E^2}{(\nu + 2)G^2}. \tag{17}$$

Higher moments have not been determined. The distribution may be expected to be nearly symmetric and slightly platykurtic. In the well-known test of Durbin and Watson (1951) information about Q is avoided by use of inequalities.

Implementation. These tests have been implemented in the function *TESTS*, described in *HOWTESTS*, listed in Appendix 3. The output should be readily intelligible. What follows the ± sign is always the relevant standard deviation (on the null hypothesis).

A few comments about details in the program may be helpful. Some quantities calculated are available as global output variables, and others would be if the user deleted them from the header. The variables A and D are repeatedly redefined. The comment at line [6] could be useful if space were short; it would be acted on after halting execution at this point with the stop control. At line [7] execution terminates if the residual mean square added to 1 is the same as 1 to the first thirteen decimal digits or so (the comparison-tolerance or fuzz feature in APL). At line [27] a test is made that the Wilson-Hilferty approximation will yield an equivalent normal deviate greater than $-\infty$; if it will not, the equivalent normal deviate for g_2 is not calculated. Similarly at line [31] a test is made that the approximate expression (16) for var($g_1 \mid g_2$) will be positive; if it will not, the equivalent normal deviate for $g_1 \mid g_2$ is not calculated. In line [37] the value reported for h will be

$$0 \pm 0 \quad (RATIO = .00)$$

if the denominator in the expression for var(h) is effectively zero. And similarly for a in line [42]. The global variable underscored-X, against which a measures progressive change in variance, is altered at line [10] through subtraction of weighted column means; other global variables invoked by *TESTS* are not altered. At line [45] Tukey's test is by-passed if either Design Condition 1 is not satisfied or the denominator in the expression for f is effectively zero.

Examples. *TESTS* is applied to the data of Chapter 11. Figure B:46 has been executed after repeating what was done in Figure B:31, linear regression of *SE*70 on the three explanatory variables. The serial test is not asked for, since the serial order of the states has little significance. There is no evidence of nonnormality. The h-test is mildly significant, and the a-test for change in variance with *PI*68 is more strongly significant. Tukey's one-degree-of-freedom term is very small.

Figure B:47 gives similar information concerning what was called the logarithmic analysis (ii), linear regression of 100 times the natural logarithm of *SE*70 on 100 times the natural logarithms of *PI*68 and *Y*69 and also on *URBAN*70 untransformed. Now all the tests, for nonnormality, heteroscedasticity and nonadditivity, show deviations that can be dismissed as not significant.

These two figures were necessarily executed in a workspace a good deal larger than the 32000-byte size that sufficed for most other computing shown in the book. There is also a difference in style of output between the figures in this appendix and

```
      ⍝ FIRST ILLUSTRATION OF EXECUTION OF 'TESTS'  ~  ~  ~  ~

      ⍝ CONTINUE FROM THE FOOT OF FIGURE B:31

      ⍴X←PI68,Y69,[1.5] URBAN70
51 3

      TESTS 1
SERIAL TEST?  Y OR N
NO
FAST COMPUTATION, EXACT SECOND AND THIRD MOMENTS.

   *DISTRIBUTION SHAPE*
SKEWNESS:  G1 = 0.089012 ± 0.36298  (RATIO =   .25)
KURTOSIS:  G2 = ¯0.20666 ± 0.73117
   (STANDARDIZED 0 3 MOMENT = 1.583, ROUGH E.N.D. FOR G2 =   ¯.11)
   (STANDARDIZED 2 1 MOMENT = 0.88451, CRUDE E.N.D. FOR G1 GIVEN G2 =
      .28)

   *HETEROSCEDASTICITY TESTS*
AGAINST FITTED VALUES:  H = 0.013523 ± 0.0063148  (RATIO =  2.14)
AGAINST COLUMN 1 OF X:  A[1;] = 0.00094732 ± 0.00038366  (RATIO =  2.47)
AGAINST COLUMN 2 OF X:  A[2;] = ¯0.0094954 ± 0.0099159  (RATIO =   ¯.96)
AGAINST COLUMN 3 OF X:  A[3;] = 0.001924 ± 0.0014196  (RATIO =  1.36)

   *NONADDITIVITY--REGRESSION OF  RY  ON SQUARED FITTED VALUES*
COEFFICIENT:  F = ¯0.00020259
TUKEY'S 1-D.F. TERM:  15.424
REMAINING SUM OF SQUARES:  33474
MEAN SQUARE (46 D.F.):  727.71
```

Figure B:46

```
      ⍝ SECOND ILLUSTRATION OF EXECUTION OF 'TESTS'  ~  ~  ~  ~

      ⍝ CONTINUE FROM LINEAR REGRESSION OF  100×⍟SE70  ON 3 VARIABLES

      ⍴X̲←(100×⍟PI68,[1.5] Y69),URBAN70
51 3

      TESTS 1
SERIAL TEST?  Y OR N
NO
FAST COMPUTATION, EXACT SECOND AND THIRD MOMENTS.

   *DISTRIBUTION SHAPE*
SKEWNESS:  G1 = 0.012559 ± 0.36339  (RATIO =    .03)
KURTOSIS:  G2 = ¯0.53686 ± 0.73241
   (STANDARDIZED 0 3 MOMENT = 1.5821, ROUGH E.N.D. FOR G2 =    ¯.74)
   (STANDARDIZED 2 1 MOMENT = 0.88399, CRUDE E.N.D. FOR G1 GIVEN G2 =
      .06)

   *HETEROSCEDASTICITY TESTS*
AGAINST FITTED VALUES:  H = 0.012992 ± 0.012264  (RATIO =  1.06)
AGAINST COLUMN 1 OF X̲:  A[1;] = 0.016884 ± 0.012157  (RATIO =  1.39)
AGAINST COLUMN 2 OF X̲:  A[2;] = ¯0.028486 ± 0.035884  (RATIO =   ¯.79)
AGAINST COLUMN 3 OF X̲:  A[3;] = 0.0013215 ± 0.0014204  (RATIO =   .93)

   *NONADDITIVITY--REGRESSION OF  RY  ON SQUARED FITTED VALUES*
COEFFICIENT:  F = ¯0.0043127
TUKEY'S 1-D.F. TERM:  348.34
REMAINING SUM OF SQUARES:  7854.6
MEAN SQUARE (46 D.F.):  170.75
```

Figure B:47

earlier figures, that when a line of numerical output exceeds the specified page width (72 characters in these figures) a number represented by more than one decimal digit may be broken at the end of the line, whereas earlier the breaks occurred only at blanks. By chance no broken numbers occur in Figures B:46 and B:47, but they do occur in Figures B:48 and B:49 below. Such details vary for no obvious reason with the implementation of APL. The breaking of numbers at output could be avoided by not using the page-width control in the system but instead using one's own outputting function, like the paragraphing function mentioned in Exercise 12(b) at the end of Part A; but that is not done here.

B. Tests Based on Residuals in a Two-Way Table

It was remarked above that for various special designs simple non-matrix expressions for the test statistics and their moments could be found. One such design is the simple homogeneous sample, for which the regression calculation reduces to estimating the mean, and $q_{ij} = \delta_{ij} - 1/n$ for all i and j. The function *SUMMARIZE* makes the nonnormality and serial-correlation tests specified above, using suitably specialized formulas. Tests for nonadditivity and heteroscedasticity are inappropriate. The matter is not discussed further here.

Consider now a two-way table. The above notation for regression is somewhat modified. The table, having k rows and l columns, is denoted by $\{y_{ij}\}$ ($i = 1, 2, \ldots, k; j = 1, 2, \ldots, l$). We assume that k and l are not less than 3, and we set $n = kl$, $\nu = (k-1)(l-1)$. We consider the "null hypothesis" that for all i and j

$$y_{ij} = \alpha_i + \beta_j + \epsilon_{ij}, \tag{18}$$

where $\{\alpha_i\}$ and $\{\beta_j\}$ are vectors of row-constants and column-constants and $\{\epsilon_{ij}\}$ is a matrix of "errors", independent identically distributed random variables having the normal distribution $N(0, \sigma^2)$. The standard analytic procedure involves calculating the grand mean $\bar{y}$, the row effects $\{y_{i.} - \bar{y}\}$ and the column effects $\{\bar{y}_{.j} - \bar{y}\}$, where

$$n\bar{y} = \Sigma_{ij} y_{ij}, \quad l\bar{y}_{i.} = \Sigma_j y_{ij}, \quad k\bar{y}_{.j} = \Sigma_i y_{ij}. \tag{19}$$

The matrix of residuals $\{z_{ij}\}$ is defined by

$$z_{ij} = y_{ij} - \bar{y}_{i.} - \bar{y}_{.j} + \bar{y}, \tag{20}$$

and the residual mean square s^2, with ν degrees of freedom, by

$$\nu s^2 = \Sigma_{ij} z_{ij}^2. \tag{21}$$

Definitions and moments. Tukey's test can be expressed in terms of regression of $\{z_{ij}\}$ on $\{(\bar{y}_{i.} - \bar{y})(\bar{y}_{.j} - \bar{y})\}$, the outer product of the row effects and the column effects. The regression coefficient is

$$\frac{\Sigma_{ij} z_{ij}(\bar{y}_{i.} - \bar{y})(\bar{y}_{.j} - \bar{y})}{\Sigma_i (\bar{y}_{i.} - \bar{y})^2 \Sigma_j (\bar{y}_{.j} - \bar{y})^2}, \tag{22}$$

and Tukey's one-degree-of-freedom term is equal to the regression coefficient multiplied by the above numerator. Note that the regression coefficient is double the one defined at (7) above, but the resulting one-degree-of-freedom terms are equivalent.

The Fisherian shape statistics are

$$g_1 = \frac{n \Sigma_{ij} z_{ij}^3}{\nu(k-2)(l-2)s^3}, \tag{23}$$

and

$$g_2 = \left(\frac{\Sigma_{ij} z_{ij}^4}{s^4} - \frac{3\nu^3}{n(\nu+2)} \right) \Big/ D, \tag{24}$$

where

$$D = \frac{\nu}{n^2} \left((k^2 - 3k + 3)(l^2 - 3l + 3) - \frac{3\nu^2}{\nu + 2} \right).$$

Nonzero moments on the null hypothesis are

$$\operatorname{var}(g_1) = \frac{6n\nu}{(k-2)(l-2)(\nu+2)(\nu+4)}, \tag{25}$$

$$\operatorname{var}(g_2) = \frac{24\nu^3}{(\nu+2)(\nu+4)(\nu+6)D}, \tag{26}$$

$$\mathcal{E}(g_1^2 g_2) = \frac{216\nu^3}{(\nu+2)(\nu+4)(\nu+6)(\nu+8)D}\left(1 - \frac{2\nu}{(\nu+2)(k-2)(l-2)}\right), \tag{27}$$

$$\mathcal{E}(g_2^3) = \frac{1728\nu^5}{(\nu+2)(\nu+4)(\nu+6)(\nu+8)(\nu+10)D^3}$$
$$\times\left[\frac{\nu}{n^3}(k^3 - 5k^2 + 10k - 7)(l^3 - 5l^2 + 10l - 7)\right.$$
$$\left. - \frac{\nu^4(\nu+8)}{n^3(\nu+2)^2} - \frac{6\nu D}{n(\nu+2)}\right]. \tag{28}$$

The heteroscedasticity statistic against fitted values is

$$h = \frac{\Sigma_{ij} z_{ij}^2(\bar{y}_{i.} + \bar{y}_{.j} - 2\bar{y})}{s^2\left[(1 - 2/k)(l-1)\Sigma_i(\bar{y}_{i.} - \bar{y})^2 + (1 - 2/l)(k-1)\Sigma_j(\bar{y}_{.j} - \bar{y})^2\right]}. \tag{29}$$

On the null hypothesis

$$\operatorname{var}(h) = \frac{2\nu}{(\nu+2)d}, \tag{30}$$

where d is the expression in square brackets in the denominator of the right side of (29).

Another sort of heteroscedasticity statistic may be of interest, detecting differences in error variance from row to row (or alternatively from column to column). Suppose that (18) holds as stated except that $\{\epsilon_{ij}\}$ are independent and normally distributed with mean 0, and ϵ_{ij} has variance σ_i^2. Then the estimated variance factors for rows may be defined as

$$\frac{\hat{\sigma}_i^2}{s^2} = \frac{1}{k-2}\left[\frac{k\Sigma_j z_{ij}^2}{(l-1)s^2} - 1\right]. \tag{31}$$

Here $\hat{\sigma}_i^2$ is an unbiased estimate of σ_i^2 (Anscombe and Tukey, 1963). These variance factors for rows sum to k. They are capable of being negative. On the null hypothesis

$$\mathcal{E}\left(\frac{\hat{\sigma}_i^2}{s^2}\right) = 1, \quad \operatorname{var}\left(\frac{\hat{\sigma}_i^2}{s^2}\right) = \frac{2(k-1)^2}{(\nu+2)(k-2)}. \tag{32}$$

As a combined test for inequality of row variances, the following expression,

$$X = \frac{A\Sigma_i\left[(\hat{\sigma}_i^2/s^2) - 1\right]^2}{k\mathrm{var}(\hat{\sigma}_i^2/s^2)}, \tag{33}$$

where

$$A = \frac{(\nu + 4)(\nu + 6)}{(l + 1)\left[\nu - 2/(k - 2)\right]},$$

may be regarded as having the tabulated χ^2 distribution with A degrees of freedom. (A will usually not be an integer.) X has the same lower bound of zero and the same first two moments as the approximating χ^2 variable. The variance of X depends on fourth moments of the variance factors $\{\hat{\sigma}_i^2/s^2\}$. Similarly for columns.

Usually the entries in a two-way table do not have an interesting serial order, and a serial test is inappropriate.

Implementation. These tests have been implemented in the function *ROWCOL* listed in Appendix 3. The function has grown since we first met it in Chapter 7. It performs the usual additive analysis of a two-way table and also the various tests of residuals just described. In many details it closely resembles *TESTS*.

Examples. The two illustrations refer to artificial data sets. (Artificial sets are chosen, both in order to see how known effects are reflected in the calculated tests, and also for brevity—real data deserve more thought.) The first, in Figure B:48, is a table with 6 rows and 8 columns, consisting of random normal deviates (zero mean and unit variance) without added row or column effects, but with two readings arbitrarily changed to the value 5.0. The table is named Y; it is first displayed, and then *ROWCOL* is applied to it. Tukey's one-degree-of-freedom term is not very different from the residual mean square. In the tests for distribution shape, g_1 is 3.29 times its standard error. Since the marginal distribution for g_1, on the null hypothesis, is probably long-tailed, this ratio should not be considered as significant as if the distribution were normal, but undoubtedly the observed value for g_1 is well beyond a mere 5% point. The kurtosis measure g_2 is more than 6 times its standard error, but after allowance for the skew distribution shape we find an equivalent normal deviate of 3.24, still strongly significant. In the rough conditional test for g_1, given g_2, we find an equivalent normal deviate of 1.36. For a combined test of distribution shape, we may take $(3.24)^2 + (1.36)^2$, or 12.3, as a χ^2 value with 2 degrees of freedom—significant at about the 0.2% level. As for the heteroscedasticity tests, h is about twice its standard error. The variance factors for rows are strongly unequal—the combined rough χ^2 test comes out at 15.73 with 5.15 degrees of freedom. The variance factors for columns are still more strongly unequal (and it may be seen that one of them is negative); the combined rough χ^2 is 25.80 with 6.59 degrees of freedom.

Thus the tests for distribution shape and for heteroscedasticity have both reacted strongly to the two outliers introduced into the otherwise homogeneous normal data set.

```
      ⍝ FIRST ILLUSTRATION OF EXECUTION OF 'ROWCOL'  ~  ~  ~  ~

      Y
 5      ¯2.72  0.18 ¯0.11  0.01 ¯0.22 ¯0.8  ¯0.01
¯0.31  5      ¯1.29  1.22  1.12 ¯1      0.42  0.72
 0.58  0.03  0.68  0.19 ¯1.19 ¯0.6   1.85  1.76
¯1.83  0.93  0.85 ¯0.52 ¯0.96  0.75 ¯0.01 ¯0.03
 0.38 ¯1.21  0.84  0.78  0.27 ¯0.23  0.39  1.12
¯0.09  1.62 ¯1.14  1.1  ¯0.31  1.46  0.12  0.81

      ROWCOL Y

GRAND MEAN (GM) IS 0.325
TOTAL SUM OF SQUARES ABOUT MEAN (TSS), DEGREES OF FREEDOM AND MEAN SQUAR
      E ARE 87.578 47 1.8634

   *ROWS*
EFFECTS (RE) ARE ¯0.15875 0.41 0.0875 ¯0.4275 ¯0.0325 0.12125
SUM OF SQUARES (SSR), D.F. AND M.S. ARE 3.1958 5 0.63916

   *COLUMNS*
EFFECTS (CE) ARE 0.29667 0.28333 ¯0.305 0.11833 ¯0.50167 ¯0.29833 0.0033
      333 0.40333
SUM OF SQUARES (SSC), D.F. AND M.S. ARE 4.6721 7 0.66744

   *RESIDUALS*
SUM OF SQUARES (RSS), D.F. (NU) AND M.S. ARE 79.71 35 2.2774
THE MATRIX OF RESIDUALS IS NAMED RY.

   *TUKEY'S TEST*
REGRESSION COEFFT. (TB) OF RY ON RE∘.×CE IS 3.4152
TUKEY'S 1-D.F. TERM IS 3.6281
REMAINING SUM OF SQUARES, D.F. AND M.S. ARE 76.082 34 2.2377

   *DISTRIBUTION SHAPE*
SKEWNESS:  G1 = 1.7729 ± 0.5395   (RATIO =  3.29)
KURTOSIS:  G2 = 7.3364 ± 1.1936
   (STANDARDIZED 0 3 MOMENT = 1.4334, ROUGH E.N.D. FOR G2 =  3.24)
   (STANDARDIZED 2 1 MOMENT = 0.79064, CRUDE E.N.D. FOR G1 GIVEN G2 =  1
      .36)

   *HETEROSCEDASTICITY TESTS*
AGAINST FITTED VALUES:  H = 1.3433 ± 0.62884  (RATIO =  2.14)
VARIANCE FACTORS FOR ROWS (VFR) = 2.7884 1.9962 0.27447 0.5117 0.17949 0
      .24976 ± 0.58124
   (ROUGH CHI-SQUARED =  15.73,  5.15 D.F.)
VARIANCE FACTORS FOR COLUMNS (VFC) = 3.0225 3.4499 0.69248 ¯0.035949 0.1
      3055 0.49817 0.2108 0.031603 ± 0.66441
   (ROUGH CHI-SQUARED =  25.80,  6.59 D.F.)
```

Figure B:48

In the second illustration, Figure B:49, the data set, named X, consists of uniformly distributed integers between 1 and 50, to which small row and column effects have been added. The kurtosis measure γ_2 (ratio of fourth cumulant to variance squared) for the uniform distribution is about ¯1.2. The estimate g_2

```
      ⍝ SECOND ILLUSTRATION OF EXECUTION OF 'ROWCOL' ~ ~ ~ ~

      ⎕X←((⍳8)∘.+⍳10)+? 8 10 ⍴50
 9 41 27 32 17 10 42 43 57 31
29 46  7  9 34 42 10 30 15 33
39 35 53 50 35 14 43 32 48 59
44 20 10 45 26 42 49 62 32 27
56 44 46 42 14 43 57 27 36 54
31 20 23 28 20 37 58 60 19 62
34 35 26 61 37 27 19 63 20 43
29 24 57 39 37 62 18 55 56 60

      ROWCOL X

GRAND MEAN (GM) IS 36.35
TOTAL SUM OF SQUARES ABOUT MEAN (TSS), DEGREES OF FREEDOM AND MEAN SQUAR
      E ARE 19114 79 241.95

   *ROWS*
EFFECTS (RE) ARE ¯5.45 ¯10.85 4.45 ¯0.65 5.55 ¯0.55 0.15 7.35
SUM OF SQUARES (SSR), D.F. AND M.S. ARE 2528 7 361.14

   *COLUMNS*
EFFECTS (CE) ARE ¯2.475 ¯3.225 ¯5.225 1.9 ¯8.85 ¯1.725 0.65 10.15 ¯0.975
      9.775
SUM OF SQUARES (SSC), D.F. AND M.S. ARE 2629.5 9 292.16

   *RESIDUALS*
SUM OF SQUARES (RSS), D.F. (NU) AND M.S. ARE 13957 63 221.54
THE MATRIX OF RESIDUALS IS NAMED RY.

   *TUKEY'S TEST*
REGRESSION COEFFT. (TB) OF RY ON RE∘.×CE IS 0.016564
TUKEY'S 1-D.F. TERM IS 22.796
REMAINING SUM OF SQUARES, D.F. AND M.S. ARE 13934 62 224.74

   *DISTRIBUTION SHAPE*
SKEWNESS:  G1 = ¯0.090818 ± 0.38034  (RATIO =    ¯.24)
KURTOSIS:  G2 = ¯1.3476 ± 0.82847
   (STANDARDIZED 0 3 MOMENT = 1.3131, ROUGH E.N.D. FOR G2 =  ¯2.45)
   (STANDARDIZED 2 1 MOMENT = 0.69664)

   *HETEROSCEDASTICITY TESTS*
AGAINST FITTED VALUES:  H = ¯0.00080806 ± 0.023377  (RATIO =   ¯.03)
VARIANCE FACTORS FOR ROWS (VFR) = 1.0973 1.2612 0.83736 0.89022 0.92346
      0.87246 0.8618 1.2562 ± 0.50128
   (ROUGH CHI-SQUARED =   .74,  6.71 D.F.)
VARIANCE FACTORS FOR COLUMNS (VFC) = 0.64815 0.97413 1.0165 0.76078 0.55
      193 1.3831 1.672 1.285 1.1566 0.55175 ± 0.55816
   (ROUGH CHI-SQUARED =  3.38,  8.19 D.F.)
```

Figure B:49

comes out at ¯1.35, and the equivalent normal deviate is ¯2.45, clearly significant. Because our crude approximation to the variance of the conditional distribution for g_1, given g_2, is negative, no equivalent normal deviate is quoted for g_1, given g_2.

These two data sets may be regenerated in a standard APL system. *X* was obtained in a clear workspace, without alteration of the link in the random-number generator. Then Y was obtained with the function *RNORMAL* (Appendix 3), the entries being rounded to two decimal places, and finally the two diagonal elements were changed to 5.

C. Kurtosis Tests Based on Time Series Innovations

The well-known methods for analysis of time series, whether in the time domain or in the frequency domain—for fitting parametric structures, for regression, for forecasting—all involve second-moment statistics. If all variables are jointly normally distributed in stationary sequences, simple first and second moments contain all the information. If not, there is the possibility that some of the needed information is not contained in the statistics used. When a random sequence is other than stationary and jointly normal, it may sometimes equally well be described and thought of as stationary but not jointly normal or as nonstationary.

Some of the series examined in Chapter 10 showed big movements in times of war, unlike their behavior at other times. A similar examination of series relating to copper production and copper prices (Anscombe, 1977) showed bigger movements during the 1920's and 1930's than at other times. Such inhomogeneity in behavior would be unusual in a jointly normal stationary random process, and may therefore be taken as evidence of nonnormality. Our technique of "harmonic regression" was largely influenced by the bigger movements, and did not reveal permanent associations between the series. In dealing with time series, a test for departure from joint normality may occasionally be useful, as signalling possible inadequacy of second-moment methods.

Definitions of kurtosis statistics. A given time series $\{x_t\}$, for some consecutive integer values for t, is supposed to be realized from a stationary sequence of random variables $\{\xi_t\}$, and we wish to test the hypothesis that the random variables are jointly normally distributed.

Univariate normality of the marginal distribution for ξ_t could be tested by making a histogram of the aggregate of given values $\{x_t\}$ or by calculating (for example) a kurtosis statistic,

$$b_2 = N\left(\Sigma_t(x_t - \bar{x})^4\right)/\left(\Sigma_t(x_t - \bar{x})^2\right)^2, \tag{34}$$

where N is the number of t-values and $N\bar{x} = \Sigma_t x_t$. To determine a significance level for b_2, proper account would have to be taken of the correlation structure of the sequence $\{\xi_t\}$.

Univariate marginal normality of a stationary random sequence does not imply joint normality, and the latter is what we are interested in here. Mardia (1970) has considered testing joint normality, given n independent observations of a p-variate distribution. His procedure involves linearly transforming the p-variate distribution so that it becomes spherical, and then he considers the n distances of the observations from their center of gravity and constructs a kurtosis statistic by comparing the sum of the fourth powers of the distances with the squared sum of the squared distances. An analogous way to treat a stationary time series would be to express the series in terms of independent identically distributed "innovations" and then calculate kurtosis statistics either from single innovations, or from pairs of consecutive innovations, or from triples of consecutive innovations, etc.

Our suggested procedure is, first of all, to try to represent the sequence in finite autoregressive form, say

$$(\xi_t - \mu) - \alpha_1(\xi_{t-1} - \mu) - \cdots - \alpha_p(\xi_{t-p} - \mu) = \eta_t \quad (t = 0, \pm 1, \pm 2, \ldots), \tag{35}$$

where $\mu, \alpha_1, \ldots, \alpha_p$ are constants and $\{\eta_t\}$ are independent identically distributed "error" random variables having a normal distribution with zero mean. For a given positive integer p, such a finite autoregressive structure can be estimated by performing ordinary linear regression of $\{x_t\}$ on $\{x_{t-1}\}, \{x_{t-2}\}, \ldots, \{x_{t-p}\}$, where we may say that $1 \leqslant t \leqslant n$ if the whole given series is of length $n + p$, that is, the given series is $\{x_t\}$ for $1 - p \leqslant t \leqslant n$. The residuals $\{u_t\}$ are the "innovations", estimating the errors $\{\eta_t\}$:

$$u_t = x_t - a_0 - a_1 x_{t-1} - \cdots - a_p x_{t-p} \quad (1 \leqslant t \leqslant n), \tag{36}$$

where $a_0, a_1, \ldots, a_p$ are the regression coefficients. The innovations depend on the choice of p. A possible method of choosing p is to calculate the empirical discrete spectrum of the given series (prewhitened and tapered), smooth it with a suitable moving average to form (after adjusting for the prewhitening) an estimated spectral density, and then try to approximate the reciprocal of the spectral density by a low-order polynomial in the cosine of the angular frequency; p is taken to be the degree of the polynomial. See Exercise 18 at the end of Chapter 10. We shall suppose that n is much larger than p. Formation of innovations has been recently discussed by Kleiner, Martin and Thomson (1979), for a different purpose.

For a given vector of innovations $\{u_t\}$, a kurtosis statistic can be defined from single innovations:

$$b_{21} = n\left(\sum_{t=1}^{n} u_t^4\right) \Big/ \left(\sum_{t=1}^{n} u_t^2\right)^2. \tag{37}$$

A kurtosis statistic defined from pairs of consecutive innovations is

$$b_{22} = n\left(\sum_{t=1}^{n-1} (u_t^2 + u_{t+1}^2)^2\right) \Big/ \left(\sum_{t=1}^{n} u_t^2\right)^2, \tag{38}$$

and one based on triples of consecutive innovations is

$$b_{23} = n\left(\sum_{t=1}^{n-2} (u_t^2 + u_{t+1}^2 + u_{t+2}^2)^2\right) \Big/ \left(\sum_{t=1}^{n} u_t^2\right)^2, \tag{39}$$

and so on.

If indeed the sequence $\{\xi_t\}$ is correctly described by an expression of the form (35), for some finite p, the left side of (35) is a sequence of independent identically distributed normal variables. Then if the correct value for p is used, the innovations $\{u_t\}$ will presumably seem to be realized from nearly independent identically distributed normal variables and the kurtosis statistics $b_{21}, b_{22}, \ldots$ should behave accordingly. In particular, if n is large, b_{21} is expected to be near to $\mathcal{E}(\eta_t^4)/(\mathcal{E}\eta_t^2)^2$, which is equal to 3, b_{22} is expected to be near to $\mathcal{E}((\eta_t^2 + \eta_{t+1}^2)^2)/(\mathcal{E}\eta_t^2)^2$, which is equal to 8, b_{23} is expected to be near to 15, etc.

If the sequence $\{\xi_t\}$ is jointly normal but, for some p, cannot be represented in the form (35), either because a larger value for p would be needed or because the sequence does not have a finite autoregressive expression, and if for the chosen value of p the parameters $\mu, \alpha_1, \ldots, \alpha_p$ are chosen to minimize the variance of the left side of (35), then the left side constitutes a stationary sequence of normal variables that are not independent. The innovations calculated for that p will presumably also seem to be realized from correlated normal variables. Correlation in the innovations may be expected to have less effect on b_{21} than on $b_{22}, b_{23}, \ldots$.

Distributions on the null hypothesis. Consider the null hypothesis that the autoregressive structure (35) is correct, with a known value for p. If the coefficients μ, $\alpha_1, \ldots, \alpha_p$ were known, the errors $\{\eta_t\}$ could be deduced from the realized series $\{x_t\}$ and used in place of the innovations $\{u_t\}$ in (37), (38), (39). Moments of the distributions for b_{21}, b_{22}, b_{23} could then be determined exactly, with some effort. The same would be true if the innovations $\{u_t\}$ were defined, not as at (36), but as residuals from regression of $\{x_t\}$ on explanatory variables that could be regarded as fixed, with independent identically distributed normal errors. Then, if the explanatory variables satisfied the mild conditions needed, the asymptotic results (13) would apply directly to b_{21}, and similar results would apply for b_{22} and b_{23}, irrespective of the values, and even of the number, of the explanatory variables:

$$\mathcal{E}(b_{21}) \sim \frac{3n}{n+2}, \quad \operatorname{var}(b_{21}) \sim \frac{24}{n+15}, \quad \gamma_1(b_{21}) \sim \frac{14.70}{\sqrt{n+29}}, \tag{40}$$

$$\mathcal{E}(b_{22}) \sim \frac{8n}{n+3}, \quad \operatorname{var}(b_{22}) \sim \frac{112}{n+15.7}, \quad \gamma_1(b_{22}) \sim \frac{14.04}{\sqrt{n+31.9}}, \tag{41}$$

$$\mathcal{E}(b_{23}) \sim \frac{15n}{n+4}, \quad \operatorname{var}(b_{23}) \sim \frac{296}{n+14.1}, \quad \gamma_1(b_{23}) \sim \frac{13.76}{\sqrt{n+40.6}}. \tag{42}$$

(Noninteger coefficients have been rounded to 1 or 2 decimal places.)

Now in fact (36) defines the innovations $\{u_t\}$ as residuals from autoregression of $\{x_t\}$. We cannot regard the explanatory variables as fixed without fixing all but one of the values of the dependent variable. Nevertheless, it seems plausible that the above asymptotic expressions are good approximations to the moments—though whether they are asymptotically correct within a factor $1 + O(n^{-2})$, as they would be if the innovations were residuals from ordinary regression, the writer has been unable to determine. In the absence of further information, let us boldly use these conjectured moments. The reader is reminded that n in the above expressions is the number of innovations, equal to the length of the given series $\{x_t\}$ minus the order of autoregression fitted.

The skewness measure (γ_1) is almost the same for b_{21}, b_{22} and b_{23}. Since the Type V (reciprocal of χ^2) approximation to the distribution of b_2 for ordinary regression residuals works quite well, the same kind of approximation may be expected to be satisfactory here.

Implementation. The tests have been implemented in the function *TSNT* (for time-series normality test) listed in Appendix 3. Equivalent normal deviates are found in the same way as in *TESTS*.

Power considerations. All the test statistics suggested above for ordinary regression residuals measured features of the data that were directly interesting, and no apology for considering them seemed called for. That is not quite so for these time-series tests. If we wish to examine joint normality in a time series, is measuring kurtosis of adjacent innovations the way to do it?

We saw in Exercise 9 at the end of Chapter 12 that Mardia's multivariate kurtosis test agreed well with visual judgment of scatterplots of samples of Plackett distributions having normal margins. As soon as the scatterplot showed one or two outliers in the emptier quadrants, Mardia's statistic responded. For time series some explicit consideration of power is not superfluous.

The tests are intended to be responsive to nonnormality in the joint distribution of the random sequence. They should preferably respond little to specification error in a jointly normal random process, that is, to choosing too low a value for the order p of the autoregressive structure fitted.

Suppose that p is chosen to be 1. For present purposes, the mean of the sequence may be set equal to 0. Then, if p is correct, the null hypothesis is that $\{\xi_t\}$ is a jointly normal stationary Markov sequence:

Hypothesis A: $\xi_t = \rho\xi_{t-1} + \eta_t$, *where* ρ *is constant* ($|\rho| < 1$) *and* η_t *is distributed* $N(0, 1 - \rho^2)$ *independently of* $\xi_{t-1}, \xi_{t-2}, \ldots$.

An alternative hypothesis involving marginal normality but not joint normality is that $\{\xi_t\}$ is a stationary Markov "jump" sequence:

Hypothesis B: *With probability* ρ/a, $\xi_t = a\xi_{t-1} + \eta_t$, *where* a *and* ρ *are constant* ($0 < \rho < a \leqslant 1$) *and* η_t *is distributed* $N(0, 1 - a^2)$ *independently of* $\xi_{t-1}, \xi_{t-2}, \ldots$; *and with probability* $1 - \rho/a$, $\xi_t = \eta_t^*$, *distributed* $N(0,1)$ *independently of* ξ_{t-1}, $\xi_{t-2}, \ldots$.

When $a = 1$, a realization of this sequence is quite unlike a realization of Hypothesis A with the same value for ρ. But when a is close to ρ, realizations differ in appearance only subtly: with Hypothesis B occasional large jumps are more frequent than with Hypothesis A. Something like this kind of joint nonnormality is sometimes observed in practice.

An alternative hypothesis involving joint normality but incorrect specification is

Hypothesis C: $\{\xi_t\}$ *is a jointly normal stationary autoregressive sequence of order greater than* 1, *or a jointly normal stationary moving-average sequence.*

In each case, if ρ is the lag-1 serial correlation coefficient, let

$$\zeta_t = \xi_t - \rho\xi_{t-1}.$$

Then as $n \to \infty$ the kurtosis statistics converge in probability:

$$b_{21} \to \frac{\mathcal{E}(\zeta_t^4)}{(\mathcal{E}\zeta_t^2)^2}, \quad b_{22} \to \frac{\mathcal{E}(\zeta_t^2 + \zeta_{t+1}^2)^2}{(\mathcal{E}\zeta_t^2)^2}, \quad b_{23} \to \frac{\mathcal{E}(\zeta_t^2 + \zeta_{t+1}^2 + \zeta_{t+2}^2)^2}{(\mathcal{E}\zeta_t^2)^2}. \tag{43}$$

Under Hypothesis A these limits are 3, 8 and 15, respectively.

Under Hypothesis B,

$$\frac{\mathcal{E}(\zeta_t^4)}{(\mathcal{E}\zeta_t^2)^2} = 3 + \frac{12\rho^3(a-\rho)}{(1-\rho^2)^2}, \tag{44}$$

$$\frac{\mathcal{E}(\zeta_t^2 + \zeta_{t+1}^2)^2}{(\mathcal{E}\zeta_t^2)^2} = 8 + \frac{4\rho(a-\rho)(1+4\rho^2+a\rho^3)}{(1-\rho^2)^2}, \tag{45}$$

$$\frac{\mathcal{E}(\zeta_t^2 + \zeta_{t+1}^2 + \zeta_{t+2}^2)^2}{(\mathcal{E}\zeta_t^2)^2} = 15 + \frac{4\rho(a-\rho)(2+a\rho+5\rho^2+a^2\rho^4)}{(1-\rho^2)^2}. \tag{46}$$

(The formidable polynomial manipulations have been computerized, and the above results are believed to be correct.)

For an asymptotic measure of power when n is large, the excesses of these expressions over the null-hypothesis values of 3, 8, 15 may be divided by the (conjectured) asymptotic standard deviations, $\sqrt{24/n}$, $\sqrt{112/n}$, $\sqrt{296/n}$, respectively. Suppose that $a > 0.9$ (say). Then b_{22} is more powerful than b_{21} when ρ is less than 0.75 about (the critical value varies a little with a); b_{22} is much more powerful when ρ is near 0; it is a little less powerful when ρ exceeds the critical value. And b_{23} is more powerful than b_{22} when ρ is less than 0.7 about.

Under Hypothesis C, ζ_t, ζ_{t+1}, ζ_{t+2} are jointly normally distributed with, in general, nonzero correlation. Let the lag-h serial correlation coefficient be δ_h. Then

$$\frac{\mathcal{E}(\zeta_t^4)}{(\mathcal{E}\zeta_t^2)^2} = 3, \quad \frac{\mathcal{E}(\zeta_t^2 + \zeta_{t+1}^2)^2}{(\mathcal{E}\zeta_t^2)^2} = 8 + 4\delta_1^2,$$
$$\frac{\mathcal{E}(\zeta_t^2 + \zeta_{t+1}^2 + \zeta_{t+2}^2)^2}{(\mathcal{E}\zeta_t^2)^2} = 15 + 8\delta_1^2 + 4\delta_2^2. \tag{47}$$

The sampling distributions for the kurtosis statistics will be affected by the lack of independence of $\{\zeta_t\}$, but the limiting value for b_{21} is not affected by the specification error. The limiting values for b_{22} and b_{23} are little affected if their excesses over the null-hypothesis values, namely $4\delta_1^2$ and $8\delta_1^2 + 4\delta_2^2$, are small compared with the respective standard deviations. The values of δ_1 and δ_2 can be estimated from the innovations $\{u_t\}$ as their lag-1 and lag-2 serial correlation coefficients.

Examples. In Figure B:50 *TSNT* is applied to the Yale enrolment series of Chapter 10. Logarithms are taken and a linear trend is subtracted, as at the top of Figure B:8. It was mentioned that the readings so detrended appeared in aggregate to be short-tailed—a kurtosis statistic such as at (34) deviates from expectation for normally distributed observations by being too low. However, the innovations tell a different story. The estimated spectrum of the series after prewhitening with the filter ($^-0.86$, 1), shown in the second column of Figure B:10, is almost flat. Entries in the first column were denoted by r_0 and in the second column by C_{00}.

```
      ⍝ ILLUSTRATIONS OF EXECUTION OF 'TSNT'  ~  ~  ~  ~

      Y←(⍳180) FIT 100×⍟ENROLMENT
MEANS ARE  90.5 732.28
REGRESSION COEFFICIENT IS 2.357
   WITH ESTIMATED STANDARD ERROR  0.032045   (178 D.F.)

      ⍴U←1,[1.5] ¯1↓Y
179 2
      ⍴U←(1↓Y)-U+.×⎕←(1↓Y)⌹U
0.13927 0.87506
179

      1↓C÷1↑C←6 AUTOCOV U
0.025168 ¯0.017813 0.026835 ¯0.0039529 ¯0.032629 0.017968

      TSNT U
B21  =  30.512
APPROXIMATE MEAN  =  2.9669,  S.E.  =  0.35173,  GAMMA1  =  1.0193
   (E.N.D.  =  8.33)
B22  =  67.651
APPROXIMATE MEAN  =  7.8681,  S.E.  =  0.75845,  GAMMA1  =  0.96678
   (E.N.D.  =  8.62)
B23  =  111.46
APPROXIMATE MEAN  =  14.672,  S.E.  =  1.2381,  GAMMA1  =  0.92854
   (E.N.D.  =  8.84)

      Y←¯60↓Y
      ⍴U←1,⍉ 0 ¯3 ↓ 2 1 0 ⌽(3,⍴Y)⍴Y
117 4
      ⍴U←(3↓Y)-U+.×⎕←(3↓Y)⌹U
0.057114 1.0365 0.072097 ¯0.16547
117

      1↓C÷1↑C←6 AUTOCOV U
¯0.054801 ¯0.020799 0.10662 0.0026017 ¯0.10615 ¯0.025097

      TSNT U
B21  =  3.9711
APPROXIMATE MEAN  =  2.9496,  S.E.  =  0.4264,  GAMMA1  =  1.2166
   (E.N.D.  =  1.95)
B22  =  10.353
APPROXIMATE MEAN  =  7.8,  S.E.  =  0.9187,  GAMMA1  =  1.1506
   (E.N.D.  =  2.16)
B23  =  19.64
APPROXIMATE MEAN  =  14.504,  S.E.  =  1.5026,  GAMMA1  =  1.0961
   (E.N.D.  =  2.47)
```

Figure B:50

Then at frequency $\lambda = r_0/160$ we estimate the spectral density of the series before prewhitening to be proportional to $C_{00}/(1 + a^2 - 2a\cos 2\pi\lambda)$, where $a = 0.86$. When the reciprocal of this estimated density is plotted against $\cos 2\pi\lambda$, the points fall close to a straight line. If a smooth curve, not quite straight, is drawn through the points, it has one point of inflexion and can be well represented by a cubic polynomial. Just what order of autoregressive relation is fitted to obtain the innovations seems to make little difference. A first-order (Markov) relation is fitted in Figure B:50, but nearly the same results would be obtained with a third-order autoregression. Obviously the three tests are saying that the detrended logarithmic enrolment could not possibly have arisen from a stationary jointly normal process.

Presumably it is the violent disturbances in the enrolment series during the two world wars that contribute most to the nonnormality. The lower part of Figure B:50 refers to the same series with the last 60 years omitted, that is, running 120 years from 1796 to 1915. If the reciprocal of estimated spectral density is plotted against $\cos 2\pi\lambda$, as before, the points do not lie so close to a straight line, and if a smooth curve is drawn through them with the same sensitivity as before it has one or two points of inflexion and can be well approximated by a cubic or quartic polynomial. There is now a more perceptible difference in results between fitting a first-order autoregression and fitting a third-order or fourth-order one. A third-order relation is fitted in Figure B:50. The kurtosis measures could be judged "significant", but are much more modest than for the whole series.

For this material the three kurtosis tests yield rather similar significance levels or equivalent normal deviates. They do not always do so, but obviously they are not independent and similar results are usual. We make no suggestion for amalgamating them into a combined test.

D. Which Residuals, How Far to Go?

We have alluded to the problem of deciding the order of autoregression to be fitted in obtaining the innovations. It is not unlike deciding where to stop in "stepwise" ordinary regression, mentioned in Chapter 11. Some problem of this kind is present in much statistical analysis. There is a dilemma. On the one hand we should be responsive to the features of the data. We do not approach examination of data with a completely open and vacuous mind, but neither should we be blinded by preconceptions. Tukey has powerfully encouraged an attitude of listening to what the data are trying to tell us. One does quite often notice unexpected things that, taken at face value, would alter our understanding. But how far should we go? Statistical theory, and computer simulation of specified random processes, warn that appearances can be deceptive. A sample has many individual features that do not reflect its source and would not persist if the sample size were much increased. Given adequate computing power, the question of how far to go outstrips available significance tests or other critical apparatus to aid our judgment so that we do not mock ourselves with falsehood. To refrain from examining the data because we do not know how to evaluate what we see, that surely is foolish. To assume without evaluation that everything seen is important, is foolish too. Fisher, so much less exposed than we are to the temptation to look rather thoroughly at the data, favored ignoring apparent effects that did not reach the 5% significance level when tested.

Sometimes with ordinary regression the question of how far to go does not seem troublesome. A factorial experiment may be designed to estimate many effects, all possibly interesting, many probably small and not interesting. The effects are usually orthogonal, and their meaning does not depend on which other effects are estimated. Provided the design stops short of being supersaturated, and there are some residual degrees of freedom for estimation of error, all the effects may be estimated and the small ones may then be ignored. Only if we wish to pool small effects with the error term to obtain a more precise estimate of error variance do we encounter a problem of some subtlety, traditionally ignored. Another of these questions is the choice of residuals for making tests of goodness of fit. Tukey's rule of thumb, to bring explanatory variables into the fitted regression relation only as far as to minimize the ratio of residual mean square to residual degrees of freedom (s^2/ν), usually disposes of the question satisfactorily. Other questions of how far to go in regression analysis can be very troublesome—as in observational studies where vast numbers of uncontrolled explanatory variables may be entertained.

To any question of how far to go in ordinary regression there are similar questions relating to autoregression and other kinds of regression of time series. Our answers must be less sure, because we have so much less knowledge concerning the properties of estimates and residuals.

This appendix concludes with an account of Tukey's rule, because of its fundamental bearing on tests of residuals in ordinary regression and its possible adaptation to tests of innovations in time-series autoregression.

Tukey's rule. (See the discussion of Anscombe, 1967, pp. 47–48.) Suppose that in a regression analysis there are not only the vectors on which regression is performed, say $\{1\}$, $\{x_i^{(1)}\}, \ldots, \{x_i^{(p-1)}\}$, but also further given vectors $\{x_i^{(p)}\}, \ldots,$ on which (at present) we do not perform regression; these vectors are all linearly independent. We suppose, in place of (1) above, that for all i

$$y_i = \left\{ \beta_0 + \beta_1 x_i^{(1)} + \cdots + \beta_{p-1} x_i^{(p-1)} \right\} + \eta_i + \epsilon_i, \tag{48}$$

where the β's are constants, $\{\eta_i\}$ is a linear combination of the further set of x-vectors, and $\{\epsilon_i\}$ is a vector of independent identically distributed "error" random variables having mean zero and variance σ^2, independent of all of the x-vectors. Let $\{q_{ij}\}$, $\{z_i\}$, ν, s^2 be defined as before. Then

$$z_i = \Sigma_j q_{ij}(\eta_j + \epsilon_j). \tag{49}$$

We are interested in the behavior of the errors $\{\epsilon_i\}$. Each error ϵ_i can be estimated by z_i/q_{ii}. The difference is the estimation error or "noise", which we should like to be small:

$$(z_i/q_{ii}) - \epsilon_i = \left[\Sigma_j q_{ij}\eta_j + \Sigma_j(1 - \delta_{ij})q_{ij}\epsilon_j\right]/q_{ii}. \tag{50}$$

Because Design Condition 1 is satisfied, $\Sigma_i q_{ij} = 0$ for all j. We shall now also impose Design Condition 2, that $q_{ii} = \nu/n$ for all i, which is usually nearly if not exactly satisfied. Then the sum of all the noises is

$$\Sigma_i\left[(z_i/q_{ii}) - \epsilon_i\right] = -\Sigma_i \epsilon_i, \tag{51}$$

and the expected value of this over the distribution of the ϵ's is 0. The mean squared noise is

$$(1/n)\Sigma_i\left[(z_i/q_{ii}) - \epsilon_i\right]^2 = (n/\nu^2)\Sigma_i z_i^2 - K, \tag{52}$$

where

$$\begin{aligned} K &= (2/\nu)\Sigma_i z_i \epsilon_i - (1/n)\Sigma_i \epsilon_i^2 \\ &= (2/\nu)\Sigma_{ij} q_{ij} \epsilon_i(\epsilon_j + \eta_j) - (1/n)\Sigma_i \epsilon_i^2. \end{aligned}$$

Taking the expectation over the distribution of the ϵ's, we find easily that $\mathcal{E}(K) = \sigma^2$. If we could assume that K was precisely equal to its expectation, we could immediately conclude from (52) that the best selection of x-vectors to include in the regression calculation would be the one that minimized $(\Sigma_i z_i^2)/\nu^2$, or s^2/ν.

Now if, for a given y-vector and a given set of x-vectors, we consider various choices of the x-vectors to be included in the regression, the rest being left to contribute to $\{\eta_i\}$, the error vector $\{\epsilon_i\}$ is always the same, the bias vector $\{\eta_i\}$ and the projection matrix $\{q_{ij}\}$ vary, and therefore K varies. It is only too likely that an unrestricted search over all possible subsets of x-vectors for inclusion in the regression, in order to minimize s^2/ν, will not lead to the minimum mean squared noise. It has therefore been suggested that the subsets examined should be restricted in whatever ways seem reasonable a priori. In the logarithmic analysis of Chapter 11 we considered only three subsets, referred to as (i), (ii), (iii). We found that analysis (ii) led to the smallest value for s^2/ν of these three, and we chose that for examination of residuals.

In the foregoing account of Tukey's rule, the x-vectors were supposed to be linearly independent, to avoid uninteresting reference to rank, but nothing was said about orthogonality or any kind of statistical independence between them. The ϵ's were asserted to be independent of each other and of the x-vectors, and that would imply approximate orthogonality of $\{\epsilon_i\}$ and $\{\eta_i\}$. The informativeness of the residuals $\{z_i\}$ concerning the errors $\{\epsilon_i\}$ has been measured (inversely) by what we have defined as the mean squared noise—this is clearly one summary measure, not necessarily perfectly attuned to all purposes.

Other criteria have been suggested for guiding "stepwise" regression, so that the response surface is well estimated and a good prediction can be made of an unobserved y, given the values of all the x-variables. At first glance, precision in prediction seems different from sensitivity in assessment of fit. However, prediction is based on fitted values, and assessment of fit on differences between the given y-values and the fitted values. Regression on a subset of x-vectors that leads to good fitted values will therefore permit, at the same time, good prediction and good assessment of fit. In fact, the criteria for prediction differ in details but do not behave very differently from Tukey's rule.

Suppose we wish to predict some future y, given the x-values. Precision will obviously depend on what x-values are entertained. Let us suppose that the set of x-values is equally likely to be any one of the n sets in the data. A rough-and-ready argument about precision now goes as follows. For the subset of x-vectors that is chosen, pretend that equation (1) holds. Then the variance of estimation of the fitted value is $p\sigma^2/n$. When σ^2 is added to this, we have the variance of the error in using the fitted value to predict the future y. The prediction variance is estimated by

$$(1 + p/n)s^2. \tag{53}$$

This will presumably be used in indicating the precision of the prediction. Thus to achieve what seems to be the most precise prediction, choose a subset of x-vectors that minimizes (53). If p is small compared with n, $(n + p)/n$ is nearly equal to n/ν, and we have approximately Tukey's rule.

To consider more carefully the errors actually incurred rather than crudely estimated, we need to start with (48) rather than (1). This has been done by

Mallows (1973), who has proposed a statistic C_p. It involves not only s^2 and p and ν for the subset chosen, but also an estimate of ultimate error variance $\hat{\sigma}^2$ obtained by doing regression on all the x-vectors. In place of (53) we obtain a linear function of C_p,

$$(1 + C_p/n)\hat{\sigma}^2 = (\nu s^2 + 2p\hat{\sigma}^2)/n. \tag{54}$$

Usually $\hat{\sigma}^2$ is only a little less than s^2. If we replace $\hat{\sigma}^2$ by s^2 in the right side of (54) we have immediately (53). Thus Mallows's C_p behaves much like Tukey's rule. What difference there is is in this direction: if C_p is minimized for a different subset than Tukey's s^2/ν, Mallows's subset will have the larger p (or smaller ν).

As for the "stepwise" fitting of an autoregression to a given time series, supposed stationary, Akaike (1969) has suggested minimizing prediction variance estimated in a similar way to (53), without expressing qualms about the unfixed state of the explanatory variables. He has subsequently extended the procedure in several directions. Parzen (1977) has proposed a more sophisticated-looking criterion, justified by asymptotic arguments; he addresses, not the error of prediction in the time domain, but the error of representation of the spectral density of the process by that of the approximating autoregressive process. Several authors have reported that Parzen's and Akaike's criteria are often minimized at the same order of autoregression.

We have conjectured that our kurtosis statistics based on time-series innovations would be distributed, on the null hypothesis, in approximately the same way as if the innovations had been residuals from an ordinary regression on fixed explanatory variables. If the conjecture is good, Tukey's simple rule may be expected to be useful in deciding the order of autoregression to be fitted. Let the given series be of length N. When a pth-order autoregression is fitted, the number of innovations $\{u_t\}$ is $n = N - p$. The residual degrees of freedom, after estimating $p + 1$ regression coefficients, are reckoned to be

$$\nu = n - (p + 1) = N - (2p + 1).$$

Then Tukey's statistic is

$$n(\Sigma_t u_t^2)/\nu^2, \tag{55}$$

where the factor n has been included because it is no longer constant. In fact, our choice in Figure B:50 of $p = 1$ for the whole detrended logarithmic enrolment series, and choice of $p = 3$ for the same when the last 60 members were dropped, are the ones that minimize this statistic for those series, and they seem to agree well with visual judgments about the behavior of the spectral density functions.

Appendix 3

ASP: A Statistical Package in A Programming Language

Designing a statistical package—a set of programs for statistical analysis of data—poses problems, some of them common to all computing, some perhaps peculiar to statistics.

Features of computing for statistical analysis, not common to all computing, are that many diverse operations are carried out on different sets of data, that several operations may be applied to any particular set of data, and that only after an operation has been applied and the output examined can the user see (usually) whether the operation was appropriate and satisfactory. Unexpected features of the output will sometimes remind the viewer of background information that he had lost sight of, or suggest a new approach or explanation. Statistical analysis is a process of trial and error, not to be specified simply and completely in advance. A statistical package is therefore naturally thought of as a kit of tools rather than as a single machine. Ingenuity may be devoted to organizing the tools into one or a few machines, but if, as with APL, the computing is conversational, flexibility is to be preferred to grandeur.

A prewritten package of programs is attractive to a user who does not consider himself expert in the language in which they are written. Programs in *VS APL*, such as those presented here, demand some knowledge of the language and especially some knowledge of the system commands and of how to manage workspaces. Given modest knowledge of that kind, the programs can be successfully used by beginners in computing who would not readily have composed equivalent programs themselves. Ultimately, however, the only good reason for using APL for statistical work is to have complete freedom to do what one wishes. These programs should be thought of as suggestions for how to proceed and as a source of examples of programming technique, rather than as prescriptions for good statistical practice. The programs are there to be altered and adapted. They are not locked. They are not proprietary. They are not sacred.

The package has been composed for the use of its author. He hopes it may prove useful to some others. Nothing has been put in merely because others might expect to find it there. The package is not intended primarily to assist persons who are beginners in both statistics and computing. For example, the function *SCATTERPLOT* (or *TSCP*) does not automatically label the axes. That is both because labeling is often unnecessary and also because when labeling is needed no completely automatic method is satisfactory. The author can quickly add labels and make other modifications to the output matrix, when these are desirable; whereas a beginner finds that troublesome. (Of the many fully labeled scatterplots that the author has made with these functions, none appears in this book, but Figures B:12 and 13 are examples of output of *TSCP* modified by two or three subsequent simple instructions.) A modified scatterplotting function with satisfactory semiautomatic labeling of axes could be composed and would obviously be needed by a less knowledgeable clientele.

APL has a rich vocabulary of primitive functions and other operations. No attempt has been made, for the benefit of a beginning user, to restrict programming to some subset of the whole language. Nearly all of the vocabulary is used somewhere in these programs, and much of it is used frequently. That in itself suggests that most of the things that a computer can do are needed sometimes in statistical work. Statistics does not call for a special subset of computer resources.

The package is written in the dialect known as *VS APL*. Some of the programs will execute in any APL system, many will not. Since the first public release of APL there have been only a few changes in the system-independent part of the language, but current implementations differ noticeably in system commands, system variables, system functions, shared variables and file access. Four system variables and one system function appear in these programs:

⎕*LC* (line counter), the statement numbers of functions in execution or halted, the most recently activated first.

⎕*PW* (printing width), the maximum number of characters in a line of output.

⎕*TC* (terminal control), a 3-element character vector consisting of the backspace, the carrier return and the linefeed.

⎕*WA* (working area), the number of bytes of available (unoccupied) space in the active workspace.

⎕*NC* (name classification), taking a character vector or matrix argument, gives the classification of the object named by the vector, or by each row of the matrix, coded: 0 = name not in use, 1 = label, 2 = variable, 3 = function, 4 = other (group name, not a usable name).

Anyone wishing to execute the programs in an implementation different from *VS APL* may have to change these as well as perhaps other elements of the language. Systematic changes are made easily through the canonical representation of functions, if something like that is available. Otherwise adaptation may require careful reading of the functions; but that is not all loss, for the user should read the functions carefully in any case.

Documentation. Programs (functions) and data sets to be kept some time or seen by others should be documented. The very least is to insert a line or two of comment in the definition of a function (as for example in *RNORMAL*—anyone who knows what is meant by random normal deviates should have no difficulty in using this function after seeing a print-out of the definition). The author's usual practice is to provide brief information about a data set under a name stored in the same workspace, consisting of the letters *WHAT* joined to the name of the data set, and information about a function under a name beginning with the letters *HOW*. Such documentation occupies least storage space, usually, if it is a variable consisting of a character vector. Examples are displayed below. Each *HOW*-variable is concluded with a date, the date of the most recent perceptible change made in it. The user may quickly discover the date of a *HOW*-variable by calling for the last 13 characters, and only display the whole if the date is later than that of his last print-out.

Such documentation does not attempt to explain statistical principles. More extended information, explanations of purpose and illustrations of use, can be given in other ways, as elsewhere in this book.

Precautions. As far as conveniently possible, functions should be written so that they will not break down (be suspended) in execution, but rather will terminate with a warning message if something is wrong. At least to begin the definition of a function by testing the arguments (if there are any) for compatibility seems to be good practice. This not only takes care of a common cause of failure in execution, but greatly assists a reader who is not fully informed about what the function does. If the user is puzzled to get the message '*NO GO.*' when he tries to execute the function, he can examine the first two or three lines of the definition to see what requirements have been placed on the arguments.

Consider for example the function *FIT*. As *HOWFIT* explains, this function is intended to subtract from a given vector (Y, say) its linear regression on another given vector (X, say). It could be used to detrend a time series by removing its linear regression on the time variable. The means, the regression coefficient and its conventionally estimated standard error, are printed out for information. The number of readings N is defined in line [1], and thereafter the desired calculations and output are specified in lines [3], [4], [5], [8], [9], [10], in a manner that should not prove difficult to read. Line [1] is primarily concerned with testing that the arguments are indeed vectors of equal length not less than 3, and if they are not line [2] causes a message to be displayed and execution to stop. Line [6] tests whether all members of the first argument (the "independent" variable) are equal. If they are, the vector X in this line consists of equal numbers almost or exactly zero, and the regression calculation in line [8] should not (perhaps cannot) be executed, and so execution of the function stops at line [7]. If execution stops at line [2], the explicit result Z is empty. If execution stops at line [7], the explicit result is just the second argument minus its mean. Only if both these traps are passed does the explicit result become the residuals from the fitted regression, as intended.

This function *FIT* has not been protected from every conceivable cause of breakdown. If the arguments were vectors of equal length not less than 3, but if either or both were vectors of characters rather than of numbers, execution would be suspended at line [3] or [4] where the sum reduction was called for. The numerical character of the arguments could have been tested in line [1]. The expression

$$(0 = 1\uparrow 0\uparrow ,X) \wedge 0 = 1\uparrow 0\uparrow ,Y$$

takes the value 1 only if both X and Y are numerical arrays. To supply a character argument to a function where a numerical argument is intended, or vice versa, is an unusual error in execution, scarcely worth guarding against. At any rate, it has not been guarded against here.

Another possible reason for breakdown in execution of a correctly written function is that there is insufficient room in the active workspace. If the space requirement were determined, as a function of size of arguments and any other relevant variables, this also could be tested for. That has generally not been done, but an example of a possibility of the sort may be seen in the function *FILTER*. The meat of the function is in the single line [9], which executes fast but needs room, sometimes a lot of room. The last three lines in the definition constitute an alternative program to line [9], slower to execute because of the loop but needing less space. In line [8] a test is made of available space, with a branch to [10] if [9] cannot be executed. (For this calculation it is supposed that the argument *X* consists of floating-point numbers, which it generally does in practice.) Breakdown in execution because of lack of space has not been guarded against completely, there may be too little room in the active workspace even for the more economical program, but the worst space problem in *FILTER* has been taken care of.

Suspensions of function execution will occur. Sometimes the user's attention is so fixed on remedying the cause or taking other action prompted by the suspension that he omits to terminate execution of the suspended function. Then the function's local variables supersede any global objects with the same names. The most annoying effect of this kind comes from statement labels, which cannot (during suspended execution of a function) be treated like ordinary names of variables and assigned new values. Preferably statement labels should be unlike any other names used. Accordingly all statement labels here begin with an underscored *L*. If the user refrains from ever assigning to a variable or function a name beginning with underscored *L* (no very irksome constraint, since underscoring is troublesome anyway), there can be no confusion.

Possibly there should be some similar distinguishing of all local variables in defined functions. That, however, is not so easy to achieve, partly because anyone developing a function should feel free to change the status of variables between local and global. For economy's sake, no local variable should have a name longer than three characters.* If global variables and functions intended for permanent storage are given descriptive names more than three characters long (the author's frequent but not invariable practice), at least there will be no confusion between them and local variables.

*There is a small unobvious economy in storage of functions to be derived from keeping the names of local variables short, mostly single characters. That is that in a collection of functions the same names of variables will occur again and again, for there are only 26 letters in the alphabet to choose from. In storage each *different* name appears just once in the symbol table, which therefore needs to be large enough to accommodate all the different names in the workspace—the names of groups, functions and variables, including the variables and statement labels in the function definitions. In the functions shown here, a handful of one-letter names account for many of the local variables; each appears several times in the definitions, but only once in the symbol table. Therefore these functions can be stored in workspaces with symbol tables reduced below the standard setting.

A particular kind of suspension occurs with quad input, or as the manual calls it, evaluated input. Only six of the functions shown here have this feature: *INTEGRATE*, *SCATTERPLOT*, *TSCP*, *DOWNPLOT*, *TDP*, *PD*3. In answering the first question in *SCATTERPLOT* or *TSCP*, the user will normally type two numbers, as suggested; but if instead he gives the name of a previously defined 2-element vector, this must be different from the names of all eleven local variables (not counting statement labels) in the function. The same applies to the third question in *TSCP*, where now there may be greater temptation to reply with the name of a previously defined character vector (*Z* would be permissible, or any name with more than one character). Similarly, if the function asked for in *INTEGRATE* is previously defined, its name must not be any one of *A*, *H*, *N*, *Z* (it could happily be *F*, or any name with more than one character). Perhaps those local-variable names should be underscored.

Architecture. The package of programs is not thought to constitute a statistical "language". The language is clearly APL; no APL primitives have been by-passed. Many of the problems facing a would-be language inventor have therefore been avoided. But as the collection of programs has grown some architectural questions have presented themselves. The author of the programs, not to mention any other person, finds remembering how to use the programs easier, on the whole, if what seem to be similar situations arising in different contexts are handled in similar ways. Two conventions are mentioned in the introductory statement, *DESCRIBE*, namely that abscissas are given before ordinates, and that when numerical vectors are stacked in a matrix they become the columns. (Because written English runs horizontally, character vectors representing names are more naturally stacked as rows of a matrix.) No attempt at a thorough analysis of similarities has been made; and indeed it is not always true that what seem to be similar situations are best handled the same way, because there are differences also. The two pairs of plotting functions, *SCATTERPLOT* and *TSCP*, and *DOWNPLOT* and *TDP*, all require information about scales and other matters, not contained in the arguments. After several revisions, the programs now resemble each other in that the user may either give the needed information in specially named global variables defined before execution of the functions or else let the functions interrogate him during execution. The latter seems to be the easier method for *SCATTERPLOT* and *TSCP*, the former for *DOWNPLOT* and *TDP*. The reason for the difference is that the interrogations in *SCATTERPLOT* and *TSCP* are easily answered, except perhaps for the last in *TSCP*, but more thought is usually required to set up the downplotting functions and one might as well specify the global control variables (if their names are remembered) before execution. There also seems to be a difference in typical multiple use of these functions. One sometimes makes several successive uses of one of the downplotting functions with the same scales and labeling (as in Figures B:18–23), whereas successive uses of one of the scatterplotting functions are likely to need different control settings. For better or worse, if the user is interrogated by *DOWNPLOT* or *TDP*, global control variables are defined that will control any subsequent execution of the function in the same workspace, unless the user resets or erases them; but when *SCATTERPLOT* or *TSCP* interrogates the user the computer does not remember the answers next time.

Algorithms. Most of the functions have been developed independently of anyone else's programming. An exception is *INTEGRATE*, based on a program by Richardson (1970) for integration by Simpson's rule. (Two-point Gaussian quadrature is often preferable to Simpson's rule on grounds of simplicity and accuracy.) The function *JACOBI* was developed from two APL programs written by others, but it is not now close to either source.

Comments are made elsewhere in the book on some of the algorithms. A few further remarks are made here.

The loop at the end of *SCATTERPLOT* could be avoided by an outer product, but that would call for considerable space. This perhaps provides a counterexample to the general principle of avoiding loops if possible.

One should beware of supposing that comparisons of alternative algorithms, executed in compiled Fortran, are relevant to apparently similar algorithms executed in APL. Because of its brevity, the function *RNORMAL*, using the algorithm of Box and Muller (1958), seems to be about the fastest possible generator of random normal deviates in APL. A peculiarity of the program may be noted, that all the radii needed are generated first and then all the angles. A consequence of this is that if on two occasions, in a clear workspace, *RNORMAL* is executed, the first time with argument 100, the second time with argument 101, all the deviates produced the first time will differ from all those produced the second time, even though almost all the radii and angles will be the same—they are paired differently.

The functions *NIF* and *INIF* use well-known approximations to the normal integral and its inverse that are masterpieces of the numerical analyst's art. In ignorance of any similar achievement for the χ^2 integral or incomplete gamma function, *CSIF* is based directly on a standard convergent series and asymptotic series, used respectively for lower and for higher values of the argument. The function is intended for a modest number of degrees of freedom, not necessarily integer, say less than 30. An absolute error bound at 10^{-7} is met, but often much greater accuracy is obtained at the cost of more than minimal computing.

The function to be evaluated, in the notation of Abramowitz and Stegun (1964), is the incomplete gamma function

$$P(a,z) = \int_0^z e^{-t} t^{a-1} dt / \Gamma(a), \tag{1}$$

where a is one-half the degrees of freedom and z is one-half the χ^2 value. We assume $a > 0$, $z > 0$. When a is an integer,

$$P(a,z) = 1 - e^{-z}\left\{ \frac{z^{a-1}}{(a-1)!} + \frac{z^{a-2}}{(a-2)!} + \cdots + 1 \right\}. \tag{2}$$

An approximation based on the asymptotic series, for a not an integer, is

$$P(a,z) \approx 1 - e^{-z}\left\{ \frac{z^{a-1}}{(a-1)!} + \frac{z^{a-2}}{(a-2)!} + \cdots + \frac{(1/2)z^{a-r}}{(a-r)!} \right\}, \tag{3}$$

where the last term in this finite expression is one-half the corresponding term in the infinite asymptotic expansion. Provided $r > a$, the error E_1 satisfies

$$|E_1| < 1/|2e^z z^{r-a}(a-r)!|. \tag{4}$$

For any a and z, this bound for $|E_1|$ is minimized when r is the least integer not less than $a + z$. If $z \geqslant 7.107$, $|E_1| < 10^{-7}$ provided r is the least integer not less than $a + 7.107$, the most unfavorable value for a being any integer plus 0.501. An approximation based on the convergent series is

$$P(a,z) \approx e^{-z}\left\{ \frac{z^a}{a!} + \frac{z^{a+1}}{(a+1)!} + \cdots + \frac{z^{a+r-1}}{(a+r-1)!}\left[1 + \frac{z}{2(a+r-z)}\right]\right\}. \tag{5}$$

The last term here has been augmented by one-half of an upper bound to the sum of the omitted terms. Provided $z < a + r$, the error E_2 satisfies

$$|E_2| < \frac{z^{a+r}}{2e^z[1 - z/(a+r)](a+r)!}. \tag{6}$$

For $z < 7.107$, $|E_2| < 10^{-7}$ if r is the least integer not less than $24.75 - a$. *CSIF* uses (3) as much as possible, down to the critical value 7.107 for z, and uses (5) only when $z < 7.107$. When a is small this is a good choice, because (3) requires fewer terms than (5), 8 instead of 25 for a near 0. When $a = 9$ (18 degrees of freedom), (3) and (5) are about equal in effort. By the time $a = 15$ (30 degrees of freedom) (3) is doing the lion's share of the work; but the number of terms to be computed would not be drastically reduced by raising the lower limit for z at which (3) was used rather than (5). One may say that for degrees of freedom up to 20 or 30, *CSIF* is roughly as economical as any procedure can be, that accepts an unrestricted vector of χ^2 values for one value of the degrees of freedom, and that is based directly on these expansions, without use of economized (trigonometric) polynomials or equivalent ingenuity. The algorithm of Bhattacharjee (1970) does not evaluate error bounds, and therefore does not necessarily attain the intended precision, as he notes.

The package. The following pages show documentary variables and function definitions. In alphabetical order, with the page number on which they appear, they are:

Documentary variables

Function definitions

```
      )LOAD 1234 ASP
SAVED 11:50:17 02/13/81
WSSIZE IS 139696

      )FNS
ANALYZE COR     DOWNPLOT        EFFECT  FIT     QUANTILES       REGR    REGRINIT
ROWCOL  ROWCOLDISPLAY   ROWCOLPERMUTE   SCATTERPLOT     SHOW    STDIZE  STRES
SUMMARIZE       VARIANCE
      )VARS
DESCRIBE        HOWDOWNPLOT     HOWFIT  HOWPLOT HOWREGRESSION   HOWSCATTERPLOT
HOWSUMMARIZE    A̲
      )GRPS
NWAYGROUP       REGRESSIONGROUP ROWCOLGROUP
      )GRP NWAYGROUP
ANALYZE EFFECT
      )GRP REGRESSIONGROUP
STRES VARIANCE REGR SHOW REGRINIT
      )GRP ROWCOLGROUP
ROWCOLDISPLAY ROWCOLPERMUTE ROWCOL

      )SYMBOLS
IS 110; 96 IN USE
      ⎕WA
98784



      DESCRIBE

          ASP - A STATISTICAL PACKAGE IN A PROGRAMMING LANGUAGE
          ↑↑↑ . ↑ ↑.::..::..: ↑.::..: .. * *::..::..:: *::..::!

   SOME PROGRAMS FOR STATISTICAL ANALYSIS BY F. J. ANSCOMBE, WHO WILL APPRECIATE
COMMENTS AND ERROR REPORTS.  (PHONE:  203-432-4752.)

   THE FUNCTIONS RESIDE IN WORKSPACES NAMED 'ASP', 'ASP2', 'ASP3'.   SOME OF THE
FUNCTIONS ARE ORGANIZED IN GROUPS.    BEGIN BY LISTING ALL FUNCTIONS, GROUPS AND
VARIABLES IN EACH WORKSPACE.   A FEW OF THE FUNCTIONS IN 'ASP3' ARE UNDER DEVEL-
OPMENT AND NOT PROPERLY DOCUMENTED.

   SOME OF THE FUNCTIONS ARE DESCRIBED IN DOCUMENTARY VARIABLES HAVING NAMES BE-
GINNING WITH 'HOW'.  FOR PROPER DISPLAY OF THESE VARIABLES SET
      ⎕PW←120
OTHER FUNCTIONS HAVE A COMMENT LINE AT THE END OF THEIR DEFINITION.   THE VECTOR
A̲  IS THE ALPHABET, FIRST PLAIN THEN UNDERSCORED.    1-ORIGIN INDEXING IS ALWAYS
SUPPOSED.

   TWO CONVENTIONS SHOULD BE NOTED:    (1) ABSCISSAS ARE ALWAYS MENTIONED BEFORE
ORDINATES (AS WITH THE ARGUMENTS OF 'REGRINIT' AND 'SCATTERPLOT'), (2) WHEN NUM-
ERICAL VECTORS ARE STACKED IN A MATRIX, THEY ARE THE COLUMNS (AS WITH THE RESULT
OF 'JACOBI' AND THE ARGUMENT OF 'DOWNPLOT').

DEPT. OF STATISTICS, YALE UNIV., NEW HAVEN, CT  06520 -- 5 FEB. 1977 ::..::..::!


      A̲
ABCDEFGHIJKLMNOPQRSTUVWXYZA̲B̲C̲D̲E̲F̲G̲H̲I̲J̲K̲L̲M̲N̲O̲P̲Q̲R̲S̲T̲U̲V̲W̲X̲Y̲Z̲
```

```
      )LOAD 1234 ASP2
SAVED 11:56:24 02/13/81
WSSIZE IS 139696

      )FNS
AUTOCOV CONTINGENCY     CW      FFT     FILTER  FOURFOLD        HARINIT HAR1
HAR1R   HAR2    HUBER   HUBER1  HUBER2  MA      MAV     MULTIPOLY       POLAR
POOL    PREHAR  TAPER   TDP     TSCP
      )VARS
DPH     DPL     DPM     DPS     FMT1    FMT2    HOWCONTINGENCY  HOWFILTER
HOWFOURIER      HOWHAR  HOWHUBER        ITS     TOL
      )GRPS
HARGROUP        HUBERGROUP
      )GRP HARGROUP
DPM POLAR DPL PREHAR MAV ITS TAPER FFT HARINIT FMT2 FMT1 TOL HAR1 HAR2 DPH TDP
DPS
      )GRP HUBERGROUP
HUBER HUBER1 HUBER2
```

```
      )LOAD 1234 ASP3
SAVED 12:02:33 02/13/81
WSSIZE IS 139696

      )FNS
BARTLETT        BIV     CCD     COLLECT CSIF    CTG2    END     GAIN    GAINLAG
HHT     INIF    INTEGRATE       ISOTROPY        JACOBI  LFACT   MARDIA  MAX
MP      NIF     PD2     PD3     PP      RLOGISTIC       RNORMAL SAMPLE  TESTS
TESTS1  TESTS2  TEXT    TSNT    T7A     T7B     T7INIT  T7LF    T7S
      )VARS
HOWBARTLETT     HOWGAIN HOWHOUSEHOLDER  HOWINTEGRATE    HOWISOTROPY
HOWJACOBI       HOWMARDIA       HOWMAX  HOWPD   HOWTESTS        HOWT7
      )GRPS
TESTSGROUP      T7GROUP
      )GRP TESTSGROUP
TESTS TESTS1 TESTS2
      )GRP T7GROUP
T7B T7A T7INIT T7LF T7S
```

```
      ∇ Z←RNORMAL N;V
[1]     →2+∨/(N≥1)∧(N=⌊N)∧1=ρN←,N
[2]     →ρ⎕←'NO GO.',Z←''
[3]     Z←(÷2147483647)×?(2,⌈N÷2)ρ2147483647
[4]     Z←N↑(, 1 2 ∘.○Z[2;]×○2)×V,V←(¯2×⍟Z[1;])*0.5
[5]   ⍝ N  RANDOM NORMAL DEVIATES BY BOX-MULLER METHOD.
      ∇
```

```
      ∇ Z←RLOGISTIC N
[1]     →2+∨/(N≥1)∧(N=⌊N)∧1=ρN←,N
[2]     →ρ⎕←'NO GO.',Z←''
[3]     Z←(÷○3*¯0.5)×⍟Z÷1-Z←(?Nρ2147483647)÷2147483647
[4]   ⍝ N  RANDOM LOGISTIC DEVIATES, MEAN 0, VARIANCE 1.
      ∇
```

```
HOWPLOT

        FUNCTIONS FOR PLOTTING
        (1)  W←U SCATTERPLOT V
        (2)  W←U TSCP V
        (3)  ROWCOLDISPLAY I
        (4)  DOWNPLOT V
        (5)  Z TDP V

   THE FIRST THREE ARE DESCRIBED IN 'HOWSCATTERPLOT',  THE LAST TWO IN 'HOWDOWN-
PLOT'.

   EXAMPLES.  TO SEE SOME OF THE POSSIBILITIES AND COMPARE THESE FUNCTIONS, COPY
THE FUNCTIONS INTO A CLEAR WORKSPACE AND TRY THE FOLLOWING.    A  IS A VECTOR OF
LETTERS OF THE ALPHABET; IT MAY BE COPIED FROM 'ASP'.

      V←¯5+⍳20
      +V←V,[1.5] V+¯3+?20⍴5

      DPL←4⍴5                               OR ALTERNATIVELY
      DPS← ¯9.5 0.5 20                      |      DPS← ¯9 1 20
      DPH←(5⍴' '), 10 0 ⍕5×¯2+⍳6            |      DPH←(5⍴' '), 10 0 ⍕ 0 10 20

      DOWNPLOT V
      'O' TDP V
      '∘O' TDP V
      (20↑A) TDP V
      ('∘',[1.5] 20↑A) TDP V

      SCPSIZE← 20 25
      SCPSCALES←1
      SCPSYMBOLS←(20⍴'∘'),20↑A

      ((⍳20),⍳20) SCATTERPLOT V[;1],V[;2]
      ((⍳20),⍳20) TSCP V[;1],V[;2]

   ALL SEVEN PLOTS ARE ESSENTIALLY THE SAME.  NOTICE THE SMALL DIFFERENCES.

19 JAN. 1977
```

```
      ∇ Y←STDIZE X
[1]     →2+∨/X≠1↑X←,X
[2]     →⍴⎕←'NO GO.',Y←''
[3]     Y←Y×((÷¯1+⍴X)×Y+.×Y←X-(+/X)÷⍴X)*¯0.5
[4]   ⍝ THE VECTOR  X  IS RESCALED TO HAVE ZERO MEAN AND UNIT VARIANCE.
      ∇
```

```
      ∇ Z←X COR Y;N
[1]     →2+(X∨.≠1↑X)∧(Y∨.≠1↑Y)∧(⍴X←,X)=N←⍴Y←,Y
[2]     →⍴⎕←'NO GO.',Z←''
[3]     X←X-(+/X)÷N
[4]     Y←Y-(+/Y)÷N
[5]     Z←(X+.×Y)÷((X+.×X)×Y+.×Y)*0.5
[6]   ⍝  Z  IS THE CORRELATION COEFFICIENT BETWEEN VECTORS  X  AND  Y .
      ∇
```

HOWSCATTERPLOT

FUNCTIONS FOR PLOTTING
(1) *W←U SCATTERPLOT V*
(2) *W←U TSCP V*
(3) *ROWCOLDISPLAY I*

'SCATTERPLOT' IS USED FOR DISPLAYING CORRESPONDING MEMBERS OF TWO VECTORS. 'TSCP' (TRIPLE SCATTERPLOT) IS A MODIFICATION OF 'SCATTERPLOT' IN WHICH A THIRD DIMENSION IS SUGGESTED BY VARYING THE SYMBOLS USED IN PLOTTING THE POINTS. 'ROW-COLDISPLAY' IS A SPECIAL-PURPOSE FUNCTION SOMEWHAT LIKE 'TSCP'. ALL THREE FUNC-TIONS GENERATE THE WHOLE CHARACTER ARRAY IN THE WORKSPACE. THE ARRAY MAY SUBSE-QUENTLY BE LABELED OR MODIFIED.

(1) *SCATTERPLOT. U AND V ARE VECTORS OF EQUAL LENGTH. CORRESPONDING MEM-BERS U[J] AND V[J] ARE PLOTTED AS ABSCISSA AND ORDINATE OF A POINT. A SINGLE POINT IS SHOWN AS* ○, *TWO COINCIDENT POINTS AS* ⊖, *THREE OR MORE COINCIDENT POINTS AS* ⊛, *UNLESS THE MULTIPLICITY CODE IS SET DIFFERENTLY (SEE BELOW).*

TO GET A SINGLE PLOT DO NOT NAME AN EXPLICIT RESULT BUT EXECUTE THUS:
U SCATTERPLOT V
TO GET A PLOT AND ALSO STORE THE CHARACTER ARRAY FOR LATER EDITING EXECUTE THUS:
⎕←W←U SCATTERPLOT V

INFORMATION ABOUT SIZE IS NEEDED. YOU MAY SPECIFY GLOBAL NUMERICAL VARIABLES SCPSIZE AND SCPSCALES BEFORE EXECUTION; OR ELSE THE INFORMATION WILL BE ASKED FOR. SCPSIZE IS A 2*-ELEMENT VECTOR LISTING THE NUMBER OF ABSCISSA VALUES AND THE NUMBER OF ORDINATE VALUES, THAT IS, THE HORIZONTAL AND VERTICAL DIMENSIONS OF THE PLOT. SCPSCALES IS A SCALAR INDICATING WHETHER THE HORIZONTAL AND VERT-ICAL SCALES ARE TO BE SET EQUAL. IF THE SCALES ARE NOT MADE EQUAL, THE ABSCISSA VALUE SET IS EQUAL-SPACED, THE LEAST EQUAL TO* ⌊/*U* , *THE GREATEST TO* ⌈/*U* , *EX-CEPT THAT IF THESE EXTREMES DO NOT HAVE THE SAME SIGN THE VALUE SET IS MINIMALLY TRANSLATED WITHOUT CHANGE OF SCALE SO THAT* 0 *BECOMES A MEMBER OF THE SET. SIMI-LARLY FOR ORDINATES. IF THE SCALES ARE MADE EQUAL, EITHER THE ABSCISSA VALUE SET OR THE ORDINATE VALUE SET MAY BE SHORTENED, TO AVOID UNNECESSARY BLANK CO-LUMNS OR ROWS IN THE PLOT. THE COORDINATES OF THE PLOTTED POINTS ARE ROUNDED TO THE NEAREST MEMBERS OF THE COORDINATE VALUE SETS. THE AXES ARE PRINTED AS* ×*'S, EXCEPT THAT IF* 0 *OCCURS IT IS SO SHOWN.*

TO CONTROL THE SIZE OF THE PLOT AND HAVE ABSCISSA AND ORDINATE UNIT STEPS E-QUAL TO NICE ROUND NUMBERS, A POSSIBLE DEVICE IS TO CATENATE ONTO EACH ARGUMENT (U AND V) TWO ADDITIONAL VALUES, BOTH 'ROUND NUMBERS', ONE GREATER, THE OTHER LESS, THAN ALL OTHER ELEMENTS OF THE ARGUMENT, AND THEN CHOOSE SCPSIZE TO GET DESIRED STEP SIZES. THE TWO SPURIOUS POINTS CAN LATER BE EDITED OUT OF THE DIS-PLAY, AND THE AXES CAN BE CHANGED TO ANY DESIRED STYLE AND LABELED.

TO CHANGE THE MULTIPLICITY CODE, SPECIFY THE GLOBAL CHARACTER VECTOR SCPMULT BEFORE EXECUTION. FOR EXAMPLE, TO SHOW MULTIPLICITIES BY NUMERALS, 1 *TO* 9, *WITH X FOR* 10 *OR MORE, ENTER:*
SCPMULT←'X987654321'
ALL THE CHARACTERS MUST BE DIFFERENT, AND NONE EQUAL TO A BLANK OR | *OR* - . *THE ORDER OF LISTING IS OF DECREASING MULTIPLICITY.*

THE TERMINAL IS SUPPOSED SET AT SINGLE SPACING. THE LAST LINE [37] *CAUSES A HORIZONTAL SPACING OF THE DISPLAY; YOU MAY DELETE THIS LINE, IN WHICH CASE THE* '2×' *IN THE MIDDLE OF LINE* [10] *SHOULD BE DELETED ALSO. THE LOCAL VARIABLE S IN LINE* [17] *SHOULD GIVE THE RELATIVE MAGNITUDES OF THE HORIZONTAL AND VERTICAL*

STEPS, AS PRINTED BY YOUR TERMINAL. WITH 6-PER-INCH LINEFEED, THE CORRECT SETTINGS FOR S ARE

```
          FOR 12-PITCH CHARACTER SPACING, LAST LINE INTACT:    S← 1 1
          FOR 10-PITCH CHARACTER SPACING, LAST LINE INTACT:    S← 6 5
          FOR 12-PITCH CHARACTER SPACING, LAST LINE DELETED:   S← 1 2
          FOR 10-PITCH CHARACTER SPACING, LAST LINE DELETED:   S← 3 5
```

(SAME FOR 'TSCP' , THE LAST LINE BEING NUMBERED [41].)

(2) TSCP. ALL COINCIDENT POINTS ARE SHOWN AS *, REGARDLESS OF MULTIPLICITY. THE SYMBOLS TO BE USED MAY BE SPECIFIED BEFORE EXECUTION AS THE GLOBAL CHARACTER VARIABLE SCPSYMBOLS , OR ELSE THEY WILL BE ASKED FOR. IF SCPSYMBOLS IS SCALAR, SUCH AS 'o', THE RESULT WILL BE LIKE THAT OF 'SCATTERPLOT' EXCEPT THAT ALL MULTIPLE POINTS ARE SHOWN BY THE ONE SYMBOL * . OTHERWISE SCPSYMBOLS MUST BE A VECTOR OF LENGTH EQUAL TO ρU AND ρV , NONE OF THE CHARACTERS BEING A BLANK OR * . FOR A VECTOR V , EXECUTING

```
      SCPSYMBOLS←Z
      (ιρV) TSCP V
```

YIELDS A PLOT EQUIVALENT TO

```
      Z TDP V
```

IF THE SCALES ARE SUITABLY MATCHED. BECAUSE THE MINUS SIGN MAY BE NEEDED AS A PLOTTING SYMBOL, ZERO LINES ARE NOT INDICATED IN THE PLOT WITH | OR - , AS THEY ARE IN 'SCATTERPLOT'.

(3) ROWCOLDISPLAY IS A SPECIAL FUNCTION, SUBSTITUTING FOR 'TSCP', THAT MAY BE USED TO DISPLAY THE OUTPUT OF 'ROWCOL'. ABSCISSAS AND ORDINATES ARE THE COLUMN EFFECTS AND THE ROW EFFECTS. THE SYMBOLS PLOTTED REPRESENT THE RESIDUALS GRADED ON A 7-POINT SCALE FROM LARGE-NEGATIVE TO LARGE-POSITIVE: M̲, M, -, ∘, +, P, P̲.

AT THE EXPENSE OF ACCURACY IN POSITIONING THE PLOTTED POINTS, COINCIDENCES OF ROWS OR COLUMNS ARE AVOIDED THROUGH A UNIT MINIMUM SEPARATION, EVEN WHEN TWO ROW EFFECTS OR COLUMN EFFECTS ARE EQUAL.

THE ARGUMENT (I) IS THE CHANGE IN COLUMN EFFECT REPRESENTED BY A UNIT HORIZONTAL DISPLACEMENT. TRY EXECUTING THE FUNCTION WITH A LARGISH VALUE FOR I, NOT LESS THAN ((⌈/CE)-⌊/CE)÷20 . IF THE DISPLAY IS TOO SMALL REPEAT WITH A SMALLER VALUE FOR I .

THE FUNCTION REFERS TO THE GLOBAL VARIABLES CE, RE, RY GENERATED BY 'ROWCOL', BUT THESE SHOULD FIRST BE REARRANGED BY EXECUTING THE FUNCTION 'ROWCOLPERMUTE' (NO ARGUMENTS OR EXPLICIT RESULT), WHICH PERMUTES THE ROWS AND COLUMNS SO THAT CE IS IN ASCENDING ORDER AND RE IN DESCENDING ORDER.

ADJUSTMENTS IN LINE [1]: TO SUPPRESS HORIZONTAL DOUBLE SPACING SET (D←1). FOR 6-PER-INCH LINEFEED, 12-PITCH CHARACTER SPACING, IF D←2 THEN S← 1 1 , BUT IF D←1 THEN S← 1 2 . FOR 10-PITCH SPACING, SET S← 6 5 AND S← 3 5 , RESP.

6 APR. 1979

```
      ∇ Z←P QUANTILES V
[1]     →2+(∧/(1>P)∧0<P←,P)∧1=ρρV
[2]     →ρ⎕←'NO GO.',Z←''
[3]     P←(ρV)⌊1⌈0.5+P×ρV←V[⍋V]
[4]     Z←V[⌊P]+(V[⌈P]-V[⌊P])×P-⌊P
[5]   ⍝ QUANTILES OF THE VECTOR  V  FOR GIVEN PROPORTIONS  P .
      ∇
```

```
    HOWDOWNPLOT

        FUNCTIONS FOR PLOTTING
        (1)  DOWNPLOT V
        (2)  Z TDP V

   'DOWNPLOT'  IS USED FOR PLOTTING MEMBERS OF ONE OR MORE VECTORS AGAINST THEIR
INDEX NUMBERS.  'TDP' (TRIPLE DOWNPLOT) IS A MODIFICATION OF 'DOWNPLOT' IN WHICH
A 3RD DIMENSION IS SUGGESTED BY VARYING THE SYMBOLS USED IN PLOTTING THE POINTS.
BOTH FUNCTIONS GENERATE AND PRINT THE CHARACTER ARRAY LINE BY LINE; THE ARRAY IS
NOT STORED IN THE WORKSPACE.

   EACH FUNCTION REQUIRES TWO GLOBAL NUMERICAL VARIABLES,  DPS  (DOWNPLOT SCALE)
AND  DPL  (DOWNPLOT LABELING).    IF THESE HAVE NOT ALREADY BEEN SPECIFIED, THEY
ARE ASKED FOR.  DPS  IS A 3-ELEMENT VECTOR INDICATING THE PLOTTING SCALE FOR VA-
LUES OF THE RIGHT ARGUMENT (V): THE LOWEST VALUE, THE STEP SIZE, AND THE HIGHEST
VALUE TO BE ACCOMMODATED.  HIGHEST MINUS LOWEST MUST BE DIVISIBLE BY STEP SIZE,
AND THE NUMBER OF STEPS MUST BE LESS THAN THE PRINTING WIDTH (⎕PW).   DPL  IS A
4-ELEMENT VECTOR INDICATING THE LABELING OF THE ROWS.   THE DPL[1]'TH ROW OF THE
PLOT IS LABELED  DPL[2] ,  AND THEREAFTER EVERY DPL[3]'TH ROW IS LABELED WITH AN
INCREMENT OF  DPL[4] .  DPL[1 3]  MUST BE POSITIVE INTEGERS, THE SECOND NOT LESS
THAN THE FIRST.  THE LABELS ARE PRINTED WITH AT MOST 4 CHARACTERS.

   THE RULINGS  ACROSS THE TOP AND BOTTOM OF THE PLOT  MARK EVERY TENTH POSITION
WITH  + ,  UNLESS BEFORE EXECUTION THE GLOBAL CHARACTER VARIABLE  DPM  (DOWNPLOT
MARKING) HAS BEEN SPECIFIED OTHERWISE THAN IN LINE [¯2+L1].  LABELING OF VALUES
ACROSS THE TOP IN ACCORDANCE WITH  DPS  (THIS IS NOT DONE AUTOMATICALLY), OR ANY
OTHER HEADING, CAN OPTIONALLY BE HAD BY SPECIFYING THE GLOBAL CHARACTER VARIABLE
DPH  (DOWNPLOT HEADER) BEFORE EXECUTION.

   THE RIGHT ARGUMENT (V) IS EITHER A VECTOR OR A MATRIX.    IF  V  IS A VECTOR,
V[J]  IS PLOTTED AGAINST  J .  OTHERWISE,  V[J;]  IS PLOTTED AGAINST  J .   MUL-
TIPLICITY IS NOT INDICATED.

   (1) DOWNPLOT.  POINTS ARE SHOWN AS  ○ .  ← OR → MEANS THAT AT LEAST ONE VALUE
WAS OUTSIDE THE PLOTTING RANGE.

   (2) TDP.    IF ANY VALUES OF  V  FALL OUTSIDE THE PLOTTING RANGE SPECIFIED BY
DPS  THEY ARE MISSED (NO WARNING ← OR →).   THE LEFT ARGUMENT (Z) IS A CHARACTER
SCALAR OR VECTOR OR MATRIX,  INDICATING THE SYMBOLS  TO BE USED  IN PLOTTING THE
POINTS.  IF  Z  IS SCALAR, THE SAME SYMBOL IS USED EVERY TIME; EXECUTING
      '○' TDP V
YIELDS THE SAME RESULT AS
      DOWNPLOT V
PROVIDED ALL MEMBERS OF  V  ARE IN THE PLOTTING RANGE.     IF  Z  IS A VECTOR OF
LENGTH  1↑⍴V , IT INDICATES THE SYMBOL TO BE USED FOR EACH ROW OF THE PLOT.   IF
V  IS A MATRIX, THEN IF  Z  IS A VECTOR OF LENGTH  ¯1↑⍴V , Z  INDICATES THE SYM-
BOL TO BE USED FOR EACH COLUMN OF  V ,   AND IF  Z  IS A MATRIX OF SIZE  ⍴V ,  Z
INDICATES THE SYMBOL TO BE USED FOR EACH INDIVIDUAL MEMBER OF  V ;  IN THE EVENT
OF COINCIDENCE THE RIGHT-MOST SYMBOL IN  Z  TAKES PRECEDENCE.   SEE THE EXAMPLES
IN 'HOWPLOT'.

17 JAN. 1977
```

```
      ∇ W←U SCATTERPLOT V;E;I;J;M;N;R;S;Z
[1]     →2+(1=ρρU)∧(1=ρρV)∧(N≥2)∧(+/ρU)=N←+/ρV
[2]     →ρ⍞←'ARGUMENTS SHOULD BE VECTORS OF EQUAL LENGTH NOT LESS THAN 2.',W←''
[3]     →4+∧/0<R←-/E← 2 2 ρ(⌈/U),(⌊/U),(⌈/V),⌊/V
[4]     →0,ρ⍞←'ONE ARGUMENT HAS ZERO RANGE.',W←''
[5]     →(2=⎕NC 'SCPSIZE')/8
[6]     'SIZE?  TYPE TWO NUMBERS, SUCH AS:  25 25'
[7]     →9,M←,⎕
[8]     M←SCPSIZE
[9]     →6×ι(∨/M≠⌊M)∨(∨/1≥M)∨2≠ρM
[10]    →11+⎕PW≥2×M[1]+1
[11]    →7,ρ⍞←'FIRST SIZE NUMBER TOO LARGE.  REENTER SIZE'
[12]    I←R÷M-J←1
[13]    →(2=⎕NC 'SCPSCALES')/16
[14]    'SAME SCALES?  Y OR N.'
[15]    →18-'Y'=1↑⍞
[16]    →18-SCPSCALES=1
[17]    E[;2]←E[;1]-I×¯1+M←1+⌈R÷I[1 2]←S×⌈/I÷S← 1 1
[18]    W←'×',((⌽M)ρ' '),[1] '×'
[19]    →L1×ι0<×/ 1 2 ↑E←E÷I,[1.5] I
[20]    E[1;2]←(1-M[1])+E[1;1]←⌊0.5+E[1;1]
[21]    W[;(ρW)[2]-E[1;1]]←(M[2]ρ'|'),'0'
[22] L1:⎕TC[2],'EXTREME ABSCISSAS ARE:  ',⍕⌽I[1]×E[1;]
[23]    'ABSCISSA UNIT STEP IS ',⍕1↑I
[24]    →L2×ι0<×/E[2;]
[25]    E[2;2]←(1-M[2])+E[2;1]←⌊0.5+E[2;1]
[26]    W[1+E[2;1];]←'0',M[1]ρ'-'
[27] L2:'EXTREME ORDINATES ARE:  ',⍕⌽I[2]×E[2;]
[28]    'ORDINATE UNIT STEP IS ',(⍕I[2]),⎕TC[2 2]
[29]    U←(ρW)[2]-⌈E[1;1]-0.5+U÷I[1]
[30]    V←⌈0.5+E[2;1]-V÷I[2]
[31]    →(2=⎕NC 'SCPMULT')/L3-1
[32]    →L3,ρZ←'⊛⊖○'
[33]    Z←SCPMULT
[34] L3:W[V[J];U[J]]←Z[1⌈¯1+ZιW[V[J];U[J]]]
[35]    →L3×ιN≥J←J+1
[36]    U←V←''
[37]    W←((2×¯1↑ρW)ρ 0 1)\W
      ∇
```

```
      ∇ W←U TSCP V;E;I;J;M;N;R;S;Z
[1]     →2+(1=ρρU)∧(1=ρρV)∧(N≥2)∧(+/ρU)=N←+/ρV
[2]     →ρ⍞←'ARGUMENTS SHOULD BE VECTORS OF EQUAL LENGTH NOT LESS THAN 2.',W←''
[3]     →4+∧/0<R←-/E← 2 2 ρ(⌈/U),(⌊/U),(⌈/V),⌊/V
[4]     →0,ρ⍞←'ONE ARGUMENT HAS ZERO RANGE.',W←''
[5]     →(2=⎕NC 'SCPSIZE')/8
[6]     'SIZE?  TYPE TWO NUMBERS, SUCH AS:  25 25'
[7]     →9,M←,⎕
[8]     M←SCPSIZE
[9]     →6×ι(∨/M≠⌊M)∨(∨/1≥M)∨2≠ρM
[10]    →11+⎕PW≥2×M[1]+1
[11]    →7,ρ⍞←'FIRST SIZE NUMBER TOO LARGE.  REENTER SIZE'
[12]    I←R÷M-J←1
[13]    →(2=⎕NC 'SCPSCALES')/16
[14]    'SAME SCALES?  Y OR N.'
[15]    →18-'Y'=1↑⍞
[16]    →18-SCPSCALES=1
[17]    E[;2]←E[;1]-I×¯1+M←1+⌈R÷I[1 2]←S×⌈/I÷S← 1 1
```

```
[18]  W←'×',((⌽M)⍴' '),[1] '×'
[19]  →L1×⍳0<×/ 1 2 ↑E←E÷I,[1.5] I
[20]  E[1;2]←(1-M[1])+E[1;1]←⌊0.5+E[1;1]
[21]  W[1↑⍴W;(⍴W)[2]-E[1;1]]←'0'
[22] L1:→L2×⍳0<×/E[2;]
[23]  E[2;2]←(1-M[2])+E[2;1]←⌊0.5+E[2;1]
[24]  W[1+E[2;1];1]←'0'
[25] L2:→(2=⎕NC 'SCPSYMBOLS')/L3-1
[26]  'SYMBOLS TO BE USED?'
[27]  →L3,⍴Z←⎕
[28]  Z←SCPSYMBOLS
[29] L3:→((~∨/,Z∊' ⋆')∧(∧/N=⍴Z)∧ 0 1 =⍴⍴Z)/L3+ 2 3
[30]  →(L3-2),⍴⍞←'SYMBOLS WON''T DO.  REENTER.'
[31]  Z←N⍴Z
[32]  ⎕TC[2],'EXTREME ABSCISSAS ARE:  ',⍕⌽I[1]×E[1;]
[33]  'ABSCISSA UNIT STEP IS ',⍕1↑I
[34]  'EXTREME ORDINATES ARE:  ',⍕⌽I[2]×E[2;]
[35]  'ORDINATE UNIT STEP IS ',(⍕I[2]),⎕TC[2 2]
[36]  U←(⍴W)[2]-⌈E[1;1]-0.5+U÷I[1]
[37]  V←⌈0.5+E[2;1]-V÷I[2]
[38] L4:W[V[J];U[J]]←1↑((W[V[J];U[J]]≠' ')/'⋆'),Z[J]
[39]  →L4×⍳N≥J←J+1
[40]  U←V←Z←''
[41]  W←((2×¯1↑⍴W)⍴ 0 1)\W
      ∇

      ∇ DOWNPLOT V;C;J;N
[1]   →(((∧/1≤⍴V)∧ 1 2 =⍴⍴V)/ 3 4),1+J←1
[2]   →0,⍴⍞←'ARGUMENT WON''T DO.'
[3]   V←((⍴V),1)⍴V
[4]   →(2=⎕NC 'DPS')/7
[5]   'DPS (DOWNPLOT SCALE)?  TYPE 3 NUMBERS SUCH AS:  ¯1.8 0.2 2'
[6]   DPS←⎕
[7]   →9-(3=⍴DPS)∧∧/0≠¯1↓1↓DPS←,DPS
[8]   →9+(N=⌊N)∧(⎕PW≥N+5)∧2≤N←1+(-/DPS[3 1])÷DPS[2]
[9]   →6,⍴⍞←'DPS WON''T DO.  REENTER.'
[10]  →(2=⎕NC 'DPL')/13
[11]  'DPL (DOWNPLOT LABELING)?  TYPE 4 NUMBERS SUCH AS:  5 5 5 5'
[12]  DPL←⎕
[13]  →15-(DPL[1 3]∧.=⌊DPL[1 3])∧4=⍴DPL←,DPL
[14]  →15+(1≤DPL[1])∧≥/DPL[3 1],⍞←''
[15]  →12,⍴⍞←'DPL WON''T DO.  REENTER.'
[16]  V←1⌈(⌊2.5+(V-DPS[1])÷DPS[2])⌊N+2
[17]  →(2≠⎕NC 'DPH')/19
[18]  DPH
[19]  →(2=⎕NC 'DPM')/21
[20]  DPM←'---------+'
[21]  '     |',(N←⌊N)⍴DPM
[22] L1:C←(N+2)⍴' '
[23]  C[V[J;]]←'o'
[24]  C[(C[1]='o')/2]←'←'
[25]  C[(C[N+2]='o')/N+1]←'→'
[26]  →(0=DPL[3]|J-DPL[1])/L3
[27]  '     |',1↓¯1↓C
[28] L2:→((⍴V)[1]≥J←J+1)/L1
[29]  →0,⍴⍞←'     |',N⍴DPM
[30] L3:→L2,⍴⍞←(¯4↑⍕DPL[2]+DPL[4]×(J-DPL[1])÷DPL[3]),'|',1↓¯1↓C
      ∇
```

```
      ∇ Z TDP V;C;J;N
[1]    →(((∧/1≤ρV)∧ 1 2 =ρρV)/ 3 4),1+J←1
[2]    →0,ρ⎕←'ARGUMENTS WON''T DO.'
[3]    V←((ρV),1)ρV
[4]    →((0 1 2 =ρρZ)/ 8 6 5),2
[5]    →((∧/(ρZ)=ρV)/9),2
[6]    →(((ρZ)=ρV)/ 8 7),2
[7]    →9,ρZ←(ρV)ρZ
[8]    Z←((1↑ρV),1)ρZ
[9]    →(2=⎕NC 'DPS')/12
[10]   'DPS (DOWNPLOT SCALE)?  TYPE 3 NUMBERS SUCH AS:  ¯1.8 0.2 2'
[11]   DPS←⎕
[12]   →14-(3=ρDPS)∧∧/0≠¯1↓1↓DPS←,DPS
[13]   →14+(N=⌊N)∧(⎕PW≥N+5)∧2≤N←1+(-/DPS[3 1])÷DPS[2]
[14]   →11,ρ⎕←'DPS WON''T DO.  REENTER.'
[15]   →(2=⎕NC 'DPL')/18
[16]   'DPL (DOWNPLOT LABELING)?  TYPE 4 NUMBERS SUCH AS:  5 5 5 5'
[17]   DPL←⎕
[18]   →20-(DPL[1 3]∧.=⌊DPL[1 3])∧4=ρDPL←,DPL
[19]   →20+(1≤DPL[1])∧≥/DPL[3 1],⎕←''
[20]   →17,ρ⎕←'DPL WON''T DO.  REENTER.'
[21]   V←1⌈(⌊2.5+(V-DPS[1])÷DPS[2])⌊N+2
[22]   →(2≠⎕NC 'DPH')/24
[23]   DPH
[24]   →(2=⎕NC 'DPM')/26
[25]   DPM←'---------+'
[26]   '     |',(N←⌊N)ρDPM
[27] L1:C←(N+2)ρ' '
[28]   C[V[J;]]←Z[J;]
[29]   →(0=DPL[3]|J-DPL[1])/L3
[30]   '     |',1↓¯1↓C
[31] L2:→((ρV)[1]≥J←J+1)/L1
[32]   →0,ρ⎕←'     |',NρDPM
[33] L3:→L2,ρ⎕←(¯4↑⍕DPL[2]+DPL[4]×(J-DPL[1])÷DPL[3]),'|',1↓¯1↓C
      ∇
```

```
      ∇ Z←NIF X
[1]    Z←0.5+(×X)×0.5-(*X×X÷¯2)×((÷1+0.2316419×|X)∘.*⍳5)+.× 319381530 ¯356563782
       1781477937 ¯1821255978 1330274429 ×3.989422804E¯10
[2]  ⍝  HASTINGS APPROXIMATION TO THE NORMAL (GAUSS-LAPLACE) INTEGRAL FUNCTION,
       NBS HANDBOOK, AMS55, 26.2.17.
[3]  ⍝  ABSOLUTE ERROR IN  Z  LESS THAN 1E¯7.  X  MAY BE ANY NUMERICAL ARRAY.
      ∇
```

```
      ∇ X←INIF P;T
[1]    →2+∧/,1E¯20≤T←P⌊1-P
[2]    →ρ⎕←'NO GO.',X←''
[3]    T←(¯2×⍟T)*0.5
[4]    X←(×P-0.5)×T-((T∘.*0,⍳4)+.× 0.322232431088 1 0.342242088547
       0.0204231210245 0.0000453642210148)÷(T∘.*0,⍳4)+.× 0.099348462606
       0.588581570495 0.531103462366 0.10353775285 0.0038560700634
[5]  ⍝  ODEH-EVANS APPROXIMATION TO THE INVERSE OF THE NORMAL INTEGRAL FUNCTION,
        AS70, APPLIED STATISTICS, 23 (1974), 96.
[6]  ⍝  ABSOLUTE ERROR IN  X  LESS THAN 1.5E¯8.  P  MAY BE ANY ARRAY OF NUMBERS
       BETWEEN 1E¯20 AND 1.
      ∇
```

HOWSUMMARIZE

SUMMARY STATISTICS OF A DATA SET
SUMMARIZE Y

THE ARGUMENT (Y) MAY BE ANY NUMERICAL ARRAY HAVING AT LEAST 4 *MEMBERS. MEASURES OF LOCATION, SCALE AND SHAPE OF DISTRIBUTION ARE PRINTED OUT. SHAPE IS INDICATED BY FREQUENCIES OF THE DATA IN THE SIX INTERVALS INTO WHICH THE REAL LINE IS DIVIDED BY THE MEAN +* ¯2 ¯1 0 1 2 × *THE STANDARD DEVIATION OF THE DATA. EXPECTED COUNTS IN THESE INTERVALS FOR A RANDOM SAMPLE OF THE SAME SIZE TAKEN FROM A NORMAL DISTRIBUTION WITH THE SAME MEAN AND VARIANCE ARE GIVEN FOR COMPARISON. SHAPE IS ALSO INDICATED BY FISHER'S SKEWNESS AND KURTOSIS MEASURES, WITH THEIR STANDARD ERRORS FOR A RANDOM SAMPLE OF THE SAME SIZE FROM A NORMAL DISTRIBUTION, AND ROUGH EQUIVALENT NORMAL DEVIATES FOR ASSESSING SIGNIFICANCE. OPTIONALLY, FOR DATA CONSTITUTING A TIME SERIES, A FIRST SERIAL CORRELATION COEFFICIENT MAY BE GIVEN, WITH STANDARD ERROR FOR A NORMAL SAMPLE.*

2 *FEB*. 1977

```
      ∇ SUMMARIZE Y;A;D;M03;M21;N;P;S;S1;S2;Z
[1]   →2+(∨/Y≠1↑Y)∧4≤N←ρY←,Y
[2]   →ρ⎕←'THE ARGUMENT MUST BE AN ARRAY WITH AT LEAST 4 MEMBERS NOT ALL EQUAL.'
[3]   'NUMBER OF READINGS:  ',(⍕N),⎕←''
[4]   'EXTREMES:  ',⍕(Z←Y[⍋Y])[1,N]
[5]   'QUARTILES:  ',⍕⌽A←0.5×(+/Z[N+1-A]),+/Z[A←(⌊1+N÷4),⌈N÷4]
[6]   'MEDIAN:  ',⍕0.5×+/Z[(⌊1+N÷2),⌈N÷2]
[7]   'MEAN:  ',⍕D←(+/Y)÷N
[8]   'INTERQUARTILE RANGE:  ',(⍕-/A),⎕←''
[9]   'VARIANCE:  ',⍕S←(+/Z←Y×Y←Y-D)÷N-1
[10]  'STANDARD DEVIATION:  ',⍕D←S*0.5
[11]  A← 2275 13591 34134 ×N÷100000
[12]  'FREQUENCIES:  ',(⍕+/(¯4+⍳6)∘.=2⌊¯3⌈⌊Y÷D),',  WITH FITTED NORMAL EXPECTATI
      ONS:  ',(1⍕A,⌽A),⎕←''
[13]  G1←(÷/(Z+.×Y),N+ ¯1 0 ¯2)÷S×D
[14]  S1←(÷/6,N+ ¯2 0 1 ¯1 3)*0.5
[15]  'SKEWNESS:  G1 = ',(⍕G1),(P←' +',⎕TC[1],'_ '),(⍕S1),'  (RATIO = ',(2⍕G1÷S1
      ),')'
[16]  G2←(((Z+.×Z)÷S×S)-3×÷/N+ ¯1 0 ¯1 1 ¯1)÷÷/N+ ¯1 0 ¯2 1 ¯3
[17]  D←G2÷S2←(÷/24,N+ ¯3 0 ¯2 ¯1 3 ¯1 5)*0.5
[18]  'KURTOSIS:  G2 = ',(⍕G2),P,⍕S2
[19]  A←6+4×A×A+4○A←2÷M03←(÷/1728,N+ ¯3 0 ¯3 ¯1 ¯2 ¯1 ¯2 ¯1)×(2+N×N-5)÷(×/N+ 3 5
      7 9)×S2×S2×S2
[20]  →(3+1↑⎕LC)×⍳0≥D←1+D×(2÷A-4)*0.5
[21]  D←(1-(2÷9×A)+((1-2÷A)÷D)*÷3)÷(2÷9×A)*0.5
[22]  →(2+⎕LC),ρ⎕←'    (STANDARDIZED 0 3 MOMENT = ',(⍕M03),', ROUGH E.N.D. FOR G2
       = ',(2⍕D),')'
[23]  '    (STANDARDIZED 0 3 MOMENT = ',(⍕M03),')'
[24]  →(2+1↑⎕LC)×⍳0≥D←1+(G2÷S2)×M21←(÷/216,N+ ¯2 0 ¯2 0 1 ¯1 3 ¯1 5)÷S1×S1×S2×N+
      7
[25]  →(2+⎕LC),ρ⎕←'    (STANDARDIZED 2 1 MOMENT = ',(⍕M21),', CRUDE E.N.D. FOR G1
       GIVEN G2 = ',(2⍕G1÷S1×D*0.5),')'
[26]  '    (STANDARDIZED 2 1 MOMENT = ',(⍕M21),')'
[27]  'SERIAL TEST?  Y OR N',⎕←''
[28]  →('Y'≠1↑⍞)/0
[29]  R1←((A←1-÷N)+((1↓Y)+.×¯1↓Y)÷S)÷D←(N⊥ 1 ¯3 2 2)÷N×N
[30]  S←((N-1+2×A×A÷D)÷D×N+1)*0.5
[31]  'SERIAL CORRELATION:  R1 = ',(⍕R1),P,(⍕S),'  (RATIO = ',(2⍕R1÷S),')'
      ∇
```

HOWFIT

SIMPLE REGRESSION
Z←X FIT Y

THE TWO ARGUMENTS MUST BE VECTORS OF EQUAL LENGTH NOT LESS THAN 3.

THE MEANS OF THE TWO ARGUMENTS, AND THE REGRESSION COEFFICIENT OF THE SECOND ARGUMENT ON THE FIRST, WITH ITS CONVENTIONAL ESTIMATED STANDARD ERROR, ARE DISPLAYED. THE EXPLICIT RESULT IS THE VECTOR OF RESIDUALS.

30 *MAY* 1973

```
      ∇ Z←X FIT Y;B;N;S;U;V
[1]    →2+(N=+/ρY)∧(3≤N←+/ρX)∧(1=ρρY)∧1=ρρX
[2]    →ρ⎕←'ARGUMENTS MUST BE VECTORS OF EQUAL LENGTH NOT LESS THAN 3.',Z←''
[3]    X←X-U←(+/X)÷N
[4]    Z←Y-V←(+/Y)÷N
[5]    'MEANS ARE  ',⍕U,V
[6]    →8-∧/X=1↑X
[7]    →ρ⎕←'NO REGRESSION CALCULATED.'
[8]    'REGRESSION COEFFICIENT IS ',⍕B←(X+.×Z)÷S←X+.×X
[9]    S←(Z+.×Z←Z-B×X)÷S×N-2
[10]   '    WITH ESTIMATED STANDARD ERROR  ',(⍕S*0.5),'   (',(⍕N-2),' D.F.)'
      ∇
```

```
      ∇ SHOW V;E;J;R;W
[1]    →3-(∨/6≤ρV)∧(J←∧/E←'')=ρρV
[2]    →3+W[6]>1↑W←V[R←(⍋V)[(⍳3),(1+ρV)-⌽⍳3]]
[3]    →0,ρ⎕←'NO GO.'
[4]   L:E←E,'(⍕W[',(⍕J),']),'' ('',(⍕R[',(⍕J),']),''), '','
[5]    →L×⍳6≥J←J+1
[6]    '3 LOWEST, 3 HIGHEST (WITH INDICES) AND R.M.S. VALUES:'
[7]    ⍎E,''' '',⍕((V+.×V)÷ρV)*0.5'
[8]    'FREQUENCY DISTRIBUTION:  ',⍕+/(0,⍳5)∘.=⌊0.5+5×(V-1↑W)÷-/W[6 1]
      ∇
```

```
      ∇ X REGRINIT Y;J
[1]    →3-(∧/1≤ρX)∧(∧/1≤ρY)∧(∨/ 1 2 =ρρY)∧2=ρρX
[2]    →3+J←(1↑ρY)=NU←1↑ρX
[3]    →0,ρ⎕←'NO GO.'
[4]    X←0↑,TRX←(P,P)ρ1,IND←(P←¯1↑ρRX←X)ρ0
[5]    →9×⍳1=ρρISS←+⌿RY×RY←Y
[6]    RB←Pρ0
[7]    'INITIAL SUM OF SQUARES (ISS), NUMBER OF READINGS AND MEAN SQUARE ARE',⎕←
       ''
[8]    →0,ρ⎕←(⍕(⌊0.5+ISS),NU),3⍕ISS÷NU
[9]    RB←(P,¯1↑ρY)ρ0
[10]   '    *Y-VARIABLE ',(⍕J),'*',⎕←''
[11]   'INITIAL SUM OF SQUARES (ISS[',(⍕J),']), NUMBER OF READINGS AND MEAN SQUAR
       E ARE'
[12]   (⍕(⌊0.5+ISS[J]),NU),3⍕ISS⌊J]÷NU
[13]   →10×(¯1↑ρY)≥J←J+1
      ∇
```

```
HOWREGRESSION

        MULTIPLE REGRESSION BY STAGES AND EXAMINATION OF RESIDUALS
        (1)  X REGRINIT Y
        (2)  REGR L
        (3)  SHOW V
        (4)  STRES
        (5)  VARIANCE

   'REGRINIT'  IS USED ONCE AT THE OUTSET TO SET UP GLOBAL VARIABLES FOR 'REGR'.
'REGR' PERFORMS REGRESSION ON ONE OR MORE DESIGNATED INDEPENDENT VARIABLES,  AND
MAY BE CALLED REPEATEDLY.  'SHOW' MAY BE USED AT ANY STAGE TO OBTAIN SUMMARY IN-
FORMATION ABOUT A VECTOR.  'STRES' GIVES STANDARDIZED RESIDUALS OF THE DEPENDENT
VARIABLE(S) FOR USE IN SCATTERPLOTS.  'VARIANCE' YIELDS THE CONVENTIONAL ESTIMA-
TED VARIANCE MATRIX OF THE REGRESSION COEFFICIENTS.

   (1) REGRINIT.  THE FIRST ARGUMENT (X) IS A MATRIX  WHOSE COLUMNS LIST VALUES
OF THE INDEPENDENT (EXOGENOUS, PREDICTOR) VARIABLES.  USUALLY ONE COLUMN IS ALL
1'S, AND ITS INDEX NUMBER IS THE ARGUMENT IN THE FIRST CALL OF 'REGR'.  EACH CO-
LUMN OF X SHOULD HAVE BEEN MULTIPLIED BY A POWER OF 10 SO THAT THE UNIT PLACE IS
THE LAST SIGNIFICANT ONE.  WHEN A COLUMN OF X-RESIDUALS HAS SUM OF SQUARES LESS
THAN (NU÷4) IT IS JUDGED TO BE TOO CLOSE TO ROUND-OFF ERROR TO BE WORTH USING.

   THE SECOND ARGUMENT (Y) IS  EITHER  A VECTOR LISTING VALUES OF ONE  DEPENDENT
VARIABLE OR A MATRIX WHOSE COLUMNS LIST VALUES OF SEVERAL DEPENDENT VARIABLES TO
BE STUDIED IN PARALLEL.  THE NUMBER OF OBSERVATIONS OR DATA SETS IS EQUAL TO
          1↑ρX ↔ 1↑ρY
JUST AS WITH  X , THE UNIT PLACE SHOULD BE THE LAST SIGNIFICANT ONE.

   GLOBAL VARIABLES IN 'REGRINIT' AND 'REGR':
P     THE NUMBER OF INDEPENDENT VARIABLES,  (ρX)[2] .
IND   A LOGICAL VECTOR OF LENGTH P SHOWING WHICH INDEPENDENT VARIABLES HAVE BEEN
      BROUGHT INTO THE REGRESSION RELATION (INITIALLY ALL 0'S).
NU    THE NUMBER OF RESIDUAL DEGREES OF FREEDOM (INITIALLY THE NO. OF OBS.).
RX    TRANSFORMED OR RESIDUAL X-VARIABLES (INITIALLY EQUAL TO X).
RY    RESIDUALS OF Y (INITIALLY EQUAL TO Y).
B     REGRESSION COEFFICIENTS (VECTOR OR MATRIX) OF Y ON COLUMNS OF X CONSIDERED
      SO FAR.  (B APPEARS ONLY IN THE OUTPUT OF 'REGR'.)
RB    DITTO WITH RX IN PLACE OF X (INITIALLY ALL 0'S).
TRX   P-BY-P MATRIX TRANSFORMING X TO RX.  RX ↔ X+.×TRX ; B ↔ TRX+.×RB .
      THE FITTED VALUES ARE  Y-RY ↔ X+.×B ↔ RX+.×RB .
ISS   INITIAL SUM OF SQUARES (APPEARING ONLY IN THE OUTPUT OF 'REGRINIT').
RSS   RESIDUAL SUM OF SQUARES (APPEARING ONLY IN THE OUTPUT OF 'REGR').

   AFTER 'REGRINIT' IS EXECUTED IT CAN BE ERASED,  AS ALSO ITS ARGUMENTS  (WHICH
SHOULD BE STORED FOR COPYING LATER WHEN NEEDED).

   (2) REGR.  THE ARGUMENT (L) IS A SCALAR OR VECTOR LISTING THE INDEX NO.(S) OF
THE INDEPENDENT VARIABLE(S) TO BE BROUGHT NEXT INTO THE REGRESSION.

   RENAME  RSS  OR ANY OTHER GLOBAL OUTPUT VARIABLE TO SAVE IT, BEFORE AGAIN EX-
ECUTING 'REGR'.

   THE POSSIBLE REDUCTIONS IN  RSS  TABULATED IN THE PRINTED OUTPUT SHOW THE EF-
FECT OF BRINGING ONE FURTHER INDEPENDENT VARIABLE INTO THE REGRESSION,  AND  MAY
BE USED TO GUIDE THE NEXT EXECUTION OF 'REGR'.

   THE SO-CALLED MODIFIED GRAM-SCHMIDT ALGORITHM IS USED.  SEE
```

ANSCOMBE: J. ROYAL STATIST. SOC. B 29 (1967), 1-52, ESPECIALLY SECTION 1.4.
BJÖRCK: BIT - NORD. TIDSKR. INFORMATIONSBEHAND. 7 (1967), 1-21.

IF YOU WANT TO FOIL THE (NU÷4) PROVISION MENTIONED ABOVE, REPLACE THE '4' IN LINE [1] BY A LARGER NUMBER. WHEN SOME COLUMNS OF X ARE INDICATORS CONSISTING OF 0'S AND 1'S ONLY, EITHER DO THIS OR ELSE MULTIPLY THOSE COLUMNS BY A FACTOR (SUCH AS 10).

(3) SHOW. THE ARGUMENT (V) IS A VECTOR, SUCH AS RY (IF A VECTOR) OR A COLUMN OF RY (IF RY IS A MATRIX) OR A COLUMN OF RX OR DIAGQ (SEE 'STRES').

THE FREQUENCY DISTRIBUTION IN THE PRINTED OUTPUT IS OVER 6 INTERVALS OF EQUAL LENGTH. THE 1ST AND 6TH INTERVALS ARE CENTERED ON THE LEAST AND GREATEST VALUES OCCURRING IN V .

'SHOW' REFERS TO NO GLOBAL VARIABLES, AND MAY BE USED OUTSIDE THIS REGRESSION CONTEXT.

(4) STRES. NO ARGUMENTS; SHOULD BE USED ONLY AFTER 'REGR' HAS BEEN EXECUTED.

GLOBAL OUTPUT VARIABLES:
DIAGQ IS A VECTOR CONSISTING OF THE MAIN DIAGONAL OF A MATRIX Q (NOT FURTHER DETERMINED HERE) THAT PROJECTS Y INTO RY , THUS RY ↔ Q+.×Y .
SRY EACH ELEMENT OF RY IS DIVIDED BY THE SQUARE ROOT OF THE CORRESPONDING ELEMENT OF DIAGQ (TIMES A CONSTANT). IF THE MEMBERS OF DIAGQ ARE ALL EQUAL, SRY IS THE SAME AS RY .

(5) VARIANCE. NO ARGUMENTS; USE ONLY AFTER 'REGR' HAS BEEN EXECUTED.

ONE GLOBAL OUTPUT VARIABLE:
XPXI IS 'X-PRIME X INVERSE'. WHEN REGRESSION HAS BEEN PERFORMED ON ALL COLUMNS OF X, XPXI IS THE INVERSE OF (⍉X)+.×X; OTHERWISE IT IS THE CORRESPONDING EXPRESSION FOR THE COLUMNS OF X THAT HAVE BEEN USED, TOGETHER WITH ZEROES FOR REGRESSION COEFFICIENTS NOT ESTIMATED.

26 JAN. 1977

```
      ∇ REGR L;D;I;J;K;S;U
[1]    →2+(0<D←4)∧J←(∧/L∊⍳P)∧∧/1≤⍴L←,L
[2]    →0,⍴⎕←'NO GO.'
[3]    →4+1≤NU←NU-0=IND[K←L[J]]
[4]    →0,⍴⎕←'NU  HAS VANISHED.'
[5]    IND[K]←1
[6]    →8×⍳(NU÷D)≤S←RX[;K]+.×RX[;K]
[7]    →⍴⎕←'DIVISOR TOO SMALL AT COLUMN ',(⍕K),' OF  RX .'
[8]    RX←RX-RX[;K]∘.×U←(RX[;K]+.×RX×(⍴RX)⍴~IND)÷S
[9]    TRX←TRX-TRX[;K]∘.×U
[10]   →11+2=⍴⍴RY←RY-RX[;K]∘.×U←(RX[;K]+.×RY)÷S
[11]   →13,RB[K]←RB[K]+U
[12]   RB[K;]←RB[K;]+U
[13]   →3×⍳(⍴L)≥J←J+1
[14]   I←(NU÷D)≤S←+⌿RX×RX
[15]   RSS←+⌿RY×RY
[16]   →21×⍳J←∨/2=⍴⍴B←TRX+.×RB
[17]   'RESIDUAL SUM OF SQUARES (RSS), D.F. (NU) AND MEAN SQUARE ARE',⎕←''
[18]   (⍕(⌊0.5+RSS),NU),3⍕RSS÷NU
```

```
[19]  'POSSIBLE REDUCTIONS IN  RSS  ARE'
[20]  →27,ρ⎕←0⍕((I×RY+.×RX)*2)÷S⌈NU÷D
[21]  '    *Y-VARIABLE ',(⍕J),'*',⎕←''
[22]  'RESIDUAL SUM OF SQUARES (RSS[',(⍕J),']), D.F. (NU) AND MEAN SQUARE ARE'
[23]  (⍕(⌊0.5+RSS[J]),NU),3⍕RSS[J]÷NU
[24]  'POSSIBLE REDUCTIONS IN  RSS[',(⍕J),']  ARE'
[25]  0⍕((I×RY[;J]+.×RX)*2)÷S⌈NU÷D
[26]  →21×⍳(1↓ρRY)≥J←J+1
[27]  →0×⍳1=∧/I
[28]  'DIVISOR TOO SMALL AT COLUMN ',(⍕(~I)/⍳P),' OF  RX .',⎕←''
      ∇
```

```
      ∇ STRES;M
[1]   →2+∧/0<DIAGQ←1-+/M÷(ρM)ρ+⌿M←(IND/RX)*2
[2]   →ρ⎕←'A MEMBER OF  DIAGQ  VANISHES.  SRY  NOT FOUND.'
[3]   SRY←RY×⍉(⌽ρRY)ρ(DIAGQ×(ρRX)[1]÷NU)*¯0.5
      ∇
```

```
      ∇ VARIANCE
[1]   XPXI←(TRX×(P,P)ρIND\÷+⌿IND/RX×RX)+.×⍉TRX
[2]   →(1 2 =ρρRY)/ 3 4
[3]   →ρ⎕←'ESTIMATED VARIANCE MATRIX OF  B  IS:  XPXI×RSS÷NU'
[4]   'ESTIMATED VARIANCE MATRIX OF  B[;J]  IS:  XPXI×RSS[J]÷NU'
      ∇
```

```
      ∇ Z←N SAMPLE CP
[1]   →2+(∧/,(N=⌊N)∧1≤N)∧(1≥ρρN)∧∧/(1≥CP)∧0≤CP←,CP
[2]   →ρ⎕←'NO GO.',Z←''
[3]   Z←+⌿CP∘.<(÷2147483647)×?Nρ2147483647
[4]  ⍝  SAMPLE OF SIZE  N  FROM A DISTRIBUTION OVER NONNEGATIVE INTEGERS HAVING
      CUMULATIVE PROBABILITIES  CP .
      ∇
```

```
      ∇ Z←MP X;C;K;MC;MR;R
[1]   →2+(TOL>0)∧(TOL<1)∧(0=ρρTOL)∧(∧/2≤ρX)∧2=ρρZ←X
[2]   →0,ρ⎕←'NO GO.',Z←''
[3]   MR←(⌊MR),⌈MR←(1+R←1↑ρX)÷2+K←0
[4]   MC←(⌊MC),⌈MC←(1+C←¯1↑ρX)÷2
[5]   M←8×⌈/|,X
[6]  L1:→L2×⍳(K≥1)∧∧/TOL≥|A←0.5×+⌿(,Z)[(⍋,Z+(⍳R)∘.×CρM)[MC∘.+C×¯1+⍳R]]
[7]   Z←Z-A∘.+Cρ0×K←K+1
[8]   →L2×⍳∧/TOL≥|B←0.5×+⌿(,Z)[(⍋,Z+(RρM)∘.×⍳C)[MR∘.+R×¯1+⍳C]]
[9]   →(L2×⍳20≤K←K+1),L1,ρZ←Z-(ρZ)ρB
[10] L2:(⍕K),' STEPS.'
[11]  A←X[;1]-Z[;1]
[12]  'ROW EFFECTS (A):  ',⍕A←A-MR←0.5×+/A[(⍋A)[MR]]
[13]  B←B-1↑B←X[1;]-Z[1;]
[14]  'COLUMN EFFECTS (B):  ',⍕B←B-MC←0.5×+/B[(⍋B)[MC]]
[15]  'ADDED CONSTANT (M):  ',⍕M←MR+MC
[16] ⍝  ADDITIVE ANALYSIS OF A TWO-WAY TABLE BY MEDIAN POLISH.
[17] ⍝  TO WATCH PROGRESS INSERT  [8.1] (⍕A),⎕TC[3],⍕B
      ∇
```

```
      ∇ ROWCOL Y;A;C̲;D;D2;D3;J;K;M03;M21;N;P;S;S1;S2;S3;S4;Z
[1]    →2+(∧/3≤K←ρY)∧2=ρρY
[2]    →0,ρ⎕←'NO GO.'
[3]    (C̲←⎕TC[2]),'GRAND MEAN (GM) IS ',⍕GM←(+/,Y)÷N←×/K
[4]    'TOTAL SUM OF SQUARES ABOUT MEAN (TSS), DEGREES OF FREEDOM AND MEAN SQUARE
        ARE ',⍕TSS,(N-1),(÷N-1)×TSS←+/,RY×RY←Y-GM
[5]    C̲,'   *ROWS*'
[6]    'EFFECTS (RE) ARE ',⍕RE←(+/RY)÷1↓K
[7]    'SUM OF SQUARES (SSR), D.F. AND M.S. ARE ',⍕SSR,(¯1+1↑K),(÷¯1+1↑K)×SSR←(1↓
       K)×RE+.×RE
[8]    C̲,'   *COLUMNS*'
[9]    'EFFECTS (CE) ARE ',⍕CE←(+⌿RY)÷1↑K
[10]   'SUM OF SQUARES (SSC), D.F. AND M.S. ARE ',⍕SSC,(¯1+1↓K),(÷¯1+1↓K)×SSC←(1↑
       K)×CE+.×CE
[11]   C̲,'   *RESIDUALS*'
[12]   'SUM OF SQUARES (RSS), D.F. (NU) AND M.S. ARE ',⍕RSS,NU,S←(÷NU←×/K-1)×RSS←
       +/,Z←RY×RY←RY-RE∘.+CE
[13]   'THE MATRIX OF RESIDUALS IS NAMED  RY .',C̲
[14]   →(∨/(SSR,SSC)=RSS+SSR,SSC)/0
[15]   →(∧/RSS<RSS+SSR,SSC)/ρ⎕←'   *TUKEY''S TEST*'
[16]   →20,ρ⎕←'TEST FAILS BECAUSE  SSR×SSC  VANISHES.'
[17]   'REGRESSION COEFFT. (TB) OF RY ON RE∘.×CE IS ',⍕TB←(A←+/,RY×RE∘.×CE)×N÷SSR
       ×SSC
[18]   'TUKEY''S 1-D.F. TERM IS ',⍕A←A×TB
[19]   'REMAINING SUM OF SQUARES, D.F. AND M.S. ARE ',⍕A,(NU-1),(÷NU-1)×A←RSS-A
[20]   C̲,'   *DISTRIBUTION SHAPE*'
[21]   G1←N÷NU÷(+/,Z×RY)÷(×/K-2)×S*1.5
[22]   S1←(÷/6,(×/K-2),N,NU+ 2 0 4)*0.5
[23]   'SKEWNESS:  G1 = ',(⍕G1),(P←' +',⎕TC[1],'_ '),(⍕S1),'  (RATIO = ',(2⍕G1÷S1
       ),')'
[24]   D2←(NU÷N×N)×(×/3+K×K-3)-3×÷/NU+ 0 2 0
[25]   G2←(((+/,Z×Z)÷S×S)-NU×÷/3,N,NU+ 0 2 0)÷D2
[26]   D←G2÷S2←(÷/24,D2,NU+ 0 2 0 4 0 6)*0.5
[27]   'KURTOSIS:  G2 = ',(⍕G2),P,⍕S2
[28]   M03←(((×/¯7+K×10+K×K-5)-NU×NU×÷/NU+ 0 2 8 2)÷N×N)-6×D2÷NU+2
[29]   A←6+4×A×A+4○A←2÷M03←(1728×NU÷N×D2×D2×D2)×(÷/NU+ 0 2 0 4 0 6 0 8 0 10)×M03÷
       S2×S2×S2
[30]   →(3+1↑⎕LC)×⍳0≥D←1+D×(2÷A-4)*0.5
[31]   D←(1-(2÷9×A)+((1-2÷A)÷D)*÷3)÷(2÷9×A)*0.5
[32]   →(2+⎕LC),ρ⎕←'   (STANDARDIZED 0 3 MOMENT = ',(⍕M03),', ROUGH E.N.D. FOR G2
        = ',(2⍕D),')'
[33]   '   (STANDARDIZED 0 3 MOMENT = ',(⍕M03),')'
[34]   →(2+1↑⎕LC)×⍳0≥D←1+(G2÷S2)×M21←(÷/216,NU+ 2 0 4 0 6 0 8)×(1-÷/2,(×/K-2),NU+
       0 2)÷D2×S1×S1×S2
[35]   →(2+⎕LC),ρ⎕←'   (STANDARDIZED 2 1 MOMENT = ',(⍕M21),', CRUDE E.N.D. FOR G1
        GIVEN G2 = ',(2⍕G1÷S1×D*0.5),')'
[36]   '   (STANDARDIZED 2 1 MOMENT = ',(⍕M21),')'
[37]   C̲,'   *HETEROSCEDASTICITY TESTS*'
[38]   D3←(SSR,SSC)+.×(×⌿ 1 0 ⌽K∘.-⍳2)÷N
[39]   H←J\(+/,Z×RE∘.+CE)÷S×(J←RSS<RSS+D3)/D3
[40]   S3←(J\(2×NU÷NU+2)÷J/D3)*0.5
[41]   'AGAINST FITTED VALUES:  H = ',(⍕H),P,(⍕S3),'  (RATIO = ',(2⍕J\H÷J/S3),')'
[42]   S4←(÷/2,(NU+2),K[1]- 1 2 1)*0.5
[43]   VFR←(¯1+(+/Z)÷S÷K[1]÷K[2]-1)÷K[1]-2
[44]   'VARIANCE FACTORS FOR ROWS (VFR) = ',(⍕VFR),P,⍕S4
[45]   A←(×/NU+ 4 6):(K[2]+1)×NU-2÷K[1]-2
[46]   '   (ROUGH CHI-SQUARED = ',(2⍕(A÷K[1]×S4×S4)×(VFR-1)+.×VFR-1),', ',(2⍕A),'
        D.F.)'
```

```
[47]  S4←(÷/2,(NU+2),K[2]- 1 2 1)*0.5
[48]  VFC←(¯1+(+/Z)÷S÷K[2]÷K[1]-1)÷K[2]-2
[49]  'VARIANCE FACTORS FOR COLUMNS (VFC) = ',(⍕VFC),P,⍕S4
[50]  A←(×/NU+ 4 6)÷(K[1]+1)×NU-2÷K[2]-2
[51]  '   (ROUGH CHI-SQUARED = ',(2⍕(A÷K[2]×S4×S4)×(VFC-1)+.×VFC-1),', ',(2⍕A),'
       D.F.)'
[52] ⍝  ADDITIVE ANALYSIS OF A TWO-WAY TABLE, WITH TESTS ON RESIDUALS.
      ∇
```

```
      ∇ ROWCOLPERMUTE
[1]   RE←RE[RP←⍒RE]
[2]   CE←CE[CP←⍋CE]
[3]   RY←RY[RP;CP]
[4]   'ROW PERMUTATION (RP) IS ',⍕RP
[5]   'COLUMN PERMUTATION (CP) IS ',⍕CP
[6]   'RE ,  CE  AND  RY  ARE NOW PERMUTED.'
[7]  ⍝  USE AFTER 'ROWCOL', BEFORE 'ROWCOLDISPLAY'.
      ∇
```

```
      ∇ ROWCOLDISPLAY I;C;D;R;S;W
[1]   →2+(0<+/,I)∧(0=⍴⍴I)∧(2=⍴S← 1 1)∧(D←2)∊⍳2
[2]   →0,⍴⎕←'NO GO.'
[3]   →4+⎕PW≥⍴C←(¯1↑W←+\D,D×1⌈⌊0.5+((1↓CE)-¯1↓CE)÷I)⍴0
[4]   →⍴⎕←'THE ARGUMENT IS TOO SMALL, TRY AGAIN.'
[5]   C[W]←1
[6]   '    *CODED DISPLAY OF PERMUTED RESIDUALS (RCD)*',⎕←''
[7]   R←(¯1↑W←+\1,1⌈⌊0.5+((¯1↓RE)-1↓RE)÷÷/I,S)⍴0
[8]   R[W]←1
[9]   'NE-SW DISPLACEMENTS ROUGHLY MEASURE CHANGES IN FITTED VALUES.',⎕TC[2]
[10]  ⎕←RCD←C\R⍀'M̲M-∘+PP̲'[⌊4.5+3×RY÷⌈/,|RY]
[11] ⍝  SEE 'HOWSCATTERPLOT'.
      ∇
```

HOWHOUSEHOLDER

REGRESSION BY HOUSEHOLDER TRANSFORMATIONS, UNCORRELATED RESIDUALS
X HHT Y

ARGUMENTS AS FOR 'REGRINIT', EXCEPT THAT THE LAST SIGNIFICANT DIGIT IN ALL COLUMNS OF X THAT HAVE OBSERVATIONAL ERROR SHOULD BE IN THE SAME PLACE, BUT NOT NECESSARILY THE UNIT PLACE. THE TOLERANCE ASKED FOR SHOULD BE SOMETHING LIKE N TIMES THE VARIANCE OF OBSERVATIONAL ERROR, WHERE N IS 1↑⍴X .

SUCCESSIVE HOUSEHOLDER TRANSFORMATIONS ARE CARRIED OUT WITH SELECTION OF PIVOTS, SO THAT THE RESIDUALS THAT ARE CONSTRAINED TO VANISH TEND TO CORRESPOND TO ROWS OF X FOR WHICH DIAGQ (SEE 'HOWREGRESSION') IS SMALL. THE REMAINING RESIDUALS ARE UNCORRELATED AND HOMOSCEDASTIC, ON THE USUAL NULL HYPOTHESIS. RESIDUALS NOT CONSTRAINED TO VANISH ARE INDICATED BY THE N-ELEMENT VECTOR ROWS . A LISTING OF COLUMN NUMBERS OF X USED IS NAMED COLS ; IGNORE ANY ZEROES.

SEE GOLUB AND STYAN: *J. STATIST. COMPUT. SIMUL.* 2 (1973), 253-274.

1 *FEB.* 1977

```
      ∇ X HHT Y;J;K;M;N;P;S;T;U
[1]    →2+((1↑ρY)=N←1↑ρX)∧(1≤×/ρY)∧(1≤×/ρX)∧(∨/ 1 2 =ρρY)∧2=ρρX
[2]    →0,ρ⎕←'NO GO.'
[3]    ROWS←NρJ←1
[4]    U←(ρX)ρCOLS←(P←¯1↑ρX)ρ0
[5]    →8×⍳2=⎕NC 'TOL'
[6]    'TOLERANCE FOR COLUMN SUM OF SQUARES IN TRANSFORMED X-MATRIX?'
[7]    TOL←⎕
[8]    →6×⍳(∨/TOL≤0)∨0≠ρρTOL
[9]   L̲1:→L̲2×⍳TOL≥M←⌈/T←ROWS+.×X×X
[10]   S←(T+.×T←ROWS×X[;COLS[J]←T⍳M])*0.5
[11]   ROWS[K←(|T)⍳M←⌈/|T]←0
[12]   U[ROWS/⍳N;J]←(ROWS/T)÷(2-4×T[K]<0)×S×U[K;J]←(0.5×1+M÷S)*0.5
[13]   X←X-U[;J]∘.×2×U[;J]+.×X
[14]   Y←Y-U[;J]∘.×2×U[;J]+.×Y
[15]   →L̲1×⍳P≥J←J+1
[16]  L̲2:(⍕J-1),' COLUMNS USED.'
[17]   'RESIDUAL SUM OF SQUARES (RSS) = ',⍕RSS←+⌿RY×RY←ROWS⍀ROWS⌿Y
[18]   'SUM OF SQUARES FOR REGRESSION = ',⍕+⌿Y×Y←(~ROWS)⌿Y
[19]   J←Pρ0
[20]   J[(COLS>0)/COLS]←1
[21]   'THE REGRESSION COEFFICIENTS ARE NAMED  B .'
[22]   B←J⍀Y⌹J/(~ROWS)⌿X
[23]   'UNCORRELATED RESIDUALS ARE NAMED  RY .'
      ∇
```

```
      ∇ ANALYZE Y
[1]    →2+(∧/2≤ρY)∧2≤ρρY
[2]    →0,ρ⎕←'NO GO.'
[3]    DF←(1,1+ρρY)ρ1
[4]    'GRAND MEAN (GM) IS ',⍕GM←(+/,Y)÷×/ρY
[5]    'TOTAL SUM OF SQUARES ABOUT MEAN (TSS), DEGREES OF FREEDOM AND MEAN SQUARE
        ARE ',⍕TSS,NU,(÷NU←¯1+×/ρY)×TSS←+/,RY×RY←Y-GM
[6]    'PROCEED BY REPEATEDLY EXECUTING THE FUNCTION ''EFFECT''.'
[7]    'THE ARRAY OF RESIDUALS IS ALWAYS NAMED  RY , WITH DEGREES OF FREEDOM  NU
       .'
[8]   ⍝  BEGINS AN ANALYSIS OF VARIANCE OF A PERFECT RECTANGULAR ARRAY.
      ∇
```

```
      ∇ EFFECT V;J;K;M;P;R;S;Z
[1]    →3-∧/∨/M←(,V)∘.=⍳R←+/ρρRY
[2]    →3+∧/0≤M←1-+⌿M
[3]    →0,ρ⎕←'NO GO.',Z←''
[4]    J←+/1↑ρDF←DF,[1] M,0
[5]    NU←NU-DF[J;1+R]←(×/(ρRY)[V])-DF[;1+R]+.×(+/DF[J;])=DF+.×DF[J;]
[6]    S←K×+/,Z×Z←(÷K)×+⌿((K←×/M/ρRY),(ρRY)[V])ρ(⍋P←(M/⍳R),V)⍉RY
[7]    RY←RY-P⍉((M/ρRY),(ρRY)[V])ρZ
[8]    V←(V≠' ')/V←⍕V
[9]    'THE EFFECTS ARE NAMED  E',V,' .'
[10]   'SUM OF SQUARES (SS',V,'), D.F. AND M.S. ARE ',⍕S,DF[J;1+R],S÷DF[J;1+R]
[11]   ⍎'E',V,'←Z'
[12]   ⍎'SS',V,'←S'
[13]  ⍝  'ANALYZE' SHOULD BE EXECUTED FIRST.
[14]  ⍝  NAMES COMPUTED AT [11] AND [12] MAY BE AMBIGUOUS IF  ρρRY  EXCEEDS 11.
      ∇
```

```
    HOWHUBER

        ROBUST REGRESSION
        (1)  ZZ←R HUBER Z
        (2)  DZ←HUBER1
        (3)  DZ←HUBER2

   THE FUNCTION  'HUBER'  PERFORMS ONE CYCLE OF ITERATION TOWARDS MINIMIZING THE
SUM OF A FUNCTION  RHO  OF THE RESIDUALS IN A REGRESSION PROBLEM, WHERE  RHO  IS
DEFINED IN TERMS OF A POSITIVE CONSTANT  K  AS FOLLOWS:
      ∇ U ← RHO Z
[1]   U←((0.5×Z*2)×K≥|Z)+(K×(|Z)-K÷2)×K<|Z  ∇
(SEE HUBER:  ANN. MATH. STATIST. 35 (1964), 73-101.)

   K MAY BE SPECIFIED IN ADVANCE, OR ALTERNATIVELY MAY BE DETERMINED BY THE CON-
DITION THAT A SPECIFIED PROPORTION OF RESIDUALS EXCEED  K  IN MAGNITUDE.  CALCU-
LATION OF ESTIMATED CHANGES IN THE REGRESSION PARAMETERS AND IN THE RESIDUALS IS
DONE BY A SUBSIDIARY FUNCTION, 'HUBER1' OR 'HUBER2'.   IF NEITHER OF THESE FUNC-
TIONS IS APPROPRIATE FOR THE DATA, INSERT A NEW LINE IN 'HUBER' AS FOLLOWS:
[8.1] →L4,⍴DZ←HUBER3
AND DEFINE A NEW SUBSIDIARY FUNCTION STARTING
      ∇ DZ←HUBER3; . . . ∇

   (1) HUBER.   THE FIRST ARGUMENT (R) MUST BE SCALAR.  IF IT IS ¯1, THE CURRENT
GLOBAL VALUE FOR  K  IS USED; OTHERWISE  R  MUST LIE BETWEEN 0 AND 1, AND IS THE
PROPORTION OF RESIDUALS TO EXCEED  K  IN MAGNITUDE.   THE SECOND ARGUMENT (Z) IS
THE ARRAY OF RESIDUALS  CORRESPONDING TO A TRIAL SETTING OF THE REGRESSION PARA-
METERS  BETA , WHICH MUST BE SPECIFIED BEFORE 'HUBER' IS EXECUTED.  THE EXPLICIT
RESULT  (ZZ)  IS A NEW SET OF RESIDUALS,  THAT CAN BE USED AS SECOND ARGUMENT IN
THE NEXT EXECUTION OF 'HUBER'.

   GLOBAL OUTPUT VARIABLES:  K, DBETA (ESTIMATED REQUIRED CHANGE IN PARAMETERS),
BETA (NEW PARAMETER VALUES).

   THE PARAMETER CHANGE  DBETA  YIELDED BY THE SUBSIDIARY FUNCTION IS DETERMINED
FROM THE FIRST AND SECOND DERIVATIVES OF THE SUM OF RHOS  AT THE INITIAL  BETA .
IT YIELDS THE TRUE MINIMUM OF THE SUM OF RHOS, FOR THE VALUE OF  K  USED,  IF NO
CHANGE OCCURS  IN THE INDEXING OF THE RESIDUALS  THAT EXCEED  K  IN MAGNITUDE --
THOSE RESIDUALS BEING REFERRED TO AS 'MODIFIED RESIDUALS'.     OTHERWISE THE TRUE
MINIMUM IS NOT OBTAINED.   THEN THE RESIDUALS ARE CALCULATED WITH THE FULL PARA-
METER CHANGE  DBETA  AND ALSO WITH CHANGES THAT ARE ONLY 0.9, 0.8,... OF  DBETA,
STOPPING WHEN THE SUM OF RHOS IS LEAST.  THE MULTIPLIER OF  DBETA  IS NAMED  AL-
PHA  IN THE DISPLAYED OUTPUT.   THE VALUE  ALPHA = 0.0  CORRESPONDS TO THE INPUT
PARAMETER SETTING  BETA .  WHAT IS FINALLY YIELDED AS THE OUTPUT VARIABLE  DBETA
IS THE FIRST  DBETA  MULTIPLIED BY THE BEST  ALPHA ;   AND THAT  DBETA  IS ADDED
TO THE INPUT  BETA  TO MAKE THE OUTPUT  BETA .

   'HUBER1' AND 'HUBER2' WILL FAIL IN EXECUTION  IF THE RIGHT ARGUMENT OF ⌹ IS
SINGULAR.   IF THAT HAPPENS, (CLEAR THE STATE INDICATOR AND) START 'HUBER' AGAIN
WITH A LARGER  K  (OR SMALLER POSITIVE  R ).

   (2) HUBER1  IS INVOKED WHEN  Z  IS A VECTOR.    THERE MUST BE A GLOBAL MATRIX
X  WHOSE COLUMNS ARE THE INDEPENDENT VARIABLES IN A REGRESSION.  FOR CONSISTENCY
          ⍴Z ↔ 1↑⍴X ;  ⍴BETA ↔ ¯1↑⍴X .
THE ORIGINAL DATA VECTOR, NOT NAMED IN THE PROGRAM, WAS EQUAL INITIALLY TO
          Z + X+.×BETA .
```

(3) *HUBER2* IS INVOKED WHEN Z IS A MATRIX. Z MUST HAVE AT LEAST 3 ROWS AND 3 COLUMNS. THE USUAL ADDITIVE STRUCTURE IS FITTED. BETA IS A VECTOR OF LENGTH 1++/ρZ , AND LISTS THE MEAN, THE ROW CONSTANTS (SUMMING TO 0), AND THE COLUMN CONSTANTS (SUMMING TO 0). THE ORIGINAL DATA MATRIX, NOT NAMED IN THE PROGRAM, WAS EQUAL INITIALLY TO

```
          Z + BETA[1] + ((ρZ)[1]↑1↓BETA)∘.+(ρZ)[1]↓1↓BETA .
```

23 JUNE 1973

```
      ∇ ZZ←R HUBER Z;DZ;I;I1;J;N;P;S;S1;ZZ1
[1]    →2+(ρρBETA←,BETA)∧(0=ρρR)∧J←3≤N←×/ρZ
[2]    →L1,ρ⎕←'HUBER NO GO.'
[3]    →(R=¯1)/6
[4]    →((1>P)∨N<P←0.5+N×1-R)/2
[5]    K←+/(1 0 + ¯1 1 ×P-⌊P)×|(,Z)[(⍋|,Z)[(⌊P),⌈P]]
[6]    'K = ',⍕K
[7]    'INITIAL NUMBER OF MODIFIED RESIDUALS = ',⍕+/,I←K<|Z
[8]    'FOR ALPHA = 0.0, SUM OF RHOS = ',⍕(0.5×+/,Z×Z×~I)+K×+/,I×(|Z)-K÷2
[9]    →((1 2 =ρρZ)/L2,L3),2
[10]  L1:→ρZZ←''
[11]  L2:→(L1×ι0=×/ρDZ←HUBER1),L4
[12]  L3:→L1×ι0=×/ρDZ←HUBER2
[13]  L4:→L6×ι∧/,I=I1←K<|ZZ←Z+DZ
[14]   'FOR ALPHA = 1.0, SUM OF RHOS = ',⍕S←(0.5×+/,ZZ×ZZ×~I1)+K×+/,I1×(|ZZ)-K÷2
[15]  L5:I1←K<|ZZ1←Z+DZ×J←J-0.1
[16]   'FOR ALPHA = ',(⍕J),', SUM OF RHOS = ',⍕S1←(0.5×+/,ZZ1×ZZ1×~I1)+K×+/,I1×(|
       ZZ1)-K÷2
[17]   →L7×ι(S≤S1)∨J=0
[18]   S←S1
[19]   →L5,ρZZ←ZZ1
[20]  L6:'MINIMIZED SUM OF RHOS = ',⍕(0.5×+/,ZZ×ZZ×~I)+K×+/,I×(|ZZ)-K÷2
[21]  L7:'DBETA = ',⍕DBETA←DBETA×1⌊J+0.1
[22]   'BETA = ',⍕BETA←BETA+DBETA
      ∇


      ∇ DZ←HUBER1;Y
[1]    →2+((ρBETA)=¯1↑ρX)∧(N=1↑ρX)∧(2=ρρX)∧1=ρρZ
[2]    →ρ⎕←'HUBER1 NO GO.',DZ←''
[3]    DZ←-X+.×DBETA←((K⌊Z⌈-K)+.×X)⌹(⍉Y)+.×Y←(~I)⌿X
      ∇


      ∇ DZ←HUBER2;A;B;C;D;M;U
[1]    →2+((ρBETA)=1++/ρZ)∧(∧/3≤ρZ)∧2=ρρZ
[2]    →ρ⎕←'HUBER2 NO GO.',DZ←''
[3]    U←(+/,U),(+/U),+⌿U←K⌊Z⌈-K
[4]    C←((ρZ)[2 2]ρ1,(ρZ)[2]ρ0)×(ρZ)[2 2]ρA←+⌿~I
[5]    D←((ρZ)[1 1]ρ1,(ρZ)[1]ρ0)×(ρZ)[1 1]ρB←+/~I
[6]    M←((+/B),-(+/I),+⌿I),[1](B,D,-I),[1] A,(⍉-I),C
[7]    A←B←C←D←''
[8]    DZ←-(1↑DBETA)+((ρZ)[1]↑1↓DBETA)∘.+(ρZ)[1]↓1↓DBETA←U⌹M
      ∇
```

```
HOWTESTS

     TESTS ON RESIDUALS AFTER LEAST-SQUARES LINEAR REGRESSION
     (1)  TESTS I
     (2)  TESTS1
     (3)  TESTS2

   TESTS FOR DISTRIBUTION SHAPE, HETEROSCEDASTICITY, NONADDITIVITY, AND (OPTION-
ALLY) SERIAL CORRELATION,  ARE CARRIED OUT ON RESIDUALS FROM A FITTED REGRESSION
RELATION.  SEVERAL GLOBAL VARIABLES PRODUCED BY 'REGR' ARE NEEDED:  RX, RY, IND,
RB, NU, RSS .  SPECIAL CASES OF SOME OF THE TESTS APPEAR IN 'ROWCOL' AND 'SUMMA-
RIZE'.  SEE
ANSCOMBE:  PROC. 4TH BERKELEY SYMP. 1 (1961), 1-36.
COMPUTING IN STATISTICAL SCIENCE THROUGH APL, APPENDIX 2.

   THE ARGUMENT (I)  IS EITHER 1 OR 2,  CONTROLLING WHICH SUBSIDIARY FUNCTION IS
USED FOR COMPUTING MOMENTS OF THE TEST STATISTICS.     IF  I  IS 1, EXACT 2ND AND
3RD MOMENTS ARE OBTAINED BY RAPID MATRIX COMPUTATIONS  REQUIRING SUFFICIENT ROOM
IN THE WORKSPACE FOR 3 OR 4 N-BY-N MATRICES, WHERE  N  IS  1↑⍴RX .  IF  I  IS 2,
EXACT 2ND MOMENTS ARE FOUND BY  N  ITERATIONS OF A LOOP REQUIRING LESS ROOM, AND
THE TWO 3RD MOMENTS CALCULATED ARE USUALLY ONLY APPROXIMATE.

   IF BEFORE EXECUTION A GLOBAL MATRIX  X  IS SPECIFIED HAVING  N  ROWS, THE HE-
TEROSCEDASTICITY TESTS WILL INCLUDE REGRESSION OF SQUARED RESIDUALS ON EACH COL-
UMN OF  X  AS WELL AS ON THE FITTED VALUES.

10 FEB. 1977
```

```
      ∇ TESTS2;C;C1;C2;D;I;Q;R
[1]     'EXACT SECOND MOMENTS, APPROXIMATE THIRD MOMENTS.  LOOP ITERATIONS:'
[2]     D1←D3←D4←D5←D6←E6←F1←I5←0
[3]     R←(1=⍳N)-U+.×U[I←1;]
[4]   L1:R←Q×Q←R
[5]     →I5/L2
[6]     I5←I5∨1≠1++/Q
[7]     D5←D5+YY[I;]×Q+.×YY
[8]   L2:D3←D3+Y[I;]×R+.×Y
[9]     →IX/L3
[10]    D4←D4+X[I;]×R+.×X
[11]  L3:D1←D1+Q+.×R
[12]    D2←D2+R+.×R
[13]    F1←F1+DIAGQ[I]×Q+.×DIAGQ
[14]    F2←F2+DIAGQ[I]×R+.×DIAGQ
[15]    →((N<I),0=10|I←I+1)/L6,L5
[16]  L4:R←(I=⍳N)-U+.×U[I;]
[17]    →I6/L1
[18]    E6←E6-Q[I]
[19]    →L1,⍴D6←D6+((1↓Q)+.×¯1↓R)+(¯1↓Q)+.×1↓R
[20]  L5:→L4,⎕←I
[21]  L6:D←NU×C×(C2←C×C)+(¯3×C÷NU+2)+(C1×C1←1-C←NU÷N)÷N-1
[22]    M03←(¯6×C×D÷NU+2)+NU×C2×(C×C2)+(3×C×C1×C1÷N-1)+(C1×C1×C1÷÷/N- 1 2 1)-C÷÷/
        NU+ 2 8 2
[23]    M03←(÷/NU+ 0 2 0 4 0 6 0 8 0 10)×1728×M03÷D×D×D
[24]    M21←(÷/216,NU+ 2 0 4 0 6 0 8)×(1-2÷(NU+2)×C-C1÷N-1)÷D
      ∇
```

```
     ∇ TESTS I;D;D1;D2;D3;D4;D5;D6;E6;F1;F2;IX;I5;I6;J;M03;M21;N;P;R;S;S1;S2;S3
        ;S4;S6;T;U;Y;YY;Z
[1]    →2+(∨/,I∊⍳2)∧(0=⍴⍴I)∧(NU≥3)∧(N≥NU)∧(1↑⍴RY)=N←(⍴RX)[1]
[2]    →0,⍴⎕←'NO GO.'
[3]    T←DIAGQ+.×DIAGQ←1-+/U×U←U÷(⍴U)⍴(+/U×U←IND/RX)*0.5
[4]    RY←(2↑(R←⍴RY),1)⍴RY
[5]    YY←Y×Y←(⍴RY)⍴Y-(⍴Y)⍴(÷NU)×DIAGQ+.×Y←RX+.×RB
[6]  ⍝  RX ,  IND ,  RB  MAY NOW BE ERASED.
[7]    →2×⍳∨/1≥1+S←RSS÷NU
[8]    →(IX←2≠⎕NC 'X̲')/11
[9]    →2×⍳(2≠⍴⍴X̲)∨N≠1↑⍴X̲
[10]   X̲←X̲-(⍴X̲)⍴(÷NU)×DIAGQ+.×X̲
[11]   'SERIAL TEST?  Y OR N'
[12]   D2←(3÷NU+2)×F2←-T×T÷NU
[13]   I6←'Y'≠1↑⍞
[14]   →L̲1×⍳I=2
[15]   TESTS1
[16]   →1+L̲1
[17]  L̲1:TESTS2
[18]   '   *DISTRIBUTION SHAPE*',⎕←''
[19]   G1←(+/RY×Z←RY×RY)÷D1×S*1.5
[20]   S1←(÷/(6+9×F1÷D1),D1,NU+ 0 2 0 4)*0.5
[21]   'SKEWNESS:  G1 = ',(⍕G1),(P←' +',⎕TC[1],'_ '),(⍕S1),'  (RATIO = ',(2⍕G1÷S1
       ),')'
[22]   G2←(((+/Z×Z)÷S×S)-3×T×÷/NU+ 0 2)÷D2
[23]   D←G2÷S2←(÷/(24+72×F2÷D2),D2,NU+ 0 2 0 4 0 6)*0.5
[24]   'KURTOSIS:  G2 = ',(⍕G2),P,⍕S2
[25]   →(5+1↑⎕LC)×⍳0.01≥M03←M03÷S2×S2×S2
[26]   A←6+4×A×A+4○A←2÷M03
[27]   →(3+1↑⎕LC)×⍳∨/0≥D←1+D×(2÷A-4)*0.5
[28]   D←(1-(2÷9×A)+((1-2÷A)÷D)*÷3)÷(2÷9×A)*0.5
[29]   →(2+⎕LC),⍴⎕←'  (STANDARDIZED 0 3 MOMENT = ',(⍕M03),', ROUGH E.N.D. FOR G2
        = ',(2⍕D),')'
[30]   '  (STANDARDIZED 0 3 MOMENT = ',(⍕M03),')'
[31]   →(2+1↑⎕LC)×⍳∨/0≥D←1+(G2÷S2)×M21←M21÷S1×S1×S2
[32]   →(2+⎕LC),⍴⎕←'  (STANDARDIZED 2 1 MOMENT = ',(⍕M21),', CRUDE E.N.D. FOR G1
        GIVEN G2 = ',(2⍕G1÷S1×D*0.5),')'
[33]   '  (STANDARDIZED 2 1 MOMENT = ',(⍕M21),')'
[34]   '   *HETEROSCEDASTICITY TESTS*',⎕←''
[35]   H←J\(J/+/Y×Z)÷(J←RSS<RSS+D3)/D3×S
[36]   S3←(J\(2×NU÷NU+2)÷J/D3)*0.5
[37]   'AGAINST FITTED VALUES:  H = ',(⍕H),P,(⍕S3),'  (RATIO = ',(2⍕J\(J/H)÷J/S3)
       ,')'
[38]   →IX/L̲3
[39]   J←1<1+D4×I←1
[40]   A←J⍀((J⌿⍉X̲)+.×Z)÷(J/D4)∘.×S
[41]   S4←(J\(2×NU÷NU+2)÷J/D4)*0.5
[42]  L̲2:'AGAINST COLUMN ',(⍕I),' OF X̲:  A[',(⍕I),';] = ',(⍕A[I;]),P,(⍕S4[I]),'
       (RATIO = ',(2⍕A[I;]÷S4[I]+S4[I]=0),')'
[43]   →L̲2×⍳(⍴A)[1]≥I←I+1
[44]  L̲3:'   *NONADDITIVITY--REGRESSION OF  RY  ON SQUARED FITTED VALUES*',⎕←''
[45]   →(I5/L̲5),(∨/RSS≥RSS+D5)/L̲6
[46]   'COEFFICIENT:  F = ',⍕F←(Z←+/RY×YY)÷D5
[47]   'TUKEY''S 1-D.F. TERM:  ',⍕Z←F×Z
[48]   'REMAINING SUM OF SQUARES:  ',⍕Z←RSS-Z
[49]   'MEAN SQUARE (',(⍕NU-1),' D.F.):  ',⍕Z÷NU-1
[50]  L̲4:→I6/L̲5-1
```

```
[51]  '    *FIRST-ORDER SERIAL CORRELATION*',⎕←''
[52]  R1←(E6+(+⌿(1 0 ↓RY)× ¯1 0 ↓RY)÷S)÷D6
[53]  S6←((NU-2×E6×E6÷D6)÷D6×NU+2)*0.5
[54]  'COEFFICIENT:   R1 = ',(⍕R1),P,(⍕S6),'  (RATIO = ',(2⍕R1÷S6),')'
[55]  →0,ρRY←RρRY
[56] L5:→L4,ρ⍞←'NO GO, ROWS OF  Q  DO NOT SUM TO 0.'
[57] L6:→L4,ρ⍞←'NO GO, DIVISOR VANISHES.'
     ∇


     ∇ TESTS1;Q;R;T1;T2;T3;T4;U1;U2;U3;U4;U5
[1]   'FAST COMPUTATION, EXACT SECOND AND THIRD MOMENTS.'
[2]   Q←((N,N)ρ1,Nρ0)-U+.×⍉U
[3]   U←''
[4]   →L1×ιI5←∨/1≠1++/Q
[5]   D5←+⌿YY×Q+.×YY
[6]  L1:F1←DIAGQ+.×Q+.×DIAGQ
[7]   F2←F2+DIAGQ+.×T4←DIAGQ+.×R←Q×Q
[8]   T4←T4+.×T4
[9]   D2←D2++/,R×R
[10]  D3←+⌿Y×R+.×Y
[11]  →IX/L2
[12]  D4←+⌿X×R+.×X
[13] L2:U1←+/,Q×R+.×R
[14]  T1←+/,R×R+.×R
[15]  U5←DIAGQ+.×Q+.×R+.×DIAGQ
[16]  D1←+/,R←R×Q
[17]  U4←+/R+.×Q+.×DIAGQ
[18]  R←''
[19]  R←(Q×(ρQ)ρDIAGQ)+.×Q
[20]  U3←DIAGQ+.×R+.×DIAGQ
[21]  T2←DIAGQ+.×(R←Q×R)+.×DIAGQ
[22]  U2←+/,R←Q×R
[23]  T3←+/,Q×R
[24]  M21← 216 216 108 144 216 +.×U1,U2,U3,U4,U5,R←''
[25]  M21←(÷/NU+ 0 2 0 4 0 6 0 8)×(M21-(÷/108,NU+ 0 4 2)×T× 2 3 +.×D1,F1)÷D2×D1×
      D1
[26]  M03← 1728 1728 3456 2592 +.×T1,T2,T3,T4
[27]  M03←(÷/NU+ 0 2 0 4 0 6 0 8 0 10)×(M03-(864×T÷×/NU+ 0 2)×((NU⊥ 5 34 60)×T×T
      ÷×/NU+ 0 2)+4×(NU+5)×D2+3×F2)÷D2*3
[28]  →I6/0
[29]  E6←-+/ 1 1 ⍉ ¯1 1 ↓Q
[30]  D6←+/,(¯1 ¯1 ↓Q)× 1 1 ↓Q
[31]  D6←D6++/,(¯1 1 ↓Q)× 1 ¯1 ↓Q
     ∇


     ∇ T←CCD S;J;N;U
[1]   →3-(1≤N←1↑ρS)∧2=ρρS
[2]   →3+(∧/(,S)=,⍉S)∧=/ρT←(ρS)ρJ←0
[3]   →ρ⍞←'ARGUMENT MUST BE A SYMMETRIC MATRIX.',T←''
[4]   →5+0<1↑U←(J↓S[J+1;])-U[;1]+.×U←(J,J-N)↑T
[5]   →ρ⍞←'ARGUMENT IS NOT POSITIVE DEFINITE.'
[6]   T[J+1;J+ιN-J]←U÷U[1]*0.5
[7]   →4×N>J←J+1
[8]  ⍝ COMPLETE CHOLESKY DECOMPOSITION:  S ↔ (⍉T)+.×T
     ∇
```

HOWGAIN

GAIN OF A LINEAR FILTER
(1) *G←F GAIN W*
(2) *Z←F GAINLAG W*

BOTH FUNCTIONS GIVE THE GAIN AT FREQUENCIES F (CYCLES PER UNIT TIME) OF A LINEAR FILTER HAVING WEIGHTS W .

(1) *GAIN. IF THE WEIGHTS ARE SYMMETRIC* (∧/*W*=⌽*W*) *THE LAG IS TAKEN TO BE CONSTANT, EQUAL TO* 0.5×¯1+ρ*W* , *AND THE GAIN G MAY BE NEGATIVE. IF THE WEIGHTS ARE UNSYMMETRIC THE LAG IS NOT COMPUTED AND G IS MERELY THE MAGNITUDE OF THE GAIN. ATTENTION IS DRAWN TO ANY ZEROES AMONG THE VALUES OF G .*

(2) *GAINLAG IS USED TO COMPUTE LAG AND GAIN WHEN W IS NOT SYMMETRIC. F IS SUPPOSED TO BE IN INCREASING ORDER, AND THE GAIN SHOULD NOT VANISH AT ANY MEMBER OF F ; 'GAIN' MAY BE RUN FIRST TO VERIFY THIS. CONSECUTIVE MEMBERS OF F SHOULD BE CLOSE ENOUGH TOGETHER FOR PHASE CHANGES TO BE SMALL. INSERT MORE VALUES IN F AND RERUN IF YOU GET A WARNING MESSAGE ABOUT LARGE PHASE CHANGES. (ABSENCE OF A WARNING MESSAGE DOES NOT GUARANTEE A PRINCIPAL SOLUTION.) THE EXPLICIT RESULT* (*Z*) *LISTS GAINS IN ITS FIRST COLUMN AND LAGS IN ITS SECOND.*

19 *APR.* 1977

```
      ∇ G←F GAIN W;J
[1]   →2+(∧/0≤F)∧(∧/0.5≥F←,F)∧2≤ρJ←¯1+⍳ρW←,W
[2]   →0,ρ⎕←'NO GO.',G←''
[3]   →(∨/W≠⌽W)/5
[4]   →0,ρG←(2○F∘.×○(2×J)-¯1↑J)+.×W
[5]   →0×⍳∧/1≠1+G←(+⌿G×G←(2 1 ∘.○F∘.×J×○2)+.×W)*0.5
[6]   'GAIN VANISHES AT FREQUENCY NO. ',⍕(1=1+G)/⍳ρF
      ∇


      ∇ Z←F GAINLAG W;A;I;J;X
[1]   →2+(∧/0<(1↓F)-¯1↓F)∧(∧/0≤F)∧(∧/0.5≥F←,F)∧2≤ρJ←¯1+⍳ρW←,W
[2]   →0,ρ⎕←'NO GO.',Z←''
[3]   X←(÷○1)×¯3○÷⌿A←(1 2 ∘.○F∘.×J×○2)+.×W
[4]   →L1×⍳∧/I←0.25>|1↓X←¯0.5+1|0.5+X-0,¯1↓X
[5]   'LARGE PHASE CHANGE AT FREQUENCY NO. ',⍕1+(~I)/⍳¯1+ρF
[6]  L1:→(0=1↑F)/L2
[7]   →(1+L2),ρZ←(+\X)÷2×F
[8]  L2:Z←((W+.×J)÷+/W),(+\1↓X)÷2×1↓F
[9]   Z←(+⌿A× 1 2 ∘.○Z×F×○2),[1.5] Z
      ∇


      ∇ C←N AUTOCOV V;J
[1]   →2+(,N≤¯1++/ρV)∧(J←2≤+/ρV)∧(1=ρρV)∧0=ρρN
[2]   →ρ⎕←'NO GO.',C←''
[3]   C←V+.×V
[4]   →4×N≥J←ρC←C,(J↓V)+.×(-J)↓V
[5]  ⍝ FIRST  N  SERIAL CORRELATIONS OF  V  ARE  1↓C÷C[1]
      ∇
```

```
HOWFILTER

          FILTERING TIME SERIES
          (1)  U←M FILTER X
          (2)  U←W MAV X
```

(1) FILTER. THE FIRST ARGUMENT (M) IS A PAIR OF INTEGERS DEFINING THE FILTER. THE SECOND ARGUMENT (X) IS A VECTOR OF DATA TO BE FILTERED. FILTERING CONSISTS OF SUBTRACTING A COSINE-WEIGHTED MOVING AVERAGE OF EXTENT M[2] FROM A SIMILAR MOVING AVERAGE OF EXTENT M[1] . THE RESULTING WEIGHTS ARE DISPLAYED, TOGETHER WITH THEIR SUM OF SQUARES. THE ELEMENTS OF M MUST BE EITHER BOTH ODD OR BOTH EVEN, OR ELSE ONE OF THEM MUST BE 0. TO GET AN 11-POINT MOVING AVERAGE, SET M EQUAL TO 11 0 . TO DETREND A SERIES BY SUBTRACTING FROM IT AN 11-POINT MOVING AVERAGE, LET M BE 1 11 . FOR A SMOOTHED DETRENDED SERIES, LET M BE (E.G.) 3 11 . SETTING M EQUAL TO 3 1 YIELDS THE SECOND DIFFERENCE OF X (DIVIDED BY 4). IF THE NONZERO ELEMENTS OF M ARE ODD, THE RESULT CORRESPONDS TO THE SAME TIME POINTS AS THE DATA (APART FROM LOSS OF SOME END VALUES). BUT IF BOTH ELEMENTS OF M ARE EVEN, THE RESULT CORRESPONDS TO TIME POINTS MIDWAY BETWEEN THOSE OF THE DATA.

IF AN ELEMENT OF M EXCEEDS 4, A LITTLE CHEATING GOES ON AT THE ENDS OF X , THE FIRST AND LAST ELEMENTS BEING REPEATED R TIMES AS DEFINED AT [7]. IF YOU DO NOT WANT THIS FEATURE, EITHER REPLACE [7] BY

```
[7]   R←0
```

OR DROP THE FIRST AND LAST R ELEMENTS OF U .

(2) MAV. A MOVING AVERAGE WITH ARBITRARY WEIGHTS IS TAKEN OF EITHER ONE VECTOR OR SEVERAL VECTORS SIMULTANEOUSLY. THE FIRST ARGUMENT (W) IS THE VECTOR OF WEIGHTS. THE SECOND ARGUMENT (X) IS EITHER A VECTOR OF DATA OR A MATRIX WHOSE COLUMNS ARE THE VECTORS OF DATA TO BE AVERAGED. THE RESULT (U) IS A VECTOR OR A MATRIX: `ρρU ↔ ρρX ; (ρU)[1] ↔ (ρX)[1]+1-ρW` .

30 JAN. 1977

```
      ∇ U←M FILTER X;P;R;W
[1]     →3-(∧/2=ρM)∧(1=ρρM)∧1=ρρX
[2]     →3+∧/(0≤M)∧(M=⌊M)∧((0=×/M)∨0=2|-/M)∧(≠/M)∧(2×¯2+ρX)≥P←⌈/M
[3]     →0,ρ⎕←'NO GO.',U←''
[4]     W←(⌊0.5×P-M[1])⌽((P-M[1])ρ0),(1-2○(⍳M[1])×○2÷1+M[1])÷1+M[1]
[5]     W←W-(⌊0.5×P-M[2])⌽((P-M[2])ρ0),(1-2○(⍳M[2])×○2÷1+M[2])÷1+M[2]
[6]     'FILTER WEIGHTS ARE  ',(⍕W),'  WITH S.S.  ',⍕W+.×W
[7]     R←⌊(P-1)÷4
[8]     →L1×⍳⎕WA≤16×P×ρX←(Rρ1↑X),X,Rρ¯1↑X
[9]     →0,ρU←W+.×(0,1-P)↓(¯1+⍳P)⌽(P,ρX)ρX
[10]  L1:R←ρρU←((ρX)+1-P)ρ0
[11]    U←U+W[R]×(R-1)↓(R-P)↓X
[12]    →(P≥R←R+1)/L1+1
      ∇


      ∇ Z←W MA X;R
[1]     →2+(R<ρX)∧(2≤R←+/ρW)∧(1=ρρW)∧1=ρρX
[2]     →ρ⎕←'NO GO.',Z←''
[3]     Z←(1-R)↓W+.×(¯1+⍳R)⌽(R,ρX)ρX
[4]   ⍝ MOVING AVERAGE OR FILTERING OF A SERIES  X  WITH WEIGHTS  W .
      ∇
```

```
HOWFOURIER

          FOURIER ANALYSIS OF TIME SERIES
          (1)  V←K TAPER U
          (2)  S←D FFT V
          (3)  T←POLAR S

     (1) TAPER.  THE FIRST ARGUMENT (K) IS A POSITIVE INTEGER, AND THE SECOND (U)
IS A VECTOR OF LENGTH GREATER THAN  2×K .  THE RESULT (V) IS A VECTOR OF LENGTH
(ρU)-K , A CIRCULARIZED VERSION OF  U  OBTAINED BY OVERLAPPING THE FIRST AND THE
LAST  K  ELEMENTS.  THE FIRST  K  ELEMENTS OF  U  MULTIPLIED BY A VECTOR OF LI-
NEAR WEIGHTS  W , PLUS THE LAST  K  ELEMENTS OF  U  MULTIPLIED BY  1-W , FORM  K
ELEMENTS OF  V , SOME AT THE BEGINNING AND SOME AT THE END.  THE REST OF  V  IS
THE SAME AS THE REST OF  U .  THIS PROCESS IS NEARLY BUT NOT QUITE WHAT IS GENE-
RALLY CALLED TAPERING.

     (2) FFT (FAST FOURIER TRANSFORM).  THE FIRST ARGUMENT (D) IS SCALAR, EITHER 0
OR 1 OR 2.  THE SECOND ARGUMENT (V) IS A THREE-DIMENSIONAL ARRAY.  THE FIRST TWO
ELEMENTS OF  ρV  SHOULD (FOR EFFICIENCY) BE NEARLY EQUAL,  AND THE THIRD ELEMENT
MUST BE 2.

     IF  D  IS 0,  THE FUNCTION YIELDS  THE COMPLEX FOURIER TRANSFORM  OF A SINGLE
COMPLEX TIME SERIES.   (,V[;;1]) IS THE REAL PART AND (,V[;;2]) IS THE IMAGINARY
PART OF THE SERIES.  DENOTE THESE BY X[T] AND Y[T], WHERE  T = 0, 1, ..., (N-1),
AND  N  IS  ×/2↑ρV .  THE RESULT (S) IS A MATRIX WITH N ROWS AND 2 COLUMNS.  THE
(R+1)TH ROW OF  S  IS (A[R], B[R]), WHERE  R = 0, 1, ..., (N-1), AND
     A[R] + I×B[R]  ↔  SUM OVER T OF  (X[T] + I×Y[T]) × EXP(2×PI×I×R×T÷N) .
HERE  I  IS ¯1*0.5 .  NOTE THAT THE SUM IS NOT DIVIDED BY A POWER OF  N.  CHECK:
          +/,S*2 ↔ N×+/,V*2

     IF  D  IS 1,  THE FUNCTION YIELDS THE REAL FOURIER TRANSFORM OF A SINGLE REAL
SERIES; THE TRANSFORM IS SCALED TO GIVE A DIRECT ANALYSIS OF VARIANCE.  THE SE-
RIES IS (,V);  DENOTE THIS BY X[T], WHERE  T = 0, 1, ..., (N-1),  AND  N  IS NOW
×/ρV .  NOTE THAT  N  IS NECESSARILY DIVISIBLE BY 2.  THE RESULT (S) IS A MATRIX
OF (N+2)÷2 ROWS AND 2 COLUMNS.  THE (R+1)TH ROW OF  S  IS (A[R], B[R]), WHERE
          A[R]  ↔  ((2÷N)*0.5) ×  SUM OVER T OF  X[T] × COS(2×PI×R×T÷N)
          B[R]  ↔  ((2÷N)*0.5) ×  SUM OVER T OF  X[T] × SIN(2×PI×R×T÷N)
EXCEPT THAT WHEN  R  IS 0 OR  N÷2  THE MULTIPLIER ON THE RIGHT SIDE IS (N*¯0.5).
CHECK THE ANALYSIS OF VARIANCE:
          +/,S*2 ↔ +/,V*2

     IF  D  IS 2,  THE FUNCTION YIELDS SIMULTANEOUSLY  THE REAL TRANSFORMS  OF TWO
REAL SERIES, EACH SCALED AS WHEN  D  IS 1.    THE FIRST SERIES IS (,V[;;1]), THE
SECOND IS (,V[;;2]); THEIR COMMON LENGTH  N  IS  ×/2↑ρV .   THE RESULT (S) IS 3-
DIMENSIONAL, HAVING ⌊(N+2)÷2 PLANES, 2 ROWS AND 2 COLUMNS.  S[;;1] IS THE TRANS-
FORM OF THE FIRST SERIES, S[;;2] THAT OF THE SECOND SERIES.  CHECK:
          +/,S[;;1]*2 ↔ +/,V[;;1]*2
          +/,S[;;2]*2 ↔ +/,V[;;2]*2

     (3) POLAR IS USED TO TRANSFORM THE OUTPUT OF 'FFT' TO POLAR FORM.  THE ARGU-
MENT (S) IS A MATRIX WITH 2 COLUMNS,  SUCH AS THE RESULT OF 'FFT' WHEN  D  IS 1,
OR HALF THE RESULT (THE THIRD INDEX BEING CONSTANT, 1 OR 2) OF 'FFT' WHEN  D  IS
2.  THE RESULT (T) IS A MATRIX OF THE SAME SIZE, THE FIRST COLUMN BEING SQUARED
AMPLITUDES AND THE SECOND COLUMN PHASES:
          T[;1] ↔ +/S*2
          S[;1] ↔ (T[;1]*0.5) × COS T[;2]
          S[;2] ↔ - (T[;1]*0.5) × SIN T[;2]

1 FEB. 1977
```

```
      ∇ W←CW R
[1]   W←(1-2○(⍳R)×○2÷R+1)÷R+1
[2]  ⍝ WEIGHTS FOR A COSINE-WEIGHTED MOVING AVERAGE OF LENGTH  R .
      ∇


      ∇ U←W MAV X;D;J;L
[1]   →3-(1=⍴⍴W)∧∨/ 1 2 =⍴⍴X
[2]   →3+(L≥1)∧(L←⍴W)≤1↑D←⍴X
[3]   →0,⍴⎕←'NO GO.',U←''
[4]   D[1]←D[1]+1-L
[5]   U←D⍴J←0
[6]   U←U+W[J←J+1]×D⍴J⊖X
[7]   →6×L>J
      ∇


      ∇ S←D FFT V;N;N1;N2;U
[1]   →3-(0=⍴⍴D)∧3=⍴⍴V
[2]   →3+(D∊ 0 1 2)∧(2=(⍴V)[3])∧(2≤N1←(⍴V)[1])∧2≤N2←(⍴V)[2]
[3]   →0,⍴⎕←'NO GO.',S←''
[4]   S← 2 1 ∘.○(¯1+⍳N1)∘.×(¯1+⍳N1)×○2÷N1
[5]   S←S+.×V
[6]   V[;;1]←S[1;;;1]-S[2;;;2]
[7]   V[;;2]←S[1;;;2]+S[2;;;1]
[8]   S←''
[9]   S← 2 1 ∘.○(¯1+⍳N1)∘.×(¯1+⍳N2)×○2÷N←N1×N2
[10]  U←(S[1;;]×V[;;1])-S[2;;]×V[;;2]
[11]  V[;;2]←(S[1;;]×V[;;2])+S[2;;]×V[;;1]
[12]  V[;;1]←U
[13]  U←S←''
[14]  V← 2 1 3 ⍉V
[15]  S← 2 1 ∘.○(¯1+⍳N2)∘.×(¯1+⍳N2)×○2÷N2
[16]  S←S+.×V
[17]  V[;;1]←S[1;;;1]-S[2;;;2]
[18]  V[;;2]←S[1;;;2]+S[2;;;1]
[19]  V←S←(N,2)⍴V
[20]  →(D=0)/0
[21]  V[;2]←0,0.5×(⌽1↓S[;1])-1↓S[;1]
[22]  V[;1]←S[1;2],0.5×(1↓S[;2])+⌽1↓S[;2]
[23]  S[;1]←S[1;1],0.5×(1↓S[;1])+⌽1↓S[;1]
[24]  S[;2]←0,0.5×(1↓S[;2])-⌽1↓S[;2]
[25]  →(D=1)/L̲
[26]  V←((1+⌊N÷2),2)↑V
[27]  S←((1+⌊N÷2),2)↑S
[28]  S←((2÷N)*0.5)×S,[2.5] V
[29]  S[1;;]←S[1;;]×2*¯0.5
[30]  →(1=2|N)/0
[31]  →0,⍴S[1+N÷2;;]←S[1+N÷2;;]×2*¯0.5
[32] L̲:V←V,[1] V[1;]
[33]  S←S,[1] S[1;]
[34]  U← 2 1 ∘.○○(0,⍳N)÷N
[35]  S[;1]←S[;1]+(U[1;]×V[;1])-U[2;]×V[;2]
[36]  S[;2]←S[;2]+(U[1;]×V[;2])+U[2;]×V[;1]
[37]  U←V←''
[38]  S←S×N*¯0.5
[39]  S[1,N+1;]←S[1,N+1;]×2*¯0.5
      ∇
```

```
      ∇ V←K TAPER U
[1]     →3-(0=ρρK)∧1=ρρU
[2]     →3+(K=⌊K)∧(K≥1)∧1≤(ρU)-2×K
[3]     →0,ρ⎕←'NO GO.',V←''
[4]     (⍕K),' LINEAR TAPER WEIGHTS WITH INCREMENT ',⍕÷K+1
[5]     U[⍳K]←((-K)↑U)+((⍳K)÷K+1)×U[⍳K]-(-K)↑U
[6]     V←(⌊K÷2)⌽(-K)↓U
      ∇
```

```
      ∇ T←POLAR S;V
[1]     →2+(2=¯1↑ρS)∧2=ρρS
[2]     →ρ⎕←'NO GO.',T←''
[3]     T←(+/S×S),[1.5]○0.5×-(×S[;2])×S[;1]=0
[4]     S←(V←S[;1]≠0)/S
[5]     T[;2]←(○2)|T[;2]+V\(○0>S[;1])-¯3○÷/S[; 2 1]
      ∇
```

HOWHAR

HARMONIC REGRESSION OF A TIME SERIES ON ONE OR MORE OTHER TIME SERIES

(1) PREHAR
(2) HARINIT U
(3) B HAR1 V
(4) Z←HAR1R
(5) B HAR2 V

SEE CHAPTER 10 OF <u>COMPUTING IN STATISTICAL SCIENCE THROUGH APL</u>, ESPECIALLY FIGURES B:8, 11, 17, AND NEIGHBORIMG TEXT.

'PREHAR' GENERATES PHASE-DIFFERENCE PLOTS BETWEEN PAIRS OF SERIES. 'HARINIT' INITIALIZES FOR THE REMAINING FUNCTIONS. 'HAR1' PERFORMS HARMONIC REGRESSION OF A 'DEPENDENT' SERIES ON ONE 'PREDICTOR' SERIES. 'HAR2' PERFORMS HARMONIC REGRESSION OF A DEPENDENT SERIES ON TWO PREDICTOR SERIES. 'HAR1R' GENERATES THE RESIDUAL SERIES AFTER EXECUTION OF 'HAR1'.

THE DATA ARE PASSED TO 'PREHAR' IN THE GLOBAL VARIABLE FT , A 3-DIMENSIONAL ARRAY. THE SECOND ELEMENT OF ρFT IS 2, THE THIRD IS THE NUMBER OF SERIES. FT[;;1] IS THE FOURIER TRANSFORM OF THE DEPENDENT SERIES, AND FT[;;2], FT[;;3], ... ARE THOSE OF THE PREDICTOR SERIES, AS GIVEN BY 'FFT' (SEE 'HOWFOURIER'). THE SERIES ARE REFERRED TO IN THE OUTPUT AS Y0 (THE DEPENDENT SERIES), Y1, Y2, ... IT IS NOT NECESSARY THAT THESE REALLY ARE THE NAMES OF THE SERIES.

BEFORE 'HARINIT' IS EXECUTED, THE VECTOR W MUST BE SPECIFIED, AND BEFORE 'HAR1' OR 'HAR2' THE SCALARS TOL AND ITS .

THE LENGTH OF THE SERIES IS NOT PASSED DIRECTLY TO THESE FUNCTIONS. THE LENGTH IS ASSUMED TO BE EVEN, IN WHICH CASE IT WILL BE EQUAL TO 2×HN , WHERE HN IS DEFINED IN 'HARINIT'. THE FUNCTIONS CAN BE ADAPTED TO HANDLE SERIES OF ODD LENGTH AS FOLLOWS. IN 'PREHAR' OMIT '¯1↓' IN [17] AND '¯1 0 ↓' IN [20]. IN HAR1[6], HAR1R[2], HAR2[6], CHANGE '2×HN' TO '1+2×HN'. OTHERWISE LET THE LENGTH BE EVEN.

27 JAN. 1981

```
      ∇ PREHAR;E;K;T
[1]   ⍝ GLOBAL  FT  IS THE TRANSFORMS OF INPUT SERIES  Y0 , Y1 ,  ..., YP ; (1↓ρ
      FT) ↔ 2,P+1
[2]    →3+(2=1↓¯1↓ρFT)∧(2≤¯1↑ρFT)∧3=ρρFT
[3]    →0,ρ⎕←'NO GO.'
[4]    PAIRS← 0 2 ρK←1
[5]    →(1↑⎕LC)×ι(2!¯1↑ρFT)>1↑ρPAIRS←PAIRS,[1](ιK-1),[1.5] K←K+1
[6]    Z←E←(ρFT)[1 3]ρ0.5+K←1
[7]   L̲1:Z[;K]← 0 ¯1 ↓T←POLAR FT[;;K]
[8]    E[;K]← 0 1 ↓T
[9]    →((¯1↑ρFT)≥K←K+1)/L̲1
[10]   ZZ←DE←((1↑ρFT),1↑ρPAIRS)ρ0.5+T←1
[11]  L̲2:ZZ[;T]←(×/Z[;PAIRS[T;]])*0.5
[12]   DE[;T]←(○2)|-/E[;PAIRS[T;]]
[13]   →((1↑ρPAIRS)≥T←T+1)/L̲2
[14]   E←1
[15]  L̲3:'PHASE DIFFERENCE PLOT FOR  Y',(1 0 ⍕PAIRS[E;1]-1),'  AND  Y',(1 0 ⍕
       PAIRS[E;2]-1),' ?  Y OR N'
[16]   →('N'=1↑⍞)/L̲4
[17]   T←1↓¯1↓ZZ[;E]*0.5
[18]   'DISTRIBUTION OF SYMBOLS (.∘○⊖⎕⍠):  ',⍕+/'.∘○⊖⎕⍠'∘.=T←'.∘○⊖⎕⍠'[⌈6×T÷⌈/T]
[19]   K←⍞
[20]   T TDP 1 0 ↓ ¯1 0 ↓DE[;E]∘.+0,○2
[21]   K←⍞
[22]  L̲4:→L̲3×(1↑ρPAIRS)≥E←E+1
      ∇
```

```
      ∇ HARINIT U
[1]    →3-(1=+/,W)∧(1=2|ρW)∧(3≤ρW)∧1=ρρW
[2]    →3+(∧/0<(1↓U)-¯1↓U)∧(∧/U∊¯1+ι¯1↑ρZ)∧2≤NV←ρVV←U←,U
[3]    →0,ρ⎕←'NO GO.'
[4]    'HN = ',⍕HN←¯1+1↑ρZ
[5]    'S = ',⍕S←+/(¯1+ρW)÷2
[6]    'EQUIVALENT D.F. = ',⍕2÷W+.×W
[7]    SSE←((S-1)+ι1+HN-2×S),((1+HN-2×S),¯2+4×NV)ρ0.5
[8]    SSE[;1+ιNV]←W MAV Z[;1+U]
[9]    'COLUMNS (CC) OF  ZZ  AND  DE  = ',⍕CC←(ι1↑ρPAIRS)+.×PAIRS∧.=1+U[⍉PAIRS[ι2
       !NV;]]
      ∇
```

```
      ∇ B HAR1 V;A;D;J;K;L;M;MM;X
[1]    →3-(NV=2)∧(0=ρρTOL)∧(0=ρρITS)∧(0=ρρB)∧(0≠1↓¯1↓V)∧(∧/V=⌊V)∧3=ρV←,V
[2]    →3+(S≤⌊/V[1 3])∧((HN-S)≥⌈/V[1 3])∧0=V[2]|-/V[1 3]
[3]    →0,ρ⎕←'NO GO.'
[4]    K←V[1]+1-S
[5]    V[3]←(+/1↓V)+1-S
[6]    MM←M,[0.5] M×M←((ιρW)-S+1)÷2×HN
[7]   L̲1:J←(ιρW)+K-1+L←0×D←2×TOL
[8]   L̲2:A←(○2)|(○0>X[2])-¯3○÷/X←(1 2 ∘.○DE[J;CC[1]]+B×M)+.×W×ZZ[J;CC[1]]
[9]    →((TOL>|D)∨ITS<L←L+1)/L̲3
[10]   →L̲2,B←B-D←÷/(MM× 1 2 ∘.○DE[J;CC[1]]+A+B×M)+.×W×ZZ[J;CC[1]]
[11]  L̲3:SSE[K;4]←(÷SSE[K;3])×X←(2○DE[J;CC[1]]+A+B×M)+.×W×ZZ[J;CC[1]]
[12]   SSE[K; 5 6 7]←A,B,X×X÷×/SSE[K; 2 3]
[13]   (L-1),D
[14]   FMT1⍕SSE[K;]
[15]   →L̲1×V[3]≠K←K+V[2]
      ∇
```

```
      ∇ Z←HAR1R
[1]     'VV = ',⍕VV
[2]     Z←((1↑SSE[;5])-(1↑SSE[;6])×(⌽⍳S)÷2×HN),SSE[;5],(¯1↑SSE[;5])+(¯1↑SSE[;6])×(
        ⍳S)÷2×HN
[3]     Z←((S⍴1↑SSE[;4]),SSE[;4],S⍴¯1↑SSE[;4]),[1.5] Z
[4]     Z←Z[; 1 1]×⍉ 2 1 ∘.○Z[;2]
[5]     Z←FT[;;1+VV[1]]-(-/Z×FT[;;1+VV[2]]),[1.5]+/Z×⌽FT[;;1+VV[2]]
      ∇
```

```
      ∇ B HAR2 V;A;C;D;E;G;J;K;L;M;MM;X
[1]     →3-(NV=3)∧(0=⍴⍴TOL)∧(0=⍴⍴ITS)∧(2=⍴B←,B)∧(0≠1↓¯1↓V)∧(∧/V=⌊V)∧3=⍴V←,V
[2]     →3+(S≤⌊/V[1 3])∧((HN-S)≥⌈/V[1 3])∧0=V[2]|-/V[1 3]
[3]     →0,⍴⎕←'NO GO.'
[4]     K←V[1]+1-S
[5]     V[3]←(+/1↓V)+1-S
[6]     MM←M×M←((⍳⍴W)-S+1)÷2×HN
[7]    L1:J←(⍳⍴W)+K-1+L←0×D←2×TOL
[8]    L2:C← 4 4 ⍴(16⍴ 1 0 0 0 0)\SSE[K; 3 3 4 4]
[9]     C[1; 3 4]←C[3 4 ;1]←(2 1 ∘.○DE[J;CC[3]]-M×-/B)+.×W×ZZ[J;CC[3]]
[10]    C[2; 3 4]←C[3 4 ;2]← ¯1 1 ×C[1; 4 3]
[11]    E←(2 1 ∘.○DE[J;CC[1]]+M×B[1])+.×W×ZZ[J;CC[1]]
[12]    E←E,(2 1 ∘.○DE[J;CC[2]]+M×B[2])+.×W×ZZ[J;CC[2]]
[13]    A←(X←POLAR 2 2 ⍴C←E⌹C)[;2]
[14]    G←X[;1]*0.5
[15]    →((TOL>⌈/|D)∨ITS<L←L+1)/L3
[16]    C← 2 2 ⍴((¯1 1 ×G[2]), 1 ¯1 ×G[1])×(2○DE[J;CC[3]]-(-/A)+M×-/B)+.×W×MM×ZZ[J
        ;CC[3]]
[17]    C[1;1]←C[1;1]+(2○DE[J;CC[1]]+A[1]+M×B[1])+.×W×MM×ZZ[J;CC[1]]
[18]    C[2;2]←C[2;2]+(2○DE[J;CC[2]]+A[2]+M×B[2])+.×W×MM×ZZ[J;CC[2]]
[19]    E←(G[2],-G[1])×(1○DE[J;CC[3]]-(-/A)+M×-/B)+.×W×M×ZZ[J;CC[3]]
[20]    E[1]←E[1]+(1○DE[J;CC[1]]+A[1]+M×B[1])+.×W×M×ZZ[J;CC[1]]
[21]    E[2]←E[2]+(1○DE[J;CC[2]]+A[2]+M×B[2])+.×W×M×ZZ[J;CC[2]]
[22]    →L2,B←B-D←E⌹C
[23]   L3:SSE[K; 5 8 6 9 7 10 11]←G,A,B,(C+.×E)÷SSE[K;2]
[24]    (L-1),D
[25]    FMT2⍕SSE[K;]
[26]    →L1×V[3]≠K←K+V[2]
      ∇
```

```
      ∇ S←N CSIF X;A;J;K;R;Z
[1]     A←N÷2
[2]     Z←X÷2
[3]     →⎕LC+1+(∧/,A>0)∧(0=⍴⍴A)∧⍴⍴Z←,Z⌈0
[4]     →0,⍴⎕←'NO GO.',S←''
[5]     →(A=⌊A)/L2
[6]     S←Z≠Z
[7]     →(0=K←+/J←Z≥7.107)/L1
[8]     S[J/⍳⍴S]←1-(*-J/Z)×((J/Z)∘.*A-⍳R)+.÷(1+R=⍳R)×!A-⍳R←⌈7.107+A
[9]    L1:→(0=K←+/J←(Z>0)∧~J)/0
[10]    →0,S[J/⍳⍴S]←(*-J/Z)×(((J/Z)∘.*¯1+A+⍳R)×1+(÷2×¯1+(A+R)÷J/Z)∘.×R=⍳R)+.÷!¯1+A
        +⍳R←1⌈⌈24.75-A
[11]   L2:S←1-(*-Z)×(Z∘.*¯1+⍳A)+.÷!¯1+⍳A
[12]   ⍝  CHI-SQUARED INTEGRAL FUNCTION WITH  N  D.F., ABS. ERROR LESS THAN 1E¯7.
[13]   ⍝  FOR INCOMPLETE GAMMA FUNCTION OMIT LINES [1]-[2] AND CHANGE HEADER TO
[14]   ⍝  ∇ S←A IGF Z;J;K;R
      ∇
```

```
      ∇ TSNT U;B;D;E;G;M;N;S
[1]     →2+3=⎕NC 'END'
[2]     →ρ⍞←'COPY END FROM 1234 ASP3'
[3]     →4+(∧/,0<D←(+/S←U×U)*2)∧(1≤N←+/ρU)∧1=ρρU
[4]     →0,ρ⍞←'NO GO.'
[5]     'B21  =  ',⍕B←(N×S+.×S)÷D
[6]     'APPROXIMATE MEAN  =  ',(⍕M←3÷1+2÷N),',   S.E.  =  ',(⍕E←(24÷N+15)*÷2),',
        GAMMA1  =  ',⍕G←14.7÷(N+29)*÷2
[7]     '   (E.N.D.  = ',(2⍕END),')'
[8]     'B22  =  ',⍕B←(N×+/((1↓S)+¯1↓S)*2)÷D
[9]     'APPROXIMATE MEAN  =  ',(⍕M←8÷1+3÷N),',   S.E.  =  ',(⍕E←(112÷N+15.7)*÷2),'
        ,  GAMMA1  =  ',⍕G←14.04÷(N+31.9)*÷2
[10]    '   (E.N.D.  = ',(2⍕END),')'
[11]    'B23  =  ',⍕B←(N×+/((2↓S)+(1↓¯1↓S)+¯2↓S)*2)÷D
[12]    'APPROXIMATE MEAN  =  ',(⍕M←15÷1+4÷N),',   S.E.  =  ',(⍕E←(296÷N+14.1)*÷2),
        ',  GAMMA1  =  ',⍕G←13.76÷(N+40.6)*÷2
[13]    '   (E.N.D.  = ',(2⍕END),')'
[14]  ⍝  TIME SERIES NORMALITY TEST.  THE ARGUMENT IS A VECTOR OF INNOVATIONS.
      ∇
```

```
      HOWMARDIA

          MARDIA'S MULTIVARIATE KURTOSIS TEST
          MARDIA X
```

THE ARGUMENT X IS A MATRIX OF 1↑ρX INDEPENDENT OBSERVATIONS OF A (ρX)[2]-DIMENSIONAL DISTRIBUTION. THE NULL HYPOTHESIS IS THAT THE DISTRIBUTION IS JOINTLY NORMAL. (BIOMETRIKA, 1970; SANKHYA, B, 1974.) THE ASYMPTOTIC THIRD MOMENT IS DUE TO MARDIA AND YAMAGUCHI.

2 SEP. 1980

```
      ∇ MARDIA X;B;E;G;M;N;P
[1]     M←(⍉X)+.×X←X-(ρX)ρ(+⌿X)÷N←(ρX)[1]
[2]     'B2',(⍕P←(ρX)[2]),'  =  ',⍕B←N×B+.×B←+/(X+.×⌹M)×X
[3]     'EXPECTATION  =  ',⍕M←(P×P+2)×÷/N+ ¯1 1
[4]     'S.E.  =  ',⍕E←(((8×P×P+2)×÷/N+ ¯3 3 ,(¯1-P),5,1-P)*0.5)÷N+1
[5]     'APPROXIMATE GAMMA1  =  ',⍕G←(÷/8,(N+30),P+ 8 0 8 2)*÷2
[6]     '   (E.N.D. = ',(2⍕END),')'
      ∇
```

```
      ∇ X←END;A
[1]     A←6+4×A×A+4○A←2÷G
[2]     →3+0<X←1+(B-M)÷E÷(2÷A-4)*÷2
[3]     →ρ⍞←'E.N.D. NOT FOUND.',X←''
[4]     X←(1-(2÷9×A)+((1-2÷A)÷X)*÷3)÷(2÷9×A)*÷2
[5]  ⍝  INVOKED BY 'TSNT' AND 'MARDIA'.
      ∇
```

HOWCONTINGENCY

ANALYSIS OF CONTINGENCY TABLES
(1) CONTINGENCY X
(2) FOURFOLD X
(3) MULTIPOLY X
(4) Y←V POOL X

THE FIRST TWO FUNCTIONS APPLY TO 2-DIMENSIONAL CONTINGENCY TABLES, THE OTHERS TO CONTINGENCY TABLES IN ANY NUMBER OF DIMENSIONS. 'CONTINGENCY' PERFORMS A CHI SQUARED TEST OF ASSOCIATION, WITH DISPLAY OF STANDARDIZED RESIDUALS. 'FOURFOLD' MAY BE APPLIED WHEN THE CATEGORIES OF EACH CLASSIFICATION ARE ORDERED. EMPIRICAL LOG CROSSPRODUCT (FOURFOLD) RATIOS ARE DISPLAYED; A PLACKETT DISTRIBUTION IS FITTED, AND GOODNESS OF FIT IS TESTED BY CHI SQUARED, WITH DISPLAY OF STANDARDIZED RESIDUALS. 'MULTIPOLY' FINDS EMPIRICAL LOG CROSSPRODUCT RATIOS, ANALOGOUS TO THOSE OF 'FOURFOLD' FOR 2-DIMENSIONAL TABLES. 'POOL' POOLS CATEGORIES IN ANY TABLE.

(1) CONTINGENCY. THE ARGUMENT (X) MUST BE A MATRIX OF NONNEGATIVE INTEGERS, HAVING NO ZERO MARGINAL TOTAL. (IF X HAS FRACTIONAL COUNTS, LINE [1] SHOULD BE BYPASSED.) GLOBAL OUTPUT VARIABLES: EF (MATRIX OF EXPECTED FREQUENCIES), SR (MATRIX OF STANDARDIZED RESIDUALS). SR IS DEFINED THUS:

```
        SR ← (M-EF)÷EF*0.5
+/,SR×SR  IS CHI SQUARED.
```

NOTE. FOR GUTTMAN PREDICTION OF THE COLUMN CATEGORY FROM THE ROW CATEGORY,

```
    LAMBDA ← ((+/⌈/X)-G)÷(+/,X)-G←⌈/+⌿X
```

(SEE GOODMAN AND KRUSKAL: JASA 49 (1954), 732-764.)

(2) FOURFOLD. SAME REQUIREMENT FOR X AS IN 'CONTINGENCY'. THE MATRIX OF LOG FOURFOLD RATIOS HAS DIMENSION VECTOR (⍴X)-1 , AND REFERS TO ALL POSSIBLE 2-BY-2 TABLES THAT CAN BE FORMED FROM X BY POOLING ADJACENT ROWS AND COLUMNS. IF

```
        N11 | N12
        N21 | N22
```

IS SUCH A TABLE, THE NATURAL-LOG FOURFOLD RATIO (LFR) CALCULATED IS

```
        ⍟÷/0.5+(N11,N12,N22,N21)
```

THE 0.5 ADDED TO EACH COUNT MAKES LFR A ROUGHLY UNBIASED ESTIMATE OF LOG PSI, WHERE PSI IS THE CORRESPONDING FOURFOLD RATIO OF PROBABILITIES, AND PREVENTS DISASTER IF ANY COUNT IS ZERO.

POSSIBLY THESE VALUES OF LFR DIFFER ONLY BY SAMPLING ERROR. A PLACKETT DISTRIBUTION HAVING CONSTANT PSI IS FITTED ITERATIVELY BY MAXIMUM CONDITIONAL LIKELIHOOD, GIVEN THAT THE MARGINAL PROBABILITIES ARE PROPORTIONAL TO THE GIVEN MARGINAL TOTALS. AT STAGE 0 THE THREE TRIAL VALUES FOR LOG PSI ARE AN AVERAGE OF THE MEMBERS OF LFR AND THE SAME ± TWO CRUDELY ESTIMATED STANDARD ERRORS. THE LOG LIKELIHOOD FUNCTION IS COMPUTED FOR THE 3 VALUES OF LOG PSI , A PARABOLA IS FITTED, AND AT STAGE 1 THE NEXT 3 TRIAL VALUES FOR LOG PSI ARE THE VALUE TO MAXIMIZE THE PARABOLA AND A VALUE ON EITHER SIDE, USUALLY CLOSER THAN BEFORE. AFTER AT MOST 4 ITERATIONS, EXPECTED FREQUENCIES (EF) ARE CALCULATED, AND HENCE STANDARDIZED RESIDUALS (SR) AND CHI SQUARED.

THE 4 ITERATIONS WILL NOT FIND THE CORRECT M.L. VALUE FOR LOG PSI IF THAT IS INFINITE. TO ADJUST THE MAXIMUM NUMBER OF ITERATIONS, CHANGE THE '4' IN LINE [1] TO ANY INTEGER NOT LESS THAN 2.

SEE

PLACKETT: JASA 60 (1965), 516-522.
TUKEY: EXPLORATORY DATA ANALYSIS, ADDISON-WESLEY, PRELIM. ED. 1971, CHAP. 29X.

(3) MULTIPOLY. THE ARGUMENT (X) MUST BE AN ARRAY OF NONNEGATIVE INTEGERS IN 2 OR MORE DIMENSIONS. GLOBAL OUTPUT VARIABLES: LCR (LOG CROSSPRODUCT RATIOS), ESE (ESTIMATED STANDARD ERRORS). LCR AND ESE HAVE DIMENSION VECTOR (ρX)-1 .

IF X IS A MATRIX, LCR IS THE SAME AS LFR YIELDED BY 'FOURFOLD'. IF X IS 3-DIMENSIONAL, LCR REFERS TO ALL POSSIBLE 2-BY-2-BY-2 TABLES THAT CAN BE FORMED FROM X BY POOLING ADJACENT PLANES, ROWS AND COLUMNS. IF

```
        N111 | N112     N211 | N212
        N121 | N122     N221 | N222
```

IS SUCH A TABLE, THE NATURAL-LOG CROSSPRODUCT (EIGHTFOLD) RATIO LCR IS

```
        ⍟÷/0.5+(N111,N112,N122,N121,N221,N211,N212,N222)
```

THE VARIANCE MAY BE ROUGHLY ESTIMATED BY

```
        +/÷0.5+(N111,N112, . . . ,N222)
```

THE SQUARE ROOTS ARE PUT OUT AS ESE . SIMILARLY IF X HAS MORE THAN THREE DIMENSIONS.

(4) POOL. THE SECOND ARGUMENT (X) IS AN ARRAY IN 2 OR MORE DIMENSIONS, TYPICALLY A CONTINGENCY TABLE OR A TABLE OF EXPECTED FREQUENCIES. THE FIRST ARGUMENT (V) IS A VECTOR WITH AT LEAST 3 ELEMENTS. V[1] SPECIFIES THE COORDINATE, AND 1↓V SPECIFIES THE INDEX-VALUES, OVER WHICH THERE IS TO BE POOLING. SECTIONS OF X CORRESPONDING TO INDEX-VALUES 2↓V OF COORDINATE V[1] ARE ADDED TO THE SECTION WITH INDEX-VALUE V[2] , AND THEN THE FORMER SECTIONS ARE DELETED. THUS IF X IS A MATRIX WITH 5 ROWS, AND IF V IS 1 4 5 1, POOLING WILL BE OVER THE 1ST COORDINATE (ROWS); THE CONTENTS OF ROWS 4, 5 AND 1 WILL BE ADDED, PLACED IN ROW 4, AND THEN ROWS 5 AND 1 WILL BE DROPPED, SO THAT THE RESULT (Y) HAS 3 ROWS, THE OLD 2ND AND 3RD AND THEN THE SUM OF THE OLD 1ST, 4TH AND 5TH ROWS. IF V HAD BEEN 1 1 4 5 THE SAME RESULT WOULD HAVE BEEN OBTAINED EXCEPT FOR A PERMUTATION OF ROWS.

9 JULY 1973

```
      ∇ CONTINGENCY X;C;R
[1]     →2+(∧/,0≤X)∧(∧/,X=⌊X)∧(∧/2≤ρX)∧2=ρρX
[2]     →ρ⎕←'NO GO, THE ARGUMENT SHOULD BE A MATRIX OF NONNEGATIVE INTEGERS.'
[3]     →4+(∧/1≤C←+⌿X)∧∧/1≤R←+/X
[4]     →ρ⎕←'NO GO, A MARGINAL TOTAL VANISHES.'
[5]     'THE MATRIX OF EXPECTED FREQUENCIES IS NAMED  EF .'
[6]     'LEAST MEMBER OF  EF  =  ',⍕⌊/,EF←R∘.×C÷+/R
[7]     'STANDARDIZED RESIDUALS (SR):'
[8]     2⍕SR←(EF*¯0.5)×X-EF
[9]     'CHI-SQUARED  =  ',(⍕+/,SR×SR),⎕←''
[10]    'DEGREES OF FREEDOM  =  ',⍕×/¯1+ρX
      ∇
```

```
      ∇ MULTIPOLY X;I;K;N
[1]     →(ρI← 1 0)+(∧/,X≥0)∧(∧/,X=⌊X)∧(∧/2≤ρX)∧2≤K←ρρX
[2]     →0,ρ⎕←'NO GO.'
[3]     →3×ι(2*K)>ρI←I,~I
[4]     →4×ι(2×K)>ρρX←((ι(ρρX)-K),((ρρX)-K)+(2↓ιK+1), 2 1)⍉X+.×(~N),[2.5] N←(ιN)∘.
        >ι¯1+N←¯1↑ρX
[5]     X←(N←1)+2×((ρI),K↓ρX)ρX
[6]     (⍕K),'-VARIABLE INTERACTION--LOG CROSSPRODUCT RATIOS (LCR):'
[7]     2⍕LCR←⍟(×⌿I⌿X)÷×⌿(~I)⌿X
[8]     'ESTIMATED STANDARD ERRORS (ESE):',⎕←''
[9]     2⍕ESE←(2×+⌿÷X)*0.5
      ∇
```

```
      ∇ FOURFOLD X;C;CCT;D;E;GT;ITS;J;L;NU;R;RCT;S;W
[1]     →2+(2≤ITS←4)∧(∧/,X≥J←0)∧(∧/,X=⌊X)∧(∧/2≤⍴X)∧2=⍴⍴X
[2]     →⍴⎕←'NO GO, THE ARGUMENT SHOULD BE A NONNEGATIVE INTEGER MATRIX WITH POSIT
        IVE MARGINAL TOTALS.'
[3]     W←+⍀+\X
[4]     GT←W[(R←(⍴X)[1]);C←(⍴X)[2]]
[5]     →2×⍳∨/(RCT,GT)=0,RCT←¯1↓W[;C]
[6]     →2×⍳∨/(CCT,GT)=0,CCT←¯1↓W[R;]
[7]     W← ¯1 ¯1 ↓W
[8]     W←0.5+W,((RCT∘.+(C-1)⍴0)-W),(W-RCT∘.+CCT-GT),[2.5](((R-1)⍴0)∘.+CCT)-W
[9]     'LOG FOURFOLD RATIOS (LFR):'
[10]    2⍕LFR←⍟÷/W
[11]    LP←(+/,E×LFR)×S←÷+/,E←÷+/÷W
[12]    LP←LP+ ¯1 0 1 ×S←2×S*0.5
[13]  L1:'STAGE ',(⍕J),':  LOG PSI  =  ',⍕LP←LP+1E¯11×GT×∨/GT>200000000000×|LP
[14]    W←GT+(¯1+PSI←*LP)∘.×RCT∘.+CCT
[15]    W←(W-((W×W)-(4×PSI×PSI-1)∘.×RCT∘.×CCT)*0.5)÷(2×PSI-1)∘.+(⍴LFR)⍴0
[16]    W←W-0,[2] 0 ¯1 0 ↓W←W-0, 0 0 ¯1 ↓W←(W,[2]((⍴LP)⍴0)∘.+CCT),((⍴LP)⍴0)∘.+RCT,
        GT
[17]    →L3×⍳J=ITS
[18]    L←+/+/((⍴W)⍴X)×⍟W
[19]    →L2×⍳0≥D←-/L[2 1 2 3]
[20]    LP←,LP[2]+E←(S×-/L[3 1])÷2×D
[21]    →L1×⍳ITS=J←J+1
[22]    ITS←(1 0 =S≥2×|E)/(J+1),ITS
[23]    →L1,⍴LP←LP+ ¯1 0 1 ×S←(S÷4)⌈(0.5×|E)⌊S
[24]  L2:'LIKELIHOOD FN CURVES WRONG WAY.  ENTER 3 INCREASING EQUAL-SPACED VALUES
         FOR LOG PSI.'
[25]    →(L1×⍳((-/¯2↑LP)=-/2↑LP)∧0>-/2↑LP←3↑,⎕),L2
[26]  L3:'   (S.E.  =  ',(⍕S÷D*0.5),')'
[27]    'THE MATRIX OF EXPECTED FREQUENCIES IS NAMED  EF .',⎕←''
[28]    'LEAST MEMBER OF  EF  =  ',⍕⌊/,EF←+⌿W
[29]    'STANDARDIZED RESIDUALS (SR):'
[30]    2⍕SR←(EF*¯0.5)×X-EF
[31]    'CHI-SQUARED  =  ',(⍕+/,SR×SR),⎕←''
[32]    'DEGREES OF FREEDOM  =  ',⍕NU←¯1+×/⍴LFR
      ∇
```

```
      ∇ Y←V POOL X;P
[1]     →3-(∧/(1↑,V)≤⍴⍴X)∧(∧/2≤⍴X)∧(2≤⍴⍴X)∧(1=⍴⍴V)∧(3≤+/⍴V)∧(∧/,1≤V)∧∧/,V=⌊V
[2]     →3+(∧/(1↓V)≤(⍴X)[V[1]])∧(¯1+⍴V)=+/,(1↓V)∘.=1↓V
[3]     →0,⍴⎕←'NO GO.',Y←''
[4]     Y←((1↑⍴X),×/1↓⍴X)⍴X←(⍋P←V[1],(V[1]≠⍳⍴⍴X)/⍳⍴⍴X)⍉X
[5]     Y[V[2];]←+⌿Y[1↓V;]
[6]     Y←P⍉((1↑⍴Y),1↓⍴X)⍴Y←Y[(∧⌿(2↓V)∘.≠⍳1↑⍴Y)/⍳1↑⍴Y;]
      ∇
```

```
      ∇ Y←LFACT X;D;I
[1]     D←⍴X
[2]     Y←⍟!56⌊X←,X
[3]     X←(I←X>56)/X
[4]     Y[I/⍳⍴Y]←((X+0.5)×⍟X)-X-(0.5×⍟○2)+(1-÷8.5+30×X×X)÷12×X
[5]     Y←D⍴Y
[6]   ⍝ LOGARITHMS OF FACTORIALS OF  X .
      ∇
```

```
      ∇ CTG2 X;C;D;E;J;NU;R;S;W
[1]    →3-∨/C←3≠⎕NC R← 3 5 ρ'CSIF INIF LFACT'
[2]    →0,ρ⍞←'COPY ',(,C⌿R),' FROM 1234 ASP3'
[3]    →4+(∧/,0≤X)∧(∧/,X=⌊X)∧(∧/2≤ρX)∧2=ρρX
[4]    →ρ⍞←'NO GO, THE ARGUMENT SHOULD BE A MATRIX OF NONNEGATIVE INTEGERS.'
[5]    →6+(∧/1≤C←+⌿X)∧∧/1≤R←+/X
[6]    →ρ⍞←'NO GO, A MARGINAL TOTAL VANISHES.'
[7]    'THE MATRIX OF EXPECTED FREQUENCIES IS NAMED  EF .'
[8]    'LEAST MEMBER OF  EF  =  ',⍕⌊/,EF←R∘.×C÷+/R
[9]    'DEGREES OF FREEDOM  =  ',⍕NU←×/¯1+ρX
[10]   '    *PROBABILITY-OF-SAMPLE TEST*',⍞←''
[11]   'STANDARDIZED RESIDUALS (SR1):'
[12]   T1←+/,SR1←2×(LFACT X)-(LFACT EF-0.5)+(X-EF-0.5)×⍟EF
[13]   2⍕SR1←(×X-EF-0.5)×(0⌈SR1)*0.5
[14]   'STATISTIC (T1)  =  ',⍕T1
[15]   →(∨/25>C,R)/L̲4
[16]   W←((J←,EF≥5)/,EF)∘.- 1.16 1.97 2.71
[17]   W←(1 2 8 ×NU)-+⌿÷W×(ρW)ρ 12 3 0.5
[18]   →(∧/J)/L̲
[19]   S←S,1-+/S←(E∘.*D)×(*-E←(~J)/,EF)∘.÷!D←0,⍳19
[20]   D←¯2×(⍟(!E-0.5)∘.÷!D)-((E-0.5)∘.-D←D,20)×(⍟E)∘.+21ρ0
[21]   W[1]←W[1]++/¯1+J←+/D×S
[22]   W[2]←W[2]++/¯2++/J←S×D×D←D-J∘.+21ρ0
[23]   W[3]←W[3]++/¯8++/J×D
[24] L̲:W[1]←W[1]-W[3]×W[2]←W[2]÷2×W[3]×W[3]←W[3]÷4×W[2]
[25]   S←1E¯12⌈(1-1E¯12)⌊W[2] CSIF(T1-W[1])÷W[3]
[26]   'ROUGH SIGNIFICANCE LEVEL  =  ',(⍕1-S),',  E.N.D.  =  ',2⍕INIF S
[27] L̲4:'    *LIKELIHOOD-RATIO TEST*',⍞←''
[28]   'STANDARDIZED RESIDUALS (SR2):'
[29]   T2←+/,SR2←2×(X×⍟(1⌈X)÷EF)-X-EF
[30]   2⍕SR2←(×X-EF)×(0⌈SR2)*0.5
[31]   'STATISTIC (T2)  =  ',⍕T2
[32] ⍝  ALTERNATIVE TO THE FUNCTION 'CONTINGENCY'.
      ∇


      ∇ P←P1 PD2 P2;C1;C2
[1]    →2+3=⎕NC 'BIV'
[2]    →ρ⍞←'COPY BIV FROM 1234 ASP3'
[3]    →4+2=⎕NC 'L'
[4]    →ρ⍞←'DEFINE L (NATURAL-LOG FOURFOLD RATIO)'
[5]    →L̲-(0=ρρL)∧(1=ρρP1)∧1=ρρP2
[6]    →L̲+(1=+/P1)∧(1=+/P2)∧(∧/0<P1,P2)∧(2≤ρP1)∧2≤ρP2
[7]  L̲:→0,ρ⍞←'NO GO.',P←''
[8]    'L = ',⍕L
[9]    P←(C1←+\¯1↓P1) BIV C2←+\¯1↓P2
[10]   P←P-0,[1] ¯1 0 ↓P←P-0, 0 ¯1 ↓P←(P,[1] C2),C1,1
      ∇


      ∇ Z←X BIV Y;K
[1]    →(L≠0)/3
[2]    →0,ρZ←X∘.×Y
[3]    Z←1+(K←¯1+*L)×X∘.+Y
[4]    Z←(Z-((Z×Z)-(X∘.×Y)×4×K×K+1)*0.5)÷2×K
[5]  ⍝  INVOKED BY 'PD2' AND 'PD3'.
      ∇
```

```
HOWPD

        DISCRETE PLACKETT DISTRIBUTIONS IN 2 AND 3 DIMENSIONS
        (1)  P←P1 PD2 P2
        (2)  PD3
        (3)  Z←X BIV Y

   (1) PD2.  THE EXPLICIT RESULT IS A TWO-DIMENSIONAL PLACKETT DISTRIBUTION HAV-
ING DISCRETE MARGINAL DISTRIBUTIONS  WITH PROBABILITIES LISTED IN THE ARGUMENTS.
EACH ARGUMENT MUST CONSIST OF AT LEAST TWO POSITIVE NUMBERS SUMMING TO 1.   ONE
GLOBAL VARIABLE  L  MUST BE DEFINED BEFORE EXECUTION,  THE NATURAL LOGARITHM  OF
THE FOURFOLD CROSSPRODUCT RATIO OF PROBABILITIES.

   (2) PD3 REQUIRES 7 GLOBAL VARIABLES TO BE DEFINED BEFORE EXECUTION:  P1 ,  P2
AND  P3  LIST THE PROBABILITIES  IN THE DISCRETE UNIVARIATE MARGINAL DISTRIBUTI-
ONS;  L12 ,  L13  AND  L23  GIVE THE LOG FOURFOLD RATIOS IN THE BIVARIATE MARGI-
NAL DISTRIBUTIONS;  TOL  IS A POSITIVE TOLERANCE SUCH AS 0.001.   IF THE FIRST 6
VARIABLES ARE INCOMPATIBLE, EXECUTION CEASES AT [21] WITH A MESSAGE.   OTHERWISE
THE USER IS ASKED FOR  L123 , THE NATURAL LOGARITHM  OF THE EIGHTFOLD CROSSPROD-
UCT RATIO OF PROBABILITIES.  THE GLOBAL OUTPUT VARIABLE  P  GIVES THE PROBABILI-
TIES IN THE 3-DIMENSIONAL DISTRIBUTION,  DETERMINED ITERATIVELY  SO THAT THE LOG
EIGHTFOLD RATIOS DIFFER FROM  L123  BY LESS THAN  TOL .   POSSIBLY NOT ALL MEMB-
ERS OF  P  ARE POSITIVE, IN WHICH CASE  L123  IS NOT COMPATIBLE WITH THE MARGIN-
AL DISTRIBUTIONS.   EXECUTION IS SUSPENDED AT [36], SO THAT THE USER CAN EXAMINE
P , POSSIBLY RESET  TOL , AND THEN TRY ANOTHER VALUE FOR  L123  BY RESUMING EXE-
CUTION WITH
      →⎕LC

   (3) BIV IS INVOKED BY 'PD2' AND 'PD3'.

28 JULY 1980


      ∇ PD3;B12;B13;B23;C;C1;C2;C3;C12;C13;C23;IND;J;L;ME;NEG;POS;W
[1]    →2+3=⎕NC 'BIV'
[2]    →0,ρ⎕←'COPY BIV FROM 1234 ASP3'
[3]    →5-∨/W←2≠⎕NC L← 7 4 ρ'P1  P2  P3  L12 L13 L23 TOL '
[4]    →0,ρ⎕←'DEFINE  ',,W⌿L
[5]    →L-(0=ρρL12)∧(0=ρρL13)∧(0=ρρL23)∧(1=ρρP1)∧(1=ρρP2)∧(1=ρρP3)∧(J←0)=ρρTOL
[6]    →L+(1=+/P1)∧(1=+/P2)∧(1=+/P3)∧(∧/0<P1,P2,P3,TOL)∧(2≤ρP1)∧(2≤ρP2)∧2≤ρP3
[7]   L:→0,ρ⎕←'NO GO.'
[8]    'P1 = ',(⍕P1),';  P2 = ',(⍕P2),';  P3 = ',⍕P3
[9]    'L12 = ',(⍕L12),';  L13 = ',(⍕L13),';  L23 = ',⍕L23
[10]   L←L12
[11]   B12←,(C12←(C1←+\¯1↓P1) BIV C2←+\¯1↓P2)∘.+C3≠C3←+\¯1↓P3
[12]   L←L13
[13]   B13←, 1 3 2 ⍉(C13←C1 BIV C3)∘.+C2≠C2
[14]   L←L23
[15]   B23←,(C1≠C1)∘.+C23←C2 BIV C3
[16]   W←((,(C12≠C12)∘.+C3)-B13+B23),[1.5] 1+B12+B13+B23-,C1∘.+C2∘.+C3
[17]   W←0,B12,B13,((,C1∘.+C23≠C23)-B12+B13),B23,((,(C1≠C1)∘.+C2∘.+C3≠C3)-B12+B23
       ),W
[18]   IND← 1 0 0 1 0 1 1 0 ,B12←B13←B23←''
[19]   POS←⌊/IND/W
[20]   →(∧/0<POS+NEG←⌊/(~IND)/W)/L1
[21]   →0,ρ⎕←'NO JOINT DISTRIBUTION.'
[22]  L1:W←W+(0.5×NEG-POS)∘.×IND-~IND
[23]   'L123 ?',⎕←''
```

```
[24]  →(0≠ρρL←⎕)/1+L1
[25] L2:'STAGE ',(⍕J),':  MAX ERROR = ',⍕ME←⌈/|C←(+/⍟IND/W)-(+/⍟(~IND)/W)+L
[26]  →(TOL≥ME)/L4
[27]  W←W-(C←C÷+/÷W)∘.×IND-~IND
[28]  J←J+1
[29] L3:→(∧/,W>0)/L2
[30]  →L3,ρW←W+(C←C÷ρ⎕←'**')∘.×IND-~IND
[31] L4:P←((ρC1),ρC23)ρW[;1]
[32]  P←((P,[1] C23),[2] C13,[1] C3),(C12,[1] C2),C1,1
[33]  P←P-0,[1] ¯1 0 0 ↓P←P-0,[2] 0 ¯1 0 ↓P←P-0, 0 0 ¯1 ↓P
[34]  'NUMBER OF NONPOSITIVE MEMBERS OF  P  = ',(⍕+/,P≤0),⎕←''
[35]  SΔPD3←1+1↑⎕LC
[36]  →L1+1+J←0
     ∇
```

```
     ∇ C←A PP B;E
[1]   →((1≤1↑ρA)∧(1=ρρA)∧1=ρρB)/4
[2]   →((0 2 =ρρA)∧(2=ρρB)∧∧/ 2 1 ≤2↑ρB)/ 5 6
[3]   →0,ρ⎕←'NO GO.',C←''
[4]   →0,ρC←+⌿(1-⍳ρA)⌽A∘.×B,1↓A≠A
[5]   →0,ρC←(A×B[1;]),[1] 1 0 ↓B
[6]   →((0=¯1↑ρA)∨(1↑(ρA)≠ρB)∨∨/(0≠1|E)∨0>E←(, 1 0 ↓A),, 1 0 ↓B)/3
[7]   C←(,A[1;]∘.×B[1;]),[1] E⊤,(E⊥ 1 0 ↓A)∘.+(E←1+1↓(⌈/A)+⌈/B)⊥ 1 0 ↓B
[8]  ⍝ POLYNOMIAL PRODUCT.  SEE EXERCISE 16 AT END OF PART A, COMPUTING IN STAT
      ISTICAL SCIENCE THROUGH APL.
     ∇
```

```
     ∇ C←COLLECT A;E;J;X
[1]   →3-(∧/2≤ρA)∧2=ρρA
[2]   →3+∧/,(0=1|X)∧0≤X← 1 0 ↓A
[3]   →0,ρ⎕←'NO GO.',C←''
[4]   X←(E←1+⌈/X)⊥X
[5]   X←(J←0≠C←(<⍀X∘.=X)+.×A[1;])/X
[6]   C←((J/C),[1] E⊤X)[;⍋X]
[7]  ⍝ USED WITH 'PP'.
     ∇
```

HOWISOTROPY

TEST OF SPHERICITY OF A MULTIVARIATE NORMAL DISTRIBUTION
N ISOTROPY L

THIS IMPLEMENTS THE TEST FOR INEQUALITY OF THE ROOTS OF A SAMPLE VARIANCE MATRIX GIVEN BY ANDERSON, INTRO. TO MULTIVARIATE STATISTICAL ANALYSIS (1958), SECTION 10.7.4. THE FIRST ARGUMENT (N) IS THE NUMBER OF DEGREES OF FREEDOM IN THE VARIANCE ESTIMATES. THE SECOND ARGUMENT (L) IS A VECTOR OF POSITIVE ROOTS OF THE VARIANCE MATRIX. IT IS ASSUMED THAT N IS MUCH LARGER THAN ρL . THE FIRST APPROXIMATION TO THE SIGNIFICANCE LEVEL IS BASED ON A SINGLE CHI-SQUARED; THE SECOND INTRODUCES A CORRECTION TERM OF ORDER N¯2 . IF THE TWO APPROXIMATIONS ARE NOT NEARLY EQUAL, PERHAPS NEITHER IS GOOD.*

29 *DEC*. 1980

```
      ∇ N ISOTROPY L;F;O2;P;Z
[1]    →2+(∧/,L>0)∧(P≥2)∧(∧/,N>P←+/ρL)∧(0=ρρN)∧1=ρρL
[2]    →0,ρ⎕←'NO GO.'
[3]    →4+3=⎕NC 'CSIF'
[4]    →ρ⎕←'COPY CSIF FROM 1234 ASP3'
[5]    O2←(P⊥ 2 4 ¯11 ¯17 10 4 8)÷288×(P×N←N-(P⊥ 2 1 2)÷6×P)*2
[6]    'STATISTIC = ',(⍕Z←-N×⍟(×/L)÷((+/L)÷P)*P),',  D.F. = ',⍕F←¯1+2!P+1
[7]    'SIGNIFICANCE LEVEL:  1ST APPROXIMATION = ',⍕1-P←F CSIF Z
[8]    (21ρ' '),'2ND APPROXIMATION = ',⍕(1-P)+O2×P-(F+4) CSIF Z
      ∇

      HOWJACOBI

           CHARACTERISTIC ROOTS AND VECTORS OF A SYMMETRIC MATRIX
           Y←C JACOBI X

   THE SECOND ARGUMENT (X) IS THE GIVEN MATRIX, TO BE TRANSFORMED TOWARDS DIAGO-
NAL FORM BY JACOBI'S METHOD.   THE FIRST ARGUMENT (C), A POSITIVE SCALAR, IS THE
TOLERANCE FOR OFF-DIAGONAL ELEMENTS IN THE TRANSFORMED MATRIX.  THE EXPLICIT RE-
SULT (Y) IS A THREE-DIMENSIONAL ARRAY WITH 2 PLANES.  Y[1;;]  IS THE TRANSFORMED
MATRIX, AND SO  1 1 ⍉Y[1;;]  IS THE VECTOR OF ROOTS, WHICH IS ALSO OUTPUT AS THE
GLOBAL VARIABLE 'ROOTS'.  THE CORRESPONDING VECTORS ARE THE COLUMNS OF  Y[2;;] .
THE GLOBAL SCALAR VARIABLE 'ITERATIONS'  IS THE NUMBER OF PERFORMED STEPS OF THE
JACOBI PROCESS.

   USUALLY (IF THERE HAVE BEEN ENOUGH ITERATIONS) THE ROOTS APPEAR IN DECREASING
ORDER.

   TO SEE THE EFFECT OF REDUCING THE VALUE OF  C , LET  C1  AND  C2  BE TOLERAN-
CES, THE FIRST GREATER THAN THE SECOND.  ENTER, FOR A GIVEN MATRIX  X ,
      Y1←C1 JACOBI X
      ROOTS
      ITERATIONS
      Y2←C2 JACOBI Y1[1;;]
      ROOTS
      ITERATIONS
THE SECOND VERSION OF  'ROOTS'  IS THE BETTER ONE.    THE CORRESPONDING IMPROVED
EIGENVECTORS ARE THE COLUMNS OF  Y1[2;;]+.×Y2[2;;] .

5 DEC. 1972

      ∇ Y←C JACOBI X;A;I;J;K;M;N;R;S;Z
[1]    →2+(∧/C>0)∧(0=ρρC)∧((+/,C)>⌈/|(,X)-,⍉X)∧(4≤+/ρX)∧(=/ρX)∧2=ρρX
[2]    →0,ρ⎕←'NO GO.',Y←''
[3]    I←,(⍳N)∘.<⍳N←+/1↑ρY←X
[4]    R←(N,N)ρ1,NρITERATIONS←0
[5]   L1:→(C≥M←⌈/Z←|I/,Y)/L4
[6]    J←1+(M-K←1+N|¯1+M←(I\Z)⍳M)÷N
[7]    →(Y[J;J]≠Y[K;K])/L2
[8]    →L3,ρA←○¯0.25××Y[J;K]
[9]   L2:A←0.5×(○Y[J;J]<Y[K;K])+¯3○2×Y[J;K]÷Y[K;K]-Y[J;J]
[10]  L3:S←(N,N)ρ1,Nρ0
[11]   S[K;K,J]← 1 ¯1 ×S[J;J,K]← 2 1 ∘.○A
[12]   Y←(⍉R)+.×X+.×R←R+.×S
[13]   →L1,ITERATIONS←ITERATIONS+1
[14]  L4:ROOTS← 1 1 ⍉Y
[15]   Y←Y,[0.5] R
      ∇
```

```
HOWT7

     REGRESSION WHEN ERRORS HAVE A TYPE 7 (OR 2) DISTRIBUTION
     (1)  T7INIT
     (2)  T7LF K
     (3)  T7S Q
     (4)  G←T7A
     (5)  H←T7B

   A LINEAR REGRESSION RELATION IS FITTED BY MAXIMUM LIKELIHOOD,  ASSUMING  THAT
THE ERRORS HAVE A PEARSON TYPE VII OR TYPE II DISTRIBUTION.  SEE
ANSCOMBE:  J. R. STATIST. SOC. B 29 (1967), 1-52, SECTIONS 2.3-4.
COMPUTING IN STATISTICAL SCIENCE THROUGH APL, CHAPTER 11.

   'T7INIT' INITIALIZES.    'T7LF' EVALUATES THE MARGINAL LIKELIHOOD FUNCTION  L
AT A TRIAL SET OF VALUES OF THE REGRESSION PARAMETERS,  AFTER INTEGRATION OF THE
LIKELIHOOD WITH RESPECT TO THE SCALE PARAMETER AND THE SHAPE PARAMETER.  IF DES-
IRED, THE FIRST AND SECOND DERIVATIVES,  DL  AND  D2L , OF  L  ARE FOUND.  'T7S'
ESTIMATES THE CHANGE IN THE REGRESSION PARAMETERS NEEDED TO REACH THE MAXIMUM OF
THE MARGINAL LIKELIHOOD.  'T7A' AND 'T7B' ARE INVOKED BY 'T7LF'.

   AT THE OUTSET 4 GLOBAL VARIABLES MUST BE SPECIFIED:  ALPHA WALPHA X XX .   AT
EACH EXECUTION OF 'T7LF' FIVE FURTHER VARIABLES MUST BE SPECIFIED, THE VECTOR OF
RESIDUALS  Z  (USUALLY DIFFERING BETWEEN EXECUTIONS), AND 4 SCALARS:  V0 ILV MNI
MIH  (USUALLY THE SAME).  'T7INIT' DEFINES GLOBAL VARIABLES:  C1 C2 C3 HN N UT .
'T7LF' DEFINES  L  AND POSSIBLY:  DL D2L .  'T7S' DEFINES:  A B DBETA .

   THE GLOBAL VARIABLES SPECIFIED BY THE USER ARE
ALPHA:  VECTOR OF VALUES OF THE SHAPE PARAMETER, POSITIVE FOR TYPE 7, 0 FOR NOR-
        MAL, NEGATIVE FOR TYPE 2.
WALPHA:  WEIGHTS TO GO WITH  ALPHA .     TO REPRESENT A UNIFORM DISTRIBUTION FOR
        ALPHA  BETWEEN ¯0.5 AND 1, AND INTEGRATION BY SIMPSON'S RULE,
        ALPHA←(¯5+⍳13)÷8
        WALPHA←1,(11⍴ 4 2),1
X:      MATRIX OF EXPLANATORY VARIABLES, THE FIRST COLUMN CONSISTING OF 1'S.
XX:     MATRIX OF PRODUCTS OF PAIRS OF COLUMNS OF  X  OTHER THAN THE FIRST COLUMN.
        FOR EXAMPLE, IF  X  HAS 4 COLUMNS,
        XX←(X[;2 2 2]×X[;2 3 4]),(X[;3 3]×X[;3 4]),X[;4]*2
        IN GENERAL, THE J-TH ROW OF  XX  SATISFIES
        XX[J;] ↔ UT/,(1↓X[J;])∘.×1↓X[J;]
        WHERE  UT  IS THE UPPER TRIANGULAR INDICATOR GIVEN BY 'T7INIT'.  IF A COL-
        UMN OF  XX  AS SO DEFINED IS THE SAME AS ONE OF THE COLUMNS OF  X , IT MAY
        BE OMITTED FROM  XX ,  BUT A CORRECTION MUST BE INSERTED JUST AFTER [6] IN
        'T7A' AND [5] IN 'T7B'.
Z:      VECTOR OF RESIDUALS  CORRESPONDING TO AN ENTERTAINED VECTOR  OF REGRESSION
        PARAMETERS  BETA .  IF  Y  IS THE ORIGINAL DATA VECTOR,
        Z←Y-X+.×BETA
V0:     LEAST-SQUARES VARIANCE ESTIMATE.  IF  Z  DENOTES THE LEAST-SQUARES RESIDS,
        V0←(Z+.×Z)÷⍴Z
ILV:    QUADRATURE INTERVAL FOR LOG VARIANCE RATIO.
MNI:    MAXIMUM NUMBER OF INTERVALS IN THE QUADRATURE.
MIH:    MINIMUM INCREMENT IN DETERMINATION OF  H .  SUGGESTED SETTINGS ARE
        ILV←0.5×N*¯0.5
        MNI←25
        MIH←1E¯6

   THE ARGUMENT OF 'T7LF' SPECIFIES THE ORDER OF DERIVATIVES NEEDED,  0 FOR  L ,
```

```
1 FOR  L  AND  DL , 2 FOR  L ,  DL  AND  D2L .   THE ARGUMENT SHOULD BE 2 AT THE
FIRST EXECUTION.  THE ARGUMENT  Q  OF 'T7S' SHOULD BE NOT LESS THAN  3÷N .   AF-
TER EXECUTING 'T7S' YOU MAY THEN PROCEED:
       Z←Y-X+.×BETA←BETA+DBETA
       T7LF 1
       B←DL÷L*1+Q
       DBETA←B⌹A

5 AUG. 1980


       ∇ T7INIT;I;M;P
[1]    →3-(2=ρρX)∧(2=ρρXX)∧(∧/ALPHA<2)∧(ρWALPHA←,WALPHA)=ρALPHA←,ALPHA
[2]    →3+(∧/1=X[;1])∧((ρXX)[2]≤2!P←(ρX)[2])∧(ρXX)[1]=N←(ρX)[1]
[3]    →0,ρ⎕←'NO GO.'
[4]    C1←(HN×⍟HN)-LFACT ¯1+HN←N÷2
[5]    M←÷(I←ALPHA≠0)/ALPHA
[6]    C1←C1+N×I\(-≠LFACT((¯0.5×M<0)+|M),[0.5](¯1.5×M<0)+|M-0.5)-0.5×⍟|M-0.5
[7]    C3←0.5×ALPHA×C2←(1-ALPHA÷2)*¯3
[8]    UT←,(⍳P)∘.≤⍳P
       ∇


       ∇ T7LF K;A;B;C;D;E;G;H;I;J;LVR;M;R;U;V;W
[1]    →2+(N=×/ρZ)∧(1=ρρZ)∧(,K∊ 0 1 2)∧0=ρρK
[2]    →0,ρ⎕←'NO GO.'
[3]    C←(1+ 1 ¯1 ×3*÷¯2)÷¯2×J←1+U←0
[4]    M←M×M←⌈/|Z
[5]    L1:'ALPHA[',(⍕J),'] = ',(⍕ALPHA[J]),⎕←''
[6]    →L5×⍳ALPHA[J]=0
[7]    A←MIH×2÷D←ILV×1+ALPHA[J]>0
[8]    R←MNI
[9]    W←C3[J]×Z×Z,ρB←H←I←0
[10]   →L2×⍳ALPHA[J]>0
[11]   →L3×⍳0≥E←⍟V0÷-M×C3[J]
[12]   →L2×⍳E≥ILV×MNI
[13]   D←E÷R←⌈E÷ILV
[14]  L2:V←V0÷*LVR←D×C+I←I+1
[15]   H←H+G←T7A
[16]   →(2+1↑⎕LC)×⍳(I≥R)∨∧/G[1]<A,B
[17]   →L2,B←G[1]
[18]   R,I
[19]   B←E←I←0
[20]  L3:V←V0÷*LVR←E-D×C+I←I+1
[21]   H←H+G←T7A
[22]   →(2+1↑⎕LC)×⍳(I≥MNI)∨∧/G[1]<A,B
[23]   →L3,B←G[1]
[24]   I
[25]   H←H×D÷2
[26]  L4:'H = ',⍕H
[27]   U←U+H×WALPHA[J]
[28]   →(L1×⍳(ρALPHA)≥J←J+1),L6
[29]  L5:→L4,ρH←T7B
[30]  L6:→0×⍳K=ρρL←U[1]
[31]   →0×⍳K=ρρDL←(ρX)[2]↑1↓U
[32]   D2L←(1+¯1↑ρX)↓U
       ∇
```

```
HOWBARTLETT

     BARTLETT'S TEST FOR HOMOGENEITY OF VARIANCE
     N BARTLETT S

  THE SECOND ARGUMENT IS A VECTOR OF UNBIASED VARIANCE ESTIMATES.    THE FIRST
ARGUMENT IS THE DEGREES OF FREEDOM, EITHER THE COMMON VALUE FOR ALL THE VARIANCE
ESTIMATES OR A VECTOR OF VALUES, ONE FOR EACH VARIANCE ESTIMATE.  BOX'S APPROXI-
MATION BY THE F-DISTRIBUTION IS GIVEN (BIOMETRIKA, 1949; MARDIA AND ZEMROCH, TA-
BLES OF THE F DISTRIBUTION, 1978), AND BARTLETT'S ORIGINAL CHI-SQUARED APPROXIM-
ATION (PROC. ROY. SOC., 1937).

28 JULY 1980


      ∇ N BARTLETT S;A;B;F;K;M;NN
[1]     →2+(∧/N>0)∧(∨/(K,1)=ρN←,N)∧(∧/S>0)∧2≤K←ρS←,S
[2]     →0,ρ⎕←'NO GO.'
[3]     M←-(N+.×⍟S)+NN×⍟(NN←+/N)÷S+.×N←KρN
[4]     B←F÷1-A-2÷F←(K+2)÷A×A←((+/÷N)-÷NN)÷3×K←K-1
[5]     'BOX APPROXIMATION:  F  =  ',(⍕F×M÷K×B-M),'  WITH  ',(⍕K,F),'  D.F.'
[6]     'BARTLETT APPROXIMATION:  CHI-SQUARED  =  ',(⍕M←M÷1+A),'  WITH  ',(⍕K),'
        D.F.'
[7]     →(3≠⎕NC 'CSIF')/0
[8]     '    SIGNIFICANCE LEVEL  =  ',⍕1-K[1] CSIF M
      ∇


      ∇ G←T7A;D;E;F;T
[1]     G←,+/F←*C1[J]+(HN×LVR)-(+⌿⍟1+W∘.÷V)÷ALPHA[J]
[2]     →0×⍳K=0
[3]     D←C2[J]×((Z,[0.5] Z)÷T←V∘.+W)+.×X
[4]     →10×⍳K=1
[5]     E←(T←(-V∘.-W)÷T×T)+.×X
[6]     E←C2[J]×E,T+.×XX
[7]     E[1;]←E[1;]+UT/,D[1;]∘.×D[1;]
[8]     E[2;]←E[2;]+UT/,D[2;]∘.×D[2;]
[9]     D←D,E
[10]    G←G,F+.×D
      ∇

      ∇ H←T7B;D;E;T
[1]     H←,(V0×T←N÷Z+.×Z)*HN
[2]     →0×⍳K=0
[3]     D←T×Z+.×X
[4]     →7×⍳K=1
[5]     E←-T×(+⌿X),+⌿XX
[6]     D←D,E+(1+÷HN)×UT/,D∘.×D
[7]     H←H,H×D
      ∇


      ∇ T7S Q;M;P
[1]     P←ρB←DL÷L×M←L*Q
[2]     A←(-D2L÷L×M)+(M×1+Q)×UT/,B∘.×B
[3]     A←A+⍉A←(1-0.5×(⍳P)∘.=⍳P)×(P,P)ρUT\A
[4]     'DBETA = ',⍕DBETA←B⌹A
      ∇
```

```
HOWINTEGRATE

     ONE-DIMENSIONAL DEFINITE INTEGRALS
     Z←H INTEGRATE A
```

THE FIRST ARGUMENT (H) IS A STEP SIZE, SCALAR. THE SECOND ARGUMENT (A) IS A VECTOR OF 2 OR MORE LIMITS OF INTEGRATION, IN ASCENDING ORDER, WITH DIFFERENCES ALL DIVISIBLE BY H . THE EXPLICIT RESULT (Z) IS A VECTOR OF LENGTH 1 LESS THAN THE LENGTH OF A , LISTING THE DEFINITE INTEGRALS FROM A[1] TO EACH OF THE OTHER MEMBERS OF A . THE FUNCTION TO BE INTEGRATED IS ASKED FOR, AND MUST BE EXPRESSED IN TERMS OF AN ARGUMENT X (LOCAL VARIABLE). TWO-POINT GAUSSIAN QUADRATURE IS USED IN EACH INTERVAL OF WIDTH H . CF. RICHARDSON, <u>APL QUOTE QUAD</u>, 1970.

FOR EXAMPLE, TO INTEGRATE SIN X FROM 0 TO EACH OF 1 2 3 AND COMPARE THE RESULT WITH 1 - COS X :

```
      0.1 INTEGRATE ¯1+⍳4
FUNCTION OF  X  TO BE INTEGRATED?
⎕:
      1○X
0.4596976835  1.416146804  1.989992451
      1-2○⍳3
0.4596976941  1.416146837  1.989992497
```

29 NOV. 1972

```
      ∇ Z←H INTEGRATE A;N;X
[1]     →3-(∧/2≤⍴A)∧(1=⍴⍴A)∧(∧/,H>0)∧0=⍴⍴H
[2]     →3+(∧/0<N-0,¯1↓N)∧∧/N=⌊N←((1↓A)-1↑A)÷H
[3]     →0,⍴⎕←'NO GO.',Z←''
[4]     'FUNCTION OF  X  TO BE INTEGRATED?'
[5]     X←(A[1]+H×⍳¯1↑N)∘.-(H÷2)×1+ 1 ¯1 ÷3*0.5
[6]     Z←(H÷2)×(+\+/⎕)[N]
      ∇
```

```
HOWMAX

     MAXIMIZATION OF A FUNCTION OF ONE VARIABLE
     Z←X MAX Y
```

THE ARGUMENTS ARE VECTORS OF LENGTH 3, THE FIRST HAVING NO TWO MEMBERS EQUAL. THE EXPLICIT RESULT IS THE COORDINATES OF THE VERTEX OF A PARABOLA WITH VERTICAL AXIS, THAT GOES THROUGH THE 3 POINTS WHOSE ABSCISSAS ARE THE FIRST ARGUMENT AND ORDINATES THE SECOND ARGUMENT. IF THE PARABOLA IS CONVEX FROM BELOW, THE VERTEX IS THE MINIMUM INSTEAD OF THE MAXIMUM POINT, AND THE WORD 'MINIMUM.' APPEARS IN THE OUTPUT.

29 NOV. 1972

```
      ∇ Z←X MAX Y;A;B;M
[1]     →3-(3=+/⍴X)∧(3=+/⍴Y)∧(1=⍴⍴X)∧1=⍴⍴Y
[2]     →3+(∧/(⍒X)=⌽⍋X)∧=/Z← 0 0
[3]     →⍴⎕←'ARGUMENTS WON''T DO.',Z←''
[4]     A←Y⌹1,X,[1.5] X×X←X-M←(+/X)÷3
[5]     →((A[3]=0),A[3]<0)/ 9 7
[6]     'MINIMUM.'
[7]     Z[1]←M+B←(÷/1↓A)÷¯2
[8]     →0,Z[2]←A[1 2]+.×1,B÷2
[9]     'NO TURNING POINT.',Z←''
      ∇
```

References

M. Abramowitz and I. A. Stegun (eds.) (1964). *Handbook of Mathematical Functions*. National Bureau of Standards, Applied Mathematics Series 55. U. S. Government Printing Office.

H. Akaike (1969). Fitting autoregressive models for prediction. *Annals of the Institute of Statistical Mathematics*, **21**, 243–247.

D. E. Amos (1969). On computation of the bivariate normal distribution. *Mathematics of Computation*, **23**, 655–659.

T. W. Anderson (1958). *An Introduction to Multivariate Statistical Analysis*. Wiley.

T. W. Anderson (1971). *The Statistical Analysis of Time Series*. Wiley.

D. F. Andrews, P. J. Bickel, F. R. Hampel, P. J. Huber, W. H. Rogers and J. W. Tukey (1972). *Robust Estimates of Location*. Princeton University Press.

F. J. Anscombe (1961). Examination of residuals. *Proceedings of the Fourth Berkeley Symposium on Mathematical Statistics and Probability*, **1**, 1–36. University of California Press.

F. J. Anscombe (1963). Tests of goodness of fit. *Journal of the Royal Statistical Society*, B, **25**, 81–94.

F. J. Anscombe (1964). Normal likelihood functions. *Annals of the Institute of Statistical Mathematics*, **16**, 1–19.

F. J. Anscombe (1967). Topics in the investigation of linear relations fitted by the method of least squares (with discussion). *Journal of the Royal Statistical Society*, B, **29**, 1–52.

F. J. Anscombe (1968a). Contribution to the discussion of a paper by D. R. Cox and E. J. Snell. *Journal of the Royal Statistical Society*, B, **30**, 267–268.

F. J. Anscombe (1968b). Use of Iverson's language APL for statistical computing. Yale University, Department of Statistics, Technical Report No. 4.

F. J. Anscombe (1971). Keynote address: a century of statistical science, 1890–1990. *Statistical Ecology* (ed. G. P. Patil, E. C. Pielou and W. E. Waters), **1** (Spatial Patterns and Statistical Distributions), xi–xvi. Pennsylvania State University Press.

F. J. Anscombe (1973). Graphs in statistical analysis. *American Statistician*, **27**, no. 1, 17–21.

F. J. Anscombe (1977). Harmonic regression. Yale University, Department of Statistics, Technical Report No. 44. (See also *Proceedings of the 1977 DOE Statistical Symposium*, 3–17.)

F. J. Anscombe, D. R. E. Bancroft and W. J. Glynn (1974). Tests of residuals in the additive analysis of a two-way table—a suggested computer program. Yale University, Department of Statistics, Technical Report No. 32.

F. J. Anscombe and H. H. Chang (1980). Tests for joint normality in time series. Yale University, Department of Statistics, Technical Report No. 47.

F. J. Anscombe and W. J. Glynn (1975). Distribution of the kurtosis statistic b_2 for normal samples. Yale University, Department of Statistics, Technical Report No. 37.

F. J. Anscombe and J. W. Tukey (1963). The examination and analysis of residuals. *Technometrics*, **5**, 141–160.

J. Arbuthnott (1710). An argument for divine providence, taken from the constant regularity observ'd in the births of both sexes. *Philosophical Transactions of the Royal Society*, no. 328, 186–190.

M. S. Bartlett (1935). Contingency table interactions. *Supplement to the Journal of the Royal Statistical Society*, **2**, 248–252.

M. S. Bartlett (1937). Properties of sufficiency and statistical tests. *Proceedings of the Royal Society*, A, **160**, 268–282.

M. S. Bartlett (1955). *An Introduction to Stochastic Processes*. Cambridge University Press. (Second ed. 1966, third ed. 1978.)

D. A. Belsley, E. Kuh and R. E. Welsch (1980). *Regression Diagnostics*. Wiley.

J. P. Benzécri (1973). *L'Analyse des Données*, **1** (La Taxinomie), **2** (L'Analyse des Correspondances). Dunod. (Second ed. 1976.)

P. Berry, J. Bartoli, C. Dell'Aquila and V. Spadavecchia (1978). *APL and Insight*. APL Press.

G. P. Bhattacharjee (1970). Algorithm AS 32: the incomplete gamma integral. *Applied Statistics*, **19**, 285–287.

P. J. Bickel (1978). Using residuals robustly I: tests for heteroscedasticity, nonlinearity. *Annals of Statistics*, **6**, 266–291.

Y. M. M. Bishop, S. E. Fienberg and P. W. Holland (1975). *Discrete Multivariate Analysis: Theory and Practice*. MIT Press.

Å. Björck (1967). Solving linear least squares problems by Gram-Schmidt orthogonalization. *BIT—Nordisk Tidskrift Informationsbehandling*, **7**, 1–21.

R. B. Blackman and J. W. Tukey (1959). *The Measurement of Power Spectra*. Dover.

C. I. Bliss (1967). *Statistics in Biology*, **1**. McGraw-Hill.

P. Bloomfield (1976). *Fourier Analysis of Time Series: an Introduction*. Wiley.

G. E. P. Box (1949). A general distribution theory for a class of likelihood criteria. *Biometrika*, **36**, 317–346.

G. E. P. Box and G. M. Jenkins (1970). *Time Series Analysis: Forecasting and Control*. Holden-Day. (Second ed. 1976.)

G. E. P. Box and M. E. Muller (1958). A note on the generation of random normal deviates. *Annals of Mathematical Statistics*, **29**, 610–611.

M. H. Brenner (1973). *Mental Illness and the Economy*. Harvard University Press.

D. R. Brillinger (1975). *Time Series: Data Analysis and Theory*. Holt, Rinehart and Winston.

D. R. Brillinger and G. C. Tiao (eds.) (1980). *Directions in Time Series*. Institute of Mathematical Statistics.

F. P. Brooks and K. E. Iverson (1963). *Automatic Data Processing*. Wiley.

J. M. Chambers (1977). *Computational Methods for Data Analysis*. Wiley.

W. S. Cleveland and B. Kleiner (1975). A graphical technique for enhancing scatterplots with moving statistics. *Technometrics*, **17**, 447–454.

W. S. Cleveland and E. Parzen (1975). The estimation of coherence, frequency response, and envelope delay. *Technometrics*, **17**, 167–172.

W. G. Cochran (1952). The χ^2 test of goodness of fit. *Annals of Mathematical Statistics*, **23**, 315–345.

D. R. Cox and H. D. Miller (1965). *The Theory of Stochastic Processes*. Wiley.

R. D'Agostino and E. S. Pearson (1973, 1974). Tests for departure from normality: empirical results for the distributions of b_2 and $\sqrt{b_1}$ (with correction). *Biometrika*, **60**, 613–622, and **61**, 647.

C. Daniel and F. S. Wood (1971). *Fitting Equations to Data*. Wiley. (Second ed. 1980.)

F. N. David (1962). *Games, Gods and Gambling*. Hafner.

R. B. Dawson (1954). A simplified expression for the variance of the χ^2-function on a contingency table. *Biometrika*, **41**, 280.

E. W. Dijkstra (1962). *A Primer of ALGOL 60 Programming*. Academic Press.

J. L. Doob (1953). *Stochastic Processes*. Wiley.

N. R. Draper and H. Smith (1966). *Applied Regression Analysis*. Wiley.

J. Durbin and G. S. Watson (1951). Testing for serial correlation in least squares regression, II. *Biometrika*, **38**, 159–178.

A. S. C. Ehrenberg (1975). *Data Reduction*. Wiley.

A. D. Falkoff and K. E. Iverson (1966). *The APL Terminal System: Instructions for Operation*. International Business Machines Corporation, Thomas J. Watson Research Center.

A. D. Falkoff and K. E. Iverson (1968). *APL\360 User's Manual*. International Business Machines Corporation.

A. D. Falkoff and K. E. Iverson (1973). The design of APL. *IBM Journal of Research and Development*, **17**, 324–334.

A. D. Falkoff and D. L. Orth (1979). Development of an APL standard. *APL Quote Quad*, **9**, no. 4, 409–453.

R. A. Fisher (1925, 1932, 1934). *Statistical Methods for Research Workers*, first ed., fourth ed., fifth ed. Oliver and Boyd/Hafner. (Fourteenth ed. 1973.)

R. A. Fisher (1930). The moments of the distribution for normal samples of measures of departure from normality. *Proceedings of the Royal Society of London*, A, **130**, 16–28. (Reprinted in *Contributions to Mathematical Statistics*, 1950, and *Collected Papers*, **2**, 1972.)

R. A. Fisher (1935). *The Design of Experiments*. Oliver and Boyd/Hafner. (Eighth ed. 1974.)

R. A. Fisher (1950). The significance of deviations from expectation in a Poisson series. *Biometrics*, **6**, 17–24.

R. A. Fisher and W. A. Mackenzie (1923). Studies in crop variation, II: the manurial response of different potato varieties. *Journal of Agricultural Science*, **13**, 311–320. (Reprinted in *Collected Papers of R. A. Fisher*, **1**, 1971, University of Adelaide.)

R. A. Fisher and F. Yates (1938). *Statistical Tables for Biological, Agricultural and Medical Research*. Oliver and Boyd. (Sixth ed. 1978.)

T. Francis *et al.* (1955). An evaluation of the 1954 poliomyelitis vaccine trials. Supplement to *American Journal of Public Health*, **45**.

G. H. Freeman and J. H. Halton (1951). Note on an exact treatment of contingency, goodness of fit and other problems of significance. *Biometrika*, **38**, 141–149.

W. H. Gilby (1911). On the significance of the teacher's appreciation of general intelligence. *Biometrika*, **8**, 94–108.

L. Gilman and A. J. Rose (1970). *APL\360, An Interactive Approach*. Wiley. (Second ed. 1974.)

D. V. Glass (ed.) (1954). *Social Mobility in Britain*. Routledge and Paul/Free Press.

G. H. Golub and G. P. H. Styan (1973). Numerical computations for univariate linear models. *Journal of Statistical Computation and Simulation*, **2**, 253–274.

I. J. Good (1969). Some applications of the singular decomposition of a matrix. *Technometrics*, **11**, 823–831.

L. A. Goodman (1972). Some multiplicative models for the analysis of cross classified data. *Proceedings of the Sixth Berkeley Symposium on Mathematical Statistics and Probability*, **1**, 649–696. University of California Press.

L. A. Goodman (1979). Simple models for the analysis of association in cross-classifications having ordered categories. *Journal of the American Statistical Association*, **74**, 537–552.

L. A. Goodman and W. H. Kruskal (1954). Measures of association for cross classifications. *Journal of the American Statistical Association*, **49**, 732–764. (Reprinted in *Measures of Association for Cross Classifications*, 1979, Springer.)

J. Graunt (1662). *Natural and Political Observations Mentioned in a Following Index, and Made upon the Bills of Mortality*. (Fourth ed., 1665.)

J. B. S. Haldane (1937). The exact value of the moments of the distribution of χ^2, used as a test of goodness of fit, when expectations are small. *Biometrika*, **29**, 133–143.

J. B. S. Haldane (1940). The mean and variance of χ^2, when used as a test of homogeneity, when expectations are small. *Biometrika*, **31**, 346–355.

E. J. Hannan (1970). *Multiple Time Series*. Wiley.

J. A. Hartigan (1975). *Clustering Algorithms*. Wiley.

H. G. Henderson (1958). *An Introduction to Haiku*. Doubleday.

D. C. Hoaglin and R. E. Welsch (1978). The hat matrix in regression and ANOVA. *American Statistician*, **32**, 17–22.

P. J. Huber (1964). Robust estimation of a location parameter. *Annals of Mathematical Statistics*, **35**, 73–101.

P. J. Huber (1972). Robust statistics: a review. *Annals of Mathematical Statistics*, **43**, 1041–1067.

S. Hunka (1967). APL: a computing language designed for the user. *British Journal of Mathematical and Statistical Psychology*, **20**, 249–260.

IBM (1965). *Proceedings of the IBM Scientific Computing Symposium on Statistics*,

October 21–23, 1963. International Business Machines Corporation, Data Processing Division.

IBM (1975). *APL Language*. International Business Machines Corporation, GC26-3847. (Third ed. 1976.)

F. R. Immer, J. J. Christensen, R. O. Bridgford and R. F. Crim (1935). *Barley in Minnesota*. University of Minnesota, Agricultural Extension Division, Special Bulletin 135.

F. R. Immer, H. K. Hayes and L. Powers (1934). Statistical determination of barley varietal adaptation. *Journal of the American Society of Agronomy*, **26**, 403–419.

K. E. Iverson (1962). *A Programming Language*. Wiley.

K. E. Iverson (1976). *Elementary Analysis*. APL Press.

K. E. Iverson (1979). The derivative operator. *APL Quote Quad*, **9**, no. 4, 347–354.

K. E. Iverson (1980). Notation as a tool of thought. *Communications of the ACM*, **23**, 444–465.

H. Jeffreys (1948). *Theory of Probability*, second ed. Clarendon Press. (Third ed. 1961.)

M. G. Kendall (1943). *The Advanced Theory of Statistics*, **1**. Griffin.

M. G. Kendall and A. Stuart (1966). *The Advanced Theory of Statistics*, **3**. Griffin/Hafner. (Third ed. 1976.)

C. V. Kiser and P. K. Whelpton (1950). Fertility planning and fertility rates by socio-economic status. *Social and Psychological Factors Affecting Fertility* (ed. P. K. Whelpton and C. V. Kiser), **2**. Millbank Memorial Fund.

B. Kleiner, R. D. Martin and D. J. Thomson (1979). Robust estimation of power spectra (with discussion). *Journal of the Royal Statistical Society*, B, **41**, 313–351.

D. E. Knuth (1968, 1969, 1973). *The Art of Computer Programming*, **1** (Fundamental Algorithms), **2** (Seminumerical Algorithms), **3** (Sorting and Searching). Addison-Wesley.

K. Larntz (1978). Small-sample comparisons of exact levels for chi-squared goodness-of-fit statistics. *Journal of the American Statistical Association*, **73**, 253–263.

W. A. Larsen and S. J. McCleary (1972). The use of partial residual plots in regression analysis. *Technometrics*, **14**, 781–790.

Library of Congress (1963). *Long Remembered: Facsimiles of the Five Versions of the Gettysburg Address in the Handwriting of Abraham Lincoln*. Library of Congress Facsimile No. 3.

R. F. Ling (1974). Comparison of several algorithms for computing sample means and variances. *Journal of the American Statistical Association*, **69**, 859–866.

R. F. Ling (1980). General considerations on the design of an interactive system for data analysis. *Communications of the ACM*, **23**, 147–154.

C. L. Mallows (1973). Some comments on C_p. *Technometrics*, **15**, 661–675.

K. V. Mardia (1970). Measures of multivariate skewness and kurtosis with applications. *Biometrika*, **57**, 519–530.

K. V. Mardia (1974). Applications of some measures of multivariate skewness and kurtosis in testing normality and robustness studies. *Sankhyā*, B, **36**, 115–128.

K. V. Mardia and P. J. Zemroch (1978). *Tables of the F and Related Distributions with Algorithms*. Academic Press.

P. McCullagh (1980). Regression models for ordinal data (with discussion). *Journal of the Royal Statistical Society*, B, **42**, 109–142.

E. E. McDonnell (1979). The socio-technical beginnings of APL. *APL Quote Quad*, **10**, no. 2, 13–18.

D. R. McNeil (1977). *Interactive Data Analysis*. Wiley.

D. R. McNeil and J. W. Tukey (1975). Higher-order diagnosis of two-way tables, illustrated on two sets of demographic empirical distributions. *Biometrics*, **31**, 487–510.

P. Meier (1972). The biggest public health experiment ever: the 1954 field trial of the Salk poliomyelitis vaccine. *Statistics: A Guide to the Unknown* (ed. J. M. Tanur and others), 2–13. Holden-Day.

R. C. Milton and J. A. Nelder (eds.) (1969). *Statistical Computation*. Academic Press.

L. C. Mitchell (1950). Report on fat by acid hydrolysis in eggs. *Journal of the Association of Official Agricultural Chemists*, **33**, 699–703.

F. Mosteller and J. W. Tukey (1977). *Data Analysis and Regression*. Addison-Wesley.

J. I. Naus (1975). *Data Quality Control and Editing*. Dekker.

J. A. Nelder and B. E. Cooper (eds.) (1967). Papers and discussion at a meeting on Statistical Programming, December 15, 1966. *Applied Statistics*, **16**, 87–151.

J. Neyman and E. S. Pearson (1931). Further notes on the χ^2 distribution. *Biometrika*, **22**, 298–305. (Reprinted in *Joint Statistical Papers* by J. Neyman and E. S. Pearson, 1967, University of California Press.)

R. E. Odeh and J. O. Evans (1974). Algorithm AS 70: the percentage points of the normal distribution. *Applied Statistics*, **23**, 96–97.

D. L. Orth (1976). *Calculus in a New Key*. APL Press.

E. Parzen (1967). On empirical multiple time series analysis. *Proceedings of the Fifth Berkeley Symposium on Mathematical Statistics and Probability*, **1**, 305–340. University of California Press.

E. Parzen (1977). Multiple time series: determining the order of approximating autoregressive schemes. *Multivariate Analysis—IV* (ed. P. R. Krishnaiah), 283–295. North-Holland.

K. Pearson (1900). On the criterion that a given system of deviations from the probable in the case of a correlated system of variables is such that it can be reasonably supposed to have arisen from random sampling. *London, Edinburgh and Dublin Philosophical Magazine*, **50**, 157–175. (Reprinted in *Karl Pearson's Early Statistical Papers*, 1956, Cambridge University Press.)

K. Pearson (1904). On the theory of contingency and its relation to association and normal correlation. *Drapers' Company Research Memoirs*. (Reprinted as above.)

K. Pearson (1913). Note on the surface of constant association. *Biometrika*, **9**, 534–537.

K. Pearson and D. Heron (1913). On theories of association. *Biometrika*, **9**, 159–315.

A. J. Perlis (1975). *Introduction to Computer Science*. Harper and Row.

R. L. Plackett (1965). A class of bivariate distributions. *Journal of the American Statistical Association*, **60**, 516–522.

R. L. Plackett (1974). *The Analysis of Categorical Data*. Griffin.

R. P. POLIVKA AND S. PAKIN (1975). *APL: The Language and Its Usage*. Prentice-Hall.

F. PUKELSHEIM (1980). Multilinear estimation of skewness and kurtosis in linear models. *Metrika*, **27**, 103–113.

J. RICHARDSON (1970). Simpson's rule. *APL Quote Quad*, **2**, no. 3, 26–27.

Rothamsted Experimental Station (1969). *Report for 1968, Part 2: The Broadbalk Wheat Experiment.* Lawes Agricultural Trust, Harpenden.

M. R. EL SHANAWANY (1936). An illustration of the accuracy of the χ^2 approximation. *Biometrika*, **28**, 179–187.

L. R. SHENTON AND K. O. BOWMAN (1977). A bivariate model for the distribution of $\sqrt{b_1}$ and b_2. *Journal of the American Statistical Association*, **72**, 206–211.

K. W. SMILLIE (1968). *STATPACK1: An APL Statistical Package*. Department of Computing Science, University of Alberta, Publication No. 9.

K. W. SMILLIE (1969). *STATPACK2: An APL Statistical Package*. Department of Computing Science, University of Alberta, Publication No. 17.

K. W. SMILLIE (1970). *An Introduction to APL\360 with Some Statistical Applications*. Department of Computing Science, University of Alberta, Publication No. 19.

K. W. SMILLIE (1974). *APL\360 with Statistical Applications*. Addison-Wesley.

R. R. SOWDEN AND J. R. ASHFORD (1969). Computation of the bi-variate normal integral. *Applied Statistics*, **18**, 169–180.

W. L. STEVENS (1939). Distribution of groups in a sequence of alternatives. *Annals of Eugenics*, **9**, 10–17.

A. STUART (1953). The estimation and comparison of strengths of association in contingency tables. *Biometrika*, **40**, 105–110.

STUDENT (1908). The probable error of a mean. *Biometrika*, **6**, 1–25. (Reprinted in *"Student's" Collected Papers*, ed. E. S. Pearson and J. Wishart, 1942, Cambridge University Press.)

M. W. TATE AND L. A. HYER (1973). Inaccuracy of the X^2 test of goodness of fit when expected frequencies are small. *Journal of the American Statistical Association*, **68**, 836–841.

D. J. THOMSON (1977). Spectrum estimation techniques for characterization and development of WT4 waveguide, I. *Bell System Technical Journal*, **56**, 1769–1815.

J. W. TUKEY (1949). One degree of freedom for nonadditivity. *Biometrics*, **5**, 232–242.

J. W. TUKEY (1967). An introduction to the calculations of numerical spectrum analysis. *Spectral Analysis of Time Series* (ed. B. Harris), 25–46. Wiley.

J. W. TUKEY (1971). *Exploratory Data Analysis* (limited preliminary edition), **3**. Addison-Wesley.

J. W. TUKEY (1977). *Exploratory Data Analysis*. Addison-Wesley.

U. S. Dept. of Commerce, Bureau of the Census (annual publication). *Statistical Abstract of the United States*.

U. S. Dept. of Commerce, Bureau of the Census (1975). *Historical Statistics of the United States, Colonial Times to 1970*. 2 volumes.

U. S. Dept. of Commerce, Bureau of Economic Analysis (1978). *1977 Business Statistics* (biennial supplement to the *Survey of Current Business*).

U. S. Dept. of Commerce, Environmental Science Services Administration (1965–1969). *Climatological Data: New England*, **77–81**.

P. C. Wakeley (1944). Geographic sources of loblolly pine seed. *Journal of Forestry*, **42**, 23–32.

L. S. Welch and W. C. Camp (1899). *Yale, Her Campus, Class-rooms, and Athletics*. Page.

C. White and H. Eisenberg (1959). ABO blood groups and cancer of the stomach. *Yale Journal of Biology and Medicine*, **32**, 58–61.

M. B. Wilk and R. Gnanadesikan (1968). Probability plotting methods for the analysis of data. *Biometrika*, **55**, 1–17.

D. A. Williams (1976). Improved likelihood ratio tests for complete contingency tables. *Biometrika*, **63**, 33–37.

F. S. Wood (1973). The use of individual effects and residuals in fitting equations to data. *Technometrics*, **15**, 677–695.

G. U. Yule (1900). On the association of attributes in statistics, with illustrations from the material of the Childhood Society, etc. *Philosophical Transactions of the Royal Society*, A, **194**, 257–319. (Reprinted in *Statistical Papers of George Udny Yule*, ed. A. Stuart and M. G. Kendall, 1971, Griffin/Hafner.)

G. U. Yule (1911). *An Introduction to the Theory of Statistics*. Griffin. (Eleventh ed., by G. U. Yule and M. G. Kendall, 1937.)

G. U. Yule (1912). On the methods of measuring association between two attributes (with discussion). *Journal of the Royal Statistical Society*, **75**, 579–652. (Reprinted without discussion as above.)

Perch'io non spero di tornar già mai,
ballatetta, in Toscana,
va tu leggera e piana
. . . . .
Se tu mi voi servire
mena l'anima teco,
molto di ciò ti preco,
quando uscirà del core.

Index

A

B

C

D

E

F

G

H

I

J

K

L

M

N

O

P

Q

R

S

T

U

V

W

Y

Z

I.J. Bienaymé: Statistical Theory Anticipated

C.C. Heyde and E. Seneta

"Helps to fill a gap in the history of probability and statistics, and gives recognition to a man whose discoveries have largely been ignored or credited to others."

—**The American Mathematical Monthly**

This is an exposition of the scientific work of the vitally important but virtually unknown statistician, Irenee Jules Bienaymé (1796-1878). His work is discussed not only for its intrinsic worth but also for the unique perspective it gives on developments in probability and statistics during the 19th century. Bienaymé, who was a prominent civil servant and academician, made substantial contributions to most of the significant problems of probability and statistics during his period. In addition to his work this volume treats Bienaymé's academic background, his contemporaries—he was involved with many major figures of the day, and his involvement with the publishing world.

Contents:

Historical Background.
Demography and social statistics—infant mortality and birth statistics, life tables, probability and the law, insurance and retirement funds.

Homogeneity and Stability of Statistical Trials.
Varieties of heterogeneity, Bienaymé and Poisson's law of large numbers, dispersion theory, Bienaymé's test.

Linear Least Squares.
Legendre, Gauss, and Laplace; Bienaymé's contribution; Cauchy's role in interpolation theory; consequences; Bienaymé and Cauchy on probabilistic least squares; Cauchy continues.

Other Probability and Statistics.
A limit theorem in a Bayesian setting, medical statistics, the law of averages, electoral representation, the concept of sufficiency, a general inequality, an historical note on Pascal, the simple branching process, the Bienaymé-Chebyshev inequality, and a test for randomness.

Miscellaneous Writings.

Bienaymé's Publications.

1977 / 172 pp. / 1 illus. / Cloth $27.00 / ISBN 0-387-90261-9
(Studies in the History of Mathematics and Physical Sciences, Volume 3)

Springer-Verlag
New York
Heidelberg
Berlin

GAUSS A Biographical Study W. K. Bühler

Carl Friedrich Gauss was one of the world's greatest mathematicians, and his work is still a great source of inspiration and scientific interest. Here is an admirable and pleasing study of the man, with emphasis on aspects of his work that are of particular interest today.

This volume is selective, not encyclopedic, and conceived as a guidebook to Gauss's ideas and life that takes into account the effects of the extraordinary political, social, and technical developments of his times. The author presents specific examples to illustrate Gauss's approach to mathematics, to show its influence on modern mathematics, and to summarize the differences between Gauss's work and that of his predecessors and contemporaries.

In addition to biographical details and separate summaries of Gauss's mathematical work, this volume contains a number of relatively unknown illustrations.

Contents

1981 / 208 pp. / 9 illus. / cloth $16.80
ISBN 0-387-10662-6

Springer-Verlag
New York
Heidelberg
Berlin